Second Edition

ANATOMY AND PHYSIOLOGY
Laboratory Manual

Eric Wise

Santa Barbara City College

Boston   Burr Ridge, IL   Dubuque, IA   Madison, WI   New York   San Francisco   St. Louis
Bangkok   Bogotá   Caracas   Lisbon   London   Madrid
Mexico City   Milan   New Delhi   Seoul   Singapore   Sydney   Taipei   Toronto

McGraw-Hill Higher Education

*A Division of The **McGraw-Hill** Companies*

ANATOMY & PHYSIOLOGY LABORATORY MANUAL:
THE UNITY OF FORM AND FUNCTION, SECOND EDITION

Published by McGraw-Hill, an imprint of The McGraw-Hill Companies, Inc., 1221 Avenue of the Americas, New York, NY 10020. Copyright © 2001, 1998 by The McGraw-Hill Companies, Inc. All rights reserved. No part of this publication may be reproduced or distributed in any form or by any means, or stored in a database or retrieval system, without the prior written consent of The McGraw-Hill Companies, Inc., including, but not limited to, in any network or other electronic storage or transmission, or broadcast for distance learning.

This book is printed on recycled, acid-free paper containing 10% postconsumer waste.

1 2 3 4 5 6 7 8 9 0 VNH/VNH 0 9 8 7 6 5 4 3 2 1 0

ISBN 0–07–038896–2

Vice president and editorial director: *Kevin T. Kane*
Publisher: *Michael D. Lange*
Sponsoring editor: *Kristine Tibbetts*
Senior developmental editor: *Connie Balius-Haakinson*
Developmental editor: *Patricia Hesse*
Marketing managers: *Michelle Watnick/Heather K. Wagner*
Media technology project manager: *Mark Christianson*
Editing associate: *Joyce Watters*
Production supervisor: *Laura Fuller*
Designer: *K. Wayne Harms*
Cover photo: *Chris Golly/SharpShooters*
Cover illustration: *Imagineering*
Photo research coordinator: *John C. Leland*
Senior supplement coordinator: *Audrey A. Reiter*
Compositor: *Carlisle Communications, Ltd.*
Typeface: *10/12 Janson*
Printer: *Von Hoffmann Press, Inc.*

The credits section for this book begins on page 547 and is considered an extension of the copyright page.

www.mhhe.com

CONTENTS

PREFACE

For the Instructor

Anatomy and physiology can be the jewel in our students' education or the bane of their college career. As an instructor of anatomy and physiology for many years, I decided to write a lab manual that was student-friendly and with a singular focus on the lab portion of the course. This lab manual was written for the undergraduate student of anatomy and physiology, and it consists of 47 exercises designed to help students learn basic human anatomy and the practical lab applications in physiology.

The diversity of interests in today's anatomy and physiology students is due, in part, to the number of majors that either require or recommend the subject. This lab manual provides a framework for understanding anatomy and physiology for students interested in nursing, radiology, physical or occupational therapy, physical education, dental hygiene, or other allied health majors.

This manual was written to be used independently from lecture texts. The illustrations are labeled, therefore students do not need to bring their lecture texts to the lab. The lab manual accompanies the lecture text and lecture portion of the course and can be used in either a one-term or full-year course. The illustrations are outstanding, and the balanced combination of line art and photographs provides effective coverage of material. The amount of lecture material in the manual is limited so there is little material included that is not part of the lab experience.

Practical lab experience is an invaluable opportunity to reinforce lecture concepts, enrich students' understanding of anatomy and physiology, and allow them to explore new dimensions in the subject area. The educational benefit of reinforcing lecture material with hands-on experiments and acquiring knowledge with a learn-by-doing philosophy makes the anatomy and physiology lab a very special educational environment. Many of us use lab experiences to present conceptually difficult material in physiology and to provide students with different learning styles another avenue for learning.

The 47 exercises in this lab manual provide a comprehensive overview of the human body. Each exercise presents the core elements of the subject matter. You can tailor this manual to match your own vision of the course or use it in its entirety. There are significant differences in the laboratories found around the country, and the advances in physiology equipment, especially computer modules, are numerous and continually evolving. The materials section in each lab is designed for a lab of 24 students and includes the amounts and types of reagents to be used. The labs generally take between 2 and 3 hours to complete.

This lab manual was written for three types of anatomy and physiology courses. For those courses that use the cat as the primary dissection animal, cat dissections or mammalian organ dissections follow the material on humans. For those courses that use models or charts, numerous cadaver photographs are included so students can see the representative structures as they exist in the cadaver material. Finally, for those courses that use cadavers, this lab manual can be used by studying the human material and omitting the cat dissection sections.

Key Features

1. **Four-color Art Program.** All of the illustrations have been newly rendered by a state-of-the-art digital illustration company. They are extremely accurate, use bold and appealing colors, and offer a unique three-dimensional look.

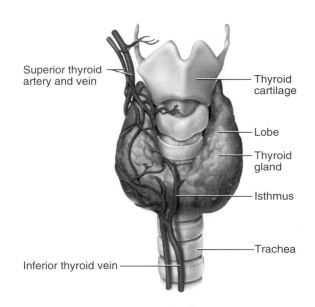

2. **Photographs.** Numerous color photographs, including detailed histological light micrographs and cadaver and cat dissections, show detailed structures.
3. **Labels.** Illustrations are labeled for students to learn the names and terminology by looking at real-life examples or models and by referring to the illustrations in the manual.
4. **Focus on the Laboratory.** This manual focuses primarily on the material necessary for the laboratory and does not repeat the material presented in the lecture text, with the expectation that students can look up material in the lecture text when necessary.
5. **Integrated Use of the Cat for Dissection Specimen.** The cat is used as the dissection animal; however, it is integrated with material on human anatomy, so that

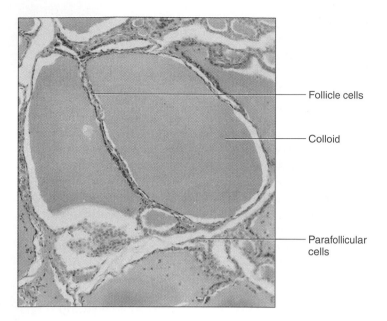

Follicle cells

Colloid

Parafollicular cells

animals do not have to be relied upon as dissection specimens if so desired.

6. **Multimedia Tie-ins.** A dancing man icon (ᛉ) appears in many of the figure legends throughout the exercises. This icon represents "The Dynamic Human" version 2.0 CD-ROM and signals the reader that information relating to this figure can be found on the CD-ROM. "The Dynamic Human" version 2.0 CD-ROM can be packaged with *Anatomy and Physiology Laboratory Manual*. A correlation guide can be found at the end of the student preface.

A videotape icon (▭) appears in many of the figure legends throughout the exercises. This icon alerts the reader that an animation related to the figure is available from the McGraw-Hill Life Science Animations Videotape series. Students can purchase the videotapes, and the tapes are available to qualified adopters. A correlation guide can be found at the end of the preface.

A CD-ROM icon (⊘) appears in several of the exercises after the materials section. This icon represents "The Virtual Physiology Lab" CD-ROM and signals the reader that a supplemental laboratory exercise can be found on the CD-ROM. "The Virtual Physiology Lab" CD-ROM can be packaged with *Anatomy and Physiology Laboratory Manual*.

7. **Safety.** Safety guidelines appear in the inside front cover for reference. The international symbol for caution (⬖) is used throughout the manual to identify material that the reader should pay close and special attention to when preparing for or performing the laboratory exercise.

8. **Clean Up.** At the end of many laboratory exercises an icon for clean up (⊛) reminds the student to clean up the laboratory. Special instructions are given where appropriate.

9. **Data Collection.** Collection of data is imbedded within each exercise as opposed to in a separate table at the back of the manual.

10. **User-Friendly Format.** Each exercise begins on a right-hand page, and the pages are perforated to allow students to more easily remove the exercises to turn them in and later store them in.

11. **Study Hints.** Study hints are located in selected chapters where additional information is provided to help students comprehend and retain the more difficult material presented.

12. **Key Terms.** Current anatomical terminology is used throughout the laboratory manual. Key terms are boldfaced.

Supplemental Materials

In addition to this laboratory manual, an extensive package of supplemental materials is available, some of them obtainable by students, and others provided to qualified instructors who adopt this book.

1. *Instructor's Manual to Accompany Anatomy and Physiology Laboratory Manual* by Eric Wise contains instructional approaches and suggestions for using the laboratory manual, compiled list of materials needed for all exercises, list of laboratory supply houses, preparation of solutions, suggested time schedule, and answers to the review exercises (0-07-038897-0).

2. *Online Learning Center*
   http://www.mhhe.com/saladin
   Students and instructors gain access to a world of opportunities through this password-protected website. Students will find quizzes, activities, links, and much more. Instructors will find all the enhancement tools needed for teaching online, or for incorporating technology in the traditional course.

3. *The Dynamic Human CD-ROM*, Version 2.0 (0-697-38935-9), three-dimensional and other visualizations of relationships between human structure and function. A *Dynamic Human icon* (ᛉ) appears in relevant figure legends in this book. A list of these correlations is on page xii.

4. *The Dynamic Human Videodisc* (0-697-38937-5) contains all of the CD-ROM animations, with a bar code directory.

5. *Virtual Physiology Lab CD-ROM* (0-697-37994-9) has 10 simulations of animal-based experiments common in the physiology component of a laboratory course; allows students to repeat experiments for improved mastery.

6. *Life Sciences Living Lexicon CD-ROM* by William Marchuk (0-697-29266-5) provides interactive vocabulary-building exercises. Includes the meanings of word roots, prefixes, and suffixes, with illustrations and audio pronunciations.

7. *Anatomy and Physiology Videodisc* (0-697-27716-X) has more than 30 physiological animations, line art, and photomicrographs, with a bar code directory.

8. *Life Science Animations* (LSA) contains 53 animations on five VHS videocassettes: Chemistry, the Cell, and Energetics (0-697-25068-7); Cell Division, Heredity, Genetics, Reproduction, and Development (0-697-25069-5); Animal Biology No. 1 (0-697-25070-9); Animal Biology No. 2 (0-697-25071-7); and Plant Biology, Evolution, and Ecology (0-697-26600-1). A videocassette icon (📼) appears in figure legends in this book to alert the reader to related animations. A list of these correlations is on page xiv. Another available videotape is Physiological Concepts of Life Science (0-697-21512-1).

9. *Anatomy and Physiology Videotape Series* consists of four videotapes, free to qualified adopters, including Blood Cell Counting, Identification, and Grouping (0-697-11629-8); Introduction to the Human Cadaver and Prosection (0-697-11177-6); Introduction to Cat Dissection: Cat Musculature (0-697-11630-1); and Internal Organs and Circulatory System of the Cat (0-697-13922-0).

10. *Human Anatomy and Physiology Study Cards, 4/e,* by Kent Van De Graaff, Ward Rhees, and Christopher Creek (0-07-290818-1) are a boxed set of 300 illustrated cards (3 × 5 in). Each of which concisely summarizes a concept of structure of function, defines a term, and provides a concise table of related information.

11. *Coloring Guide to Anatomy and Physiology* by Robert and Judith Stone (0-697-17109-4) consists of outline drawings and text that emphasize learning through color association. Students retain information through a meditative exercise in color-coding structures and correlated labels. This can be an especially effective aid for students who remember visual concepts more easily than verbal ones.

12. *Atlas of the Skeletal Muscles, 3/e,* by Robert and Judith Stone (0-07-290332-5) is a guide to the structure and function of human skeletal muscles. The illustrations help students locate muscles and understand their actions.

13. *Laboratory Atlas of Anatomy and Physiology, 2/e,* by Douglas Eder et al. (0-697-39480-8) is a full-color atlas including histology, skeletal and muscular anatomy, dissections, and reference tables.

14. *Case Histories in Human Physiology, 3/e,* by Donna Van Wynsberghe and Gregory Cooley (0-697-34234-4) is a web-based workbook that stimulates analytical thinking through case studies and problem solving; includes an instructor's answer key. (www.mhhe.com/biosci/ap/vanwyn/)

15. *Survey of Infectious and Parasitic Diseases* by Kent Van De Graaff (0-697-27535-3) is a booklet of essential information on 100 of the most significant infectious diseases.

16. *Life Science Animations* 3D—forty-two key biological processes all narrated and animated in vibrant color with dynamic three-dimensional graphics. (0-07-290652-9)

17. *Life Science Animations 3D CD-ROM*—More than 120 animations that illustrate key biological processes are available at your fingertips on this exciting CD-ROM. This CD contains all of the animations found on the Essential Study Partner and much more. The animations can be imported into presentation programs, such as PowerPoint. Imagine the benefit of showing the animations during lecture. (0-07-234296-X)

18. *Web-Based Cat Dissection Review for Human Anatomy and Physiology* is an online multimedia program that contains high-quality cat dissection photographs. The photos depict an accurately dissected cat and contain labels for easy identification of pertinent structures and relationships. This product helps students easily identify and review corresponding body structures and functions between the cat organism and the human organism. (0-07-232157-1)

Acknowledgments

Many people have been involved in the production of this lab manual, and I would like to thank the editorial staff at McGraw-Hill, including Kris Tibbetts, Connie Haakinson, Darlene Schueller, Joyce Watters, Wayne Harms, and Sandy Hahn, for their input and encouragement.

I would like to dedicate this book to my wife, Mary, and to our two daughters, Sarah and Caitlin, for their understanding and love.

Please feel free to write me or e-mail me with your comments, suggestions, and criticisms. I value your input and hope that your comments will lead to an even better third edition of this laboratory manual.

Eric Wise
Santa Barbara City College
721 Cliff Drive
Santa Barbara, CA 93109
wise@sbcc.net

Reviewers

The writing of a laboratory manual is a collaborative effort and I was impressed by the dedication and insight given to me by the reviewers. I would like to gratefully acknowledge them as they made significant contributions regarding improving the content and clarity of the manual. I appreciate their time and energy in making this a better lab manual, however, I alone, take responsibility for any remaining errors.

Mary Suzzanne Allen
North Harris College

Michael J. Harman
North Harris College

Carol Haspel
LaGuardia Community College

Charles B. Leonard, Jr.
Gallaudet University

Charmayne Jesik Mack
Dominican University

Brian K. Paulson
California University of Pennsylvania

Donna Ritch
University of Wisconsin-Green Bay

Ed W. Thompson
Winona State University

Louis Trombetta
St. John's University

Robert Vick
Elon College

Norman Wintrip
Humber College of Applied Arts and Technology

PREFACE

For the Student

This laboratory manual was written to help you gain experience in the lab as you learn human anatomy and physiology. The 47 exercises explore and explain the structure and function of the human body. You will be asked to study the structure of the body using the materials available in your lab which may consist of models, charts, mammal study specimens such as cats, preserved or fresh internal organs of sheep or cows, and possibly cadaver specimens. You may also examine microscopic sections from various organs of the body. You should familiarize yourself with the microscopes in your lab very early so you can take the best advantage of the information they can provide.

The physiology portion of the course involves experiments that you will perform on yourself or your lab partner. They may also involve the mixing of various chemicals and the study of the functions of live specimens. Because the use of animals in experiments is of concern to many students, a significant attempt has been made to reduce (but unfortunately not eliminate) the number of live experiments in this manual. Until there is a sound replacement for live animals, their use will continue to be part of the college physiology lab. Your instructor may have alternatives to live animal experimentation exercises. It is important to get the most out of what live specimen experimentation there is. Coming to the lab unprepared and then sacrificing a lab animal while gaining little or no information is an unacceptable waste of life. Use the animals with care. Needless use or inhumane treatment of lab animals is not acceptable or tolerated.

As a student of anatomy and physiology you will be exposed to new and detailed information. The time it takes to learn the information will involve *more* than just time spent in the lab. You should maximize your time in the lab by reading the assigned lab exercises before you come to class. You will be doing complex experiments, and if you are not familiar with the procedure, equipment, and time involved you could end up ruining the experiment for yourself and/or your lab group. The exercises in physiology are written so you can fill in the data as you proceed with the experiment. At the end of each exercise are review sheets that your instructor may wish to collect to evaluate the data and conclusions of your experiments. The illustrations are labeled except on the review pages. All review materials can be used as study guides for lab exams or they may be handed in to the instructor.

The anatomy exercises are written for cat and human study though these exercises can be used with or without cats or cadavers. Get *involved* in your lab experience. Don't let your lab partner do all of the dissections or all of the experimentation; likewise, don't insist on doing everything yourself. Share the responsibility and you will learn more.

Safety

Safety guidelines appear in the inside front cover for reference. The international symbol for caution (⚠) is used throughout the manual to identify material that you should pay close and special attention to when preparing for or performing laboratory exercises.

Clean Up

Special instructions are provided for clean up at the end of appropriate laboratory exercises and are identified by this unique icon (✊).

How to Study for This Course

Some people learn best by concentrating on the visual, some by repeating what they have learned, and others by writing what they know over and over again. In this course, you will have to adapt your particular learning style to different study methods. You may use one study method to learn the muscles of the body and a completely different method for understanding the function of the nervous system. Some students need only a few hours per week to succeed in this course, while others seem to study far longer with a much less satisfactory performance.

You need to come to class. Come to lab on time. The beginning of the lab is when most instructors go over the material and point out what material to omit, what to change, and how to proceed. If you do not attend lab, you do not get the necessary information.

Read the material ahead of time. The subject matter is very visual, and you will find an abundance of illustrations in this manual. Record on your calendar all of the lab quizzes and exams listed on the syllabus provided by your instructor. Budget your time so you study accordingly.

Work hard! There is absolutely no substitute for hard work to achieve success in a class. Some people do math easier than others, some people remember things easier, and some people express themselves better. Most students succeed because they work hard at learning the material. It is a rare student who gets a bad grade because of a lack of intelligence. Working at your studies will get you much farther than worrying, partying, or thinking that you can succeed by attending some of the time and reviewing your notes the night before the lab exam.

Be *actively* involved with the material and you will learn it better. Outline the material after you study it for awhile. Read your notes, go over the material in your mind, and then make the information your own. There are several ways that you can get actively involved.

Draw and doodle a lot. Anatomy is a visual science, and drawing helps. You do not have to be a great illustrator. Visualize the material in the same way you would draw a map to your house for a friend. You do not draw every bush and tree, but rather create a *schematic* illustration that your friend could use to get the *pertinent* information. As you know, there are differences in maps. Some people need more practice than others, but anyone can do it. The head can be a circle, which can be divided into pieces representing the bones of the skull. Draw and label the illustration after you have studied the material and without the use of your text! Check yourself against the text to see if you really know the material. Correct the illustration with a colored pen so you highlight the areas you need work on. Go back and do it again until you get it perfect. This does take some time, but not as much as you might think.

Write an outline of the material. Take the mass of information to be learned and go from the general to the specific. Let's use the skeletal system as an example. You may wish to use these categories:

1. Bone composition and general structure
2. Bone formation
3. Parts of the skeleton
   a. Appendicular skeleton
      (1) Pectoral girdle
      (2) Upper extremity
      (3) Pelvic girdle
      (4) Lower extremity
   b. Axial skeleton
      (1) Skull
      (2) Hyoid
      (3) Ribs
      (4) Vertebral column
      (5) Sternum

An outline helps you organize the material in your mind and lets you sort the information into areas of focus. If you do not have an organizational system, then this course is a jumble of terms with no interrelationships. The outline can get more detailed as you progress, so you eventually know that the specific nasal bone is one of the facial bones and the facial bones are skull bones, which are part of the axial skeleton, which belongs to the skeletal system!

Test yourself before the exam or quiz. If you have practiced answering questions about the material you have studied, then you should do better on the real exam. As you go over the material, jot down possible questions to be answered later, after you study. If you compile a list of questions as you review your notes, then you can answer them later to see if you have learned the material well. You can also enlist the help of friends, study partners, or family (if they are willing to do this for you). You can also study alone. Some people make flash cards for the anatomy portion of the course. It is a good idea to do this for the muscle section of the class, but you may be able to get most of the information down by using the preceding technique. Flash cards take time to fill out, so use them carefully.

Use memory devices for complex material. A mnemonic device is a memory phrase that has some relationship to the study material. For example, there are two bones in the wrist right next to one another, the trapezium and the trapezoid. The mnemonic device used by one student was that trapez*ium* rhymes with th*umb* and it is the one under the thumb.

Use your study group as a support group. A good study group is very effective in helping you do your best in class. Hang out with people who will push you to do your best. If you get discouraged, your study partners can be invaluable support people. A good group can help you improve your test scores, develop study hints, encourage you to do your best, and let you know that you are not the *only* one living, eating, and breathing anatomy and physiology.

Just as a good study group can really help, a bad group can drag you down farther than you might go on your own. If you are in a group that constantly complains about the instructor, that the class is too hard, that there is too much work, that the tests are not fair, and that you don't really need to know this much anatomy and physiology for your own field of study and that this isn't medical school, then get yourself out of that group and into one that is excited by the information. Don't listen to people who complain constantly and make up excuses instead of studying. There is a tendency to start believing the complaints, and that begins a cycle of failure. Get out of a bad situation early and get with a group that will move forward.

Do well in the class and you will feel good about the experience. If you set up a study time with a group of people and they spend most of the time talking about parties, sports, or personal problems, then you aren't studying. There is nothing wrong with parties, sports, or helping someone with personal problems, but you need to address the task at hand, which is learning anatomy and physiology. Don't feel bad if you must get out of your study group. It is *your* education, and if your partners don't want to study, then they don't really care about your academic well-being. A good study partner is one who pays attention in class, who is prepared ahead of the study time session, and who can explain information that you may have gotten wrong in your notes. You may want to get the phone number of two or three such classmates.

Supplemental Materials

A variety of materials can be purchased separately to supplement this laboratory manual. Please see the instructor's preface for a list and description of these items, or call the McGraw-Hill Customer Service Department at 1–800–338–3987.

Test-Taking

Finally, you need to take quizzes and exams in a successful manner. By doing practice tests, you can develop confidence. Do well early in the semester. Study extra hard early (there is no such thing as overstudying!). If you fail the first test or quiz, then you must work yourself out of an emotional ditch. Study early and consistently, and then spend the evening before the exam going over the material in a general way and solving those last few problems. Some people do succeed under pressure and cram before exams; however, the information is stored in short-term memory and does not serve you well in your major field! If you study on a routine basis, then you can get up on the morning of a test, have a good breakfast, listen to some encouraging music, maybe review a bit, and be ready for the exam.

Your instructor is there to help you learn anatomy and physiology, and this laboratory manual was written with you in mind. Relate as much of the material as you can to your own body and keep an optimistic attitude.

Please feel free to write me or e-mail me with your comments, suggestions, and criticisms. I value your input and hope that your comments will lead to an even better third edition of this laboratory manual.

Eric Wise
Santa Barbara City College
721 Cliff Drive
Santa Barbara, CA 93109
wise@sbcc.net

Dynamic Human Version 2.0 CD-ROM Correlation Guide

Figure	Correlation
2.1	Human body/Clinical concepts/Clinical imaging
2.3	Human body/Explorations/Anatomical orientation
2.4	Human body/Explorations/Anatomical orientation/Directional terminology
2.5	Human body/Explorations/Anatomical orientation/Planes
4.1–4.9	Human body/Anatomy/Cell components
4.10, 4.11	Human body/Explorations/Cell cycle
6.1	Human body/Histology/Simple squamous epithelium
6.3	Human body/Histology/Simple columnar epithelium
6.4	Human body/Histology/Pseudostratified ciliated columnar epithelium
6.5	Human body/Histology/Stratified squamous epithelium
6.6	Human body/Histology/Transitional epithelium
6.7	Muscular/Anatomy/Skeletal muscle
6.8	Muscular/Anatomy/Cardiac muscle
6.9	Muscular/Anatomy/Smooth muscle
6.12	Human body/Histology/Dense irregular connective tissue
6.18	Human body/Histology/Hyaline cartilage
6.19	Human body/Histology/Fibrocartilage
6.20	Human body/Histology/Elastic cartilage
8.1, 8.2	Skeletal/Anatomy/Gross anatomy
8.3–8.5	Skeletal/Explorations/Cross section of a long bone
8.6	Skeletal/Histology/Compact bone
9.1	Skeletal/Anatomy/Gross anatomy
9.2, 9.3	Skeletal/Anatomy/Gross anatomy/Appendicular skeleton/Pectoral girdle
9.4–9.6	Skeletal/Anatomy/Gross anatomy/Appendicular skeleton/Upper limb
9.7–9.11	Skeletal/Anatomy/Gross anatomy/Appendicular skeleton/Lower limb
10.1	Skeletal/Anatomy/Gross anatomy
10.2–10.4, 10.6, 10.8, 10.10	Skeletal/Anatomy/Gross anatomy/Axial skeleton/Vertebral column
10.12, 10.13	Skeletal/Anatomy/Gross anatomy/Axial skeleton/Thoracic cage
11.1–11.8	Skeletal/Anatomy/Gross anatomy/Axial skeleton/Skull
12.1, 12.2, 12.4	Skeletal/Explorations/Fibrous joints
12.3, 12.6	Skeletal/Explorations/Cartilaginous joints
12.7–12.17, 12.19	Skeletal/Explorations/Synovial joints
13.1, 13.2	Muscular/Anatomy/Body regions/Abdomen and back
14.1–14.4	Muscular/Anatomy/Body regions/Forearm
14.5	Muscular/Anatomy/Body regions/Hand
15.1–15.6	Muscular/Anatomy/Body regions/Thigh
16.1–16.3	Muscular/Anatomy/Body regions/Lower leg
	Muscular/Anatomy/Body regions/Foot
17.1–17.7	Muscular/Anatomy/Body regions/Head and neck
18.1–18.4	Muscular/Anatomy/Body regions/Abdomen and back
21.5, 21.9	Nervous/Anatomy/Gross anatomy of the brain
21.6	Nervous/Explorations/Motor and sensory pathways
22.2	Nervous/Anatomy/Spinal cord anatomy

Continued

Dynamic Human Version 2.0 CD-ROM Correlation Guide—*Continued*

Figure	Correlation
22.3	Nervous/Histology/Spinal cord
25.1	Nervous/Explorations/Taste
	Nervous/Histology/Vallate papillae
25.2	Nervous/Explorations/Zones of taste
25.3	Nervous/Explorations/Olfaction
26.3	Nervous/Explorations/Vision
26.5	Nervous/Histology/Retina
27.4	Nervous/Explorations/Hearing
	Nervous/Explorations/Dynamic equilibrium
	Nervous/Explorations/Static equilibrium
27.5, 27.6	Nervous/Explorations/Hearing
27.8	Nervous/Explorations/Static equilibrium
27.9	Nervous/Explorations/Dynamic equilibrium
28.1	Endocrine/Anatomy/Gross anatomy
28.2	Endocrine/Anatomy/Gross anatomy/Hypothalamus and pituitary gland
28.3, 28.4	Endocrine/Histology/Pituitary gland
28.6, 28.8	Endocrine/Anatomy/Gross anatomy/Thyroid gland
28.7	Endocrine/Histology/Thyroid gland
28.10	Endocrine/Histology/Pancreas
28.11	Endocrine/Anatomy/Gross anatomy/Adrenal glands
28.12	Endocrine/Histology/Adrenal cortex
28.13	Reproductive/Histology/Testis
28.14	Reproductive/Histology/Ovary
30.2	Lymphatic/Clinical concepts/Blood type
31.1–31.4	Cardiovascular/Anatomy/Gross anatomy
32.1	Cardiovascular/Explorations/Heart dynamics
32.2	Cardiovascular/Explorations/Heart dynamics
	Cardiovascular/Clinical concepts/Electrocardiogram
32.3	Cardiovascular/Clinical concepts/Electrocardiogram
32.4	Cardiovascular/Clinical concepts/Heart murmur
34.1	Cardiovascular/Histology/Vasculature
37.1	Lymphatic/Explorations/Lymph formation and movement
37.5	Lymphatic/Histology/Lymph node
37.9, 37.10	Lymphatic/Anatomy/Gross anatomy
39.1, 39.7	Respiratory/Anatomy/Gross anatomy
39.8	Respiratory/Anatomy/3D Viewer
39.10	Respiratory/Histology/Lung
40.1	Respiratory/Histology/Alveoli
42.1	Digestive/Anatomy/Gross anatomy
42.3	Digestive/Clinical concepts/Dental cavity
42.9	Digestive/Histology/Fundic stomach
42.18	Digestive/Histology/Liver
44.1–44.3	Urinary/Anatomy/Gross anatomy

Continued

Dynamic Human Version 2.0 CD-ROM Correlation Guide—*Continued*

Figure	Correlation
44.4	Urinary/Histology/Renal cortex
	Urinary/Histology/Renal medulla
44.6	Urinary/Histology/Penile urethra
44.7	Urinary/Histology/Bladder
46.1	Reproductive/Anatomy/3D Viewer: Male
	Reproductive/Anatomy/Male reproductive anatomy
46.2, 46.3	Reproductive/Anatomy/Male reproductive anatomy
46.6	Reproductive/Anatomy/3D Viewer: Male
	Reproductive/Anatomy/Male reproductive anatomy
46.7	Reproductive/Histology/Testis
46.8	Reproductive/Explorations/Spermatogenesis
46.9	Reproductive/Clinical concepts/Vasectomy
47.1	Reproductive/Anatomy/3D Viewer: Female
	Reproductive/Anatomy/Female reproductive anatomy
47.2	Reproductive/Anatomy/3D Viewer: Female
47.3	Reproductive/Explorations/Menstrual cycle
	Reproductive/Explorations/Ovarian cycle
47.6	Reproductive/Explorations/Oogenesis
47.7	Reproductive/Histology/Ovary

Life Science Animations Correlation Guide, Videotape Series, and 3–D Videotape

Figure	Correlation		
4.1–4.9	Tape 1	Concept 2	Journey into a Cell
4.9	3–D	Concept 13	Structure of DNA
4.10, 4.11	3–D	Concept 10	Mitosis
4.10, 4.11	Tape 2	Concept 12	Mitosis
20.5	Tape 6	Concept 61	Synaptic Transmission
23.3	Tape 3	Concept 25	Reflex Arcs
27.8	Tape 3	Concept 26	Organ of Static Equilibrium
28.1	3–D	Concept 41	Hormone Action
30.2	Tape 4	Concept 40	A, B, O Blood Types
32.1	Tape 3	Concept 31	Regulation of Muscle Contraction
32.1	3–D	Concept 40	Muscle Contraction
32.2, 32.3	Tape 4	Concept 38	Production of Electrocardiogram
32.4	Tape 3	Concept 32	The Cardiac Cycle and Production of Sounds
33.1	3–D	Concept 39	Action Potential
42.1	3–D	Concept 36	Digestion Overview
43.1	Tape 6	Concept 54	Lock-and-Key Model of Enzyme Action
43.2	Tape 3	Concept 34	Digestion of Carbohydrates
44.1–44.3	3–D	Concept 38	Kidney Function
46.8	Tape 2	Concept 19	Spermatogenesis
47.6	Tape 2	Concept 20	Oogenesis
47.13	Tape 2	Concept 21	Human Embryonic Development

LABORATORY EXERCISE 1

Measurement in Science

Introduction

Science is the study of physical phenomena. The human body and its structure and function fall within the realm of scientific investigation. The science of **human anatomy** involves the understanding of the structure of the human body, while **human physiology** is the study of the function of the body. There are many different ways that people can achieve an understanding of anatomy and physiology. Some of these have been described in a process known as the **scientific method.**

The beginning of the scientific understanding frequently begins with a question. An example of a question might be how the body digests food. The next part in this understanding frequently involves the development of an **hypothesis,** which is a testable proposal that seeks to explain a scientific question. The test done to prove or disprove the hypothesis is usually done in the form of an **experiment.** A more lengthy discussion of the scientific method is provided in your lecture text or in most science textbooks.

Much of science involves measurement and collecting **data.** Experimental data are the pieces of information or 'facts' obtained, which are later examined to support or reject the proposed hypothesis. In this exercise you collect data and examine how to graph data so that the information becomes more comprehensible. You will also look at what kind of measurement is done in science.

Objectives

At the end of this exercise you should be able to

1. describe the advantages of the metric system over the U.S. customary system;
2. define independent variable and dependent variable;
3. list four base units of the metric system;
4. convert fractions into decimal equivalents;
5. collect and graph data taken in class;
6. determine some factors involved in experimental error.

Materials

Meter stick
Small metric ruler (30 cm)

Procedure

Measurements in Science

Members of the scientific community and people of many nations of the world use the **metric system** to record quantities such as length, volume, mass (weight), and time. This is because the metric system is based on units of 10, and conversion to higher or lower values is relatively easy when compared to using the U.S. customary system. For example, assume you are working on a bicycle and are using a ½-inch wrench. If you need to go up in size you move to a ⁹⁄₁₆-inch, then a ⅝-inch, then an ¹¹⁄₁₆-inch, or perhaps as large as a ¾-inch wrench. This requires a bit of computation as you move from one size to the next. On the other hand, if you are using the metric system and a 12-millimeter (mm) wrench is too small, you progressively move to a 13 mm, 14 mm, or 15 mm wrench.

The same idea can be applied to volume or weight. In the case of volume, there are 8 ounces per cup, and 128 ounces per gallon. The calculation for the number of ounces in 7 gallons is a little cumbersome (7 gallons × 128 ounces). In the metric system, there are 1,000 milliliters in 1 liter so there are 7,000 milliliters in 7 liters. The conversions are much easier. Medical dosages are given frequently in milliliters or cubic centimeters (cc). One milliliter occupies one cubic centimeter and so these values are interchangeable.

You can use the metric system to measure four quantities—length, volume, mass, or time. Examine table 1.1 and compare the quantity, base unit, and U.S. equivalent.

If the quantity measured is much larger or smaller than the base unit, then the base unit can be expressed in multiples of 10. For example, if you had one thousand grams (1,000 grams), then you would have a **kilo**gram. If you had one thousandth of

Table 1.1	Metric System and Equivalents	
Quantity	**Base Unit**	**U.S. Equivalent**
Length	Meter (m)	1.09 yards (39.4 inches)
Volume	Liter (l)	1.06 quarts
Mass	Gram (g)	.036 ounces (¹⁄₄₅₄ of a pound)
Time	Second (s)	Second

a gram ($\frac{1}{1,000}$ gram), you would have a **milli**gram. Examine table 1.2 as you answer the following questions:

What is $\frac{1}{100}$ gram?   _____

What is 1,000 seconds?   _____

What is 10 meters?   _____

What is $\frac{1}{1,000,000,000}$ liter?   _____

As you can see from table 1.2, some measurements in science are very small. The amounts of hormones circulating in the blood are very minute indeed. To provide a shortened notation of numbers that are very large or small we use scientific notation. A number such as 60,000 is written as 6×10^4. You move the decimal point four places to the left and thus the superscript above the ten is a four. Write 6,000 in scientific notation _____. For very small numbers, the superscript is written as a negative number. The number 0.00006 is written as 6×10^{-5}, as you move the decimal point five places to the right. Convert the following numbers into scientific notation:

4,300,000   _____

0.000034   _____

2,200   _____

0.0019   _____

Working in the science lab requires you to focus on the procedures and materials at hand. You may work as part of a group in some labs, and it is important that you read your lab material before coming into the lab. Some of the materials you work with may be dangerous, and a thorough, prior knowledge of the lab exercise will ensure a safer lab.

Pay attention to the experiment and what is to be done and when. Casual observation and carelessness may lead to incorrect results. Establish a procedure for conducting experiments. If you are working with one or more lab partners, divide up responsibilities before the experiment begins. If you are responsible for a particular portion of the experiment, make sure your lab partners see the results. Make a careful record of the results of your experiment.

Table 1.2 Decimals of the Metric System			
Name	**Description**	**Multiple/Fraction**	
Kilo	One thousand times greater	1,000	
Deca	Ten times greater	10	
Base Unit			
Deci	One-tenth as much	$\frac{1}{10}$	0.1
Centi	One-hundredth as much	$\frac{1}{100}$	.01
Milli	One-thousandth as much	$\frac{1}{1,000}$	.001
Micro	One-millionth as much	$\frac{1}{1,000,000}$	.000001
Nano	One-billionth as much	$\frac{1}{1,000,000,000}$	.000000001

Be honest. Fudging data is not tolerated in the scientific community. Record your data as you measure it. If your results do not seem to be what they should, then discuss this with your instructor. Never record data that you think you *should* get but rather *record the observed data.*

Correlation Between Leg Length and Height

In this part of the exercise, you will look for a correlation between the length of the leg and the overall height of the individual.

Make an hypothesis about any relationship between leg length and height.

Record your hypothesis in the following space.

Figure 1.1   Measurement of leg length.

By graphing data you can more easily see trends present in a sample size. One problem in sampling is that you need to have a large enough number to have a valid sample. Let's suppose that one-half of the basketball team is enrolled in your lab section. This might have a rather unusual effect on your graph. On the other hand, if you are able to sample your entire school, the effects of the basketball players' sizes would be minimized.

Your lab partner should measure the length of your leg from the back of the knee to the bottom of your heel. To do this, kneel on a chair with the thigh vertical and the back of the calf at a 90-degree angle to the thigh (figure 1.1). Place a meter stick along the outside (lateral) edge of the back of your calf. You should feel a tendon, which is the tendon of the biceps femoris muscle. Make sure that the meter stick is just at the edge of the tendon, and record the distance, in centimeters, from this tendon to the bottom of your heel.

Length (cm) between tendon and heel: _____

Now record your overall height in centimeters. Determine this value directly or by multiplying your height in inches by 2.54 cm/in. Have your lab partner measure your height against a wall in the lab, after removing your shoes.

Overall height in centimeters: _____

Determine the distribution of the entire class in terms of leg length. Give your data to other members of the class to record in the space to the left. Plot the length of the leg on the x-axis (the independent variable is plotted on the horizontal x-axis) and the overall height of the individual on the y-axis (the dependent variable) to see if leg length determines overall height. Is there a correlation?

Plot the distribution of leg length and overall height on the following graph on page 4.

For every 5-centimeter change in leg length, what is the approximate change in height?

Leg Length and Height Data		
Student Number	Leg Length (cm)	Height (cm)
1		
2		
3		
4		
5		
6		
7		
8		
9		
10		
11		
12		
13		
14		
15		
16		
17		
18		
19		
20		
21		
22		
23		
24		

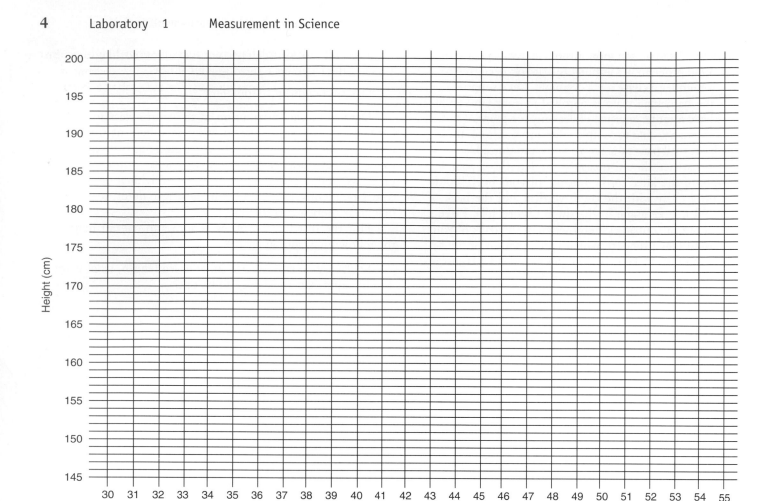

Range and Mean in Measurements

The extremes of measurement, which represent the longest and shortest measurements, represent the **range** of the measurements. Using the values obtained by members of your lab class, record the range of the leg length in centimeters for the entire class.

Longest value: _____

Shortest value: _____

The **mean** leg length of the class is the average length for the group. Using the data obtained for the leg lengths, add all of the values together (the sum of the leg lengths) and divide that number by the number of individuals in the class. This is the mean leg length. Record the number in centimeters.

Mean leg length: _____

Measurement and Experimental Error

In this part of the lab, all members of the lab will take a measurement of a single individual's hand. You should perform the next exercise by following the directions below.

1. Select *one* person in your lab class who will serve as the experimental subject for this section of the lab.

2. Everyone in the lab should take a metric ruler from the lab table and measure the length of the hand of the selected subject. In order to obtain individual results, the measurements should be conducted alone and the results recorded without any discussion with classmates.

3. With a ruler, measure the test subject's hand from the wrist to the tip of the middle finger and record this value in millimeters.

Length of the hand in millimeters:_____

4. After everyone has made a measurement and entered the data in his or her own lab book, every individual in the class should write his or her results on the board. Record all of the data from the class below.

Length of hand in millimeters:

_____	_____	_____	_____
_____	_____	_____	_____
_____	_____	_____	_____
_____	_____	_____	_____
_____	_____	_____	_____
_____	_____	_____	_____

If everyone measured the same person, all of the values should be the same. Was this the case in your lab?

If you were to do this experiment again, how would you change the procedure to try to get a more accurate recording of lengths among the people measuring the subject's hand?

If this was not the case, what are some reasons why the measurements were different?

Name _____

1. In terms of base units

   a. What is the base unit of length in the metric system?

   b. What is the base unit of volume in the metric system?

2. How many cubic centimeters are there in 200 milliliters?

3. Assume a pill has a dosage of 350 mg of medication. How much medication is this in grams?

4. How would you write 0.000345 liters in scientific notation?

5. How many milligrams are there in 4.5 kilograms?

6. How many meters is 250 millimeters?

7. If given a length of 1/10,000 of a meter

   a. convert this number into a decimal:

   b. convert it into scientific notation:

8. Use a word to describe

   a. one thousandth of a second:

   b. one thousand liters:

   c. one hundredth of a meter:

9. According to your graph, predict how tall a person would be if his or her leg length was 37 cm. How tall do you predict they would be if the individual had a leg length of 48 cm?

10. If you were to do a new experiment, what do you estimate the relationship between the *length of the arm* and the *overall height* of an individual to be?

11. In the case of arm length and overall height, which is the independent variable and which is the dependent variable?

12. What was the range in values for the hand length measured in lab?

Organs, Systems, and Organization of the Body

Introduction

The study of the human body requires an understanding of the orientation of the body or organ of study and the detail of presentation and knowledge of body regions. In this exercise, you examine the major organ systems of the body, directional terms, levels of organization of the body (from the subatomic level to the whole organism), and describe the major regions of the body.

Objectives

At the end of this exercise you should be able to

1. list the 11 organ systems;
2. place major organs such as the heart, lungs, and stomach in the proper organ system;
3. determine from an illustration whether a section is in the frontal, transverse, or sagittal plane;
4. give directional terms that are equivalent to up, down, front, back, toward the midline or edge, and toward the core or surface of the body;
5. explain what is meant by anatomical position;
6. identify the quadrants and nine abdominal regions.
7. list the major regions of the body and provide a common name if known.

Materials

Models of human torso
Charts of human torso

Procedure

Levels of Organization

The human body can be studied from a number of perspectives. The earliest study involved gross anatomy, or cutting up part or all of the body and examining its details. As more sophisticated equipment was developed, other levels of organization became apparent. Today, the manipulation of the atomic nuclei under magnetic fields has led to magnetic resonance imaging (MRI) studies that do not depend on dissection of the body (figure 2.1). Examine the following list for the numerous levels of study along with cited examples:

Level	Example
Subatomic	Neutrons
Atomic	Oxygen
Molecular	Proteins
Cellular	Squamous epithelium
Tissue	Epithelial
Organ	Stomach
Organ System	Digestive system
Organism	*Homo sapiens*

Organ Systems

Anatomy can be studied in many ways. **Regional anatomy** is the study of particular areas of the body such as the head or leg. Most undergraduate college courses in anatomy and physiology (and the format of this lab manual) involve **systemic anatomy,** which is the study of **organ systems** such as the skeletal system and nervous system. Although organ systems

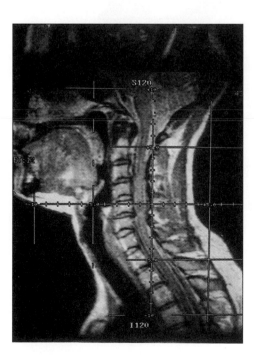

Figure 2.1 MRI. Image of the neck.

are studied separately, it is important to realize the intimate connections between the systems. If the heart fails to pump blood as part of the cardiovascular system, then the lungs do not receive blood for oxygenation and the intestines do not transfer nutrients to the blood as fuel. The brain is no longer capable of functioning, and the result is death of the body. From a clinical standpoint, the failure of one system has impacts on many other organ systems.

Examine the torso models and charts in the lab and locate various organs. Using figure 2.2, find the following organ systems:

Reproductive   Respiratory   Digestive
Urinary   Skeletal   Endocrine
Nervous   Lymphatic   Cardiovascular
Muscular   Integumentary

A quick way to remember all 11 systems is to remember this phrase: "Run Mrs. Lidec." Each letter of the phrase represents the first letter of one of the names of the organ systems.

Reproductive The gonads (testes and ovaries) contain the sex-producing cells of the body, and the accessory organs such as the uterus, vagina, penis, and seminal vesicles play a part in the transport of the sex cells and the development of the fetus.

Urinary The kidneys serve as filters of the body, and the urinary bladder serves as a storage organ. The ureters connect the kidneys to the bladder and the urethra is the exit tube from the body. The urinary system plays an important role in ridding the body of nitrogenous wastes while maintaining the chemical balance of body fluids.

Nervous The brain and spinal cord as well as the numerous nerves make up the nervous system. The nervous system coordinates body regions, interprets environmental cues, and integrates information.

Muscular Individual muscles are the organs of this system. Muscles move and strengthen joints, generate heat, and serve other functions such as abdominal compression.

Respiratory The nose, larynx, trachea, and lungs are part of the respiratory system. The function of the system is the exchange of gases (oxygen and carbon dioxide) between the blood and the air.

Skeletal Each bone is considered an organ, with blood vessels and nerves commonly found in each bone. The skeletal system functions to support the body, protect delicate organs, and produce blood.

Lymphatic The lymph nodes, spleen, and tonsils are all involved as part of the lymphatic system. A major function of the lymphatic system is to protect the body from foreign particles such as bacteria, viruses, and fungi.

Integumentary The skin is the largest organ of the body and makes up most of the integumentary system. The system also contains associated structures such as hair follicles,

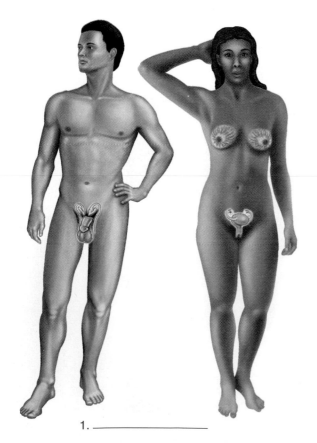
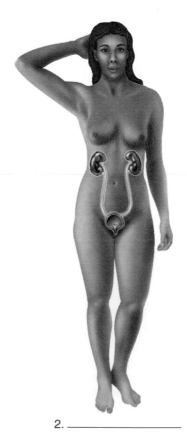
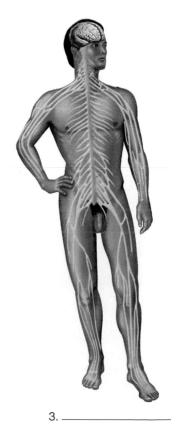

1. _____   2. _____   3. _____

Figure 2.2 Organ systems of the human body.

Continued

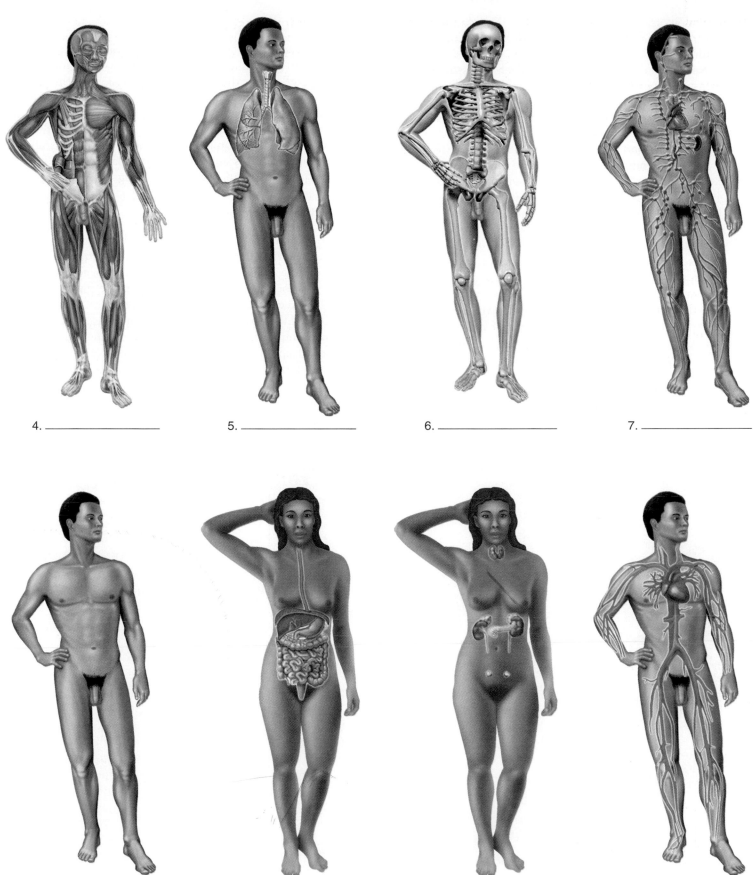

4. _____

5. _____

6. _____

7. _____

8. _____

9. _____

10. _____

11. _____

Figure 2.2—Continued.

hair, nails, and the glands of the skin. The skin protects the body against microorganisms, drying out, and produces vitamin D.

Digestive The mouth, esophagus, stomach, intestines, and liver are all parts of the digestive system, which functions to provide nutrients and water to the body.

Endocrine This system is determined by function, since endocrine organs produce hormones. Organs such as the thyroid gland and the adrenal glands are primarily endocrine glands (glands that secrete hormones without the use of ducts). Organs such as the pancreas and the gonads have dual functions; one is endocrine in nature and the other is exocrine (secretion of material through ducts). Hormones are vital in regulating growth, development, and maintaining a constant internal body condition.

Cardiovascular The heart and blood vessels make up this system, which is sometimes included with the lymphatic system as the circulatory system. The heart is the pump of the system and the blood vessels are the delivery and return portion of the system. The cardiovascular system is primarily involved in transporting oxygen, carbon dioxide, and other materials throughout the body.

You should examine models or charts in lab and locate the organs of the various organ systems.

Figure 2.3   Anatomical position. ✗

Anatomical Position

In clinical settings it is vital to have a proper orientation when dealing with patients. If two physicians are operating on a patient and one tells the other to make an incision to the left, the physician making the cut does not have to ask, "My left or your left?" because the cut is always to the *patient's* left side. When referring to the human body you will orient the body in **anatomical position.** In this position, the body is upright, facing forward, arms and legs straight, palms facing forward, feet flat on the ground, and eyes open (figure 2.3).

Directional Terms

With the body in anatomical position, there are specific terms used to describe the location of one part with respect to another. These terms are somewhat different for four-footed animals and for humans, so both examples are given. Table 2.1 lists the directional terms used for humans.

Locate the terms in figure 2.4. In quadrupeds (four-footed animals) the directional terms are somewhat different. Note that quadrupeds do not have a superior/inferior designation. Dorsal refers to the back, and ventral is on the belly side, while anterior (or craniad) is the front, or head-end, of the animal and posterior (or caudad) is the rear, or tail-end, of the animal. There are other terms for location that have unique meanings. For the digestive system, **proximal** refers to regions closer to

Table 2.1	Directional Terms Used for Humans	
Term	**Meaning**	**Example**
Superior	Above	The nose is superior to the chin
Inferior	Below	The stomach is inferior to the head
Medial	Toward the midline	The sternum is medial to the shoulders
Lateral	Toward the side	The ears are lateral to the nose
Superficial	Toward the surface	The skin is superficial to the heart
Deep	Toward the core	The lungs are deep to the ribs
Ventral (or Anterior)	To the front	The toes are ventral/anterior to the heel
Dorsal (or Posterior)	To the back	The spine is dorsal/posterior to the sternum
Proximal	For extremities meaning near the trunk	The elbow is proximal to the wrist
Distal	For extremities meaning away from the trunk	The toes are distal to the knee

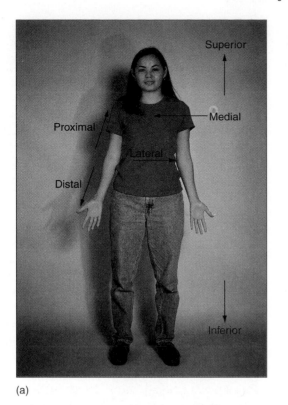

(a)

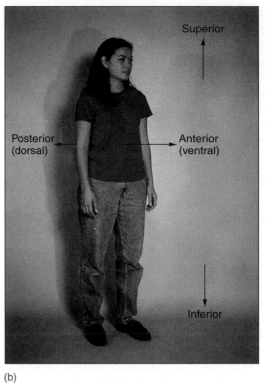

(b)

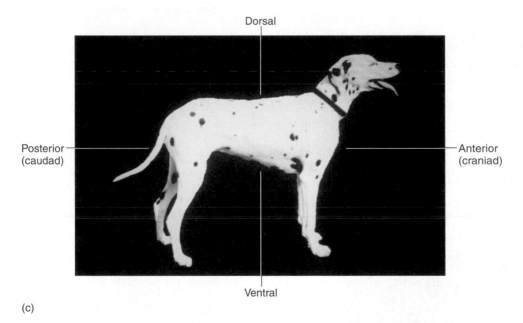

(c)

Figure 2.4   Directional terms. (a) For humans, anterior view; (b) for humans, 3/4 view; (c) for quadrupeds, lateral view. ✗

the mouth while **distal** is in reference to regions closer to the anus. **Parietal** is in reference to the body wall when compared to **visceral** which refers to areas closer to the internal organs. The heart, for example, has a visceral layer that is closer to the heart proper, while it also has a parietal layer that is further away from the heart. **Ipsilateral** refers to being on the same side of the body and **contralateral** refers to being on the other half (left side/right side). The right hand and right arm are ipsilateral, while the ears are contralateral.

Planes of Sectioning

When viewing a picture of an organ that has been cut it is very important to understand how the cut was made. Just as an apple looks very different when cut crosswise as opposed to lengthwise, so do some organs. Examine figure 2.5 for the following sectioning planes. A cut that divides the body or organ into superior and inferior parts is in the **transverse,** or **horizontal, plane.** A cut that divides the body into anterior and posterior portions is in the **coronal,** or **frontal, plane.** A cut that divides the body into left and right portions is in the **sagittal plane.** A cut that divides the body equally into left and right halves is called **midsagittal,** while one that divides the body into unequal left and right parts is called a **parasagittal plane.**

Body Cavities

The body contains two major body cavities, the **dorsal cavity** and the **ventral cavity.** These can be further divided into smaller cavities. The dorsal cavity is composed of the **cranial** and the **vertebral cavities** (figure 2.6). Compare the models and charts in

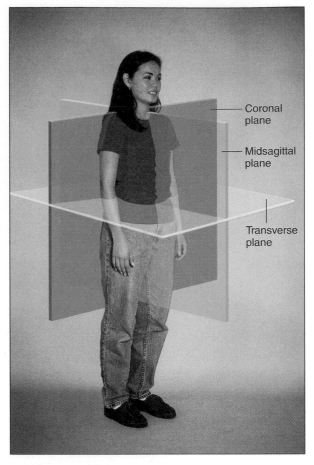

Figure 2.5   Sectioning planes.

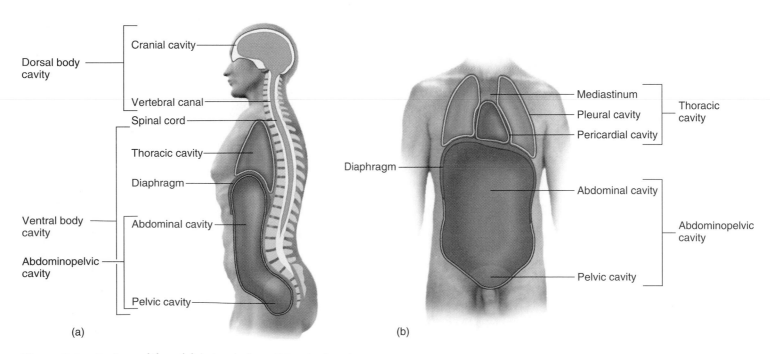

Figure 2.6   Body cavities. (a) Lateral view; (b) anterior view.

the lab with this figure. The ventral cavity is divided into (1) the thoracic cavity, above the diaphragm, which is further divided into the mediastinum and the pleural and pericardial cavities, and (2) the abdominopelvic cavity, below the diaphragm, which is divided into the abdominal and pelvic cavities.

Regions of the Body

Overview

Examine figure 2.7 for the various regions of the body. You will refer to these areas throughout this lab manual, so a complete study of these regions here is essential. Locate these regions on a torso model in lab.

Examine a muscle model in lab and name the region where the following muscles are found.

Pectoralis major _____

Trapezius _____

External oblique (front) _____

Rectus femoris _____

Gastrocnemius _____

Gluteus maximus _____

Flexor carpi radialis _____

Triceps brachii _____

Latissimus dorsi _____

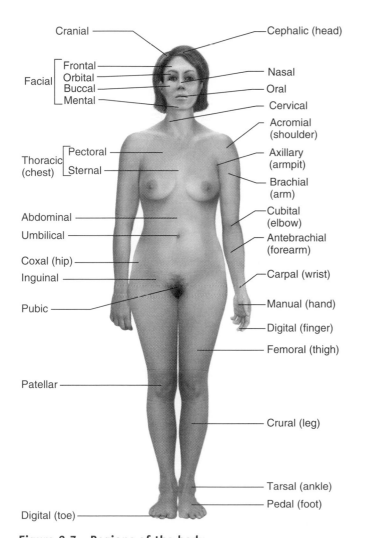

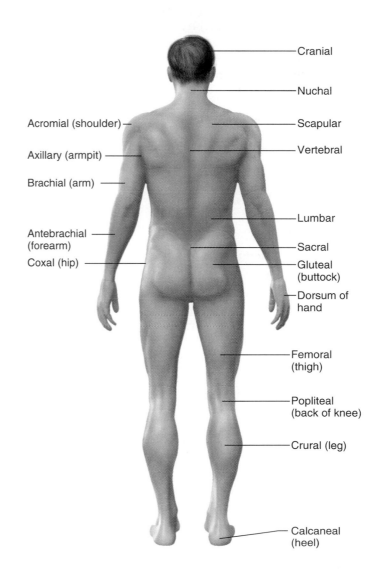

Figure 2.7   Regions of the body.

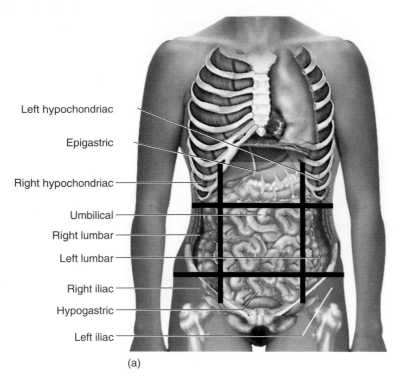

Left hypochondriac

Epigastric

Right hypochondriac

Umbilical

Right lumbar

Left lumbar

Right iliac

Hypogastric

Left iliac

(a)

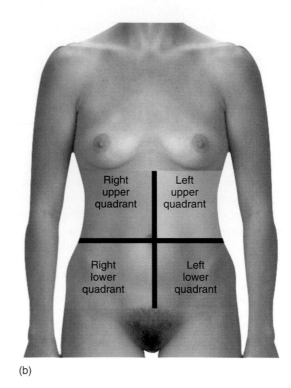

Right upper quadrant

Left upper quadrant

Right lower quadrant

Left lower quadrant

(b)

Figure 2.8 Abdominal regions. (a) Nine regions; (b) four quadrants.

Abdominal Regions

The abdomen can be further divided into either four quadrants or nine regions (figure 2.8). Anatomists generally use the nine regions approach, while clinicians typically use the four quadrants terminology.

Right upper quadrant
Left upper quadrant
Right lower quadrant
Left lower quadrant

Right hypochondriac
Left hypochondriac
Epigastric
Right lateral abdominal (lumbar)
Left lateral abdominal (lumbar)
Umbilical
Hypogastric
Right iliac (inguinal)
Left iliac (inguinal)

Name _____

1. What kind of section plane is illustrated in figure 2.6*a*?

2. What kind of section plane is illustrated in figure 2.6*b*?

3. In anatomical terms, referring to front and back, the pectoral region is _____ to the scapular region.

4. In terms of nearness to the trunk, the antebrachium is _____ to the carpal region.

5. In terms of front to back, the nipples are _____ to the shoulder blades.

6. The lungs are found in what specific cavities?

7. The lungs are found in what main body cavity (dorsal/ventral)?

8. In terms of up and down, the chin is _____ to the nose.

9. In terms of up and down, the umbilicus is _____ to the sternum.

10. In terms of nearness to the surface, the liver is _____ to the skin.

11. The kidneys belong to the _____ system.

12. The liver belongs to the _____ system.

13. The spleen belongs to the _____ system.

14. If you were to sit on a horse's back you would be on the _____ aspect of the horse.
    a. anterior   b. ventral   c. posterior   d. dorsal

15. If a hairline fracture occurred in the proximal humerus (arm bone) would the injury be closer to the shoulder or to the elbow? Why?

16. In a clinical report there is a note of a laceration (cut) on the posterior crural region. Where does this occur in layman's terms?

17. Pain in the appendix would be felt in what quadrant of the abdomen?

18. What is the difference between the abdomen and the abdominal cavity?

19. The brain is located in what specific cavity?

20. What is another name for the armpit region?

21. The region of the abdomen directly under the right side of the rib cage is the _____ region.

22. The plane that divides the body into a top half and bottom half is the _____.

23. In anatomical position the palms of the hand are facing _____.

24. The portion of the head other than the facial region is the _____.

25. What common name is used for the antebrachial region?

Microscopy

Introduction

Originally the study of anatomy and physiology was based on macroscopic, or gross, observation. With the invention and use of the light microscope much greater detail was seen, and thus began the study of cells and tissues. Light microscopy involves the use of visible light and glass lenses to magnify and observe a specimen. Further advances in microscopy led to the development of electron microscopy, which uses electrons passing through or bouncing off of material. Electron microscopy has revealed much greater detail than observable with the light microscope. This exercise involves the use of the light microscope, how to examine prepared slides under the microscope, and how to make slides of fresh material for study.

Objectives

At the end of this exercise you should be able to

1. list the rules for proper microscope use;

2. name the parts of the microscope presented in this exercise;

3. demonstrate the proper use of the light microscope;

4. place a microscope slide on the microscope and observe the material, in focus, under all magnifications of the microscope;

5. calculate the total magnification of a microscope based on the specific lenses used;

6. prepare a wet mount for observation.

Materials

Light microscope
Prepared slide with the letter *e* (or newsprint and razor blades)
Transparent ruler or sections of overhead acetates of rulers
Glass microscope slides
Coverslips
Lens paper
Kimwipes or other cleaning paper
Lens cleaner
Small dropper bottle of water
1% methylene blue solution
Toothpicks
Histological slides of kidney, stomach, or liver

Procedure

Before you begin to examine tissue slides, you first need to learn about the use of the microscope. Microscopes are very expensive pieces of equipment, and you should always take great care handling them. There are a few rules concerning microscopes that you should observe:

1. When carrying the microscope, hold it securely with two hands—one hand under the base and one on the arm.
2. Keep the microscope upright at all times.
3. Keep microscope lenses clean with lens cleaner and softened lens paper.
4. Use only the fine-focus knob when using the high-power objective lens.
5. Remove slides from the microscope before putting it away.
6. Secure the cord with a rubber band or wrap the cord around the base of microscope.
7. Store the microscope with the *low-power* objective lens in place.
8. Put the microscope away in its proper location.

To use the compound light microscope properly you need to learn some of the parts of the instrument. There are differences among microscopes in use in various labs, so the microscope you are using may not look exactly like the one described here (figure 3.1). Take one of the microscopes from the cabinet and carry it back to your table. Place a check mark next to the appropriate space when you locate the part of the microscope.

Microscope Parts		
_____ Base	_____ Arm	_____ Objective lens
_____ Condenser	_____ Iris diaphragm lever	_____ Body tube
_____ Nosepiece	_____ Coarse-focus knob	_____ Fine-focus knob
_____ Stage	_____ Ocular (eyepiece) lens	_____ Light source

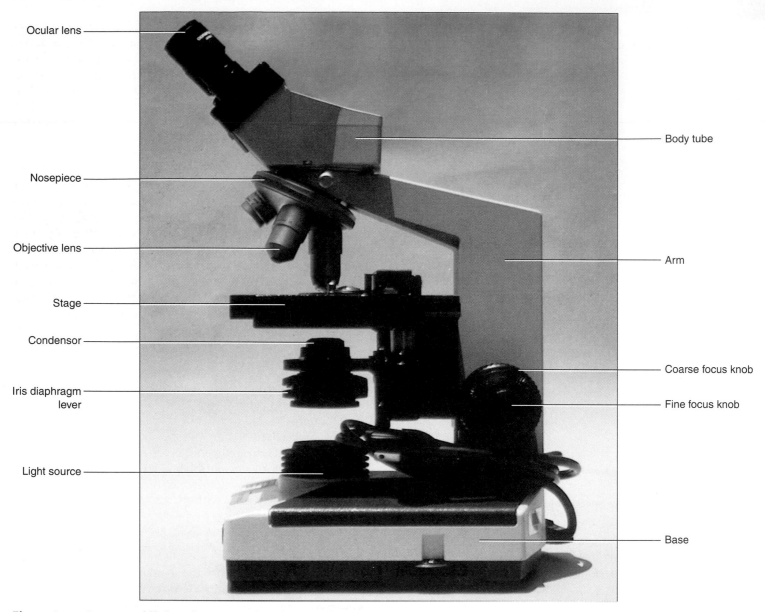

Figure 3.1 Compound light microscope. (Courtesy of Olympus America, Inc.)

To begin your observations under the microscope select a prepared slide with the letter *e* on it or cut out a letter from newsprint and place it on a glass microscope slide. Bring the slide back to your table and perform the following:

1. Place the microscope on the table before you with the ocular lens or lenses facing you.
2. Plug in the microscope. Make sure the cord is not hanging over the counter or in the aisle where someone might trip on it or pull the microscope off the table.
3. If the microscope has an illuminator (light source) dial make sure that the setting is on the lowest level.

4. Turn the nosepiece until the low-power objective lens (the one with the lowest number or the shortest barrel) clicks into place. When first looking at microscope slides always use the low-power objective lens.
5. Locate and turn on the light switch.
6. Place a prepared slide with the letter *e* on the microscope stage so that the letter is facing you.
7. Position the material on the slide so the light from the illuminator shines through the material.
8. Use the coarse-focus knob to bring the specimen into focus.

9. Adjust the iris diaphragm for brightness.
10. Examine the slide and draw what you see in the space below.

If you are having a difficult time seeing anything or seeing things in focus, there could be several reasons. Use the following troubleshooting list to help you.

Problem	Solution
Nothing is visible in the lens.	Plug in microscope. Turn power supply on. Rotate objective lens so that it is in place. Bulb is burned out; replace bulb.
You see a dark crescent.	The objective lens is not in proper position; click the lens into place.
All you see is a light circle.	Microscope is out of focus; adjust coarse focus knob. Light is up too high, turn down light. Iris diaphragm is open too much, close it down.

Focusing the Microscope

Focusing the microscope requires a little bit of patience. When first looking at microscope slides always use the low-power objective lens. Make sure the specimen is on the stage and centered in the open circle on the stage. There should be light coming through the specimen. The coverslip should be very close to the objective lens. Look at the microscope stage from the side, adjust the coarse-focus knob so that the coverslip is almost touching the objective lens. Look through the ocular lens and rotate the coarse-focus knob slowly so the objective lens and the slide begin to move away from each other. This should bring the object into focus in the field of view. The field of view is the circle that you see as you look into the microscope.

After you get the specimen in focus under low power, you should examine the material under higher powers. This is done by centering the image that you observe in the field of view and then switching the objective lens to the next higher power. Do not move the mechanical stage out of the way. The next higher objective lens should clear the slide. Once you rotate the lens, adjust the focus by using the *fine-focus knob*. You can look at the subject under high power by the same procedure by turning the objective lens to the high power lens. If you cannot focus on the high power or have lost the image that you were looking for, you should return to low power and try the process again. If you still cannot find the object under high power you should ask your instructor to help you.

Proper Lighting

Too much or too little light makes a specimen difficult to see. You can change the light level by either adjusting the light from the illuminator or adjusting the iris diaphragm lever. On some faintly stained specimens the material may be difficult to see on low power. One trick is to locate the edge of the coverslip and turn the coarse-focus knob up and down until the edge is in sharp focus. This lets you know that you are in the approximate focal plane for examining the material on the slide. Move the slide to where the specimen should be, adjust the light, and reexamine it.

Magnification and Field of View

You can determine the size of the object under view if you know the diameter of the field of view. The field of view can be measured directly when under low magnification by using a clear ruler. If higher magnifications are used, rulers won't work and you have to calculate the field of view. There is a relationship between the diameter of the field of view and the magnification used. You can first calculate the total magnification using the following procedure. Look at the barrels of your microscope and determine the magnification of each lens.

Eyepiece (ocular) magnification: _____

Low-power objective magnification: _____

Total magnification (= ocular magnification × objective magnification): _____

Place a transparent section of ruler or a ruled section of acetate on the stage of the microscope. The space between each dark line that runs vertically is 1 millimeter (mm). Count the number of millimeters at the broadest part of the field of view and enter this number as the diameter of the field of view in the following space.

Diameter of the field of view (mm): _____

You can calculate the length of an object by determining how much of the diameter of the field of view it occupies. Let's say that the diameter of the field of view is 10 mm. If an object takes up one-half of the field of view, then you can estimate its size at 5 mm. If the object took up only one-third of the field of view, how large would it be? Record your answer in the following space.

Object size (mm): _____

As the magnification increases, the field of view decreases proportionally. Thus, if the diameter of the field of view is 10 mm at one magnification and you double the magnification by changing lenses, the field of view is reduced to a diameter of 5 mm. If you switch to a new lens and increase the magnification by 10 times, then the field of view is reduced to one-tenth of the original field of view. Look at figure 3.2 for a representation of this proportional change in the field of view.

Keep the clear ruler or ruled acetate sheet under the microscope and increase the magnification to the next higher power by moving the next larger objective lens in place. Record the total magnification of your microscope with this objective lens.

Total magnification: _____

Examine the ruler under the microscope and record the diameter of the field of view in millimeters.

Diameter of the field of view in millimeters: _____

Has the increase in magnification produced a decrease in the field of view? Is the decrease in the field of view proportional to the magnification?

Now calculate the total magnification of the microscope using the high-power objective lens.

Magnification with high-power objective lens: _____

You will not be able to measure the field of view accurately under high power with a ruler or acetate sheet; however, you should be able to calculate the diameter of the field of view. For example, if the diameter of the field of view is 2.5 mm at 40 power (40×), then it would be 0.25 mm at 400× (10 times more magnified yet one-tenth the field of view). Calculate the diameter of the field of view under high power.

Diameter of the field of view under high power: _____

Preparation of a Wet Mount

You can make relatively quick and easy observations under the microscope as long as the material is thin enough and small enough. One technique for cell observation is to exam-

Figure 3.2   Increasing magnification and decreasing field of view.

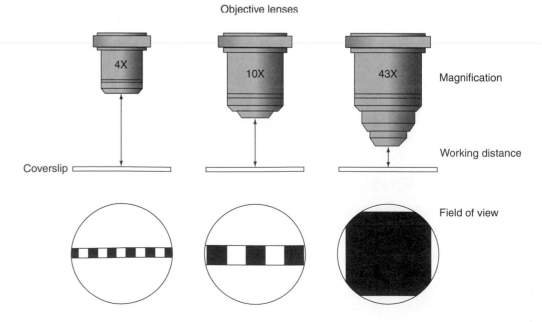

ine cells from the inside of the oral cavity. Do the following procedure.

1. Take a toothpick and *gently* scrape the inside of your cheek.
2. Smear the cheek material from the toothpick on a clean microscope slide.
3. Place a drop of methylene blue on the smear.
4. Place a coverslip on the slide and examine it under the microscope (figure 3.3). The small oval structures inside the cells are the nuclei.
5. Draw your observations in the space provided.

Your illustration of a cheek cell:

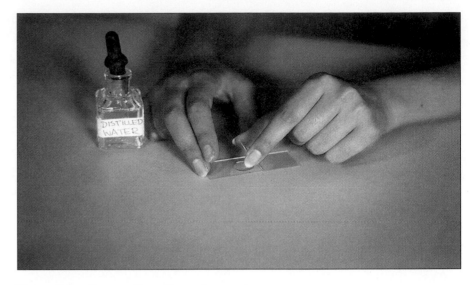

Figure 3.3   Preparation of a wet mount.

For another observation, remove a hair from your head (preferably one with split ends) and examine it by making a wet mount. Place the hair in the center of the slide and add a drop of water. Place the coverslip on one edge of the drop and slowly lower it. Try to avoid trapping any air bubbles in the process.

Observation of a Prepared Slide

Examine a prepared slide of tissue provided by your instructor. Examine the entire sample using the low-power objective lens. You should scan the entire sample looking for areas you want to observe closer. Move to the next higher power and adjust the focus using the fine-focus knob. Finally, examine the material with the high-power objective lens and draw what you see in the space provided.

Your illustration of material from a prepared slide:

Name of the sample drawn:

Oil Immersion Lens

The objective lenses you have used so far are called "dry" lenses. Your lab may be equipped with microscopes that have oil immersion lenses. The techniques for using these lenses

are somewhat different from those for dry lenses. Once you have examined the specimen using the high-power dry lens, find the spot you want to examine and center it in the field of view. Add a drop of immersion oil on top of the coverslip and carefully swing the oil immersion lens into place. Use the fine focus only or you may drive the oil immersion lens through the slide and break it. Once you have examined the slide, swing the lens away and remove the slide, carefully wiping away the immersion oil with a clean piece of lens paper (do not use your shirt or a paper towel; these can scratch the lens). Use only lens paper to clean the oil from the oil immersion lens. Use another paper to remove any remaining oil.

Cleaning the Microscope

Smudges on the images you view through the microscope may be due to several things. There may be makeup or dirt on the ocular lens (or lenses). There may be dirt, oil, salt, stains, or other material on the objective lenses. To clean a lens, place a small amount of lens-cleaning fluid on a clean sheet of lens paper. Make one circular pass on the lens and throw away the paper. If you continue to clean the lens with the same lens paper, you can grind dirt or dust into it. Use a fresh piece of lens paper and repeat the procedure if further cleaning is needed.

You may want to clean the microscope slide before you examine it. Use a cleaning paper such as a Kimwipe to clean oil or dust from the slide.

Finally, dust may have collected inside the microscope over the years or the lenses may be scratched. There is nothing you can do about this, though you may want to bring this to your instructor's attention.

Name _____

1. Ocular lenses are typically 10×. If the objective lenses of a microscope are 5×, 17×, and 35×, what would all of the total magnifications be if you used these lenses sequentially?

    a. _____

    b. _____

    c. _____

2. What is the function of the iris diaphragm of the microscope?

3. If the diameter of the field of view is 5.6 mm at 40×, what would the diameter be at 80×?

4. Label the parts of the microscope illustrated.

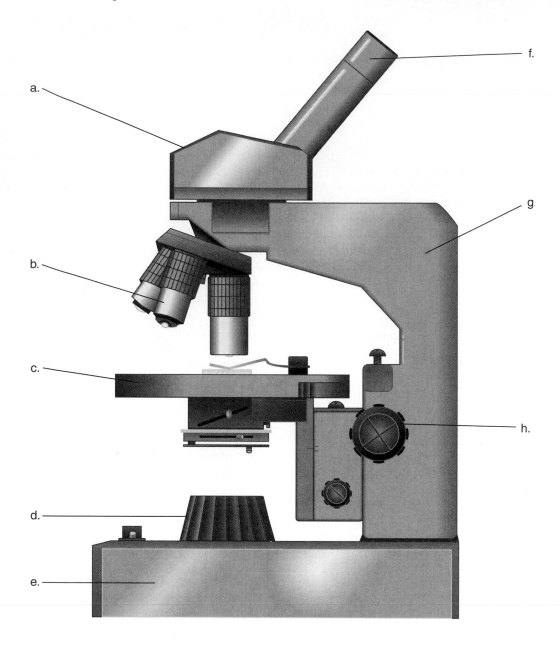

LABORATORY EXERCISE 4

Cell Structure and Function

Introduction

The cell is the structural and functional unit of the human body and is the fundamental unit of living organisms. Most diseases that produce obvious physical dysfunction can be traced to some type of cellular change. In this exercise, you examine the structure of cells and learn how cells of the body divide to make new cells. Cells perform many functions, some of which are unique to the particular organ where they are found. Generally speaking, however, cells grow, divide, acquire nutrients, release wastes, and respond to local stimuli.

Objectives

At the end of this exercise you should be able to

1. describe the importance of cells in the makeup of the body;
2. list the functions of the nucleus;
3. list all of the organelles and their functions;
4. describe the structure and functions of the mitochondria;
5. draw a representation of each of the organelles;
6. describe the three main events of the cell cycle;
7. name the four phases of mitosis and the events occurring in each phase.

Materials

Models or charts of animal cells
Electron micrographs of cells or textbook with electron micrographs
Prepared slides of whitefish blastula
Microscopes
Modeling clay (Plasticine)—two colors
Marbles

Procedure

Overview of the Cell

Cells consist of three main parts, the **plasma (or cell) membrane,** the **nucleus,** and the **cytoplasm.** The plasma membrane is the outer boundary of the cell, and though it cannot be seen using the light microscope, its location can usually be determined by the difference in color between the cytoplasm and the surrounding liquid on the microscope slide. The cyto-plasm is the portion of the cell in which water, dissolved materials, and small cellular **organelles** (organelle = small organ) are found. The portion in which the organelles are suspended is called the **cytosol.** Small filaments and tubules make up the **cytoskeleton** (also considered part of the cytoplasm). The nucleus directs the cell's activities and stores its genetic information. It is visible with the light microscope and frequently appears as a spherical or oblong structure. Figure 4.1 shows the plasma membrane, cytosol, and nucleus. Locate these on the model or charts in the lab.

Plasma Membrane

The plasma membrane is composed of a **phospholipid bilayer.** This consists of an outer and inner sheet of phosphate molecules each with hydrocarbon (lipid) tails directed toward the middle of the membrane (figure 4.2). Interspersed among the phospholipid molecules are **cholesterol molecules,** which provide stability to the membrane or, in larger concentrations, can make the membrane more fluid. Two major types of proteins are also found in the membrane. **Peripheral proteins** are found on the inner or outer surface of the membrane, while those passing through the membrane are known as **integral (transmembrane) proteins.** Peripheral proteins may have carbohydrates or other molecules associated with them and frequently serve as cell markers. Integral proteins function as channels by which some materials can pass through the membrane. The plasma membrane is important in establishing electrochemical differences, which allow for nerve impulse conduction and muscle contraction. It is also a selectively permeable membrane in that it provides an entrance or exit to the cell for some materials while excluding other material from entering or exiting the cell's interior. Proteins may serve to anchor one cell to another, provide a place for metabolic reactions to take place, act as cell markers that identify a particular cell, and act as receptors or channels.

Cytoplasm

Most of the inside of the cell is cytoplasm, which consists of the inner fluid portion of the cell known as the cytosol, the inner framework of the cell called the cytoskeleton, and the small specialized units of the cell called organelles. The cytosol is composed of water with dissolved materials such as sugars, ions, colloidal proteins, and amino acids.

Figure 4.3 illustrates the cytoskeleton, which consists of microtubules, microfilaments, and intermediate filaments, all of

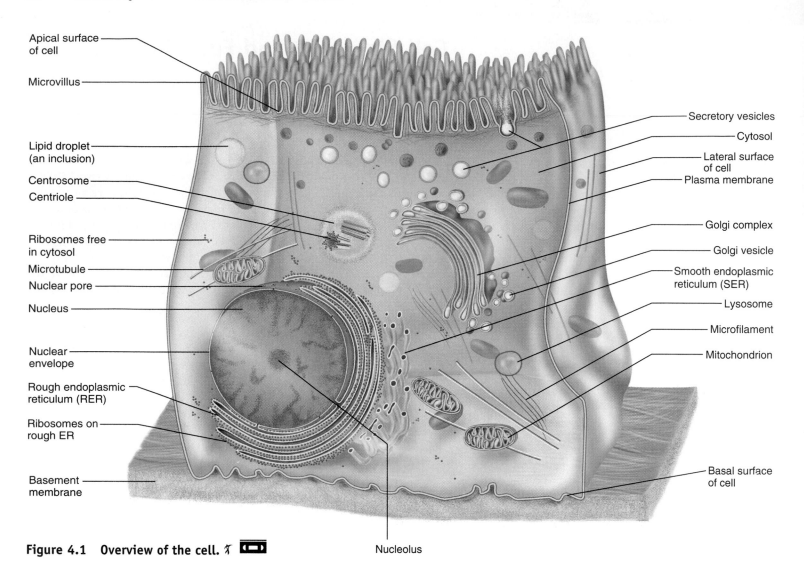

Apical surface of cell

Microvillus

Lipid droplet (an inclusion)

Centrosome

Centriole

Ribosomes free in cytosol

Microtubule

Nuclear pore

Nucleus

Nuclear envelope

Rough endoplasmic reticulum (RER)

Ribosomes on rough ER

Basement membrane

Secretory vesicles

Cytosol

Lateral surface of cell

Plasma membrane

Golgi complex

Golgi vesicle

Smooth endoplasmic reticulum (SER)

Lysosome

Microfilament

Mitochondrion

Basal surface of cell

Nucleolus

Figure 4.1   Overview of the cell.

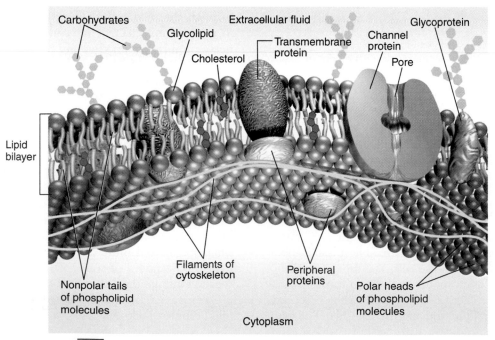

Carbohydrates

Extracellular fluid

Glycoprotein

Glycolipid

Cholesterol

Transmembrane protein

Channel protein

Pore

Lipid bilayer

Filaments of cytoskeleton

Peripheral proteins

Nonpolar tails of phospholipid molecules

Polar heads of phospholipid molecules

Cytoplasm

Figure 4.2   Plasma membrane.

Figure 4.3   Cytoskeleton. ✗ 📼

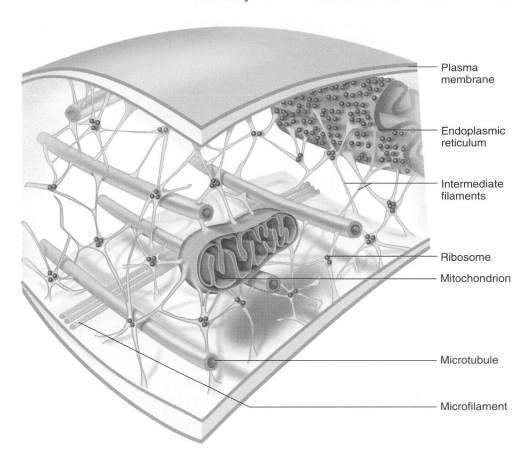

Plasma membrane

Endoplasmic reticulum

Intermediate filaments

Ribosome

Mitochondrion

Microtubule

Microfilament

which provide shape to the cell, a place to anchor organelles, and resistance to gravitational and other forces acting on the cell. Microtubules are made of the protein tubulin and are approximately 25 nanometers (nm) in diameter. Intermediate filaments are composed of fibrous proteins and are approximately 10 nm in diameter. Microfilaments are made of actin and are approximately 6 nm in diameter.

Organelles

The term *organelle* literally means "small organ." Each organelle serves a particular function in the cell. There are two types of organelles—membranous and nonmembranous. Organelles represent a wonderful example of **specialization** on a microscopic scale—individual organelles have specific structural characteristics reflecting their specific functions. Look at the illustrations of the various organelles as you read the following text.

Mitochondria

The major function of the mitochondrion (plural, *mitochondria*) is to convert the stored chemical energy in a sugar molecule to stored chemical energy in molecules of **adenosine triphosphate (ATP).** Mitochondria are elongated organelles approximately 0.2 to 5.0 micrometers (μm) in length. Mitochondria

have two membranes that are similar in structure to the plasma membrane in that they are composed of phospholipid bilayers. Examine the illustration of a mitochondrion in figure 4.4. Note the separate outer and inner membranes. The inner has folds called **cristae.** These cristae increase the surface area of

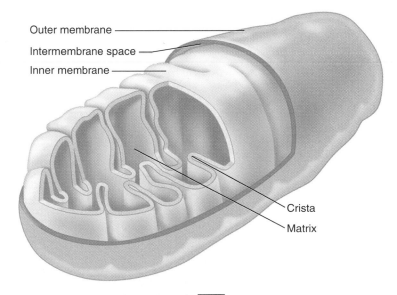

Outer membrane

Intermembrane space

Inner membrane

Crista

Matrix

Figure 4.4   Mitochondrion. ✗ 📼

the mitochondrion. The inner region of the mitochondrion is known as the mitochondrial matrix (stroma) and it is the area where the citric acid cycle occurs.

Ribosomes

Ribosomes are the smallest of the organelles (about 25 nm in diameter), are nonmembranous, and function to produce proteins. Ribosomes are composed of two subunits that come together during protein synthesis. Ribosomal subunits vary among some organisms, but humans have the typical vertebrate pattern.

Ribosomes come in two forms, **free ribosomes** (cytoplasmic ribosomes) and **bound ribosomes** (endoplasmic ribosomes.) Free ribosomes make proteins for use inside the cell, while bound ribosomes make proteins for use outside the cell. An example of externally used proteins are the digestive enzymes of the stomach. In this case, specialized cells in the inner lining of the stomach wall have bound ribosomes that secrete proteins that eventually become digestive enzymes when they are activated near the ingested food. Figure 4.5 shows examples of ribosomes.

Endoplasmic Reticulum

The **endoplasmic reticulum** is an organelle composed of a network of enclosed channels. The name *endoplasmic reticulum* literally means the "little net inside the goo." There are two types of endoplasmic reticulum (figures 4.1 and 4.6): **rough endoplasmic reticulum,** which contains bound ribosomes, and **smooth endoplasmic reticulum,** which does not have ribosomes on its surface. The rough endoplasmic reticulum is associated with the nucleus and functions to produce proteins for transport and use outside the cell. The smooth endoplasmic reticulum is often a distal extension of the rough endoplasmic reticulum and serves to produce lipid and steroid compounds and detoxify material. **Cisternae** are open spaces in the endoplasmic reticulum that receive proteins and secrete them in vesicles (fluid-filled sacs) for modification.

Golgi Complex

The **Golgi complex** (or Golgi body) receives material from the endoplasmic reticulum and other parts of the cytoplasm and serves as an assembly and packaging organelle. Examine the Golgi complex in figure 4.7 and note the vesicles that are released from this organelle. Proteins from the endoplasmic reticulum are joined with carbohydrates, lipids, metals (such as the iron in hemoglobin), or other materials in the Golgi complex and then secreted in **vesicles** for transport from the cell.

Vesicles

These fluid-filled sacs inside the cell protect the integrity of the plasma membrane. If a substance was pushed out of the cell by opening a hole in the plasma membrane, the likely consequence would be that the cell would burst. There is a limit to the size of a molecule permitted to pass through the phospholipid bilayer or through the protein channels in the plasma membrane. Larger molecular weight materials are assembled inside the Golgi complex and enclosed in vesicles. The membrane of the vesicle can fuse with the cell membrane and the larger molecules are ejected without disrupting the plasma membrane. Two common types of vesicles are discussed next.

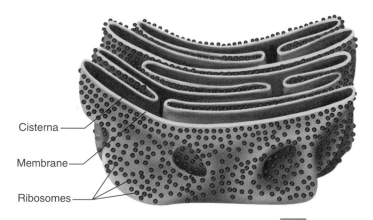

Cisterna

Membrane

Ribosomes

Figure 4.6   Rough endoplasmic reticulum.

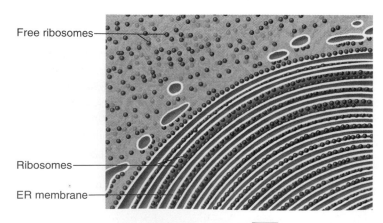

Free ribosomes

Ribosomes

ER membrane

Figure 4.5   Free ribosomes of a cell.

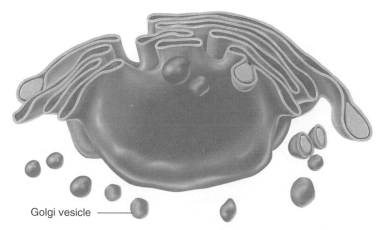

Golgi vesicle

Figure 4.7   Golgi complex.

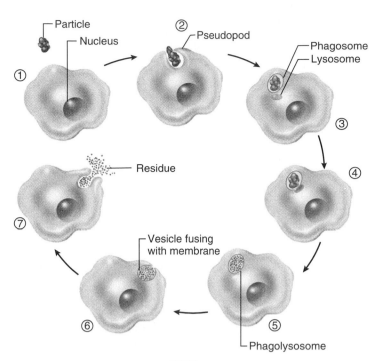

Figure 4.8 Lysosomes. ℐ ▭

Organelle	Membrane	Function
Mitochondrion	Double	ATP production; fatty acid oxidation
Ribosome	None	Protein production
Rough endoplasmic reticulum	Single	Protein production for export
Smooth endoplasmic reticulum	Single	Lipid and steroid synthesis; detoxification
Golgi complex	Single	Assembly of macromolecules; transport from cell for secretion
Lysosome	Single	Digestion of material
Peroxisome	Single	Conversion of H_2O_2 to $H_2O + O_2$

Table 4.1 Organelles and Their Functions in Cells

Lysosomes

Lysosomes are vesicles filled with digestive enzymes. Some cells, such as white blood cells, fuse with foreign particles and release the contents of the lysosome near the foreign particles. In this way digestive enzymes hydrolyze the substance, which can later be removed from or incorporated into the body. Figure 4.8 illustrates lysosomes.

Peroxisomes

Peroxisomes are vesicles that contain peroxidase, an enzyme that converts hydrogen peroxide (H_2O_2) to water (H_2O) and oxygen ($\frac{1}{2} O_2$). Hydrogen peroxide is very damaging to tissue, and removal of peroxide lessens potential damage to cells.

Go back to figure 4.1 and locate the organelles listed in table 4.1 in that illustration. Be able to recognize the variances in structure and function of the organelles.

Nucleus

The **nucleus** has two major functions: one is to house the genetic information of the cell and the other is to control the various tasks of the cell. These two functions are carried out by **DNA (deoxyribonucleic acid),** which combines with proteins to form a material called **chromatin** in the nucleus. The nucleus is bounded by a **nuclear membrane** (nuclear envelope), which is a double membrane, each one resembling the plasma membrane (in that it is composed of a phospholipid bilayer). The nuclear membrane has holes in it called **nuclear pores,** which allow movement of materials out of the nucleus.

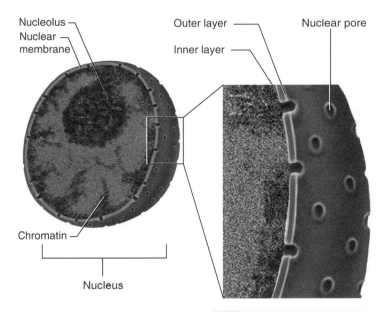

Figure 4.9 Nucleus of a cell. ℐ ▭

One or more structures known as the **nucleoli** (singular, *nucleolus*) are inside the nucleus. The nucleoli consist of **RNA (ribonucleic acid)** and protein and function to make ribosomes, the protein-producing organelles in the cytoplasm of the cell. Figure 4.9 shows the structures of the nucleus.

The Cell Cycle

One of the great wonders of science is the mechanism by which a single cell, the result of the fusion of egg and sperm, develops into a complex, multicellular organism such as a human. Various estimates put the number of cells in the human body in the trillions. All of these cells came from the first cell, or **zygote.** In this part of the lab exercise we examine the mechanism by which this occurs.

Most cells produce more cells by a process known as the cell cycle. The cell cycle can be divided into three particular events known as **interphase, mitosis,** and **cytokinesis.**

Interphase

Interphase is the time when a cell undergoes growth and duplication of DNA in preparation for the next cell division. If a cell is not going to divide any further (such as brain cells and some muscle cells), then interphase is regarded as the time when a cell carries out normal cellular function.

Interphase has three separate phases known as the G_1 **phase, S phase,** and G_2 **phase.** In the G_1 phase (G stands for growth), cells are in the process of growing in size and producing organelles. In the S phase (S stands for synthesis) the DNA of the cell is duplicated. The **double helix** of the DNA molecule unzips and two new, identical DNA molecules are produced. In the final phase of interphase, the G_2 phase, the cell continues to grow and prepares for the process of mitosis. Some cells do not undergo further division and are said to be in the G_0 (G-zero) phase.

Mitosis

Mitosis is a continuous event that has been divided into four distinct phases. Mitosis is **nuclear division** and it involves the division of genetic information to produce two identical nuclei. In order for mitosis to occur, the chromatin in the nucleus of the cell must condense into compact units called **chromosomes.** Chromosomes consist of two **chromatids** held at the center by a **centromere.** Examine figure 4.10 for the structure of a chromosome. You should also note the structure of chromosomes as you study the cells undergoing mitosis. The four phases of mitosis, **prophase, metaphase, anaphase,** and **telophase,** are described next. Refer to figure 4.11 as you read the descriptions.

Prophase
Cells in interphase have a distinct nuclear membrane, and the genetic information is dispersed in the nucleus as chromatin. The first indication that a cell is undergoing mitosis is the condensation of chromatin into chromosomes. Each chromosome consists of two elongated arms known as **chromatids,** which are connected to each other by a centromere.

In addition to the thickening of the chromosomes, the nucleolus disappears and the nuclear membrane begins to disassemble. In order for the chromosomes to separate and migrate apart, the nuclear membrane must not be present. Additionally the **mitotic apparatus** begins to become apparent. The mitotic apparatus consists of **spindle fibers,** which attach to the chromosomes at regions of the centromere known as

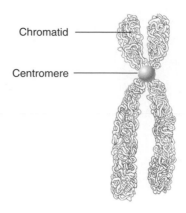

Figure 4.10   Structure of an isolated chromosome. ⚞ ▭

the **kinetochores,** and two **asters,** which are points of radiating fibers at each end (poles) of the cell. In the center of the asters are two small structures known as **centrioles.**

Metaphase
In this phase the chromosomes align between the poles of the cell in a region known as the **metaphase plate.**

Anaphase
In anaphase, the chromosomes separate at the centromere and each chromatid is now known as a **daughter chromosome.** The spindle fibers have attached to the region of the centromere known as the kinetochore and apparently pull the daughter chromosomes toward opposite poles of the cell. The centromere region moves first and the arms of the chromosomes follow.

Telophase
Once the daughter chromosomes reach the poles, telophase begins. The daughter chromosomes begin to unwind into chromatin, the nucleolus reappears, and the nuclear envelope begins to re-form. The mitotic apparatus disassembles, thus terminating mitosis.

Cytokinesis

The splitting of the cell's cytoplasm into two parts is known as **cytokinesis.** Although cytokinesis is a distinct process, it frequently begins during late anaphase or early telophase. In late anaphase, as the chromosomes are moving to the poles, the plasma membrane begins to constrict at a region known as the **cleavage furrow.** This begins the process of dividing the cytoplasm (figure 4.11), ending as the cell splits into two separate daughter cells. The cytoplasm and the organelles are effectively divided into two parts.

Examine a slide of whitefish blastula and look for the cells in the slide. Most of the cells that you see are in a particular part of the cell cycle. What is this phase and why are most of the cells in this phase?

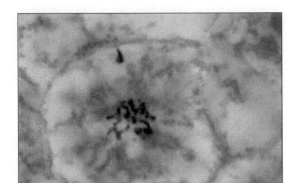

Prophase

(a)

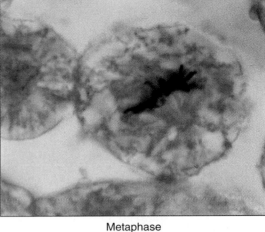

Metaphase

(b)

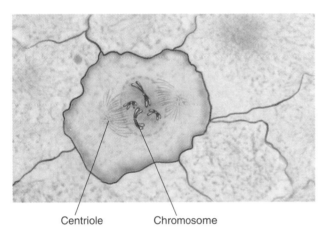

Centriole   Chromosome

Metaphase plate   Spindle fiber

Figure 4.11   Phases of mitosis (1,000× . (a)Prophase; (b)metaphase; (c)anaphase; (d)telophase with cytokinesis. ⚡ 📼

Continued

Compare the slide with the figure 4.11. Draw representative cells in each phase of mitosis in the following space.

Review the phases of mitosis in table 4.2. Take two colors of modeling clay and make chromosomes from the clay. You should have a long chromosome and a short chromosome of each color, for a total of four chromosomes. Each chromosome should have two chromatids, and the chromosomes should be joined by a marble, which represents the centromere. Draw a large circle on a sheet of paper to represent a cell. Manipulate the clay chromosomes to show how mitosis occurs. After you have done this, describe the process of mitosis in your own words. List each stage and what happens in that stage.

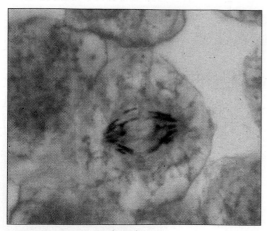

Anaphase

(c)

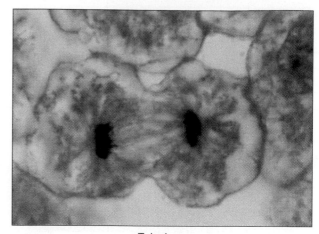

Telophase

(d)

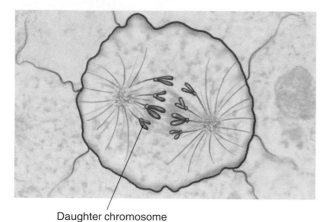

Daughter chromosome

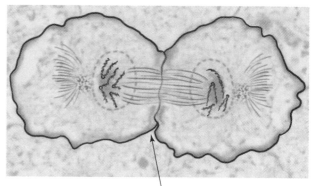

Cleavage furrow

Figure 4.11—Continued.

Table 4.2	Major Events of Mitosis
Prophase	Chromatin condenses to form chromosomes
	Nuclear envelope disappears
	Spindle apparatus forms
	Nucleolus disappears
Metaphase	Chromosomes align on the equator
Anaphase	Chromosomes split and daughter chromosomes migrate to poles; cytokinesis often begins
Telophase	Chromosomes reach poles; nuclear envelope re-forms
	Chromosomes unwind to chromatin; cytokinesis divides the cytoplasm
	Nucleolus reappears

Name _____

1. Name the phases of the cell cycle as illustrated.

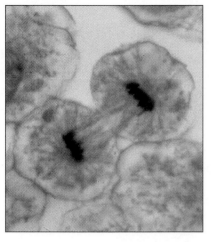

(a) _____

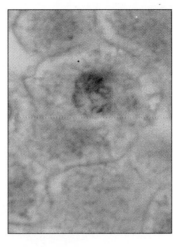

(b) _____

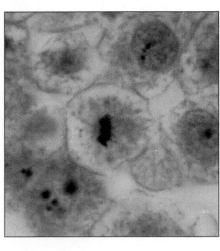

(c) _____

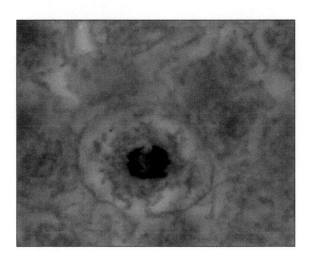

(d) _____

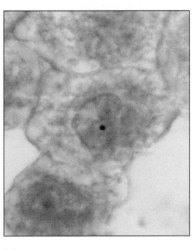

(e) _____

2. Name the organelles as illustrated.

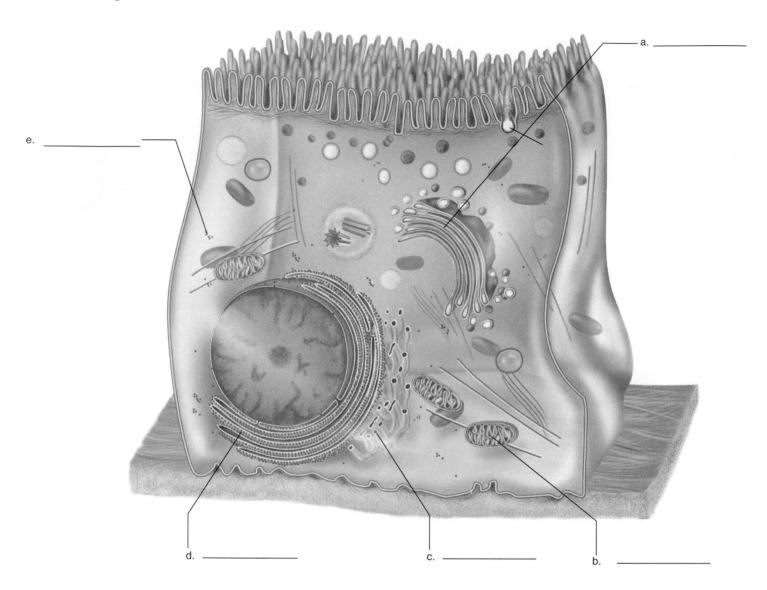

3. Fill in the name, structure, or function of the organelles.

Organelle	Function	Structure
1. _____	protein production for use inside of the cell	not membrane-bound
2. Smooth ER	_____	no attached ribosomes
3. _____	packages and transports material	vesicles on organelle
4. Mitochondrion	produces energy for the cell	_____
5. Rough ER	protein production for use outside of the cell	_____

4. How is the rough endoplasmic reticulum related to the Golgi complex in terms of function?

5. What two organelles produce proteins?

6. What is the function of the cytoskeleton?

7. The cytoplasm has a liquid portion. What is it called?

8. What structure in the cell is mostly composed of a phospholipid bilayer?

9. In what cellular object is cholesterol vital for the structure?

10. Name the stages of interphase and what happens during those phases.

Some Functions of Cell Membranes

Introduction

The membrane of the cell is the dynamic interface between the internal, living environment of the cell and the external environment. In humans, most cells are bathed in a liquid medium called extracellular fluid (ECF), which provides nutrients, oxygen, hormones, water, and other materials to the cell. From the interior of the cell, the cell releases ammonia, carbon dioxide, and other metabolic products into this same liquid. The cell membrane is vitally important in the exchange of materials between the internal fluid of the cell and the environment surrounding the cell. The exchange of materials between the cell and the ECF maintains the homeostatic balance the cell must have to survive. Even small changes in the concentration of certain materials in the cell might lead to cellular death, and so the constant adjustment of water, ions, and other metabolic products is extremely important. In large part, the cell membrane actively regulates what enters and what exits the cell. In this exercise, we look at the physical processes that influence cell membrane dynamics and study the nature of membrane transport.

Objectives

At the end of this exercise you should be able to

1. describe the processes by which substances move across membranes;
2. define the terms *hypertonic*, *hypotonic*, and *isotonic*;
3. define the terms *diffusion*, *osmosis*, and *filtration*;
4. describe the movement of water across a selectively permeable membrane;
5. compare and contrast diffusion and osmosis.

Materials

Brownian Motion
Whole milk
Fat testing solution (1% Sudan III solution)
Microscopes
Microscope slides
Coverslips
Vegetable oil or light household oil in dropper bottles
Hot plate

Diffusion Demonstration
Potassium permanganate crystals
100 ml beaker (one per table)
Water

Diffusion
Agar plates (three per table)
0.01 M potassium permanganate solution in dropper bottles
0.01 M methylene blue solution in dropper bottles
0.01 M potassium dichromate solution in dropper bottles
Plastic drinking straws
Millimeter ruler
Fine probe or small forceps
Warming tray

Osmosis Demonstration
Thistle tube
Dialysis tubing
Rubber band
Molasses or concentrated sucrose solution (20%)
1% starch solution
Ring stand and clamp
250 ml beaker
Distilled water
China marker

Osmosis Experiment
Four strips of 20 cm long dialysis tubing (one set of four per table)
Four 200 ml beakers
String or "W" clamp with wooden applicator stick
Scissors
Four solutions (2 L each) of 0%, 5%, 15%, and 30% sucrose
Balances
Towels
Pipettes (10 ml)
Pipette pumps

Osmosis and Living Cells
Clean glass microscope slides
Coverslips
5 ml of mammal blood (available at local veterinarian's office)
Distilled water in dropper bottle (one per table)
0.9% sodium chloride solution in dropper bottle (one per table)
5% sodium chloride solution in dropper bottles (one per table)
Latex or plastic gloves

Filtration
Filter paper
Funnel
Ring stand with ring clamp
10 ml graduated cylinder
500 ml beaker

Iodine solution in dropper bottles (one per table)

Filtration solution (500 ml of 1% starch, 1% charcoal, and 1% copper sulfate) consists of 5 g each of starch, charcoal, and copper sulfate in one bottle or flask

Stopwatch or clock with second hand

 Virtual Physiology Lab 9: Diffusion, Osmosis, and Tonicity

Procedure

This lab is most efficiently done if the timing of experiments overlaps. While you are waiting for the results of one experiment, begin another.

Brownian Motion

All atoms and molecules in living organisms move due to the kinetic energy that they possess. Kinetic energy is the energy of motion and is the driving force of the movement of atoms and molecules. As the temperature increases, so does the kinetic energy. The only time kinetic energy is absent is if materials are at absolute zero (–273° C).

At room temperature, atoms and molecules move about but are too small to be seen even with the light microscope. Larger particles can be seen in random, zig-zag motion, and we interpret this as movement due to the collisions of smaller particles against those that we can see. This process is known as **Brownian motion,** first described in the nineteenth century by the botanist Robert Brown.

You can examine the effects of kinetic energy indirectly by studying Brownian motion under the microscope. Individual movements of atoms are too small to see, yet you can observe the collective collision of atoms as they strike large, visible molecules in solution (in this case fat droplets in milk).

To observe Brownian motion do the following activity:

1. Place a drop of milk on a clean glass microscope slide and add a drop of fat testing solution (1% Sudan III solution in alcohol).
2. Carefully set a coverslip on the slide and gently place a thin line of vegetable oil or household oil around the edge of the coverslip to prevent the slide from drying out.
3. Examine the slide under high power and record the movement of the fat droplets in the milk in the space provided.

Movement of fat droplets:

1. Remove the slide from the microscope and place it on a warm surface (such as a hot plate on the low temperature setting) for a few seconds until the slide becomes warm.
2. Quickly return the slide to the microscope and note the movement of the particles compared to the initial observation.
3. Record any difference in the space provided.

Observation of warm slide:

How might kinetic energy play a part in the differences between the first observation and the second observation?

Diffusion

Knowing that kinetic energy moves particles in solution (or in a gas for that matter), we can see that a concentration of particles would be struck by chance collisions and that some of those particles would begin to disperse. This process is known as **diffusion** and can be defined as the movement of particles from regions of high concentration to regions of low concentration. If a bottle of perfume is poured into a dish in a room, the perfume molecules diffuse from the dish into the air of the room. The concentration of particles is higher in the dish and lower in the air, and so we say that the molecules move down the **concentration gradient.** If the particles become uniformly dispersed, then the system has reached **equilibrium.**

If dye particles are placed in water, the same effect happens as the particles begin to spread out in the water. Water is the liquid into which material dissolves, and so water is called a **solvent.** The material that becomes dissolved in water is known as the **solute,** and the combination of solvent and solute is the **solution.** Water is one of many different kinds of solvents, yet it is the solvent vital to the life of the cell. Water is a polar molecule that dissolves ions and carries sugars, amino acids, and other materials to and from cells.

Diffusion, driven by kinetic energy, is a process equally important to the cell. An example of the essential nature of diffusion is the movement of oxygen into the blood vessels of the lungs. Oxygen in the air is at a higher concentration than in the blood of the lung capillaries, consequently oxygen moves from the air to the blood.

You can demonstrate diffusion by placing a crystal of potassium permanganate in a 100 ml beaker of water. Do this as a group at each table.

1. Fill a 100 ml beaker almost to the top with tap water.
2. Place the beaker on your table and drop a small crystal of potassium permanganate into it.
3. Leave the beaker undisturbed, but note the changes that occur during 1 hour.
4. Record your observations in the place provided. While you are waiting, continue with the other experiments.

Description of potassium permanganate movement:

Many factors can affect the rate of diffusion, including changes in temperature, changes in concentration of the solute, size or weight of the solute particles, and interactions between the solute and the solvent. In the following two experiments you will examine the effects of the weight of the particle and of temperature on diffusion rate.

Diffusion Rates and Particle Weight

You may want to do this experiment as a group of 3 to 4 students.

Agar, a liquid that forms a gel, can be used as a medium in which to measure diffusion rates of materials. Agar consists of water and algal polysaccharides. The liquid nature of agar is such that diffusion occurs in the gel at a slow rate.

1. Using a plastic drinking straw and a petri dish filled with agar, gently make three stabs into the dish so they are approximately equidistant from one another (figure 5.1). Do not twist the straw or you will break the agar and leave a crack into which fluid will run.
2. Remove the small plug of agar with a fine probe if needed so a well is left in the agar.
3. Into one of the wells place a drop of 0.01 M potassium permanganate solution (molecular weight 158).
4. Into another well place a drop of about the same size of 0.01 M methylene blue solution (molecular weight 320).
5. In the last well place a small drop of 0.01 M potassium dichromate solution (molecular weight 294). Potassium permanganate is a purple solution, methylene blue is very blue, and potassium dichromate is yellow.

Figure 5.1   Placement of the wells in agar.

Be careful—these dyes stain clothes, skin, and lab notebooks!

Record the diameter of the diffusion in chart 1 and continue with the following experiments.

Diameter of Diffusion (in mm)				Chart 1
	20 min	40 min	60 min	80 min
Potassium permanganate	_____	_____	_____	_____
Methylene blue	_____	_____	_____	_____
Potassium dichromate	_____	_____	_____	_____

Effects of Temperature on Diffusion Rates

1. Prepare two more petri dishes in the same way, except this time take one of the petri dishes from a refrigerator and another from a warming tray (such as an electric warming tray on the lowest setting).
2. Make three wells in the cold petri dish and place three drops of the respective dyes as you did previously.
3. This time, however, replace the petri dish in the refrigerator and examine after 80 minutes. You only need to record the diameter of diffusion at the end of the 80 minutes.
4. Follow the same procedure for the petri dish on the warming tray.
5. Record your results in chart 2.

Diameter of Diffusion (in mm)		Chart 2
	80 min (cold)	80 min (warm)
Potassium permanganate	_____	_____
Methylene blue	_____	_____
Potassium dichromate	_____	_____

Compare the diffusion rates of the cold and warm temperatures with the dish left at room temperature for 80 minutes. How does an increase or a decrease in temperature affect the diffusion rate?

What can you say about temperature and the kinetic energy of the system?

Osmosis

In the preceding diffusion experiments there was no barrier to the movement of the particles. Cell membranes are barriers to certain molecules, while they allow other molecules to pass through. This type of membrane is called a **selectively permeable membrane.** Water, some alcohols, oxygen, and carbon dioxide move easily across the cell membrane while larger molecules, such as proteins, or charged particles (ions) are prevented from crossing the membrane.

The movement of water across a selectively permeable membrane from solutions of higher water concentration (more pure water) to lower water concentration (less pure water) is known as **osmosis.** Osmosis is a particular kind of diffusion, and the process can be viewed from the perspective of the solvent or the perspective of the solute.

The Solvent Perspective of Osmosis

In diffusion, material moves from higher concentrations to lower concentrations. The same can be seen with osmosis. If you have two solutions separated by a selectively permeable membrane, one a 10% sugar solution (for example, 90% water) and one a 5% sugar solution (for example, 95% water), then the water will move from higher water concentration (95% water) to lower water concentration (90% water). The greater the difference between the two solutions, the greater the concentration gradient, which, in this case, is known as the **osmotic potential.**

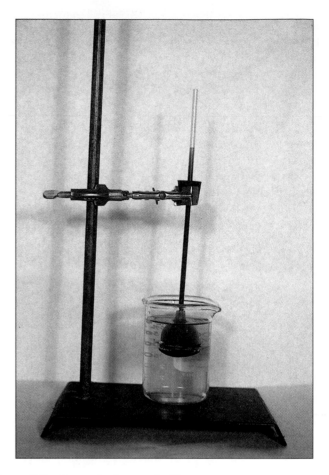

Figure 5.2   Osmosis demonstration.

The Solute Perspective of Osmosis

A 10% sugar solution has more solutes than a solution of 5% sugar. The 10% sugar solution is said to be **hypertonic** to the 5% sugar solution. The 5% sugar solution is said to be **hypotonic** to the 10% sugar solution. If these two solutions are separated by a selectively permeable membrane, then water flows *from* the hypotonic solution *to* the hypertonic solution. If enough water flows across the membrane and the two solutions reach the same concentration of sugar, then **equilibrium** is established and the net movement of water stops. If solutions have the same concentration of solutes, they are said to be **isotonic** to one another.

Demonstration of Osmosis

Observe osmosis with a thistle tube osmometer. This device consists of a hollow bell attached to a long tube. The tube is filled with molasses or sugar solution, and the large opening of the tube is covered with dialysis tubing and secured with a rubber band. The thistle tube osmometer is placed in a beaker of water and clamped to a ring stand (figure 5.2). The level of the molasses or sugar solution is indicated with a wax mark from a china marker.

Examine the setup throughout the lab period. You may find that the liquid in the thistle tube eventually stops rising. This occurs when the gravitational pressure equals the force

(a)

(b)

Figure 5.3   Dialysis bags. (a)Filling; (b)clamping.

exerted by the process of osmosis. The amount of force to balance or equilibrate osmosis is the **osmotic pressure.**

> Examine (or prepare one if your lab instructor indicates for you to do so) a thistle tube that contains a starch solution. Does the liquid move up the tube?

> Why or why not?

> What does this say about the osmotic activity of starch?

Osmosis and the Concentration Gradient

As stated earlier, the osmotic potential varies depending on the concentration gradient between the two solutions. In this experiment, you determine the effects of various concentrations of a sucrose solution on the rate of osmosis.

1. Fill four dialysis bags that have been soaking in water with 10 ml of a 15% sucrose solution. You can do this by placing a pipette pump (or bulb) on the end of a 10 ml pipette, drawing liquid to reach the 10 ml mark, and filling the dialysis bag.
2. Secure the bottom end of the bag prior to filling it with sucrose solution from a pipette and pipette pump and then tie or clamp both ends of the bag (figure 5.3).
3. Rinse each bag in distilled water and blot with a towel.

4. Weigh each bag to the nearest tenth of a gram and record the weights in chart 3. This weight is the initial weight of the bag.

Initial Weight of Dialysis Bags	Chart 3
Bag 1 _____ grams	
Bag 2 _____ grams	
Bag 3 _____ grams	
Bag 4 _____ grams	

5. Place bag 1 in a beaker filled about two-thirds to the top with a 0% sugar solution. Make sure the bag is covered with the solution, and leave it there for 20 minutes.
6. Place bag 2 in a 5% sugar solution and leave it for 20 minutes.
7. Place bag 3 in a 15% sugar solution and leave it for 20 minutes.
8. Place bag 4 in a 30% sugar solution and leave it for 20 minutes.
9. Remove the bags from the beakers after 20 minutes, and blot and weigh each bag.
10. Remember to place each bag back in its proper solution! Record the weight of the bags each 20 minutes for a total of 80 minutes in chart 4.

Calculate the **change of weight** of each bag from the initial weight for each of the time periods. For example, let's as-

Weight of Bags (in g) for Each Time Period				Chart 4
	Time			
Bag	20 min	40 min	60 min	80 min
1	_____	_____	_____	_____
2	_____	_____	_____	_____
3	_____	_____	_____	_____
4	_____	_____	_____	_____

sume a bag weighed 20.5 grams as the initial weight and the recorded weights are as follows:

20 min	40 min	60 min	80 min
23.5 g	24.2 g	25.0 g	25.6 g

The change in weight would be as follows:

3.0 g	3.7 g	4.5 g	5.1 g

Graph the change in weight for each bag of your experiment in chart 5 and connect the points with a line. Indicate which line represents which solution.

Which of the bags (if any) gained weight?

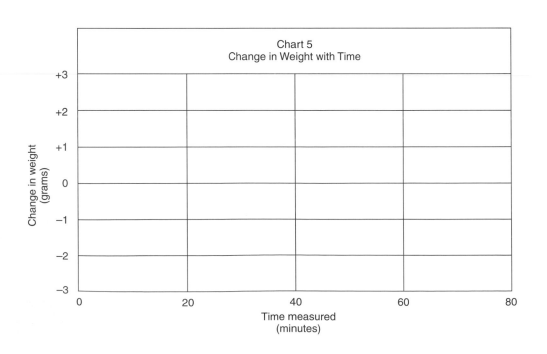

Chart 5
Change in Weight with Time

Which of the bags (if any) lost weight?

Determine the osmotic relationship (hypertonic, hypotonic, isotonic) of the bag to the solution in the beaker.

Bag 1: The bag solution is _____ to the beaker solution.

Bag 2: The bag solution is _____ to the beaker solution.

Bag 3: The bag solution is _____ to the beaker solution.

Bag 4: The bag solution is _____ to the beaker solution.

Does the change in weight correlate to what you know about osmosis? If so, how does it correlate?

Osmosis and Living Cells

The importance of isotonic solutions can be demonstrated by the following procedure:

1. Place a drop of fresh mammalian blood on a slide. Make sure you wear protective gloves while conducting this experiment.
2. To this slide add a drop of physiological saline (0.9% sodium chloride) and place a coverslip on the slide.
3. Observe the cells on high power under the microscope and note their shape.
4. Record this in the space provided.

    Shape of the cells:

1. Take a new slide and place another drop of blood on it.
2. To this slide add a drop of 5% sodium chloride (NaCl) solution.
3. Place a coverslip on the slide, and record your observations.
4. Examine the slide for at least a few minutes or until a change of shape becomes obvious.

    Shape of the cells in 5% NaCl:

Finally, with a third slide repeat the previous procedure except add a drop of distilled water instead of the sodium chloride. Immediately observe this slide and then continue to look at it for a few minutes. Record your observations.

Cells in distilled water:

When red blood cells lose water they undergo a process known as crenation. Did any of the cells show crenation?

Which solution might produce this?

When water moves into a cell at a rapid rate the cell becomes inflated and sometimes bursts in a process known as hemolysis.

Did any of the cells undergo hemolysis?

Which solution might produce this effect?

Examine figure 5.4 for the various effects of solutions on red blood cells.

Filtration

The process of **filtration** is very important in certain cells of the body and results as the pressure of a fluid forces particles through a filtering membrane. Filtration is a major component of kidney function. Hydrostatic pressure from the blood forces urea, ions, sugars, and other materials from the blood through the membranes of the kidney cells. In this way, small particles in the blood are forced into kidney tubules while larger molecules such as proteins remain in the blood. In this exercise, you learn the basic principles of filtration, such as the **selectivity** of the filtration membrane and the **filtration rate** of a system.

1. Fold a piece of filter paper in half and then fold it in half again so that it forms a cone (figure 5.5).
2. Place the cone in a funnel mounted on a ring stand over a beaker.
3. Into this cone filter add the filtration solution, which is a mixture of copper sulfate, powdered charcoal, and starch in water.

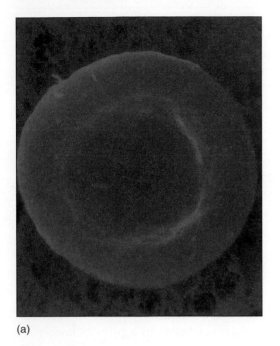
(a)

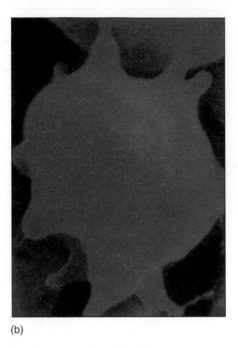
(b)

(c)

Figure 5.4   Stages of red blood cells. (a)Normal; (b)crenated; (c)inflated.
David M. Phillips/Visuals Unlimited.

4. Fill the funnel to near the top of the brim and let the filtrate (the material passing through the filter) collect in the beaker.
5. When the funnel is approximately half full, place a 10 ml graduated cylinder under it and record the time it takes to fill the cylinder to the 2 ml mark.
6. Divide the time by two and calculate the filtration rate expressed as ml/minute.
7. Record this value.

Number of seconds to produce 2 ml: _____

Filtration rate: _____ ml/minute

You should be able to determine what material passes through the membrane by the following method: Remove the funnel from the beaker. If the solution in the beaker has a blue cast to it, then copper sulfate passed through the membrane. If black particles are found in the filtrate, then charcoal passed through the membrane. If you add a few drops of iodine to the beaker and the solution turns black, then starch passed through the membrane. Record what material passed through the membrane in the space provided.

Substance	Yes	No
Copper sulfate	_____	_____
Activated charcoal	_____	_____
Uncooked starch	_____	_____

The force that drives filtration in the funnel is the force of gravity on the liquid. In the kidney, the force that drives filtration is blood pressure.

Figure 5.5   Filtration apparatus.

Name _____

1. A solution with 5% sugar is _____ (isotonic/hypertonic/hypotonic) when compared to a 3% sugar solution.

2. If these two solutions were separated by a selectively permeable membrane, which solution would lose water?

3. What is a solute?

4. Two solutions that are in equilibrium in terms of their concentrations are known as: _____.

5. What effect would lowering the temperature to absolute zero (–273°C) have on Brownian motion?

6. You have two different molecules, one of 230 molecular weight and one of 415 molecular weight. Which one would diffuse the farthest if they were both allowed to diffuse for the same length of time?

7. If a dialysis bag of 20% sugar solution was placed in a beaker of 40% sugar solution and allowed to remain there for a few minutes, would the solution in the bag be hypertonic or hypotonic to the solution in the beaker?

Would the water flow into or out of the bag?

8. What would happen to the filtration rate if you were to apply pressure from a water hose to the filtration system?

What might happen to the filtration membrane if the water pressure was too high?

Why might this be of concern to people who have both kidney disease and high blood pressure?

LABORATORY EXERCISE 6

Tissues

Introduction

The study of tissues, called **histology,** is very important in that many organic dysfunctions of the human body can be seen specifically at the tissue level. Surgical specimens are routinely sent to pathology labs so that accurate assessment of the health of the tissue, and consequently the health of the individual, can be made.

There are four main tissue types found in the human body, **epithelial tissue, connective tissue, muscular tissue,** and **nervous tissue.** In this exercise, you examine numerous slides of organ tissue and begin an introduction to histology. In later exercises, we revisit histology as we examine various organ systems.

Objectives

At the end of this exercise you should be able to

1. recognize the various cell types of epithelial tissue;
2. associate a particular tissue type with an organ such as kidney or bone;
3. examine a slide under the microscope or a picture of a tissue and name the cell type or specific tissue represented;
4. distinguish between cartilage and other connective tissues;
5. list the three parts of a neuron;
6. describe the muscle cell types according to location and structure.

Materials

Microscope
Colored pencils

Epithelial Tissue Slides
Simple squamous epithelium
Simple cuboidal epithelium
Simple columnar epithelium
Pseudostratified ciliated columnar epithelium
Stratified squamous epithelium
Transitional epithelium

Muscular Tissue Slides
Skeletal muscle
Cardiac muscle
Smooth muscle
All three muscle types

Nervous Tissue Slides
Spinal cord smear

Connective Tissue Slides
Dense connective tissue
Elastic connective tissue
Reticular connective tissue
Loose (areolar) connective tissue
Adipose tissue
Ground bone
Hyaline cartilage
Fibrocartilage
Elastic cartilage
Blood

Procedure

Before you begin this exercise you should be thoroughly familiar with the particular microscopes in your lab. If you need a review, go back to Laboratory Exercise 3. As you examine various tissues look for distinguishing features that will identify the tissue. It is a good idea to examine more than one slide of a particular tissue so you see a range of samples of that tissue. Frequently the material you see in the lab is from a slice of an organ and as such includes more than one tissue type. For example, a sample of cartilage frequently taken from the trachea contains epithelial tissue, fat, and other connective tissues in addition to the cartilage you want to study. If you are looking for smooth muscle from the digestive tract, you may also find both epithelial tissue and connective tissue in the slide. Use the figures in this exercise to help you locate tissues on the prepared slides. Examine each slide by holding the slide up to the light and visually locating the sample. Then put the slide on the microscope and examine it on low power, scanning around the slide. Move to progressively higher powers after you have identified the tissue. If you cannot identify the tissue after some searching, then ask your lab partner or your instructor for help.

Epithelial Tissue

Epithelial tissue is a highly cellular tissue, covering or lining parts of the body (such as the skin on the outside or the digestive tract on the inside) or found in glandular tissue such as the sweat glands or the pancreas. In most cases, epithelial tissue adheres to the underlying layers by way of a **basement membrane,** which is a noncellular adhesive layer. In a region like the skin, the epidermis is made of epithelial tissue and the basement membrane connects the epidermis to the underlying dermis. Epithelial tissue is classified according to the shape of the cells and the number of layers present. The cell shapes are **squamous** (meaning flattened), **cuboidal, columnar,** and **pseudostratified.** The number of layers are **simple** (cells in a single layer) or **stratified** (cells stacked in more than one layer). Epithelial tissue is listed by cell type in the following discussion.

Simple Epithelia

Simple epithelium is only one cell layer thick. The cells are located on the basement membrane, which may adhere to connective tissue, muscle, or even other epithelial tissue. Each epithelial cell type is further classified according to shape.

Simple Squamous Epithelium This cell type is composed of flat cells that lie on the basement membrane like floor tiles. Seen from a surface view the cells appear relatively equal on all sides and may initially look cuboidal. If these cells are seen from a side view, their resemblance to floor tiles becomes apparent. Examine a prepared slide of simple squamous epithelium, which is usually seen as either a surface view or side view. Simple squamous epithelium is seen in the air sacs of

lungs, it lines blood vessels, and it can be found as the surface layer of many membranes. It functions to provide a slippery surface (as found on the inside of blood vessels) or to allow for diffusion (as in the lungs). Compare your slide to the photograph in figure 6.1. Draw what you see under the microscope in the space provided. Note whether your slide shows the cell as a surface or side view.

Illustration of simple squamous epithelium:

Simple Cuboidal Epithelium Simple cuboidal epithelium consists of cube-like or pie-shaped cells that are mostly uniform in diameter. These cells form many of the major glands

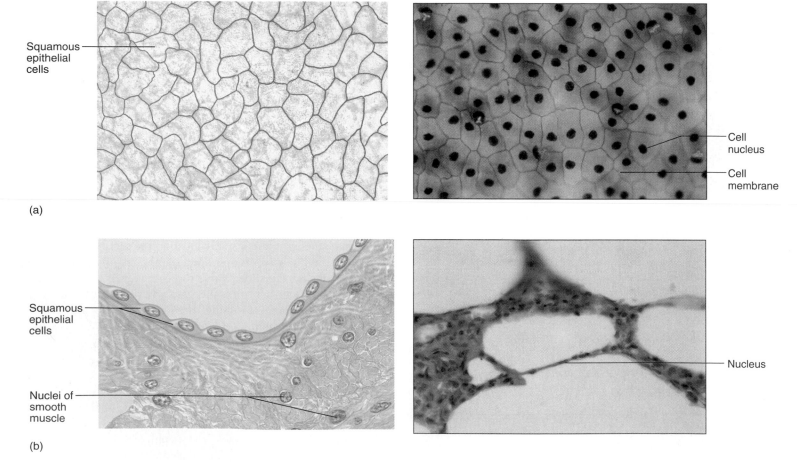

Squamous epithelial cells

(a)

Cell nucleus

Cell membrane

Squamous epithelial cells

Nuclei of smooth muscle

(b)

Nucleus

Figure 6.1 Simple squamous epithelium (400×). (a) Top view; (b) side view.

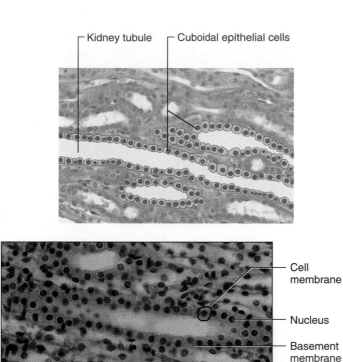

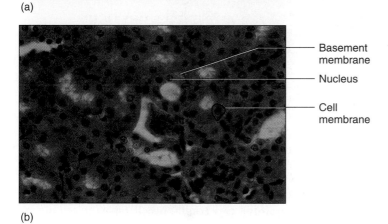

(a)

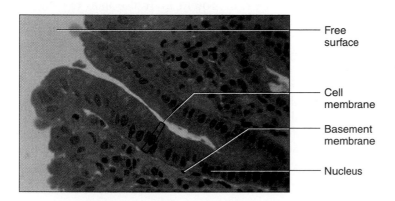

(b)

Figure 6.2   Simple cuboidal epithelium (400×). (a) Long section; (b) cross section of kidney tubules.

and glandular organs of the body and are the major cell type of the kidney. Simple cuboidal epithelium is frequently found lining tubules such as sweat ducts. Simple cuboidal epithelium often is involved in the secretion of fluids (sweat, oil) or in filtration (as in the kidneys). Examine a prepared slide of simple cuboidal epithelium and compare it to figure 6.2. Draw what you see under the microscope in the space provided and label the nucleus of the cell and the basement membrane.

Illustration of simple cuboidal epithelium:

Simple Columnar Epithelium   This cell type resembles tall columns anchored to the basement membrane. Simple columnar epithelium can be ciliated on the free edge or it may be smooth. The nonciliated type lines the inner portion of the digestive tract and provides an absorptive area for digested food. Simple columnar epithelium also lines the uterine tubes and is ciliated in this case. Examine a prepared slide of simple columnar epithelium and compare it to figure 6.3. Draw a representation of what you see in the space provided.

Illustration of simple columnar epithelium:

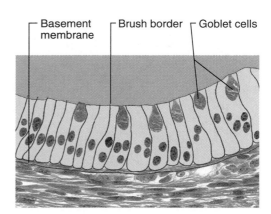

Figure 6.3   Simple columnar epithelium (400×).

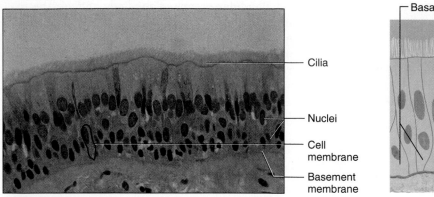

Figure 6.4   Pseudostratified ciliated columnar epithelium (400×). ⊤

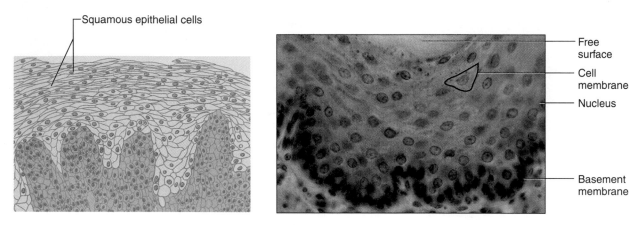

Figure 6.5   Stratified squamous epithelium (400×). ⊤

Pseudostratified Ciliated Columnar Epithelium   This cell type may first look like it occurs in a few layers, but all of the cells originate from the basement membrane, thus it is known as a simple epithelium. Pseudostratified ciliated columnar epithelium lines some portions of the respiratory passages, where it serves a protective function by trapping dust particles in a mucus sheet and moving the particles away from the lungs. Examine a prepared slide of pseudostratified ciliated columnar epithelium and compare it to figure 6.4. Draw a representative sample of what you see in the space provided.

    Illustration of pseudostratified ciliated columnar
        epithelium:

Stratified Epithelia

Stratified epithelium is so named because many epithelial cells occur in layers on a basement membrane. The term *stratified* comes from the word *strata* and refers to the layering of these cells. You will examine two common cell types belonging to this group.

Stratified Squamous Epithelium   Stratified squamous epithelium is a cell type that covers the outside of the body and forms the skin around us. It also lines the vaginal canal and mouth. The multiple layers of this cell type protect the underlying tissue from mechanical abrasion. This epithelium may have cuboidal cells at the basement layer, but it derives its name from the cell shape at the free surface. Stratified squamous epithelium comes in two distinct types, keratinized and nonkeratinized (keratin is a tough protein that hardens cells in the outer layer of the skin). Compare your slide to figure 6.5 and draw what you see under the microscope in the space provided. Locate the basement membrane, and count the layers of cells that occur between it and the surface of this cell type.

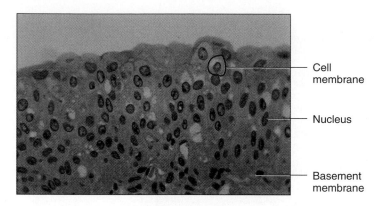

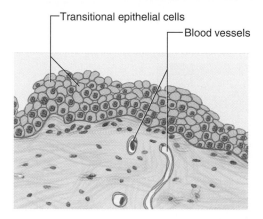

Figure 6.6   Transitional epithelium (400×). ⫶

Illustration of stratified squamous epithelium:

Record the numbers of layers of cells here: _____

Transitional Epithelium   Transitional epithelium is an unusual cell type in that it has some remarkable stretching capabilities. Transitional epithelium lines the urinary bladder and the ureter and allows these organs to expand as urine collects within them. Examine a prepared slide of transitional epithelium. Look at figure 6.6 and draw the cell type in the space provided.

Illustration of transitional epithelium:

Muscular Tissue

Muscular tissue is considered a cellular tissue in that the individual muscle fibers are actually cells. These cells are contractile and shorten in length due to the sliding of protein filaments across one another. There are three cell types of muscular tissue found in the body, and these are listed in the following discussion.

Cell Types

Skeletal Muscle   When you refer to the muscles of your body you are referring to organs made mostly of skeletal muscle. Skeletal muscle is sometimes known as striated muscle because it has obvious **striations** (what appear to look like crosshatchings) in the fiber. Skeletal muscle is voluntary in that you have control over this type of muscle in your body. The individual muscle cells have many nuclei and are thus called multinucleate or syncytial. The nuclei are somewhat elongated and occur on the periphery of the cell. Examine a prepared slide of skeletal muscle under high power. Compare this slide to figure 6.7. Draw a representation of the muscle in the space provided. How many widths of muscle cells fit across the diameter of your microscope when viewed at high power?

Illustration of skeletal muscle:

How many muscle fiber widths do you count? _____

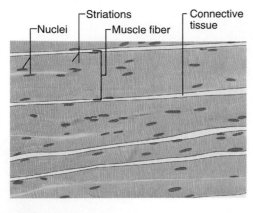
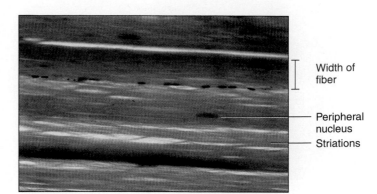

Figure 6.7   Skeletal muscle (400×).

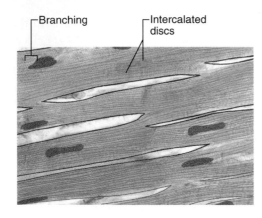
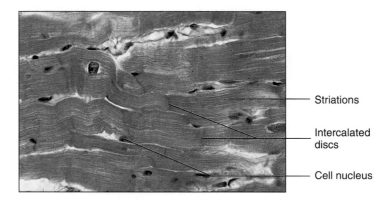

Figure 6.8   Cardiac muscle (1,000×).

Cardiac Muscle   Cardiac muscle is found only in the heart, and is the main tissue making up that organ. Cardiac muscle is somewhat similar to skeletal muscle in that it is striated, but the striations are much less obvious and the individual cell diameter is less than those of skeletal muscle cells. Place a prepared slide of cardiac muscle under the microscope, and count the number of widths of cardiac muscle fibers across the diameter of the microscope under high power. Cardiac muscle functions as an involuntary muscle in that it does not require conscious thought to function. The cells are mostly uninucleate, with the nucleus centrally located and oval in shape. Cardiac muscle cells are joined together by intercalated discs, which serve to tie the cardiac muscle cells together electrically. In this way, as one muscle receives an impulse it sends it on to the next cell. Look at a prepared slide of cardiac muscle and compare it to figure 6.8, noting the striations, intercalated discs, and nuclei of the cells. Draw the cells in the space provided.

Illustration of cardiac muscle:

Record the number of cell widths that occupy the diameter of your field of view at high power: _____

Smooth Muscle   Smooth muscle is nonstriated in that the fibers do not have crosshatchings perpendicular to the length of the fiber. Smooth muscle is involuntary, as is cardiac muscle, and is found in the digestive tract, where it propels food along by a process known as peristalsis. Smooth muscle is also found in abundance in the uterus and is the driving force of uterine contractions during labor. The cells of smooth muscle are spindle-shaped, uninucleate, and the nucleus is centrally located and elongated in shape when the muscle is relaxed. When the muscle is contracted the nucleus appears corkscrew-shaped. Examine a prepared slide of smooth muscle and note the narrow diameter of the fibers. How many fiber widths do you count across the diameter of the field of view of your microscope under high power? Compare your slide to figure 6.9 and draw an illustration of smooth muscle in the space provided.

Illustration of smooth muscle:

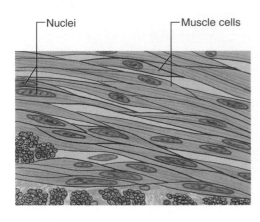

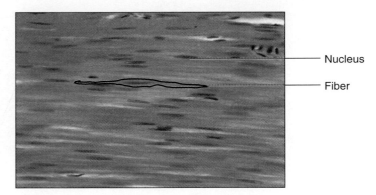

Figure 6.9   Smooth muscle (400×). ✗

Number of cell widths under high power: _____

Examine a prepared slide of all three muscle types, if available. Work with your lab partner and identify each muscle cell.

Nervous Tissue

Nervous tissue, as with epithelial and muscular tissues, is also very cellular. Nervous tissue is found in the brain, spinal cord, and peripheral nerves of the body. The conductive cell of this tissue is the **neuron,** which receives and transmits electrochemical impulses. A neuron is a specialized cell with three major regions, the **dendrites,** the **nerve cell body** (or **soma**), and the **axon.** Examine figure 6.10 for the regions of a typical neuron.

Examine the neurons of a smear of the spinal cord of an ox. Look for purple star-shaped structures, which are the nerve cell bodies. Compare them to figure 6.11.

Special support cells of the nervous system are called **glial cells** or **neuroglia.** These cells do not conduct impulses, but they act to support the neuron in its function either directly or indirectly. You will study glial cells in greater detail in Laboratory Exercise 20.

Connective Tissue

Unlike the previous tissues, most connective tissue is not very cellular at all. There are cells in connective tissue, but it mostly consists of numerous **fibers** and a background material called the **matrix.** The matrix of connective tissue can be fluid such as the plasma of blood, it can be solid like the mineral salts of bone, or it can be rubbery as in the cartilage at the tip of your nose. Due to the lesser abundance of cells in this tissue, connective tissue is classified by *specific tissue* rather than by *cell type.*

Connective tissue is diverse, and specific tissues do not seem to have much in common with one another; however, all

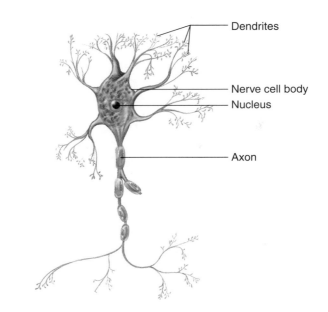

Figure 6.10   Regions of a typical neuron.

arise from an embryonic tissue known as **mesenchyme.** Connective tissue can be divided into several subgroups for easier identification. These subgroups are connective tissue proper, osseous tissue, cartilage, and vascular tissue. These are described in more detail next.

Connective Tissue Proper

Dense Connective Tissue   Dense connective tissue or white fibrous connective tissue is one of five specific tissues that make up connective tissue proper. This specific tissue is composed of collagenous fibers that may either be parallel, in which case it is known as **dense regular connective tissue,** or run in many directions, in which case it is known as **dense irregular connective tissue.** Dense regular connective tissue is found in tendons and ligaments. Dense irregular connective tissue is found in the deep layers of the skin. The fibers in

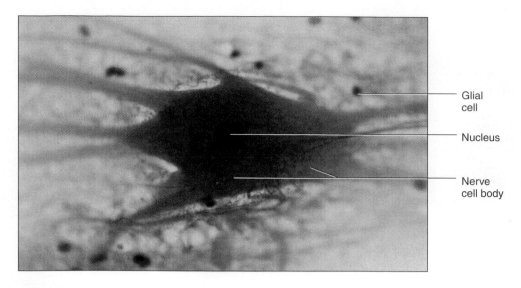

Figure 6.11 Spinal cord smear (400×).

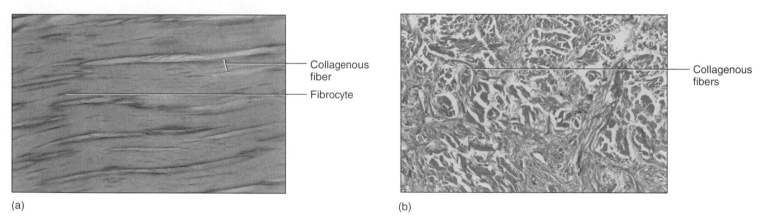

(a)

(b)

Figure 6.12 Dense connective tissue. (a) Dense regular connective tissue (400×); (b) dense irregular connective tissue (100×). ⫪

these tissues are made of a protein called **collagen** and are called **collagenous** fibers. **Fibrocytes** and **fibroblasts** may be found in the tissue, in addition to the collagenous fibers. Fibroblasts secrete fibers and then mature into fibrocytes. Examine a slide of dense connective tissue and compare it to figure 6.12. Draw a section of dense regular or dense irregular connective tissue in the space provided.

Illustration of dense connective tissue:

Elastic Connective Tissue This tissue has a cellular component of fibroblasts and fibrocytes. The fibroblasts actively secrete fibers and subsequently develop into fibrocytes. The fibers that occur in this tissue are made of the protein **elastin** and are called **elastic fibers.** Elastic tissue occurs in the walls of arteries and can be identified as dark squiggly lines in the wall of an artery. Examine figure 6.13 as you look at a prepared slide of elastic tissue. Then draw the specific tissue in the space provided.

Illustration of elastic connective tissue:

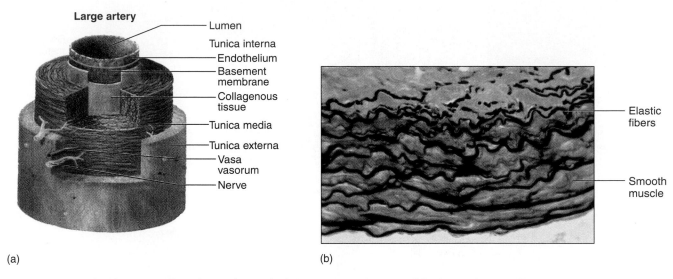

(a)   (b)

Figure 6.13   Elastic connective tissue (400×). (a) Overview of artery; (b) photomicrograph.

Reticular Connective Tissue   Reticular connective tissue also has fibroblasts and fibrocytes, but the fibers are **reticular fibers.** Reticular connective tissue is found in soft internal organs such as the liver, spleen, and lymph nodes. Reticular connective tissue provides an internal framework for the organ. If your slide is stained with a silver stain, the reticular fibers appear black. If your slide is stained with Masson stain, then the reticular fibers appear blue. In either case, look for fibers that appear branched among small round cells. The other cells are the cells of the organ that the reticular connective tissue is holding together. Examine a prepared slide of reticular tissue and compare it to figure 6.14. Draw a sample of what you see in the space provided.

Illustration of reticular connective tissue:

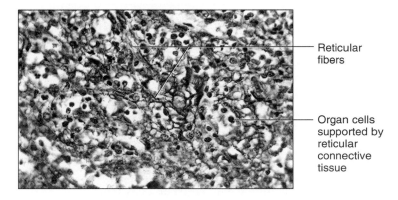

Figure 6.14   Reticular connective tissue (400×).

collection of cells including **fibroblasts, fibrocytes, mast cells,** and **macrophages.** Examine a prepared slide of loose connective tissue and compare it to figure 6.15. Make a drawing of it in the space provided.

Illustration of loose connective tissue:

Loose (Areolar) Connective Tissue   Loose (areolar) connective tissue is a complex collection of fibers and cells that has a very distinctive look. Areolar connective tissue is found as a wrapping around organs, as sheets of tissue between muscles, and in many areas of the body where two different tissues meet (such as the boundary layer between fat and muscle). Loose connective tissue consists of large pink-stained **collagenous fibers,** smaller dark **elastic** and **reticular fibers,** and a

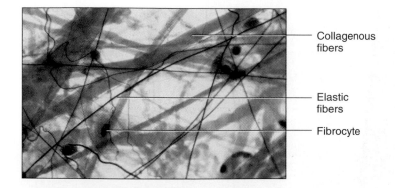

Collagenous fibers

Elastic fibers

Fibrocyte

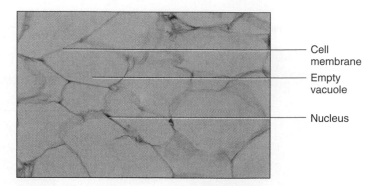

Cell membrane

Empty vacuole

Nucleus

Figure 6.16   Adipose tissue (400×).

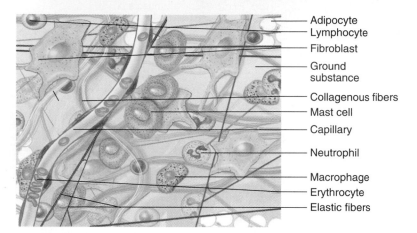

Adipocyte
Lymphocyte
Fibroblast
Ground substance
Collagenous fibers
Mast cell
Capillary
Neutrophil
Macrophage
Erythrocyte
Elastic fibers

Figure 6.15   Loose (areolar) connective tissue (400×).

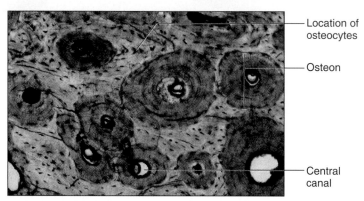

Location of osteocytes

Osteon

Central canal

Figure 6.17   Ground bone (100×).

Adipose Tissue   Adipose tissue is unusual as a connective tissue in that it is highly cellular. The cells that make up adipose tissue are called **adipocytes** (which is a technical way to say *fat cells*). Adipocytes store lipids in large vacuoles, while the nucleus and cytoplasm remain on the outer part of the cell near the plasma membrane. In prepared slides of adipose tissue the fat typically has been dissolved in the preparation process. You should be able to locate the large empty cells, with the nuclei on the edge of some of the cells. Compare what you see in the microscope to figure 6.16 and make a drawing in the space provided.

Illustration of adipose tissue:

Osseous Tissue

Osseous tissue, or bone, is a type of connective tissue in which the matrix is made of a mineral component called hydroxyapatite crystal, or bone salt. **Collagenous fibers** occur in the tissue along with osteocytes or bone cells. Look at a prepared slide of ground bone and compare it to figure 6.17. Locate the large, circular regions called **osteons,** the **central canal,** and the **osteocytes** in the tissue. A more detailed examination of bone appears in Laboratory Exercise 8. After you study the slide, record your observations in the space provided.

Illustration of ground bone:

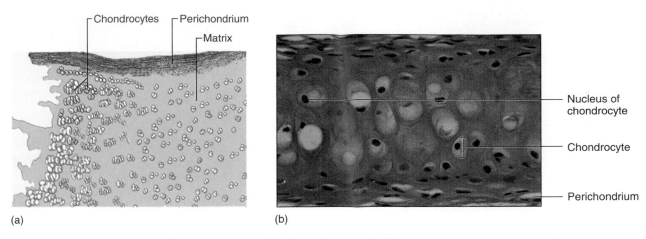

Figure 6.18   Hyaline cartilage (400×). (a) Overview; (b) photomicrograph. ✗

Cartilage (Cartilage and Perichondrium)

Cartilage is a type of connective tissue in which the matrix is composed of a pliable material that allows for some degree of movement. The matrix of cartilage is made of protein and carbohydrates and forms a semisolid gel in which fibers and cells are found. In prepared slides, frequently the cells have come out of the matrix and left a cavity. This cavity is known as a lacuna and is reasonably diagnostic of cartilage tissue. The **perichondrium** is dense, irregular connective tissue that is found on the surface of cartilage. There are three types of cartilage found in the human body, and they vary by the number of and type of protein fibers. These three types are hyaline cartilage, fibrocartilage, and elastic cartilage.

Hyaline Cartilage   The most common cartilage in the body is hyaline cartilage, which is found at the apex of the nose, at the ends of many long bones, between the ribs and the sternum, and in other locations as well. Hyaline cartilage is clear and glassy in fresh tissue but will probably appear light pink in prepared slides. The cells that occur in hyaline cartilage are called **chondrocytes,** and the fibers are **collagenous fibers.** Find the chondrocytes in a prepared slide of hyaline cartilage, but the fibers will not appear distinct because they blend into the color of the matrix. The chondrocytes of hyaline cartilage frequently occur in pairs. Look around the edge of the sample of hyaline cartilage and locate the perichondrium. Compare the material under the microscope to figure 6.18. Draw what you see in the space provided.

Illustration of hyaline cartilage:

Fibrocartilage   Fibrocartilage is similar to hyaline cartilage in that it has chondrocytes and collagenous fibers, but it differs from hyaline cartilage in that it has many more collagenous fibers. These can be seen in the prepared slides. Fibrocartilage is found in areas where more stress is placed on the cartilage, such as in the intervertebral discs, the symphysis pubis, and the menisci of each knee. One characteristic of fibrocartilage is that the chondrocytes are frequently found in rows of fours or fives in the tissue. Examine a prepared slide of fibrocartilage, and locate the collagenous fibers and chondrocytes. Compare the slide to figure 6.19 and make a drawing of fibrocartilage in the space provided.

Illustration of fibrocartilage:

Elastic Cartilage   Elastic cartilage is unique in that it contains **elastic fibers,** which gives the tissue its flexible nature. Elastic cartilage is found in the external ear and in a part of the larynx called the epiglottis. If the prepared slide is stained with a silver stain, the elastic fibers appear black. If the slide is not stained in this way, the fibers may be hard to distinguish. Examine a prepared slide of elastic tissue and look for the chondrocytes and elastic fibers. Compare your slide to

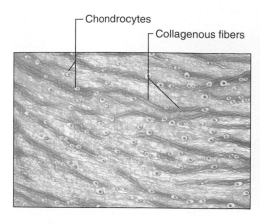

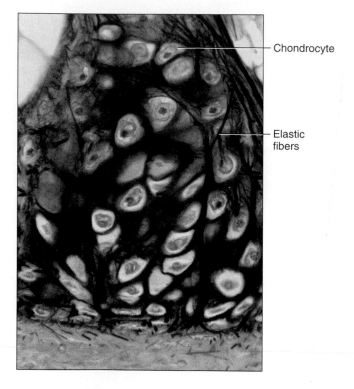

Figure 6.19   Fibrocartilage (400×). ✶

figure 6.20 and make a drawing of elastic cartilage in the space provided.

Illustration of elastic cartilage:

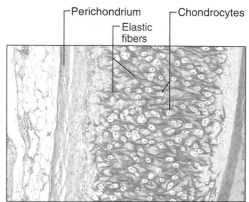

Figure 6.20   Elastic cartilage (400×). ✶

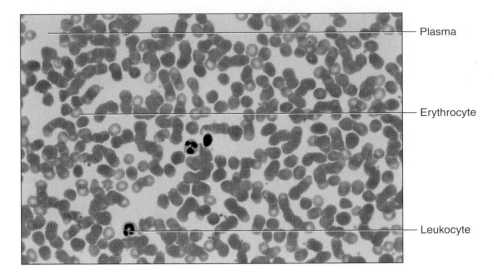

Figure 6.21 Blood (400×).

Vascular Tissue

Blood is a connective tissue in which the matrix is fluid and no fibers are present. The matrix of the blood is called **plasma,** and the cells of blood are either **erythrocytes** (red blood cells) or **leukocytes** (white blood cells). Examine a prepared slide of blood and look for the numerous pink disks in the sample. These are the erythrocytes. You may find larger cells with lobed, purple nuclei. These are the leukocytes. The details of blood are studied in Laboratory Exercise 29. Examine the slide under high power and compare it to figure 6.21. Make a drawing of blood in the space provided.

Illustration of blood:

Study Hints

Examine as many different images of tissues as you can. Drawing the material observed in the lab is a good memory tool. Try to make associations between the abstract image of tissue and something common. For example, you could think of bone as looking like a series of tree rings. Adipose tissue may look like a series of angular balloons. If you have difficulty distinguishing between two specimens, try to find a characteristic that separates the two. For example, smooth muscle and dense connective tissue may look similar in certain stained preparations. You will note that smooth muscle has nuclei inside the fibers, while dense connective tissue has nuclei of the fibrocytes between adjacent fibers.

Name _____

1. Label the following photomicrographs by using the terms provided.

   a. Collagenous fibers, elastic fibers, macrophages, fibroblasts

a. _____

b. _____

c. _____

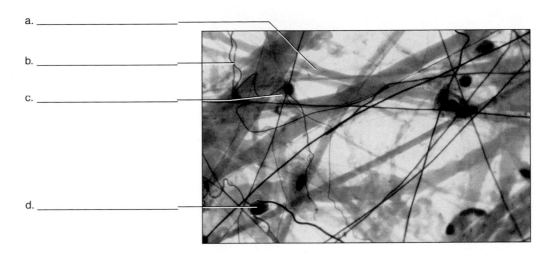

d. _____

Loose areolar connective tissue (400×).

   b. Cilia, basement membrane, nucleus

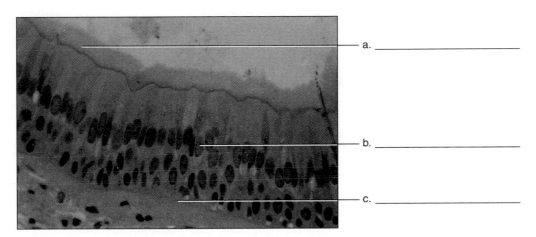

a. _____

b. _____

c. _____

Pseudostratified ciliated columnar epithelium (400×).

c. Intercalated discs, striations, nucleus

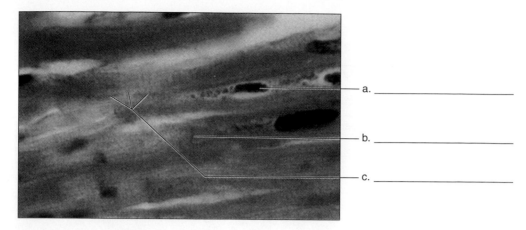

a. _____

b. _____

c. _____

Cardiac muscle (1,000×).

d. Nerve cell body, glial cells

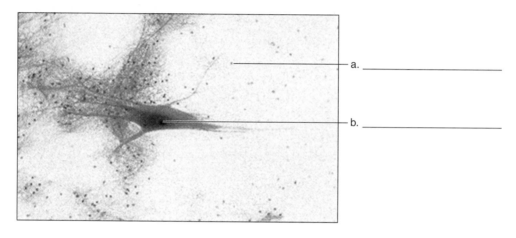

a. _____

b. _____

Nervous tissue (100×).

2. How do the groupings of cells differ between hyaline cartilage and fibrocartilage? In which specific tissue are the cells in pairs? In which specific tissue are the cells in fours or fives?

3. How many layers are in simple epithelium?

4. What shape are the cells that are squamous?

5. If cells have many layers, they are referred to as being:

6. The digestive tract has what kind of muscle?

7. The heart is made of what kind of muscle?

8. The muscle in your arm is composed of what cell type?

9. What are the three main parts of a neuron?

10. How do you distinguish the difference between epithelial cells and connective tissue?

11. There are numerous cell types and specific tissues in the body. You have seen a representative sample of the more common ones. Your instructor may wish you to know other cell types or specific tissues. Make sure to draw and label any additional tissues in the space provided.

Integumentary System

Introduction

The integumentary system consists of the skin and associated structures such as hair, nails, and various glands. The integument consists primarily of a cutaneous membrane composed of a superficial epidermis and a deeper dermis. Below these two layers is an underlying layer called the hypodermis, which anchors the integument to structures deep to the integument. The skin is the largest organ of the body and is vital in terms of water and temperature regulation and protection against microorganisms.

Objectives

At the end of this exercise you should be able to

1. list the two layers of the integument;
2. list all of the layers of the epidermis from superficial to deep;
3. describe the structure and function of sudoriferous glands and sebaceous glands;
4. draw a hair in a follicle in longitudinal and cross section;
5. discuss the protective nature of melanin.

Materials

Models and charts of the integumentary system
Prepared microscope slides of:

Thick skin (with Pacinian corpuscles)
Hair follicles
Pigmented and nonpigmented skin

Microscopes

Procedure

Overview of the Integument

Examine models or charts of the **integumentary system** in the lab and locate the two major regions of the integument, the **epidermis** and the **dermis.** Note how the dermis is the deeper layer and the epidermis is located on the surface. Also locate the **hypodermis,** which is not a layer of the integument but anchors the integument to underlying bone or muscle. The hypodermis can be distinguished by the abundant adipose tissue. Compare the models in the lab to figure 7.1.

Locate the **hair follicles** in the dermis, **sebaceous (oil) glands,** and **sudoriferous (sweat) glands.** The sudoriferous glands are connected to the surface by **sweat ducts.**

Epidermis

The epidermis is composed of **stratified squamous epithelium** with the superficial layer infused with the protein **keratin.** Keratin is a tough, water-resistant material that also serves to protect the lower layers of the epidermis from ultraviolet radiation. The epidermis has a number of layers known as **strata** (singular, *stratum*). The epidermis does not have blood vessels between the cells but is nourished by the vessels in the dermis.

Microscopic Examination of the Integument

Examine a prepared slide of thick skin. The slide will probably show three layers—one that is relatively clear due to a significant amount of adipose tissue, a pink layer, and a multicolored layer. The clear layer is the hypodermis, the pink layer is the dermis, and the region with the multicolored layers is the epidermis. The colors may vary in the slides used in your particular lab. Compare these layers in the prepared slide to figure 7.2.

Strata of the Epidermis

In thick skin, the deepest layer of the epidermis is known as the **stratum basale.** This single layer of cells is adhered to the dermis by the **basement membrane,** a noncellular layer. Cells in the stratum basale divide repeatedly and push up toward the superficial layers as the cells increase in age. This process is illustrated in figure 7.3.

In the stratum basale are specialized cells known as **melanocytes.** These cells produce a brown pigment known as **melanin.** Melanin protects the stratum basale from the damaging effects of ultraviolet radiation. The number of melanocytes is relatively constant among people of all skin colors, but the amount of melanin produced by the melanocytes is variable. People of darker pigmentation produce more melanin, while people of lighter pigmentation have melanocytes that produce less melanin. People of lighter pigmentation are more susceptible to the effects of ultraviolet radiation and skin cancer.

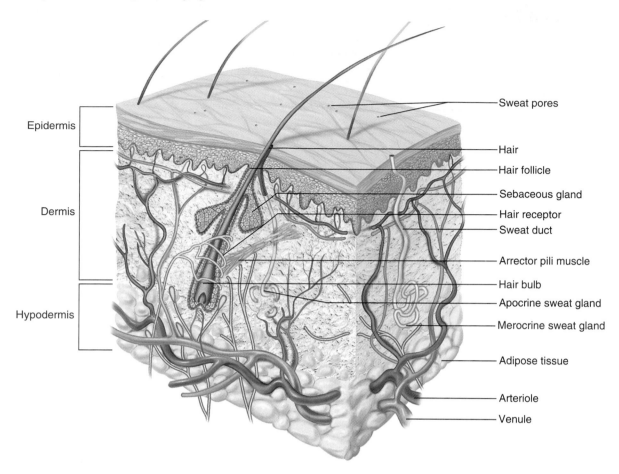

Figure 7.1   Diagram of the integument.

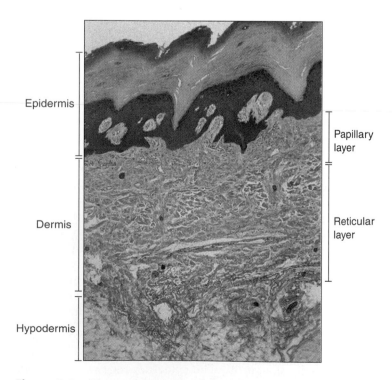

Figure 7.2   Photomicrograph of the integument (40×).

The layer of numerous cells superficial to the stratum basale is the **stratum spinosum.** In prepared slides, the cells appear star-shaped because the cell membrane pulls away from the other cells except in areas where the cells are joined at microscopic junctions called **desmosomes.** The stratum basale and stratum spinosum are frequently classified as the stratum germinativum.

A layer of granular cells is present in the next layer. This is known as the **stratum granulosum.** The cells have purple-staining granules in their cytoplasm. The granules are precursors of keratin that is found in the outermost layer of the epidermis.

Superficial to this layer is the **stratum lucidum,** so named (lucid = light) because this layer appears translucent in fresh specimens. This layer generally appears clear or pink in prepared and stained slides. It is found only in the palms of the hand and soles of the feet.

The most superficial layer of the integument is the **stratum corneum,** which is a tough layer consisting of dead cells that are flattened at the surface. Cells of the stratum corneum have been toughened by complete keratinization, and thus the epidermis is said to be made of **keratinized stratified squamous epithelium.** Examine a prepared slide of thick skin and locate the various strata of the epidermis as illustrated in figure 7.4.

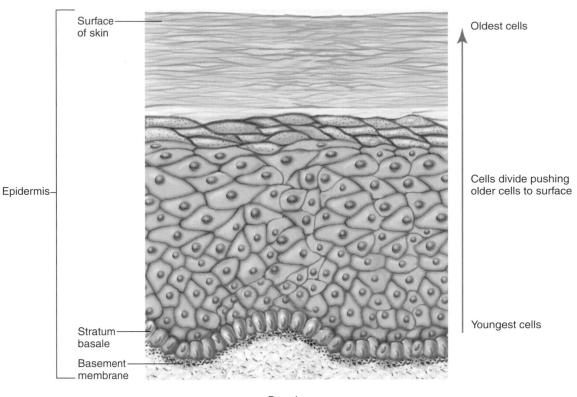

Surface of skin

Oldest cells

Epidermis

Cells divide pushing older cells to surface

Youngest cells

Stratum basale

Basement membrane

Dermis

Figure 7.3   Proliferation of the epidermis.

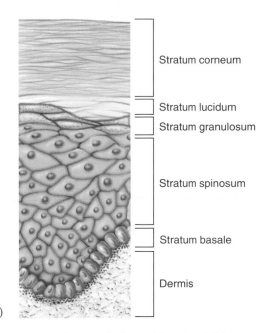

Stratum corneum

Stratum lucidum

Stratum granulosum

Stratum spinosum

Stratum basale

Dermis

(a)

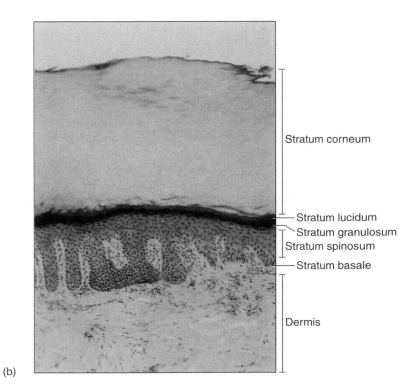

Stratum corneum

Stratum lucidum

Stratum granulosum

Stratum spinosum

Stratum basale

Dermis

(b)

Figure 7.4   Strata of the epidermis. (a) Diagram; (b) photomicrograph (100×).

Cells in the epidermis go through three major phases. In the stratum basale, the cells are involved in rapid cell division. Adjacent to this layer cells undergo a process of producing precursor molecules that leads to the general waterproofing of the cells. The final phase is the completion of the water-proofing process, which leads to the death of the epithelial cells. The most superficial cells of the epidermis are tough. They protect the skin from microorganisms and desiccation and eventually slough off. The epidermis renews itself about every 6 weeks.

Dermis

The dermis is responsible for the structural integrity of the integumentary system. It is composed primarily of closely apposed fibers along with blood vessels, nerves, sensory receptors, hair follicles, and glands. The dermis consists of two major regions, the superficial **papillary layer** and a deeper **reticular layer.** The papillary layer is so named for the **papillae** (bumps) that interdigitate with the epidermis, providing good adhesion between the two layers. The reticular layer composes most of the dermis and can be seen as an irregular arrangement of fibers. Examine a slide of thick skin and locate the two layers of the dermis as seen in figure 7.5.

The majority of the fibers of the dermis are **collagenous** along with lesser numbers of **elastic** and **reticular fibers.** The collagenous fibers provide strength and flexibility. The elastic fibers, as their name implies, provide elasticity to the skin.

Blood vessels pass into the dermis, bringing nutrients to both the dermis and epidermis. The blood vessels do not enter into the epidermis, but the nutrients and oxygen that these vessels carry diffuses from the dermis into the nearby cells of the epidermis. Blood vessels are also important here in that they release heat to the external environment. As the internal temperature of the body core rises, the capillaries of the dermis expand and allow for heat radiation from the skin. In cold weather the precapillary sphincters close, decreasing the amount of blood to the dermis, which conserves heat in the body core.

Nerves also receive sensory information from the dermis and transmit impulses to the central nervous system. Sensory information is received by numerous structures such as the **light touch corpuscles** of the dermis. Two of these light touch corpuscles are the **Meissner's corpuscles,** in the upper portion of the dermis, and **Merkel discs,** located in the upper dermis and lower epidermis. These two receptors allow for the perception of very slight touch stimuli (such as a fly lightly walking over your cheek!). In addition, deep touch or pressure receptors are found in the dermis, further away from the epidermis. **Pacinian** (lamellated) **corpuscles** sense pressure, such as when you lean against a wall or feel a vibration. Other receptors in the skin are **warm receptors** and **cool receptors.** When you are at a comfortable temperature both of these receptors are firing. If you increase or decrease the skin temperature beyond these receptors, **pain receptors** are stimulated. Pain receptors are naked nerve endings in the dermis that respond to numerous environmental stimuli. Locate the Meissner's corpuscles and Pacinian corpuscles in a prepared slide of skin and compare them to figure 7.6.

Integumentary Glands

A significant number of glands occur in the dermis and hypodermis. These include **sudoriferous (sweat) glands, lactiferous (milk) glands, sebaceous (oil)**

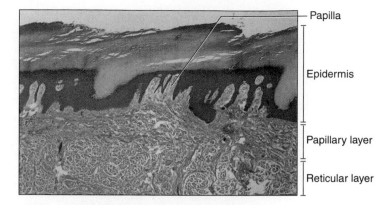

Figure 7.5   Layers of the dermis (40×).

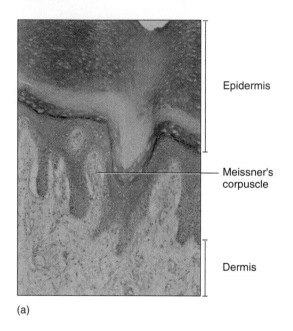

(a)

(b)

Figure 7.6   Skin receptors (100×). (a) Meissner's corpuscle; (b) Pacinian corpuscle.

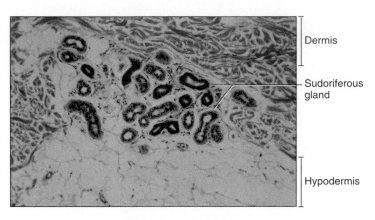

Figure 7.7   Sudoriferous glands (100×).

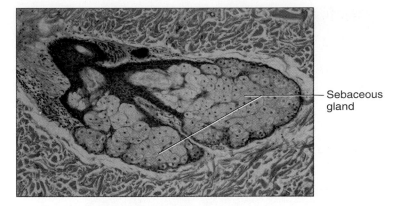

Figure 7.8   Sebaceous glands (100×).

glands, and **ceruminous (earwax) glands.** The sudoriferous glands are composed of simple cuboidal epithelium and come in two different types. **Eccrine (merocrine) glands** are the most common, are sensitive to temperature, and produce normal body perspiration. Perspiration functions to reduce the temperature of the body by **evaporative cooling. Apocrine glands** secrete water and a higher concentration of organic acids than eccrine glands. The organic acids, along with bacterial action, produce the characteristic pungent body odor. These glands are concentrated in the region of the axilla and groin and are associated with hair follicles. Examine a prepared slide of skin and look for clusters of cuboidal epithelial cells with purple nuclei in the dermis or hypodermis. These are the sudoriferous glands illustrated in figure 7.7.

Sebaceous glands are typically associated with hair follicles and are larger than sudoriferous glands. You may not see a hair follicle with the sebaceous gland if the section was made through the gland and not through the gland and follicle. These glands secrete an oily material called **sebum,** which lubricates the hair and decreases the wetting action of water on the skin. Examine a slide of hair follicles and compare the slide to figure 7.8.

Hair

Hair is considered an accessory structure of the integumentary system and consists of keratinized cells that are produced in **hair follicles.** The follicle encloses the hair in the same way a bud vase encloses the stem of a rose. Hair follicles are actually projections of the epidermal layers into the dermis. Hair can thus be considered an epidermal derivative. The hair consists of the **shaft,** a portion that erupts from the skin surface, the **root,** which is enclosed by the follicle, and the **hair bulb,** which is the actively growing portion of the hair. There are two different types of hair. **Determinate hair** grows to a specific length and then stops. **Indeterminate hair** continues to grow without regard to length. Examine models and charts in the lab and compare them to figure 7.9.

Look at a prepared slide of hair and locate the root, follicle, and hair bulb. At the center of the bulb is a small structure

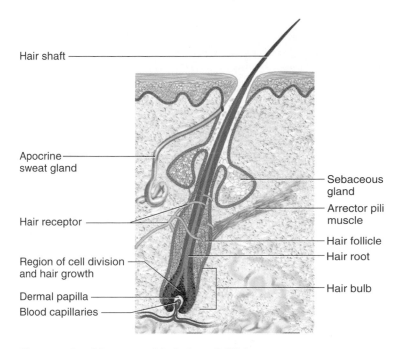

Figure 7.9   Diagram of hair in a follicle.

known as the **dermal papilla,** a region with blood vessels and nerves that reach the hair bulb. The follicle is composed of an outer dermal layer of connective tissue and an inner epithelial layer. These form the **root sheath** of the hair. Locate these structures and compare them to figure 7.10.

Arrector Pili

When the hair "stands on end" it does so by the contraction of **arrector pili** muscles. An arrector pili muscle is a cluster of parallel smooth muscle fibers that connects the hair follicle to the upper regions of the dermis. In response to cold conditions the arrector pili muscles contract, producing goose bumps. The arrector pili also contract when a person is suddenly frightened. Find an arrector pili muscle in a prepared

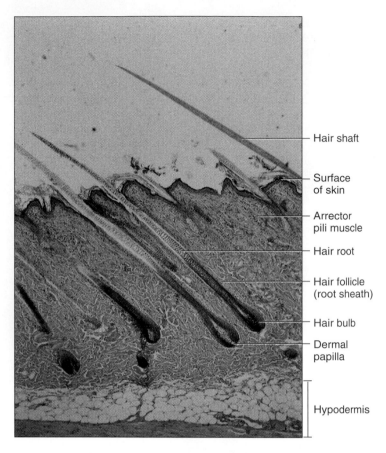

Figure 7.10   Photomicrograph of hair in a follicle (40×).

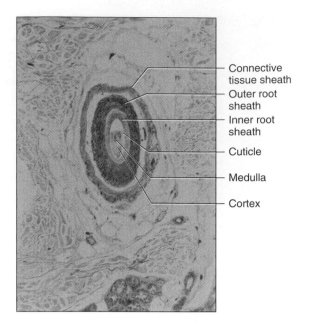

Figure 7.11   Hair root and a follicle, cross section (100×).

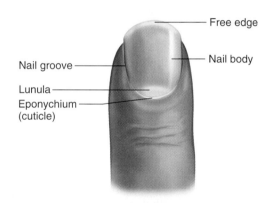

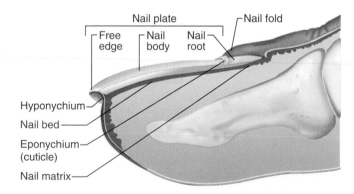

Figure 7.12   Fingernail, top and midsagittal view.

slide and compare it to figures 7.9 and 7.10. If you have trouble finding an arrector pili muscle, look for slender slips of smooth muscle that angle from the follicle to the superficial regions of the dermis. You may have to look at more than one slide to find them.

Cross Section of Hair

The central portion of the hair is known as the **medulla,** which is enclosed by an outer **cortex.** The cortex may contain a number of pigments, which give the hair its particular color. The brown and black pigments are due to varying types and amounts of **melanin.** Iron-containing melanin pigments produce red hair. Gray hair is the result of a lack of pigment and air in the hair root and shaft. The layer superficial to the cortex is the **cuticle** of the hair and looks rough in microscopic sections. Figure 7.11 shows a cross section of hair.

Nails

Nails are keratinized cells of the stratum corneum of the skin that occur on the distal ends of the fingers and toes. The nail itself is called the **nail body** and, as it grows, the **free edge** of the nail is the part you clip. The cuticle of the nail is known as the **eponychium,** and the **nail root** is underneath the eponychium and is the area that generates the nail body. The **nail bed** is the layer deep to the nail body while the **lunula** is the small

white crescent that occurs at the base of the nail. The **hyponychium** is the region under the free edge of the nail. Examine figure 7.12 and compare your fingernails to the diagram.

On each side of the nail is the **nail groove,** a depression between the body of the nail and the skin of the fingers. The ridge above the groove is the **nail fold.**

Name _____

1. Distinguish between Pacinian corpuscles, Meissner's corpuscles, and pain receptors in the skin.

2. Place labels on the following illustration.

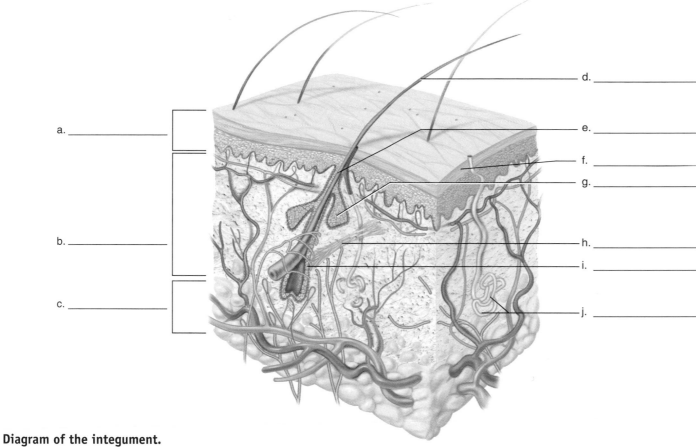

a. _____

b. _____

c. _____

d. _____

e. _____

f. _____

g. _____

h. _____

i. _____

j. _____

Diagram of the integument.

3. Hair of the axilla can be considered determinate/indeterminate hair (circle the correct answer).

4. Electrolysis is the process of hair removal by using electric current. Explain how this might destroy the process of hair growth in relation to the hair bulb.

5. Since hair color is determined by pigment in the cortex and the hair shaft is dead, explain the fallacy of a person's hair turning white overnight.

6. Which specific strata is the deepest layer of the epidermis?

7. How long does it take for the epidermis to renew itself?

8. What cell type produces a pigment that darkens the skin?

9. Which one of the three layers (epidermis, dermis, hypodermis) is not part of the integument?

10. Which layer of the integument is vascular?

11. What integumentary gland secretes oil?

12. What part of the hair is found on the outside of the skin?

13. What does the arrector pili muscle do?

14. The outermost portion of the hair is known as the
    a. cuticle   b. lunula   c. root   d. medulla

Introduction to the Skeletal System

Introduction

The skeletal and muscular systems are intimately related. The skeletal system has numerous functions, including support of the body, protection of soft tissues such as the brain and lungs, a structure for locomotion, storage of calcium, and blood formation (hematopoiesis).

The interaction between the skeletal and muscular systems is one in which the skeletal system provides the stable framework on which the muscles can work, resulting in movement. In this exercise, you examine the structure, composition, and development of the skeletal system.

Objectives

At the end of this exercise you should be able to

1. describe the composition of bone tissue;
2. discuss how the skeletal system provides for efficient movement;
3. list the two major types of bone material found in long bones;
4. draw or describe the microscopic structure of compact bone;
5. relate the anatomy of the bone to its growth.

Materials

Chart of the skeletal system
Cut sections of bone showing compact and spongy bone
Articulated human skeleton
Articulated cat skeleton (if available)
Disarticulated human skeleton
Prepared slide of ground bone
Prepared slide of decalcified bone
Bone heated in oven at 350°F for 3 hours or more
Bone placed in vinegar for several weeks or 1 N nitric acid for a few days.
Microscopes

Procedure

Divisions of the Skeletal System

Examine the skeletons or charts in the lab and locate the division of the skeletal system called the **axial skeleton.** The axial skeleton consists of the skull, hyoid bone, vertebral column, ribs, and sternum. All of the other parts of the skeletal system including the pectoral girdle, upper extremity bones, pelvic girdle, and lower extremity bones belong to the **appendicular skeleton.** Compare the material in the lab to figure 8.1 and table 8.1 (page 77).

Major Bones of the Body

Figure 8.1 shows the major bones or bone regions of the body. You should be familiar with the major bones of the body before you study the specifics of the individual bones in later exercises. It helps to take a bone from a disarticulated skeleton and compare it to an articulated skeleton. Locate the major bones in the figure as you examine the bones in the lab.

The typical young adult has around 206 bones, with more found in younger individuals and fewer in older individuals. Numerous developing bones fuse as a person ages and form larger bones, thus reducing the overall number.

Bone Features

Bones have many surface features that have specific names. These are listed in table 8.2 and are grouped by projections, depressions, or holes in the bone. You should study these terms and be able to find them on bones in lab.

Composition of Bone Tissue

Bone consists of **organic material** and **inorganic material.** The organic portion is composed mostly of cells and collagenous fibers and makes up about 30% of the bone by weight. The inorganic portion makes up the remaining 70% of the bone and mostly consists of **hydroxyapatites,** which are complex salts of calcium phosphate and calcium carbonate. If available in the lab, examine bones that have been soaked in acid. The acid has dissolved the mineral salts, leaving the collagenous fibers as a flexible framework. Now examine the bones that were baked in an oven for a few hours. In this preparation the structural characteristics of proteins have been destroyed, and though the bones remain hard because of the mineral salts, they are brittle due to the destruction of the protein fibers by the heat. Notice how easily the bones break.

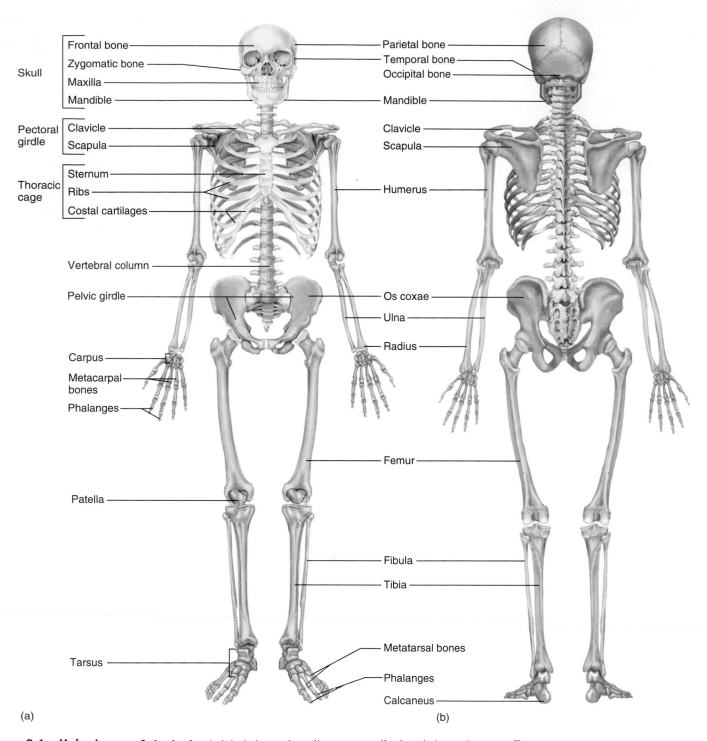

Figure 8.1   Major bones of the body. Axial skeleton in yellow, appendicular skeleton in gray. ⚡

Bone Morphology

The general structure of a long bone consists of the proximal and distal ends of the bone, called the **epiphyses,** and the shaft of the bone, called the **diaphysis** (figure 8.2). The epiphyses of bones have **articular cartilage** at the ends of the bone.

The articular cartilage is composed of hyaline cartilage and helps reduce friction as the joint moves.

In between the diaphysis and each epiphysis in older individuals is an **epiphyseal line.** The epiphyseal line can be considered a fusion plate, where the growing regions of the bone finally united. This region is the site of the old **epiphyseal**

Table 8.1 Axial and Appendicular Skeleton

Axial Skeleton

Skull	Ribs
Hyoid bone	Sternum
Vertebral column	

Appendicular Skeleton

Pectoral girdle	Pelvic girdle
Clavicle	Os coxa
Scapula	Lower extremity
Upper extremity	Femur
Humerus	Patella
Radius	Tibia
Ulna	Fibula
Carpals	Tarsals
Metacarpals	Metatarsals
Phalanges	Phalanges

plate, which occurs in individuals who are growing in height. The epiphyseal plate is made of **hyaline cartilage** and increases in thickness by the division of chondrocytes. As the cartilage grows the individual is literally pushed upwards by the increase in the cartilage plate. Remodeling of the cartilage takes place, and the cartilage is eventually replaced by bone.

Examine a cut section of bone. Note the hard, **compact bone** on the outside of the bone and the inner **spongy** (or cancellous) **bone.** The spongy bone is made of **trabeculae,** which are thin threads of bone that frequently run in the same direction as the stress applied to the bone. Stress on the bone may be in the form of gravity, or it may occur due to a common force applied to the limb.

The innermost section of bone is hollow and is called the **marrow** (or **medullary**) **cavity.** Marrow is the material that occupies this cavity and can be of two types, a hematopoietic (blood-forming) **red marrow** and an adipose-containing **yellow marrow.** In flat bones the outer and inner compact bone encases a material called **diploe,** which is spongy bone. Compare the cut bones in the lab to figures 8.2 and 8.3 and note the features listed in the discussion.

Table 8.2 Bone Features

Projections from the Bone Surface	Example
Process—a general term for a projection from the surface of the bone	Styloid process of ulna
Tubercle—a relatively small bump on a bone	Greater tubercle of humerus
Tuberosity—a relatively larger bump on a bone	Deltoid tuberosity of humerus
Spine—a short, sharp projection	Vertebral spine
Condyle—an irregular, smooth surface that articulates with another bone	Lateral condyle of femur
Epicondyle—a bump on a condyle	Medial epicondyle of humerus
Head—a hemispheric projection that articulates with another bone	Head of the femur
Neck—a constriction below the head	Neck of a rib
Crest—an elevated, ridge of bone	Crest of the ilium
Line—a smaller elevation than a crest	Gluteal line of ilium
Facet—a smooth, flat face	Articular facets of vertebrae
Trochanter—a large bump (on femur)	Greater and lesser trochanter of femur
Ramus—a branch	Ramus of the mandible
Depressions or Holes in the Bone Surface	**Example**
Foramen—a shallow hole	Foramen magnum of skull
Sinus—a cavity in a bone	Maxillary sinus
Meatus or canal—a deep hole	External auditory meatus
Fossa—a shallow depression in a bone	Iliac fossa
Notch—a deep cut-out in a bone	Greater sciatic notch
Groove or sulcus—an elongated depression	Intertubercular groove of humerus
Fissure—a long deep cleft in a bone	Inferior orbital fissure

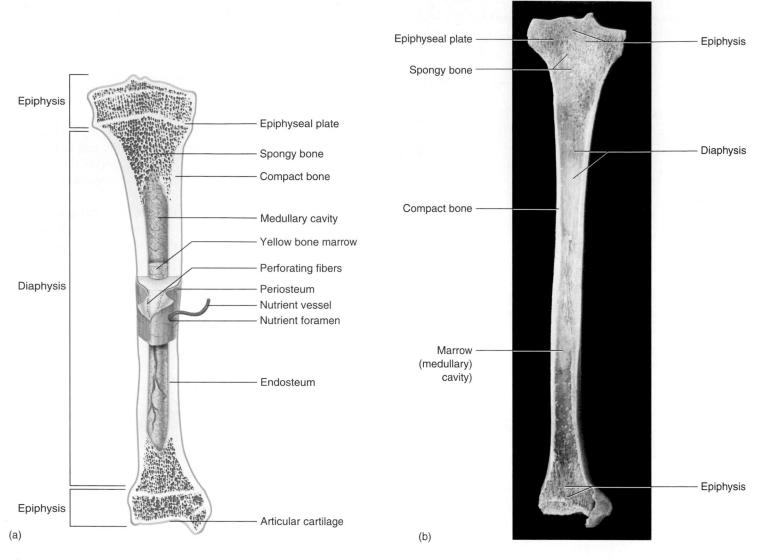

Figure 8.2   Long bone, longitudinal section. (a) Diagram; (b) photograph. ⚡

Note how the compact bone in the middle of the long section of bone is thicker than the walls at the ends of the bone. Many long bones have this feature and resemble an archery bow in this regard. This reinforcement decreases breakage at the region of bone that would be most prone to break, the middle.

On the surface of the diaphysis, locate the **nutrient foramina.** These are small holes that allow for the passage of blood vessels into and out of the bone. The nutrient foramina lead to **perforating** (Volkmann's) **canals** that pass through compact bone.

The outer surface of the bone is covered with a dense connective tissue sheath called the **periosteum.** The periosteum houses **osteoblasts,** which function to produce more bone mass. It is also the site where nerves and blood vessels occur on the outer surface of the bone. The periosteum serves as an anchoring point for **tendons** and **ligaments.** Tendons attach muscles to bone at the periosteum, and this attachment is strengthened by **perforating** (Sharpey's) **fibers** that provide greater surface

area between the tendon and the periosteum. **Ligaments** are parallel straps of connective tissue that connect one bone to another, and they are also secured to the bone by the periosteum.

The inner surface of long bones, near the marrow cavity, is lined with a layer known as the **endosteum.** The endosteum has specialized cells known as **osteoclasts** that contain lysosomal enzymes that remove bone tissue.

Bone Classification

Bones may be classified in a number of ways. Frequently bones are classified according to shape, location, and development.

Bone Shape

Bones are classified according to four main shapes, **long, short, flat,** and **irregular.** Long bones are those that are longer than broad. Bones such as those of the arm, forearm,

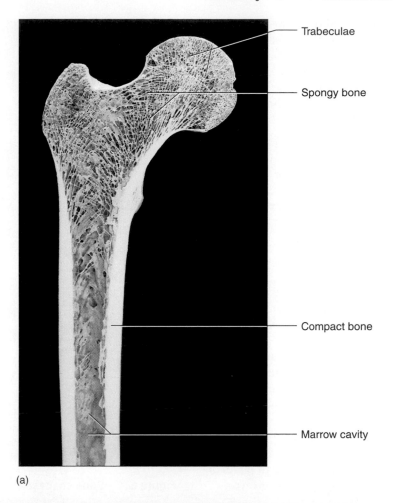

Trabeculae

Spongy bone

Compact bone

Marrow cavity

(a)

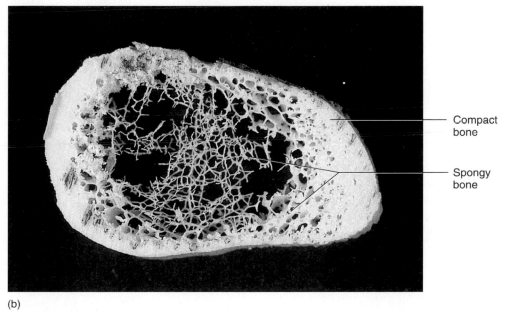

Compact bone

Spongy bone

(b)

Figure 8.3   Cut bone. (a) Long section; (b) cross section. ✶

fingers, thigh, and leg are long bones. Short bones are more or less equal in all dimensions. Bones of the wrist and ankle are examples of short bones. Flat bones, such as bones of the cranium and sternum, appear compressed in one dimension. The ribs are also considered flat bones. Irregular bones are hard to place into any of the other categories. Bones such as those in the floor of the skull, facial bones, vertebrae, pelvic and pectoral girdle bones are considered irregular. Examine bones in the lab and compare them to figure 8.4.

Bone Location

Bones can be classified not only as a part of the axial and appendicular skeleton but also as part of the upper extremity, lower extremity, pelvic girdle, and skull.

Bone Development

There are two main ways that bones develop. The first is **endochondral ossification,** which is the way most bones develop. This type of bone formation begins with a hyaline cartilage template, which is later replaced by bone.

Endochondral ossification begins with the initiation of calcified regions in the hyaline cartilage called **primary ossification centers.** Later, **secondary ossification centers** develop and expand the mineralization of the cartilage. An early, developing fetus has an almost translucent skeleton composed of hyaline cartilage. Ossification begins before birth, but much cartilage remains even at birth. At the time of birth an infant has no carpal bones and no fusion of any of the growth plates of bones. Figure 8.5 illustrates the pattern of long bone growth.

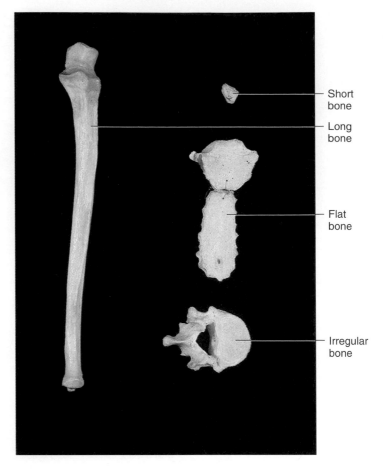

Figure 8.4   Bone shapes. ✗

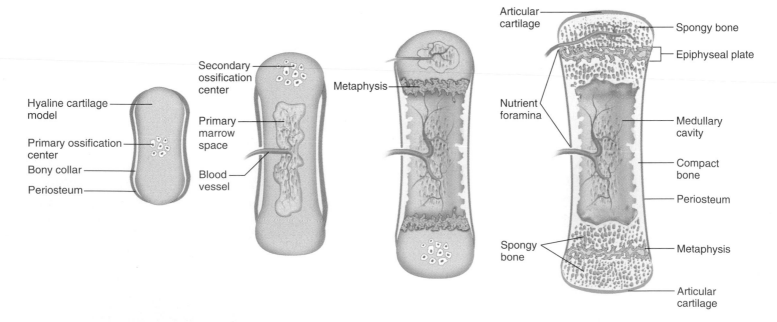

Figure 8.5   Endochondral ossification. ✗

The second pattern of bone development, **intramembranous ossification,** is different from endochondral ossification in that the bone material does not replace cartilage but rather a fibrous membrane is laid down and ossification occurs between the fibers. This type of ossification is characteristic of bones in the skull and clavicles. Collagenous fibers make up the membrane, and osteoblasts secrete **spicules** or threads of bone into the tissue. A specialized type of intramembranous bone is a **sesamoid bone.** This kind of bone is named because it looks like a sesame seed. The most common example of a sesamoid bone is the patella, which begins development at about the age of two in the tendon that stretches from the anterior thigh muscles (quadriceps femoris) to the shin bone (tibia).

Microscopic Structure of Bone

Ground Bone

Examine a prepared slide of ground bone and locate the circular structures in the field of view. These modular units of bone are called **osteons.** Each osteon has a hole in the middle called a **central** (Haversian) **canal** that serves to house blood vessels and nerves in the dense bone tissue. Around the central canal are rings of dark spots. These are the **lacunae,** (singular; *lacuna*) which are spaces that the bone cells, or **osteocytes,** occupied in living tissue. In the prepared slide the lacunae have filled with bone dust from the grinding process. The lacunae are connected by thin tubes called **canaliculi.** The role of canaliculi is to provide a passageway through the dense bone material. Osteocytes are living and need to receive oxygen and nutrients and remove wastes. They do this by extending cellular strands through the canaliculi between osteocytes. The canaliculi connect to the central canal. The osteon has layers of dense tissue that form concentric rings between the lacunae. These are the **lamellae.** Locate these structures in figure 8.6.

Decalcified Bone

Examine a slide of decalcified bone. In this preparation the bone salts have been dissolved away and the remaining thin section of bone has been sliced, mounted, and stained. In this

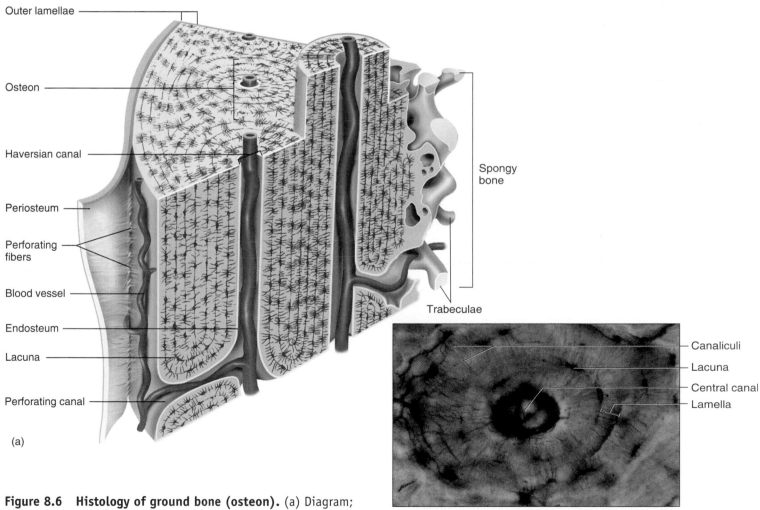

Figure 8.6   Histology of ground bone (osteon). (a) Diagram; (b) cross section (400×).

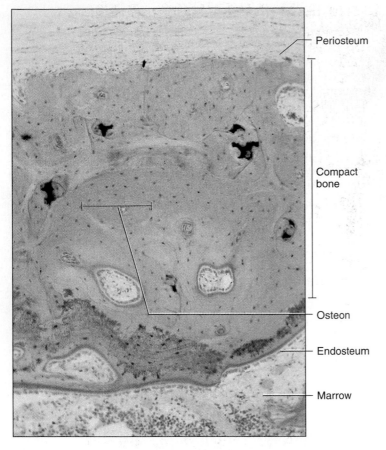

Figure 8.7   Decalcified bone (100×).

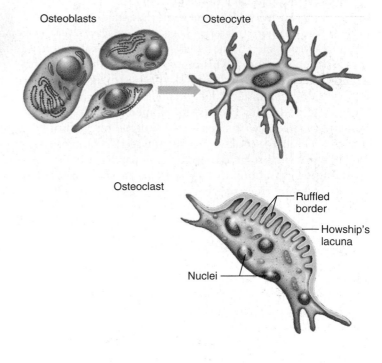

Figure 8.8   Three types of bone cells.

way the periosteum is visible on the surface of the bone, with the compact bone in the middle and the endosteum on the inside of the bone (figure 8.7).

There are three types of bone cells in osseous tissue. These are the bone-forming osteoblasts, osteocytes, and osteoclasts. **Osteoblasts** are active bone-forming cells. They are found near the periosteum (figure 8.8). Examine a slide of decalcified bone and compare your sample to the figure.

The cells that are the most common in compact bone are the **osteocytes.** In ground bone only the spaces where the os-

teocytes occur are visible (the lacunae), but in decalcified bone the actual osteocytes are visible. Examine the slide of decalcified bone for osteocytes and compare it to figure 8.8.

At the very inner margin of bone is the endosteum, and in this layer you can find cells, each of which has numerous nuclei. These cells are the **osteoclasts** and are responsible for the remodeling of bone tissue. Osteoclasts secrete lysosomal enzymes that digest osseous material. Examine the inner portion of the same slide and locate the osteoclasts. Compare them to figure 8.8.

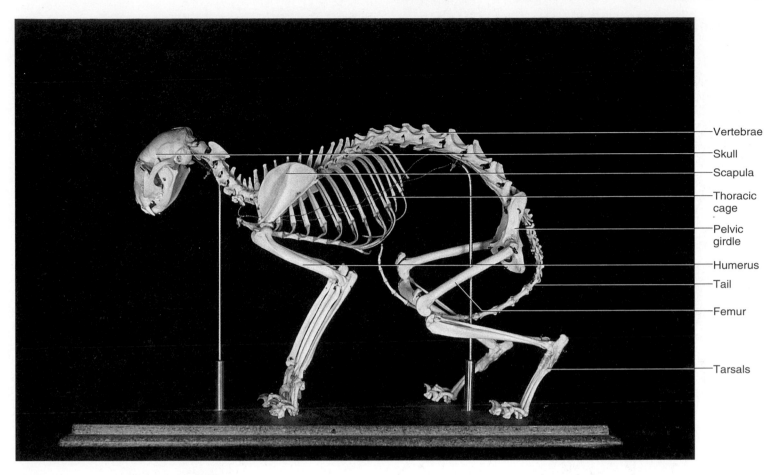

Vertebrae
Skull
Scapula
Thoracic cage
Pelvic girdle
Humerus
Tail
Femur
Tarsals

Figure 8.9   Cat skeleton, lateral view.

Skeletal Anatomy of the Cat

The cat skeleton differs from the human skeleton in several ways. Humans are bipedal, and this can be seen in the vertebral column. In humans the bodies of the vertebrae become much larger as you move from superior to inferior. Cats are quadrupeds, and so the weight of the vertebral column is distributed more evenly and their vertebrae are more evenly matched in terms of size. In humans the absence of a tail is reflected in a corresponding absence of tail vertebrae. In cats the tail vertebrae are present and fully functional.

Another difference is that humans walk on the entire inferior surface of the foot while cats walk on their toes. Examine a cat skeleton in the lab and note the general features of the skeleton (figure 8.9).

Name _____

1. The hyoid bone belongs to the

   a. appendicular skeleton   b. axial skeleton   c. upper extremity   d. skull

2. The clavicle belongs to the

   a. axial skeleton   b. pectoral girdle   c. pelvic girdle   d. upper extremity

3. In the disease osteoporosis there is a significant loss of spongy bone. Explain how the loss of this specific bone material could weaken a bone.

4. Label the following illustration.

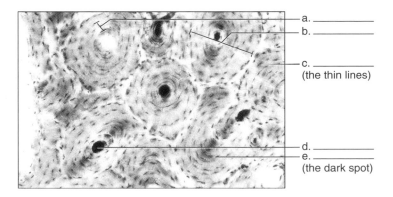

a. _____
b. _____

c. _____
(the thin lines)

d. _____
e. _____
(the dark spot)

5. The ends of a long bone are known as the _____.

6. A carpal bone is classified as a _____ bone in terms of shape.

7. The radius is part of the what skeletal division?

   a. axial   b. appendicular

8. The ribs are part of what skeletal division?

   a. axial   b. appendicular

9. The clavicle is part of what skeletal division?

   a. axial   b. appendicular

10. The patella is part of what skeletal division?

    a. axial   b. appendicular

11. A young adult has how many bones on average?

12. The inorganic portion of bone tissue is made of what mineral salt?

13. In terms of shape, what are the bones of the cranium?

    a. long   b. short   c. flat   d. irregular

14. In terms of shape, what is the sternum?

    a. long   b. short   c. flat   d. irregular

15. In terms of shape, what are the metacarpals?

    a. long   b. short   c. flat   d. irregular

16. In terms of shape what are the tarsals?

    a. long   b. short   c. flat   d. irregular

17. The long bones of the body develop by _____ ossification.

18. What is an osteon?

19. Synthetic bone material known as hydroxyapatite is frequently molded into the shape of bone. This procedure is done to replace bone that has been diseased, damaged, or surgically removed. Cells in the body remodel the synthetic material and produce new bone. What bone cells do you predict to be involved in this remodeling and what role do these cells have?

Appendicular Skeleton

Introduction

The appendicular skeleton consists of approximately 126 bones in four major groupings. The first group is the **pectoral girdle,** which is composed of the scapula and the clavicle on each side. The next is the **upper extremity,** which consists of the humerus, radius, ulna, carpals of the wrist, metacarpals of the palm, and phalanges of the fingers. The lower portion of the appendicular skeleton is composed of the **pelvic girdle** (the os coxae or hip bones), each of which consists of three fused bones, the ilium, the ischium, and the pubis. The **lower extremity** is the final group of the appendicular skeleton and consists of the femur, patella, tibia, fibula, tarsals of the ankle, metatarsals (of the foot), and phalanges of the toes.

Objectives

At the end of this exercise you should be able to

1. locate and name all of the bones of the appendicular skeleton;
2. name the significant surface features of the major bones;
3. place bones into the four major groupings, such as upper extremity bones;
4. name the individual carpal bones and tarsal bones in articulated hands and feet respectively;
5. determine whether a femur, tibia, humerus, radius, or ulna is from the left side or right side of the body;
6. place a clavicle with the sternal end pointing medially;
7. locate the major surface features of the scapula and os coxa;
8. determine whether a scapula or pelvic bone is from the left side or right side of the body.

Materials

Articulated skeleton or plastic cast of articulated skeleton
Disarticulated skeleton or plastic casts of bones
Charts of the skeletal system
Plastic drinking straws cut on a bias or pipe cleaners for
    pointer tips
Foam pads of various sizes (to protect real bones from hard
    countertops)

Procedure

Locate the bones of the appendicular skeleton on the material in the lab and on figure 9.1. In this exercise, we study the bones of the appendicular skeleton by examining an isolated (or disarticulated) bone in conjunction with the articulated skeleton (one that is joined together). Hold the individual bone up to the skeleton to see how it is positioned in relation to the other bones of the body. When examining bones in the lab, do not use your pen or pencil to locate a structure. Use a cut drinking straw or a pipe cleaner to point out structures. Your instructor may want you to place real bone material on foam pads to cushion the bone from the tabletop.

Pectoral Girdle

The pectoral girdle consists of the right and left **scapulae** (singular, *scapula*) and the right and left **clavicles.** The scapula is commonly known as the shoulder blade and is found on the posterior, superior portion of the thorax. The scapula is roughly triangular in shape and has three borders, a **superior border,** a medial or **vertebral border,** and a lateral or **axillary border.** On the superior border of the scapula is an indentation known as the scapular notch. The anterior surface of the scapula is a smooth, hollowed depression known as the **subscapular fossa.** The posterior surface of the scapula is divided by the **scapular spine** into a **supraspinous fossa** and an **infraspinous fossa.** The spine of the scapula is actually a crest that runs from the vertebral border to the edge of the scapula, where it expands to form the **acromion process.** Another projection from the scapula is the **coracoid process,** which projects anteriorly from the scapula. The scapula also has two sharp angles known as the **inferior angle** and **superior angle.** The humerus moves in the socket of the scapula known as the **glenoid cavity,** or glenoid fossa. Just below the glenoid fossa is a bump known as the **infraglenoid tubercle,** which is an attachment point for the triceps brachii muscle. The features of the scapula are important because numerous muscles attach to it. You should also be able to distinguish a right scapula from a left one. The spine of the scapula is posterior, the inferior angle is less than 90°, and the glenoid fossa is lateral for articulation with the humerus. Locate the features of the scapula on figure 9.2.

Clavicle

The **clavicle** is a small bone that serves as a strut between the scapula and the sternum. The blunt, vertically truncated end of the clavicle is the **sternal end** and the horizontally flattened end is the **acromial end.** Note the **conoid tubercle** on the inferior surface of the clavicle. The clavicle is the most

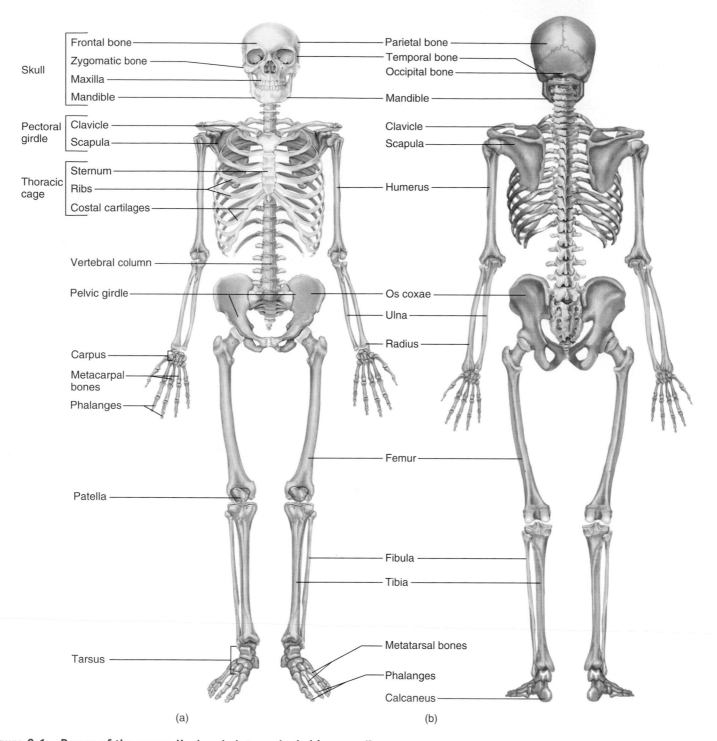

Skull
- Frontal bone
- Zygomatic bone
- Maxilla
- Mandible

Pectoral girdle
- Clavicle
- Scapula

Thoracic cage
- Sternum
- Ribs
- Costal cartilages

Vertebral column

Pelvic girdle

Carpus

Metacarpal bones

Phalanges

Patella

Tarsus

Parietal bone
Temporal bone
Occipital bone
Mandible

Clavicle
Scapula

Humerus

Os coxae
Ulna
Radius

Femur

Fibula
Tibia

Metatarsal bones
Phalanges
Calcaneus

(a)   (b)

Figure 9.1   Bones of the appendicular skeleton, shaded in gray. ⚲

frequently fractured bone in the body. Examine the clavicles in the lab and compare them to figure 9.3.

Upper Extremity

The **humerus** is a long bone that articulates with the scapula at its proximal end and with the radius and ulna at its distal end. The **head** of the humerus is a hemispheric structure that ends in a rim known as the **anatomical neck.** The anatomical neck is the site of the **epiphyseal line.** The humerus has two large processes, the **greater tubercle,** which is the largest and most lateral process, and the **lesser tubercle,** which is medial and smaller. These processes serve as attachment points for muscles originating from the scapula. Between the tubercles is the **intertubercular groove,** which is an elongated depression through which one of the biceps brachii tendons passes. Ex-

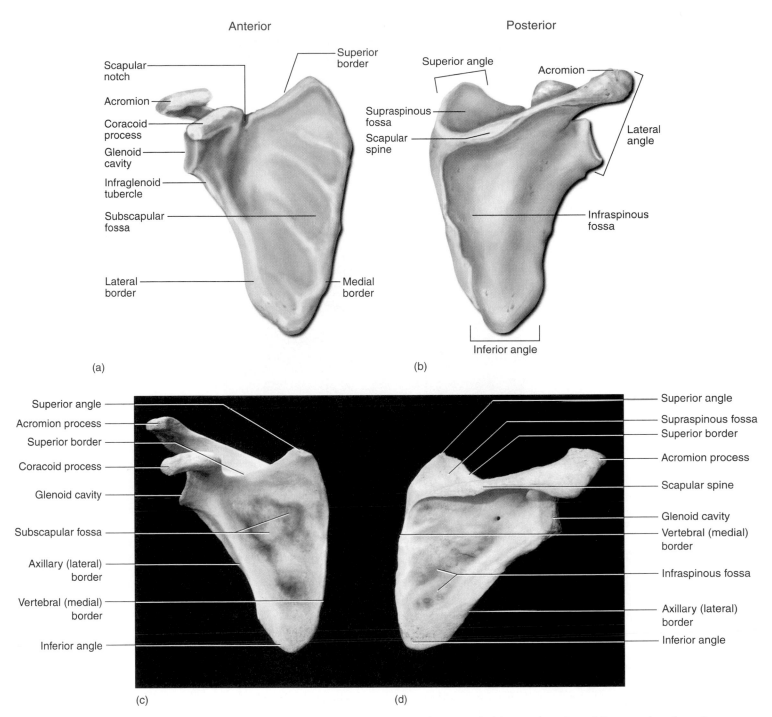

Anterior

Scapular
notch

Acromion

Coracoid
process

Glenoid
cavity

Infraglenoid
tubercle

Subscapular
fossa

Superior
border

Lateral
border

Medial
border

(a)

Posterior

Superior angle

Acromion

Supraspinous
fossa

Scapular
spine

Lateral
angle

Infraspinous
fossa

Inferior angle

(b)

Superior angle

Acromion process

Superior border

Coracoid process

Glenoid cavity

Subscapular fossa

Axillary (lateral)
border

Vertebral (medial)
border

Inferior angle

Superior angle

Supraspinous fossa

Superior border

Acromion process

Scapular spine

Glenoid cavity

Vertebral (medial)
border

Infraspinous fossa

Axillary (lateral)
border

Inferior angle

(c)

(d)

Figure 9.2   Right scapula. Diagram (a) anterior view; (b) posterior view. Photograph (c) anterior view; (d) posterior view. ⚲

amine figure 9.4 and locate these features. Below the tubercles of the humerus is a constricted region known as the **surgical neck.** The surgical neck is so named because it is a frequent site of fracture.

A roughened area on the lateral surface of the humerus is the **deltoid tuberosity,** named for the attachment of the deltoid muscle. The majority of the length of the humerus is called the **shaft,** and a small hole penetrating the shaft is the

nutrient foramen. At the distal end of the humerus are the regions of articulation with the bones of the forearm. On the lateral side is a round hemisphere known as the **capitulum.** This is where the head of the radius fits into the humerus. On the medial side is an hourglass-shaped structure called the **trochlea.** The trochlea is where the ulna attaches to the humerus. The capitulum and trochlea represent the **condyles** of the humerus. To the sides of these condyles are

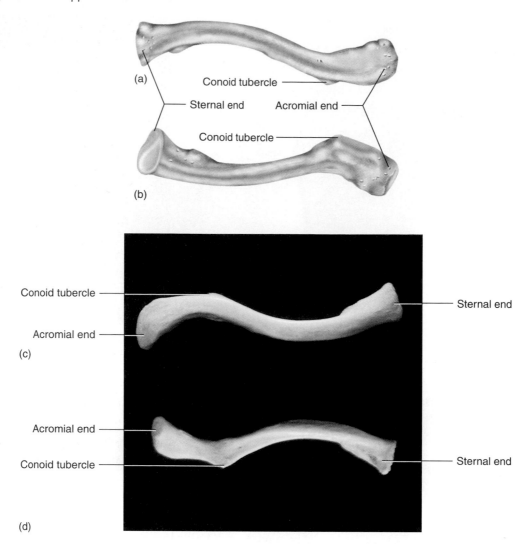

Figure 9.3   Clavicle. Diagram (a) superior view; (b) inferior view. Photograph (c) superior view; (d) inferior view. ⚥

the **epicondyles.** The medial epicondyle of the humerus is commonly known as the "funny bone," and the ridge that forms a wing of bone, proximal to the epicondyle, is the **medial supracondylar ridge.** The lateral epicondyle also has a **lateral supracondylar ridge.** The epicondyles and supracondylar ridges are points of attachments for muscles that manipulate the forearm and hand. At the distal end are two depressions in the humerus. The one on the anterior surface is the **coronoid fossa** and the one on the posterior surface is the **olecranon fossa.** Locate these structures on specimens in the lab and on figure 9.4.

Bones of the Forearm

Two bones make up the skeletal portion of the forearm. These are the **radius,** which is lateral (on the thumb side) and the **ulna,** a medial bone. The radius has a proximal **head** that looks like a wheel. Distal to this is the **radial tuberosity,**

where the biceps brachii muscle inserts, and at the most distal end is the **styloid process** of the radius. On the distal end of the radius there is a medial depression called the **ulnar notch.** This is where the distal part of the ulna joins with the radius. Examine figure 9.5 for these structures and compare them to the bones in the lab.

The ulna has a U-shaped depression (some students remember this as "U for ulna") on the proximal portion, which is the **trochlear notch** (pronounced TRŌK-lee-ur). The most proximal portion of the ulna is the **olecranon process** (pronounced oh-LEC-ruh-non), commonly known as the elbow. The process distal to the trochlear notch is the **coronoid process,** which ensures a relatively tight fit with the humerus and also provides an area for muscle attachment. The ulna has a *distal* **head,** unlike the radius (the head of the radius is proximal), and a **styloid process** near the head. Locate these structures in figure 9.5. On the proximal part of the ulna, where the head of the radius fits into the ulna, is a depression known as

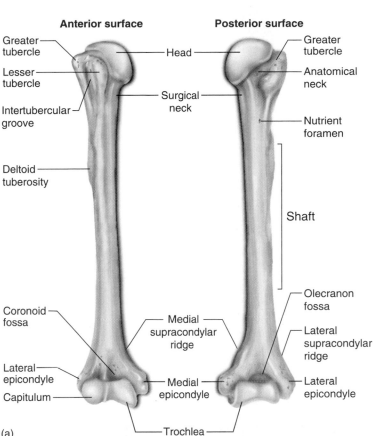

Anterior surface

Greater tubercle

Lesser tubercle

Intertubercular groove

Deltoid tuberosity

Coronoid fossa

Lateral epicondyle

Capitulum

Head

Surgical neck

Medial supracondylar ridge

Medial epicondyle

Trochlea

Posterior surface

Greater tubercle

Anatomical neck

Nutrient foramen

Shaft

Olecranon fossa

Lateral supracondylar ridge

Lateral epicondyle

(a)

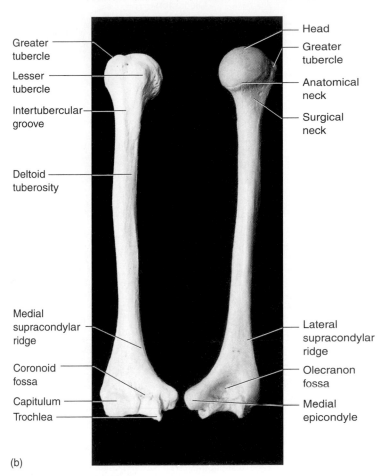

Greater tubercle

Lesser tubercle

Intertubercular groove

Deltoid tuberosity

Medial supracondylar ridge

Coronoid fossa

Capitulum

Trochlea

Head

Greater tubercle

Anatomical neck

Surgical neck

Lateral supracondylar ridge

Olecranon fossa

Medial epicondyle

(b)

Figure 9.4   Humerus. (a) Diagram; (b) photograph.

the **radial notch.** If you know that the trochlear notch is anterior and the radial notch is lateral, then you can determine if you are looking at a left or right ulna.

Bones of the Hand

Each hand consists of three groups of bones, the **carpals,** the **metacarpals,** and the **phalanges** (singular, phalanx). The human hand is composed of eight carpal bones (figure 9.6). These bones can be roughly aligned in two rows. The proximal row from lateral to medial includes the **scaphoid** (navicular), **lunate, triquetrum** (pronounced tri-KWE-trum) (triangular), and the **pisiform.** The distal row includes the **trapezium** (greater multangular), the **trapezoid** (lesser multangular), the **capitate,** and the **hamate.** On the hamate is a hook-like projection called the hamulus of the hamate. The carpal bones were named for their shape, with the scaphoid resembling a boat (*scaphos* = boat), the lunate named for the crescent moon, triquetrum (a triangle), pisiform being pea-shaped, the trapezium (rhymes with thumb and named for a shape), the trapezoid (named for a shape), the capitate being "headlike," and the hamate (*hamus* = hook). Another way to put the bones in proper order is to use a mnemonic device

that uses the first letter of each carpal bone (S for scaphoid, L for lunate, etc.) in a sentence such as, "Say Loudly To Pam, Time To Come Home." The carpal bones serve as regions of attachment for forearm muscles and also for intrinsic muscles of the hand.

The **metacarpals** are the bones of the palm. Each metacarpal consists of a proximal **base,** a **shaft,** and a distal **head.** The metacarpals are labeled by Roman numerals I to V, with I being proximal to the thumb and V being proximal to the little finger. Metacarpals II to V have little movement, but metacarpal I allows for significant movement of the thumb. Locate metacarpal I on the skeletal material and on your own hand and note the substantial range of movement. This type of movement is covered in greater detail in Laboratory Exercise 12. Examine the carpals and metacarpals in the lab and in figure 9.6.

Distal to the metacarpals are the **phalanges.** There are 14 bones that make up this group. Each finger of the hand has three phalanges, a **proximal, middle,** and **distal** phalanx, except for the thumb. The thumb has only a proximal and distal phalanx. You should be able to identify each individual bone by noting the digit it comes from and whether it is proximal, middle, or distal. Each phalanx has a **base, shaft,**

Anterior surface   **Posterior surface**

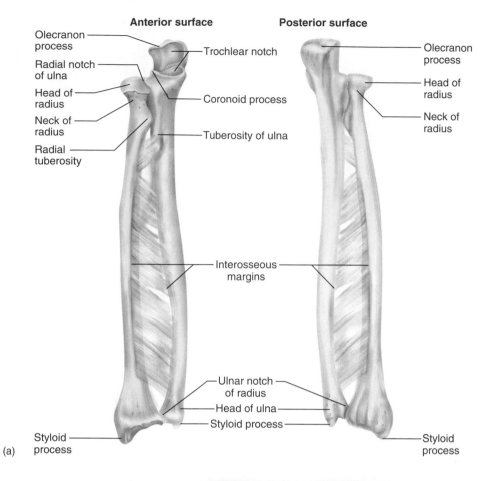

Olecranon process
Radial notch of ulna
Head of radius
Neck of radius
Radial tuberosity

Trochlear notch
Coronoid process
Tuberosity of ulna

Olecranon process
Head of radius
Neck of radius

Interosseous margins

Ulnar notch of radius
Head of ulna
Styloid process

Styloid process

Styloid process

(a)

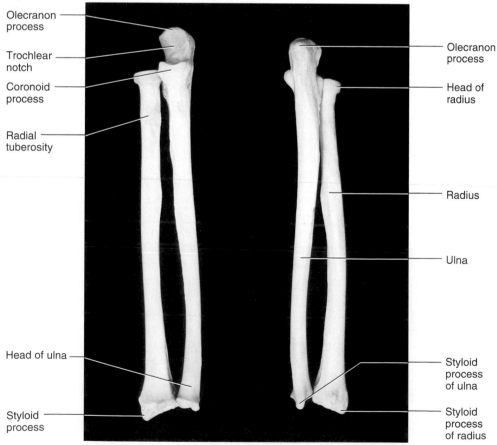

Olecranon process
Trochlear notch
Coronoid process
Radial tuberosity

Olecranon process
Head of radius

Radius

Ulna

Head of ulna
Styloid process

Styloid process of ulna
Styloid process of radius

Figure 9.5   Radius and ulna.
(a) Diagram; (b) photograph. ϯ   (b)

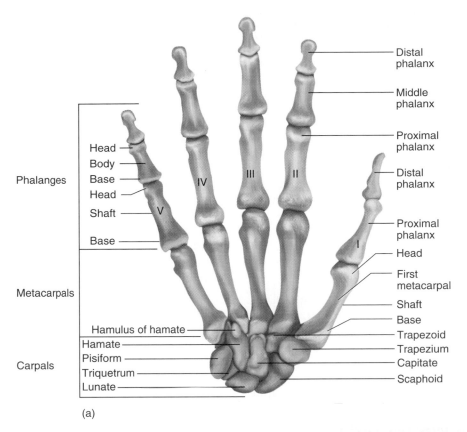

Distal
phalanx

Middle
phalanx

Proximal
phalanx

Phalanges

Head
Body
Base
Head

Shaft

Base

Distal
phalanx

Proximal
phalanx

Head

First
metacarpal

Shaft

Base

Metacarpals

Trapezoid

Trapezium

Capitate

Scaphoid

Hamulus of hamate

Hamate
Pisiform
Triquetrum
Lunate

Carpals

(a)

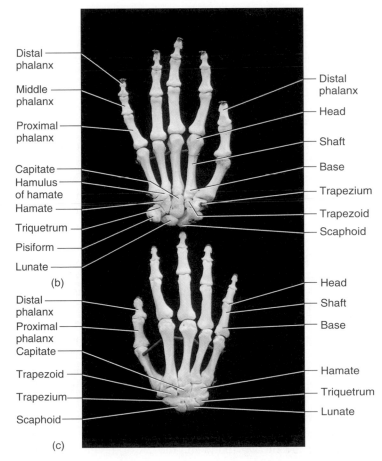

Distal
phalanx

Middle
phalanx

Proximal
phalanx

Capitate

Hamulus
of hamate

Hamate

Triquetrum

Pisiform

Lunate

Distal
phalanx

Head

Shaft

Base

Trapezium

Trapezoid

Scaphoid

(b)

Distal
phalanx

Proximal
phalanx

Capitate

Trapezoid

Trapezium

Scaphoid

Head

Shaft

Base

Hamate

Triquetrum

Lunate

(c)

Figure 9.6   Bones of the hand. Diagram (a) anterior view.
Photograph (b) anterior view; (c) posterior view. ⚘

and **head** and, like the metatarsals, phalanges are classified as long bones. Locate the various bones of the fingers in material in the lab and in figure 9.6.

Pelvic Girdle

The pelvic girdle consists of the two **coxal,** or **innominate, bones,** the **os coxae.** Each os coxa results from the fusion of three bones, the **ilium,** the **ischium** (pronounced ISS-kee-um), and the **pubis.** The os coxae are joined in the front by the **symphysis pubis,** which is a fibrocartilage pad. The **sacrum,** a wedge-shaped section of the vertebral column, attaches to the os coxae in the posterior but it is *not* considered a bone of the pelvic girdle.

Ilium

The most superior coxal bone is the ilium. If you feel the top part of your hip you are feeling the **iliac crest,** which is a long, crescent-shaped ridge of the ilium. The lateral side is the outer surface of the ilium or the **iliac blade.** The two processes that jut from the ilium in the front are the **anterior superior iliac spine** and the **anterior inferior iliac spine.** These are attachment points for some of the thigh flexor muscles. The posterior ilium has the **posterior superior iliac spine** and the **posterior inferior iliac spine.** Below the posterior inferior iliac spine is a large depression known as the **greater sciatic notch.** The outer surface of the ilium serves as an attachment point for the gluteal muscles. The interior, shallow depression of the ilium is known as the **iliac fossa.** Locate these areas in figure 9.7.

Ischium and Pubis

The ischium is inferior to the ilium and is the part of the coxal bone on which you sit. The **ischial spine** is a sharp projection on the ischium, and the **ischial tuberosity** is an attachment site for the hamstring muscles. Anterior to the ischial tuberosity is the **obturator foramen,** a large hole underneath a cup-like depression known as the **acetabulum** (a region where all three coxal bones are joined). The acetabulum is the socket into which the femur fits. The ischium connects to the pubis by an elongated portion of bone known as the **ischial ramus.** This ramus connects to the **inferior pubic ramus,** which is posterior to the **symphysis pubis.**

If the os coxae are articulated, you can easily see the upper basin of the pelvis, which is known as the **false pelvis.** The false pelvis is medial to the iliac fossa. There is a rim of bone that separates the upper basin from a deeper, smaller basin known as the **true pelvis.** This rim is known as the **pelvic brim** in an intact pelvis or **arcuate line** on an isolated pelvic bone. The true pelvis is medial to the obturator foramen.

Examine the disarticulated os coxa in lab. You can determine if you have a left or right bone and the sex of the individual. In a female pelvis the greater sciatic notch has a fairly broad angle, which approximates the arc inscribed by your outstretched thumb and index finger. In a male the greater sciatic notch is less than 90° and approximates the angle between the index finger and the middle finger. As you examine the bones in the lab look at figure 9.7 and locate the various terms listed on the illustration.

Lower Extremity

The **femur** is the longest and heaviest bone in the body and it is the only bone of the thigh. It articulates proximally with the hip and inferiorly with the **tibia.** The femur has a proximal, medial spherical **head,** which inserts into the **acetabulum** of the os coxa, and an elongated, constricted **neck.** Below the neck are two large processes called the **trochanters,** which are areas of muscle attachment. The larger, anterior process is the **greater trochanter** and the smaller, posterior process is the **lesser trochanter.** Locate these structures in figure 9.8. You should also find the **intertrochanteric crest,** which is a posterior ridge between the two trochanters and the anterior **intertrochanteric line.** The femur is curved and bows anteriorly with the posterior shaft marked by the **linea aspera** (meaning rough line). At the proximal portion of the linea aspera is a roughened area called the **gluteal tuberosity,** an attachment site for the gluteus maximus muscle. The distal portion of the femur consists of **medial** and **lateral condyles.** These two smooth processes come into contact with the condyles of the tibia. To the sides of the condyles are bulges known as the **epicondyles.** Locate the **lateral epicondyle** and the **medial epicondyle** on the femur. Also note the location of the **adductor tubercle,** a small triangular process proximal to the medial epicondyle. This is one of the attachment points for one of the adductor muscles.

Patella

The **patella** is a bone formed in the tendon that runs from the quadriceps muscles on the anterior thigh to the tibia. The patella is a specific kind of bone known as a **sesamoid bone.** The patella begins to form around the age of two to three and usually is complete by age six. When the leg is flexed, as when kneeling, the patella protects the ligaments of the knee joint. Examine the patella in the lab and compare it to figure 9.9.

Bones of the Leg

Below the femur are the bones of the leg. The largest of these bones is the **tibia,** which is the major weight-bearing bone of the leg. It is the medial bone of the leg. The condyles of the femur articulate with the **tibial condyles.** The process that separates the condyles of the tibia is the **intercondylar eminence.** The tibia is roughly triangular in cross section. One of the angles of this triangle runs down the anterior tibia and is known as the **anterior tibial crest.** This crest is occasionally

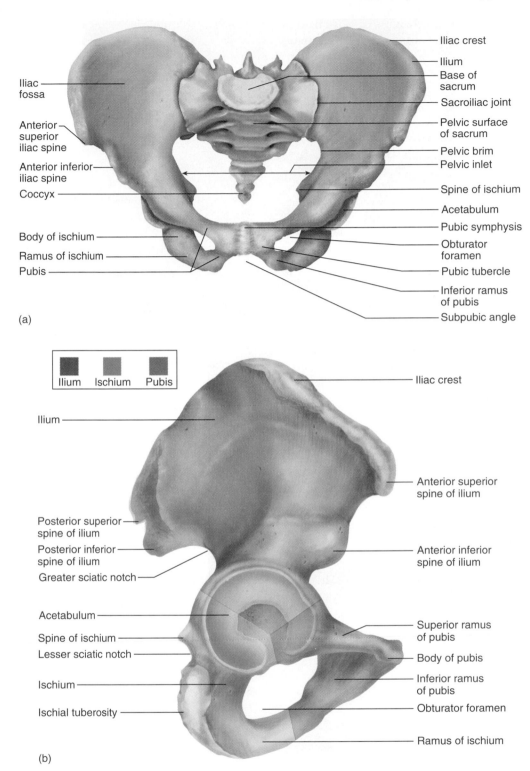

Figure 9.7 Bones of the pelvic girdle. Diagram (a) anterior view; (b) lateral view, right hip. Photograph (c) anterior view; (d) lateral view, right hip. *Continued*

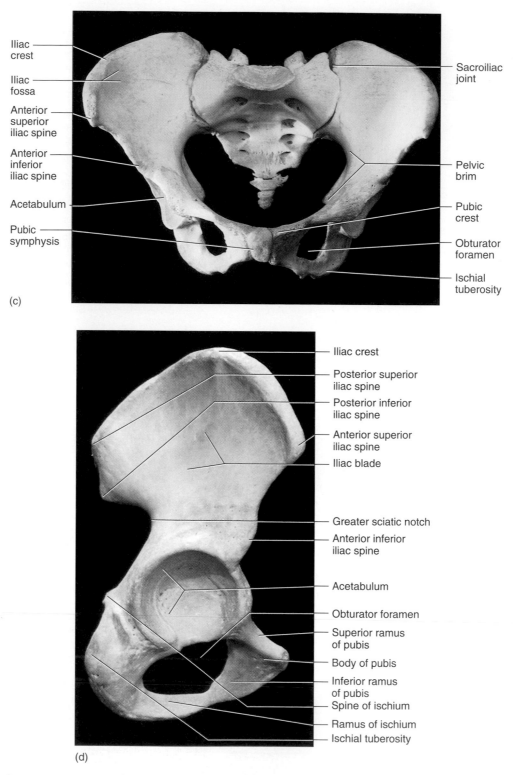

Iliac crest

Iliac fossa

Anterior superior iliac spine

Anterior inferior iliac spine

Acetabulum

Pubic symphysis

Sacroiliac joint

Pelvic brim

Pubic crest

Obturator foramen

Ischial tuberosity

(c)

Iliac crest

Posterior superior iliac spine

Posterior inferior iliac spine

Anterior superior iliac spine

Iliac blade

Greater sciatic notch

Anterior inferior iliac spine

Acetabulum

Obturator foramen

Superior ramus of pubis

Body of pubis

Inferior ramus of pubis

Spine of ischium

Ramus of ischium

Ischial tuberosity

(d)

Figure 9.7—*continued*.

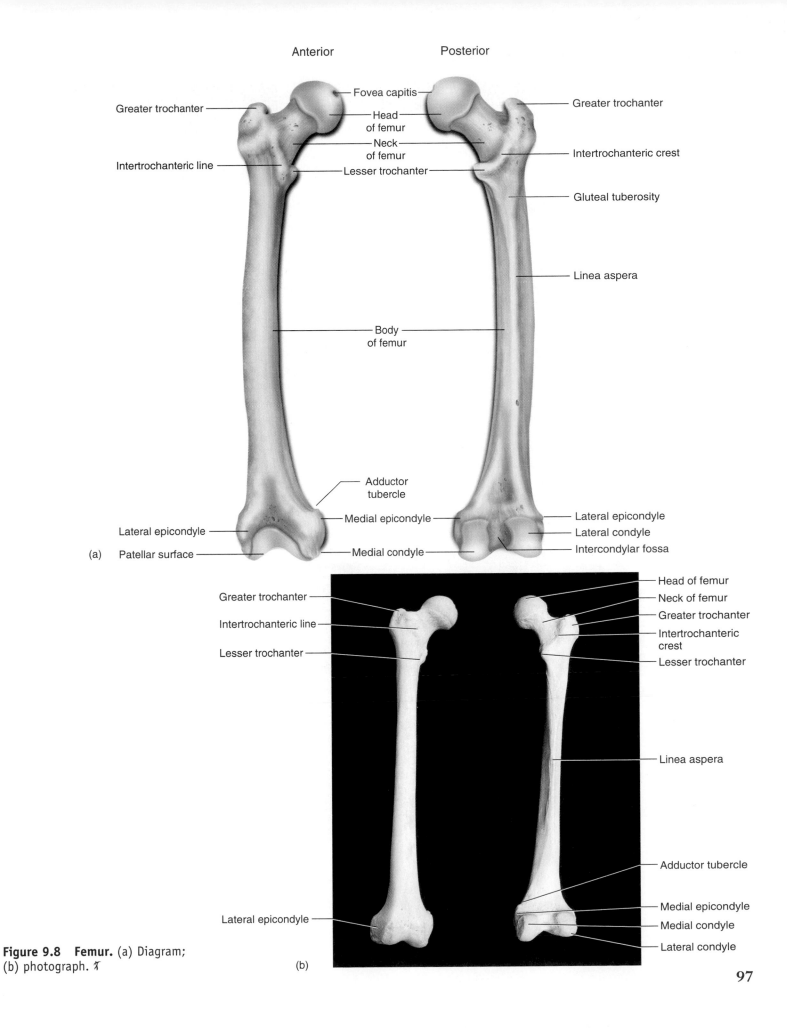

Anterior Posterior

Greater trochanter

Fovea capitis

Head of femur

Neck of femur

Greater trochanter

Intertrochanteric line

Lesser trochanter

Intertrochanteric crest

Gluteal tuberosity

Linea aspera

Body of femur

Adductor tubercle

Lateral epicondyle

Medial epicondyle

Lateral epicondyle

Lateral condyle

(a) Patellar surface

Medial condyle

Intercondylar fossa

Greater trochanter

Head of femur

Neck of femur

Greater trochanter

Intertrochanteric line

Intertrochanteric crest

Lesser trochanter

Lesser trochanter

Linea aspera

Adductor tubercle

Lateral epicondyle

Medial epicondyle

Medial condyle

Lateral condyle

(b)

Figure 9.8 Femur. (a) Diagram; (b) photograph.

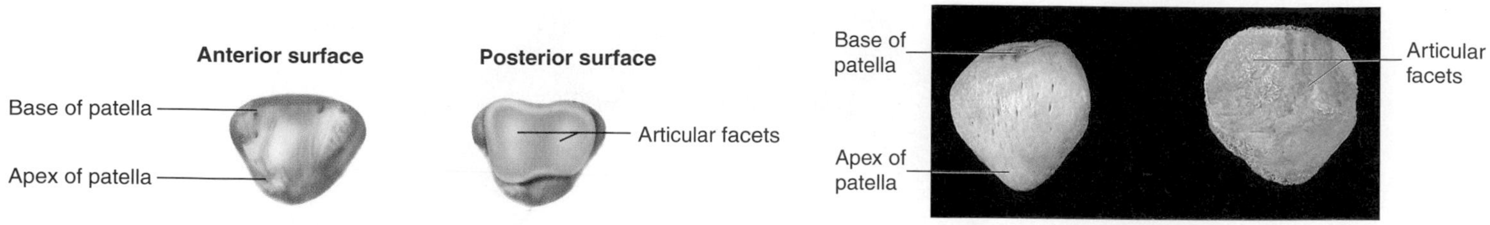

Anterior surface   **Posterior surface**

Base of patella ———

Apex of patella ———

——— Articular facets

Base of patella ———

Apex of patella ———

——— Articular facets

Figure 9.9   Patella, anterior and posterior views. ⚸

injured when you run into a coffee table or other low object in the middle of the night. There is a proximal process on the anterior surface of the tibia known as the **tibial tuberosity.** This is the attachment point for the patellar ligament. At the distal end of the tibia is an extension of bone known as the **medial malleolus.** This bump is one part of the ankle joint as it articulates with the talus of the foot. Locate the structures of the tibia in the lab and in figure 9.10.

The **fibula** is smaller than the tibia (you can remember that it is smaller than the tibia because you "tell a little fib"). The fibula is lateral to the tibia. The fibula has a proximal **head** and a distal process called the **lateral malleolus.** The lateral malleolus is the other part of the leg that forms a joint with the talus. The fibula is not a weight-bearing bone but rather serves as an attachment point for muscles. Examine the bones in the lab and compare the fibula to figure 9.10.

Foot

There are seven **tarsal bones** of the foot, including the **talus,** which serves as the bone of articulation with the leg. Directly below the talus is the **calcaneus,** which is commonly known as the heel bone. The bone anterior to both the talus and the calcaneus is the **navicular.** The foot has three **cuneiform** (pronounced cue-NEE-ih-form) bones, which are known as the first, or **medial, cuneiform;** the second, or **intermediate,**

cuneiform; and the third, or **lateral, cuneiform.** Locate these on specimens in the lab and in figure 9.11. Notice that the lateral cuneiform bone is *not* the most lateral bone in the foot but is the most lateral of the cuneiform bones. The bone that is lateral to the third cuneiform is the **cuboid.**

The **metatarsals** are similar to the metacarpals in that the first metatarsal is under the largest digit, the big toe, and the fifth metatarsal is under the smallest digit. The pattern of the phalanges of the foot is the same as in the hand, with toes 2 to 5 having a **proximal, middle,** and **distal phalanx** each and the big toe, or **hallux,** having a proximal and distal phalanx. Examine the material in the lab and compare the metatarsals and phalanges to figure 9.11.

When you have finished locating the major features of the bones of the appendicular skeleton, test yourself and your lab partner by pointing to various structures and seeing how accurate you are at identifying the structures.

Study Hints

Use your time in lab to *find* the material at hand. Once you are at home, draw or trace material from your manual and be able to name all of the parts.

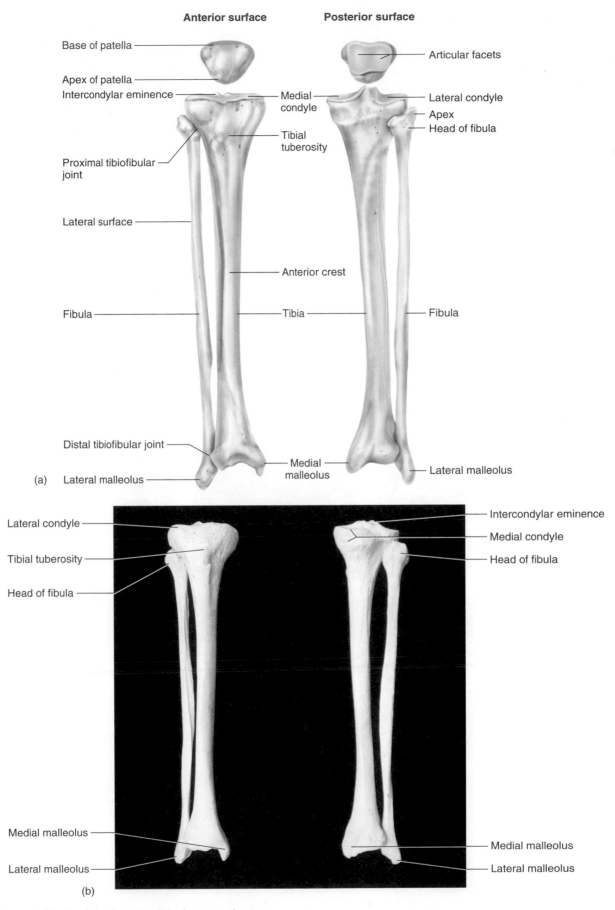

Anterior surface **Posterior surface**

Base of patella

Apex of patella

Intercondylar eminence

Medial condyle

Tibial tuberosity

Proximal tibiofibular joint

Lateral surface

Anterior crest

Fibula

Tibia

Fibula

Distal tibiofibular joint

Medial malleolus

(a) Lateral malleolus

Articular facets

Lateral condyle

Apex

Head of fibula

Lateral malleolus

Lateral condyle

Tibial tuberosity

Head of fibula

Intercondylar eminence

Medial condyle

Head of fibula

Medial malleolus

Lateral malleolus

Medial malleolus

Lateral malleolus

(b)

Figure 9.10 Tibia and fibula. (a) Diagram; (b) photograph.

Superior   Inferior

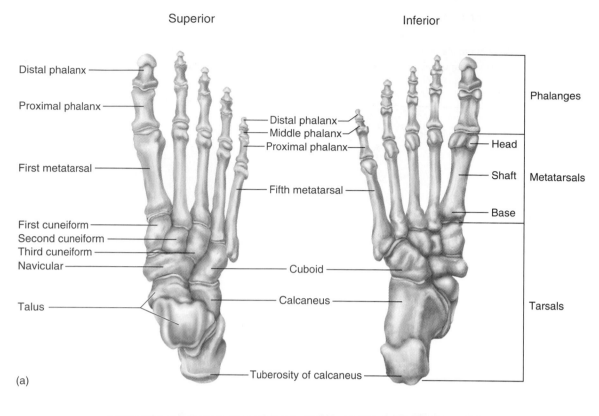

(a)

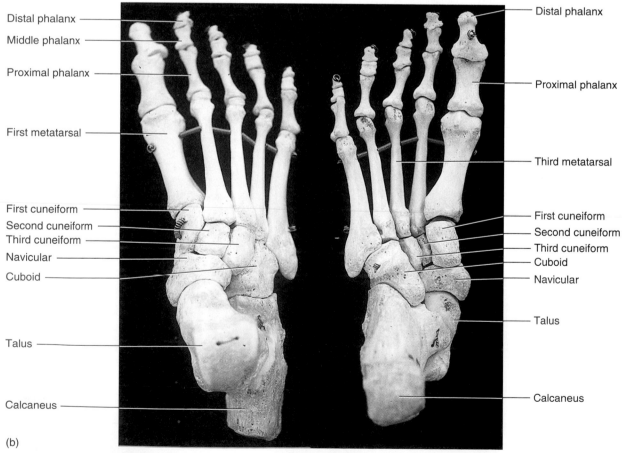

(b)

Figure 9.11   Bones of the foot. (a) Diagram; (b) photograph. ☆

Name _____

1. The acromion is on what bone?

2. In terms of position, how does the head of the ulna differ from the head of the radius?

3. How many bones are in the ankle versus the number of bones in the wrist?

4. Which bone of the hip is the most superior in position?

5. On what bone are the trochanters found?

6. Name the carpal bone at the base of the thumb.

7. In terms of types of bone markings, what is the spine of the scapula?

8. The clavicle joins what two bones?

9. What part of the femur inserts into the acetabulum?

10. Label the parts of the scapula in the following illustration.

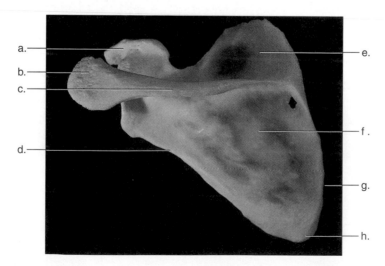

a.

b.

c.

d.

e.

f .

g.

h.

11. Label the parts of the ulna as illustrated and determine if the bone is from the left or right side of the body.

12. A wedding band is typically placed on what phalanx of what digit?

13. Label the following illustration of the foot.

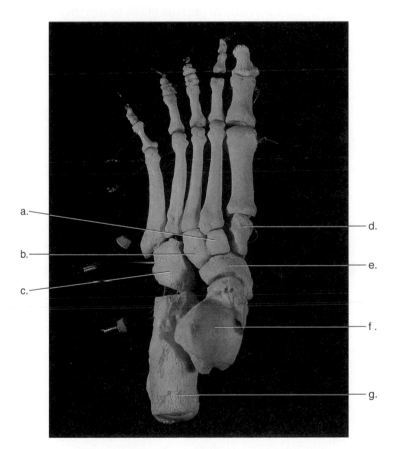

14. Label the following illustration of the hand.

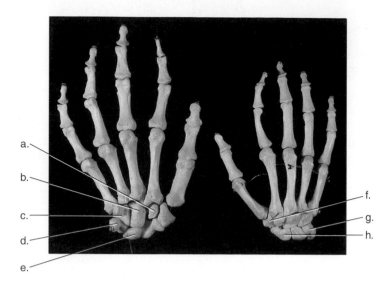

15. On the preceding illustration place an "X" on the proximal phalanx of the fourth digit and a "Y" on the middle phalanx of the second digit. Write a "Z" on the distal phalanx of the first digit.

LABORATORY EXERCISE 10

Axial Skeleton, Vertebrae, Ribs, Sternum, Hyoid

Introduction

The axial skeleton consists of 80 bones, including the skull, hyoid bone, vertebral column, sternum, and ribs. In this exercise, you examine all parts of the axial skeleton except for the skull. Details of the skull are covered separately in Laboratory Exercise 11. The vertebral column consists of the 33 individual vertebrae, some of which are fused into larger structures such as the sacrum. The thoracic cage consists of the 12 pairs of ribs and the sternum, which both serve to protect the lungs and heart yet provide for flexibility during breathing. The hyoid bone is a small floating bone that occurs between the floor of the mouth and the upper anterior neck.

Objectives

At the end of this exercise you should be able to

1. locate and name all of the bones of the vertebral column on an articulated skeleton;
2. name the significant surface features of individual vertebrae;
3. describe the differences between a cervical, thoracic, and lumbar vertebra and be able to name and distinguish between the atlas and the axis;
4. locate the major features of the hyoid bone;
5. distinguish a left rib from a right rib;
6. point out the three major regions of the sternum.

Materials

Articulated skeleton or plastic casts of a skeleton
Disarticulated skeleton or plastic casts of bones
Charts of the skeletal system
Plastic drinking straws cut on a bias or pipe cleaners for pointer tips
Foam pads of various sizes (to protect bone from hard countertops)

Procedure

Locate the bones of the axial skeleton in the lab and on figure 10.1. In this exercise, we study the bones of the axial skeleton by examining both the isolated bones and the articulated skeleton. Hold each individual bone up to the skeleton to see how it is positioned in relation to the other bones of the body. When examining bones in the lab, do not use your pen or pencil to locate a structure. Use a cut drinking straw or a pipe cleaner to point out structures. Your instructor may want you to place real bone material on foam pads to cushion the bone from the tabletop.

Vertebral Column

The **vertebral column** is significantly different in humans than in other mammals in that we are the only habitually bipedal mammal. The vertebral column in a cat, for example, has relatively uniform **vertebrae** from the cervical region to the lumbar region. In humans, on the other hand, the vertebrae increase in size from the cervical to the lumbar vertebrae. This is due to the increase in weight on the lower vertebrae. Between the vertebrae are fibrocartilaginous pads known as **intervertebral discs.** Obtain a real vertebra or plastic cast of an individual vertebra and examine it for the features represented in a typical vertebra (figure 10.2). Place the vertebra in front of you so you can see through the large hole known as the **vertebral foramen.** Note the large **body** of the vertebra, which functions to support the weight of the vertebral column and is in contact with the intervertebral discs. Note also the vertebral foramen, a hole for the passage of the spinal cord and, the **neural,** or **vertebral, arch,** which consists of two **pedicles** and two **laminae.** The pedicles are the parts of the arch that extend from the body of the vertebra to the two lateral projections called the **transverse processes.** Each lamina is a broad, flat structure between the transverse process and the dorsal **spinous process.**

If you rotate the vertebra as illustrated in figure 10.2, you should be able to see the vertebral body in lateral view with the **superior articular process** and the **superior articular facet** visible. The superior articular facet is a smooth face that articulates with the vertebra above. By comparison, there is also an **inferior articular process** and an **inferior articular facet** that articulates with the vertebra below. If you put two adjacent vertebrae together you can see the **intervertebral foramina,** holes that allow spinal nerves to exit from the vertebral column. These features are seen in figure 10.3

After examining a representative vertebra, look at the vertebrae from each region of the spinal column and note the specific features of each region. Compare these with an articulated vertebral column (figure 10.4).

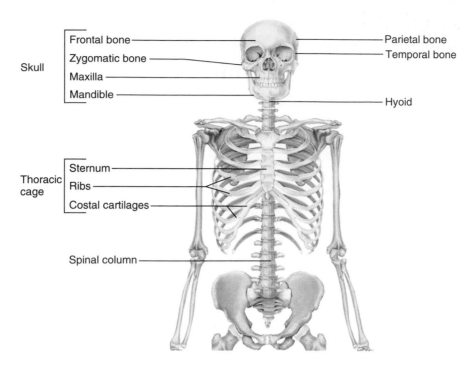

Figure 10.1   Major bones of the axial skeleton. ✗

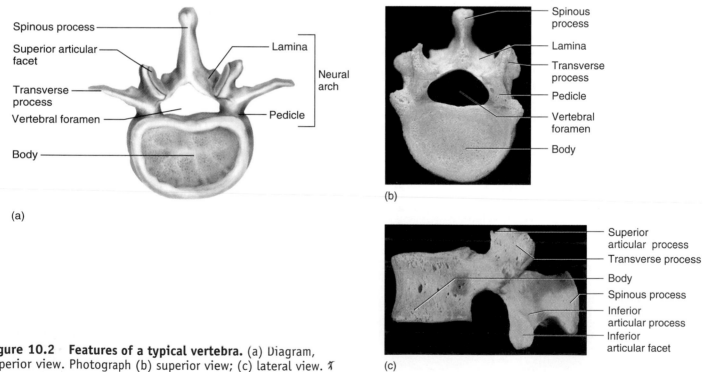

Figure 10.2   Features of a typical vertebra. (a) Diagram, superior view. Photograph (b) superior view; (c) lateral view. ✗

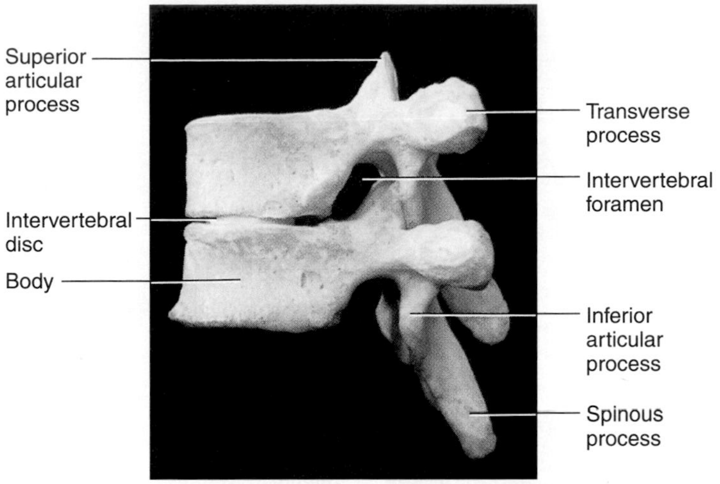

Superior articular process

Transverse process

Intervertebral foramen

Intervertebral disc

Body

Inferior articular process

Spinous process

Figure 10.3   Articulated vertebrae, lateral view. ⅄

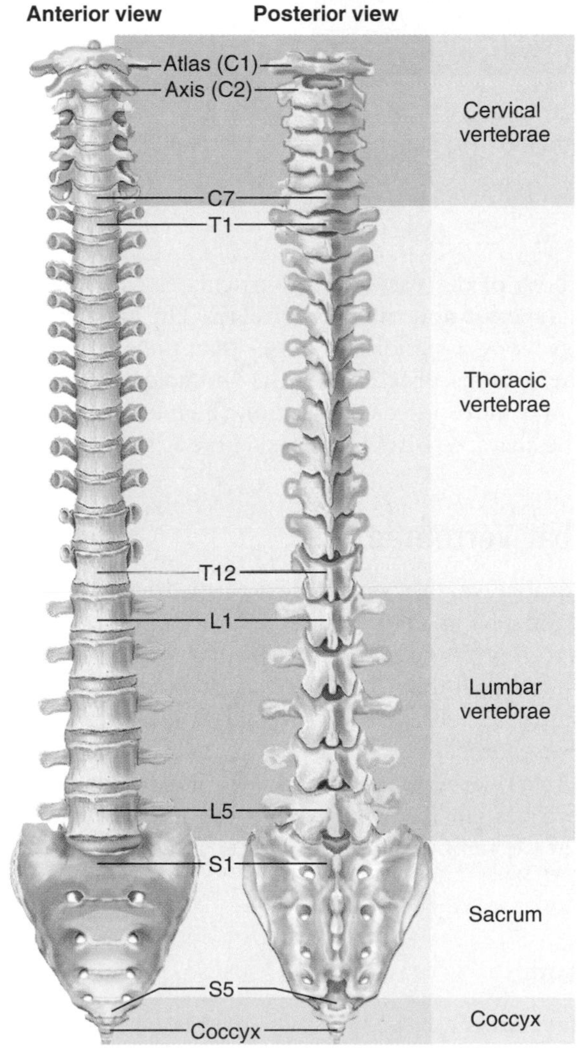

Anterior view   Posterior view

Atlas (C1)
Axis (C2)

Cervical vertebrae

C7
T1

Thoracic vertebrae

T12

L1

Lumbar vertebrae

L5

S1

Sacrum

S5

Coccyx

Coccyx

Figure 10.4   Overview of the vertebral column. ⅄

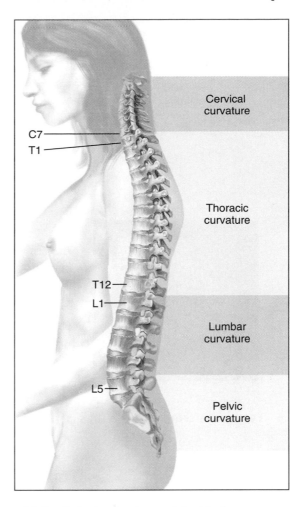

C7
T1

Cervical curvature

Thoracic curvature

T12
L1

Lumbar curvature

L5

Pelvic curvature

Figure 10.5   Spinal curvatures, lateral view.

Spinal Curvatures

The spine has four curvatures, which alternate from superior to inferior. The **cervical curvature** is convex (bowed forward) when seen from anatomical position. The **thoracic curve** is concave, while the **lumbar curve** is convex and the **sacral (pelvic) curve** is concave. These allow for balance in an upright posture. Locate the curvatures in figure 10.5.

Cervical Vertebrae

There are seven **cervical vertebrae,** and these can be distinguished from all other vertebrae in that each vertebra has three foramina. The **vertebral foramen** is the largest opening, and the two **transverse foramina** are specific to the cervical vertebrae. The transverse foramina house the vertebral arteries and vertebral veins. Some of the cervical vertebrae have **bifid spinous processes** (meaning split in two), and the bodies of the cervical vertebrae are less massive than those below. Examine the isolated cervical vertebrae in the lab and compare them to figure 10.6.

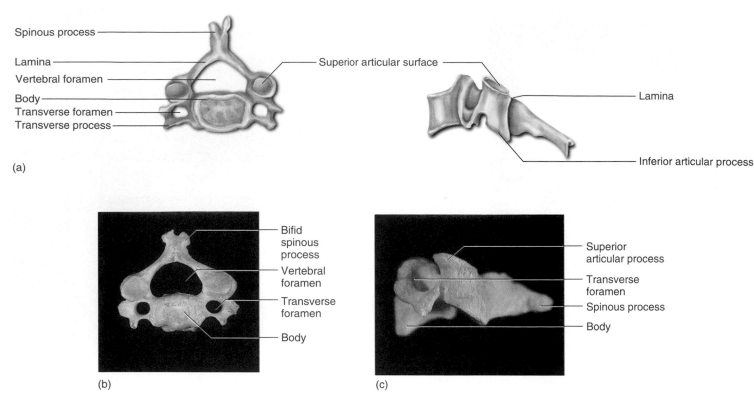

Figure 10.6 Cervical vertebra. Diagram (a) superior and lateral views. Photograph (b) superior view; (c) lateral view. 𝒳

There are some specific vertebrae that you should know. The first cervical vertebra (C1) is known as the **atlas** and is the only cervical vertebra without a body. The atlas joins with the head and provides for range of motion, as when you nod your head to indicate "yes." The second cervical vertebra (C2) is the **axis,** and it has a unique process called the **dens** or **odontoid process,** that runs superiorly through the atlas. The odontoid process allows the atlas to rotate on the axis and provides for the range of motion in your head when you rotate your head to indicate "no." The seventh cervical vertebra (C7) is known as the **vertebra prominens.** It has a **spinous process** that projects sharply in a posterior direction and can be palpated (felt) as a significant bump at the posterior base of the neck. Look at an atlas, axis, and vertebra prominens in the lab and compare them to figure 10.7.

Thoracic Vertebrae

The **thoracic vertebrae** are distinguished from all other vertebrae by their markings on the lateral posterior surface of the body of the vertebrae, which serve as attachment points for the ribs. Sometimes a rib attaches to just one vertebra, in which case the marking on the vertebral body is known as a **rib facet.** In some sections of the thoracic region the head of a rib is attached to two vertebra. Each vertebra has a smooth attachment point known as a **demifacet.** The head of the rib

spans both of the vertebra and articulates with the demifacet of the superior and inferior vertebra. The thoracic vertebrae also have longer spinous processes than the cervical vertebrae, and the spinous processes of the thoracic vertebrae tend to angle in a more inferior direction. Examine figure 10.8 and note the characteristics of the vertebrae as seen in the lab.

Lumbar Vertebrae

The **lumbar vertebrae** are distinguished from the other vertebrae by having neither transverse foramina nor rib facets. The spinous processes of the lumbar vertebrae tend to be more horizontal than the thoracic vertebrae, and the bodies of the lumbar vertebrae are larger, since they carry more weight than the thoracic vertebrae, yet the twelfth thoracic and the first lumbar vertebra are remarkably similar. The last thoracic vertebra has rib facets on it, and the first lumbar does not. Look at the lumbar vertebrae in the lab and compare them to figure 10.9.

Sacrum

The **sacrum** is a large, wedge-shaped bone composed of five fused vertebrae. The lines of fusion are called **transverse lines,** and these may be seen on both the anterior and poste-

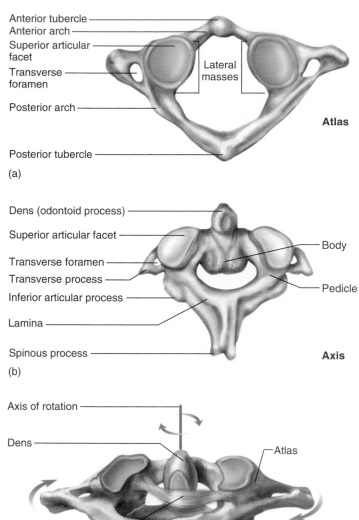

(a)

(b)

(c)

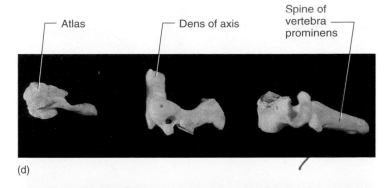

(d)

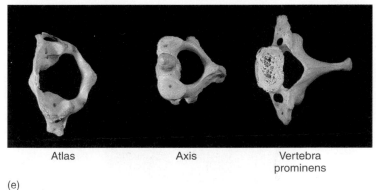

Atlas   Axis   Vertebra
prominens

(e)

Figure 10.7   Atlas, axis and vertebra prominens. (a) Atlas, superior view; (b) axis, supero-posterior view; (c) atlas and axis joined. Photographs of three vertebrae (d) lateral view; (e) superior view.

it wedges itself into the ilium. If you examine the posterior surface of the sacrum you will notice the posterior sacral foramina, the **median** and **lateral sacral crests,** and the superior and inferior openings of the **sacral canal.** As with the other vertebrae, the **superior articular process** and the **superior articular facets** join with the next most superior vertebra (the fifth lumbar vertebra). These features can be seen in figure 10.10.

Coccyx

The **coccyx** is the terminal portion of the vertebral column, and it consists of usually four fused vertebrae. The coccyx may be fused with the sacrum in some individuals if they have fallen backward and landed on a hard surface. Examine the coccyx in the lab and compare it to figure 10.10.

Hyoid

The **hyoid** is a floating bone found at the junction of the floor of the mouth and the neck. The hyoid may be anchored by muscles from the anterior, posterior, or inferior directions and serves to aid tongue movement and swallowing. Locate the central **body**

rior sides. Notice how the sacrum is shaped like a shallow, triangular bowl. If you were to place the sacrum in front of you as you would a cereal bowl, the shallow depression is the **anterior surface.** The two rows of holes you see are the **anterior sacral foramina.** The same holes from the back are known as the **posterior sacral foramina.** Compare a bone in the lab to figure 10.10.

The **sacral promontory** is a rim on the anterior superior part of the sacrum, and the **ala** are two expanded regions of the sacrum lateral to the promontory. The roughened areas lateral to the ala are the **auricular,** or **articular, surfaces** of the sacrum; each joins with the ilium to form the **sacroiliac joint.** As additional force is applied to the sacrum,

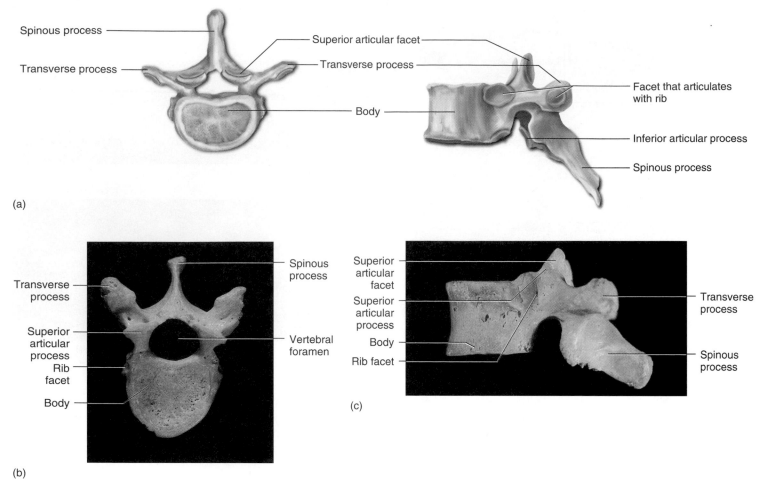

Figure 10.8   Thoracic vertebrae. Diagram (a) superior and lateral views. Photograph (b) superior view; (c) lateral view. ⚡

of the hyoid, the **greater cornua** (singular, *cornu*), and the **lesser cornua** on the material in the lab and in figure 10.11.

Ribs

There are 12 pairs of **ribs** in the human, and these, along with the sternum, make up the thoracic cage. Each rib has a **head** that articulates with the body of one or more vertebrae. On the head is a **facet,** which is the site of articulation. A constricted region near the head is the **neck** of the rib. Locate this on figure 10.12. The process near the neck is the **tubercle** of the rib. This tubercle articulates with the transverse process of the vertebra. Examine the articulated skeleton in the lab and notice that the ribs bend at about the level of the inferior angle of the scapula. This can be seen on individual ribs and is known as the **angle of the rib** or the costal angle. Some ribs have a truncated **sternal end** that attaches to costal cartilage prior to joining the sternum. The superior edge of the rib is more blunt than the inferior edge. The depression that runs along the inferior side of each rib is known as the costal groove. With the costal groove in an inferior position

and the blunt sternal end toward the midline, determine whether you are looking at a left rib or a right rib.

The first seven pairs of ribs are **true ribs.** A true rib is one that attaches to the sternum by its own cartilage. These are known also as **vertebrosternal ribs,** as they attach to the vertebra posteriorly and to the sternum anteriorly. Ribs 8 to 12 are **false ribs,** because they do not attach to the sternum by their own cartilage. Ribs 8 to 10 attach to the sternum by way of the cartilage of rib 7. These ribs are also called **vertebrochondral ribs** for their posterior attachment to the vertebrae and their anterior attachment to the cartilage of rib 7. Ribs 11 and 12 do not attach to the sternum at all and are known as **floating ribs** (a specific type of false rib). These ribs are also known as **vertebral ribs.** Examine the articulated skeleton in the lab and compare it to figure 10.13.

Sternum

The **sternum** is composed of three fused bones. The superior segment is the **manubrium** and the depression at the top of the manubrium is the suprasternal or **jugular notch.** The

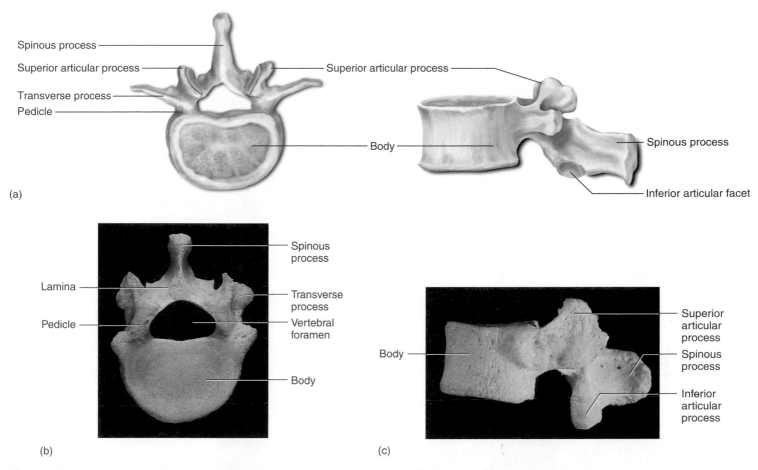

(a)

(b)   (c)

Figure 10.9   Lumbar vertebrae. Diagram (a) superior and lateral views. Photograph (b) superior view; (c) lateral view.

lateral indentations are the **clavicular notches.** The main portion of the sternum is the **body,** and between the body and the manubrium is the **sternal angle.** This is a landmark for finding the second rib when using a stethoscope to listen to heart sounds. Locate the **costal notches** on the body of the sternum. These are where the cartilages of the ribs attach. The narrow, bladelike part that is the most inferior segment of the sternum is the **xiphoid process.** Care must be taken when performing CPR (cardiopulmonary resuscitation) so that pressure is applied to the body of the sternum and not to the xiphoid process. If the force is applied to the xiphoid process it could be fractured and driven into the liver. Examine the structures of the sternum on lab specimens and in figure 10.13.

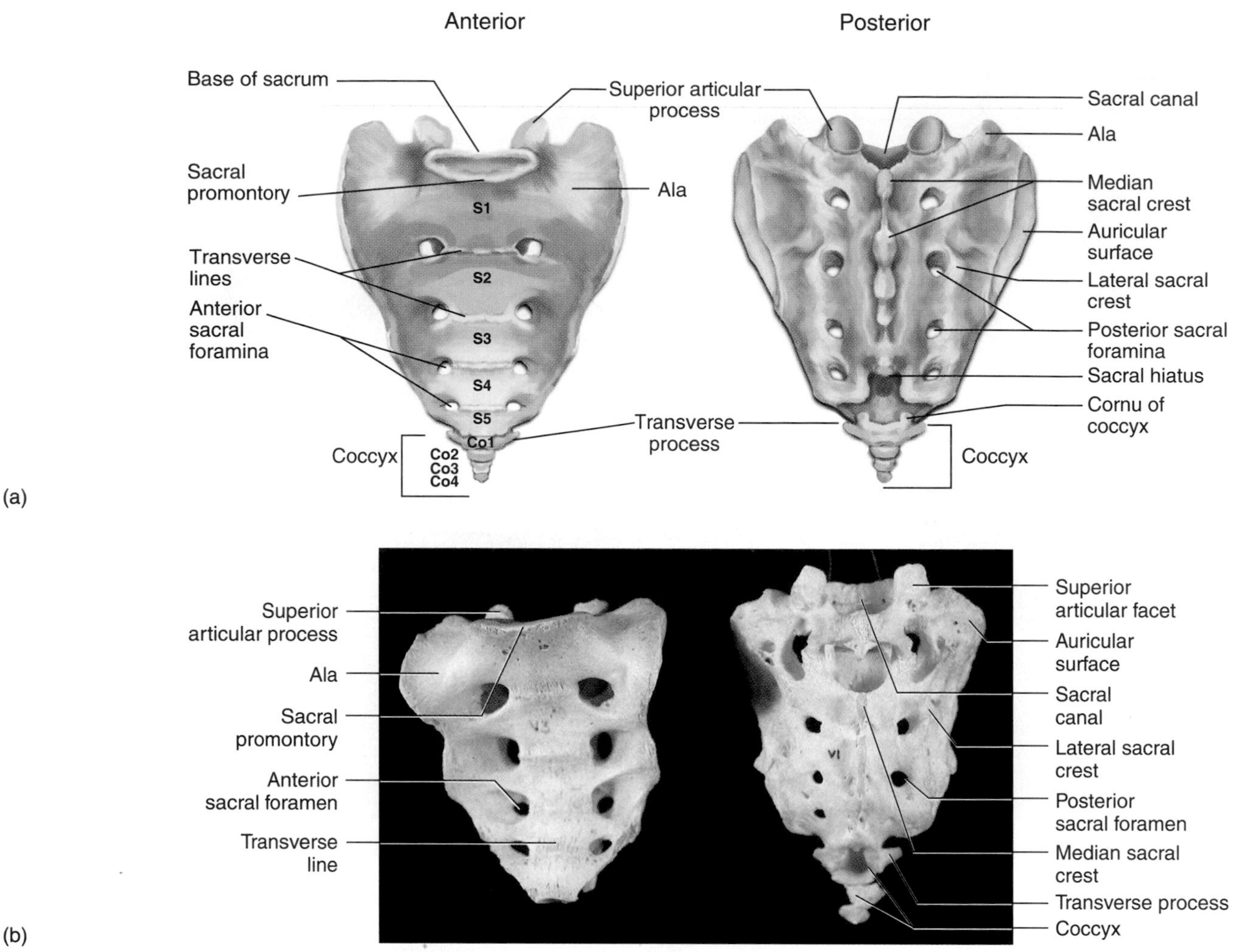

Anterior

Posterior

Base of sacrum

Superior articular process

Sacral canal

Sacral promontory

Ala

Ala

Median sacral crest

S1

Auricular surface

Transverse lines

S2

Lateral sacral crest

Anterior sacral foramina

S3

Posterior sacral foramina

S4

Sacral hiatus

S5

Cornu of coccyx

Co1

Coccyx

Co2
Co3
Co4

Transverse process

Coccyx

(a)

Superior articular process

Superior articular facet

Ala

Auricular surface

Sacral promontory

Sacral canal

Anterior sacral foramen

Lateral sacral crest

Transverse line

Posterior sacral foramen

Median sacral crest

Transverse process

Coccyx

(b)

Figure 10.10   Sacrum and coccyx. (a) Diagram; (b) photograph.

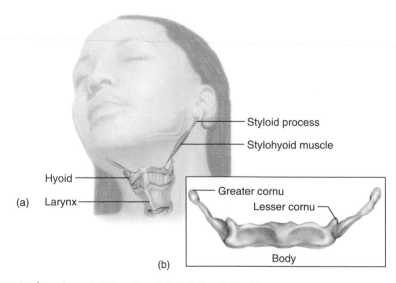

Styloid process

Stylohyoid muscle

Hyoid

Greater cornu

(a)   Larynx

Lesser cornu

Body

(b)

Figure 10.11   Hyoid bone, anterior view. (a) In situ; (b) details of hyoid.

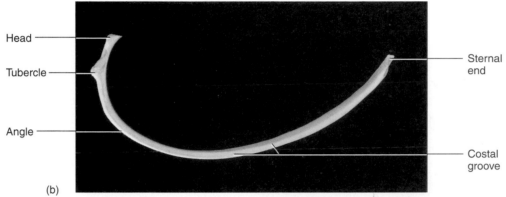

(a)

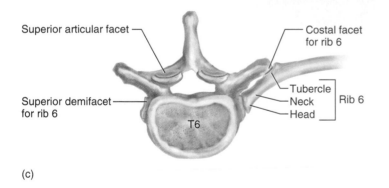

(b)

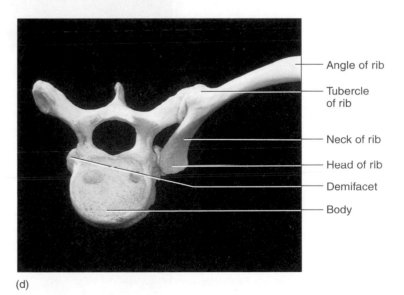

(c)

(d)

Figure 10.12   Individual and articulated rib. (a) Diagram of an individual rib; (b) photograph of an individual rib; (c) diagram of an articulated rib; (d) photograph of an articulated rib.

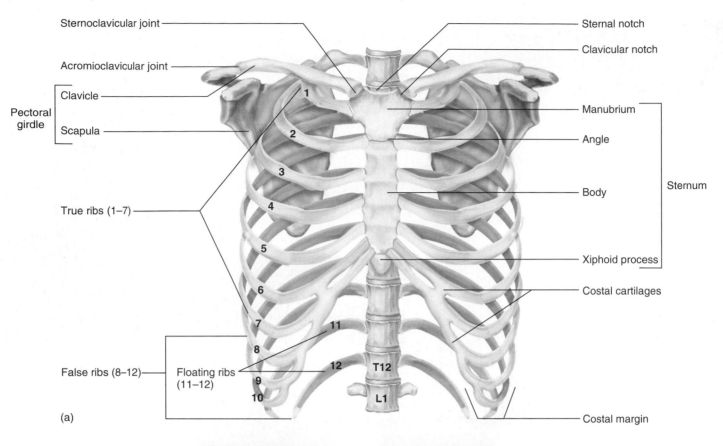

Sternoclavicular joint

Acromioclavicular joint

Pectoral
girdle
Clavicle

Scapula

True ribs (1–7)

False ribs (8–12)

Floating ribs
(11–12)

(a)

Sternal notch

Clavicular notch

Manubrium

Angle

Body

Xiphoid process

Costal cartilages

Costal margin

Sternum

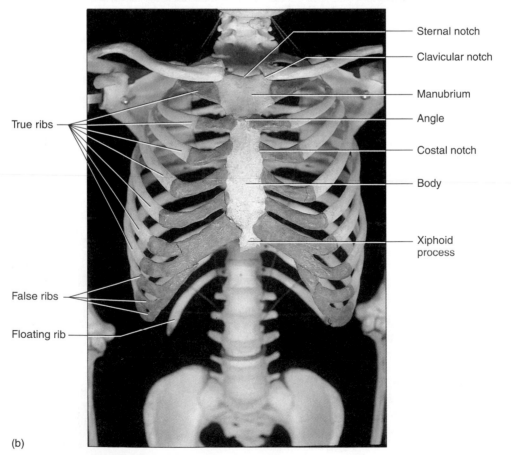

True ribs

False ribs

Floating rib

Sternal notch

Clavicular notch

Manubrium

Angle

Costal notch

Body

Xiphoid
process

(b)

Figure 10.13 The thoracic cage. (a) Diagram; (b) photograph.

Name _____

1. Label the following illustration with the appropriate terms.

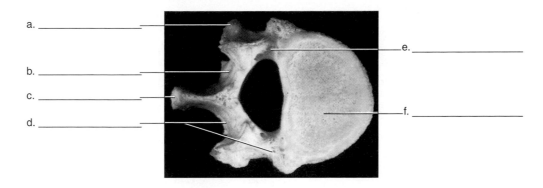

a. _____

b. _____

c. _____

d. _____

e. _____

f. _____

2. What part of a rib articulates with the transverse process of a vertebra?

3. The most inferior portion of the sternum is the

    a. body   b. manubrium   c. angle   d. xiphoid

4. Distinguish between the posterior sacral foramina and the sacral canal.

5. What two features do cervical vertebrae have that no other vertebrae have?

6. Determine from what part of the spinal column the vertebra in the following photograph comes.

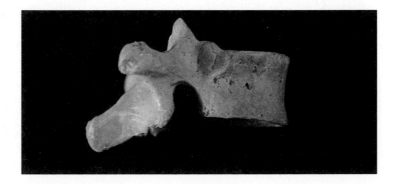

7. The superior portion of the sternum has what particular name?

8. What structures make up the vertebral arch?

9. The superior articular process articulates with what specific structure?

10. How many lumbar vertebrae are found in the human body?

11. Name the second cervical vertebra.

12. Which spinal curvature is the most superior one?

13. Does the hyoid bone have any solid bony attachments?

14. A rib that attaches to the sternum by the cartilage of another rib has what name?

15. Is the angle of the rib on the anterior or posterior side of the body?

Axial Skeleton—Skull

Introduction

The skull consists of two major regions, the **cranium** and the **face.** The skull is the most complex region of the skeletal system and not only houses the brain but contains a significant number of the sense organs as well. The cranium consists mostly of flat bones, while the face is composed of irregular bones. In this exercise, you learn the details of the skull and the anatomy of the fetal skull.

Objectives

At the end of this exercise you should be able to

1. list all of the bones of the cranium and the major features of those bones;
2. list all of the bones of the face and the major features of those bones;
3. name all of the bones that occur singly or in pairs in the skull;
4. locate the major foramina of the skull;
5. name the major sutures of the skull;
6. list all of the fontanels of the fetal skull.

Materials

Disarticulated skull, if available
Articulated skulls with the calvaria cut
Foam pads to cushion skulls from the desktop
Fetal skulls
Pipe cleaners

Procedure

Familiarize yourself with the bones of the skull in the cranium and the face. The skull bones in the cranium are listed with a number after the name of the bone indicating whether the bone occurs *singly* or as a *pair*. Take a skull back to your lab table and locate the bones listed in figure 11.1.

Frontal (1)	Parietal (2)
Occipital (1)	Temporal (2)
Sphenoid (1)	Ethmoid (1)

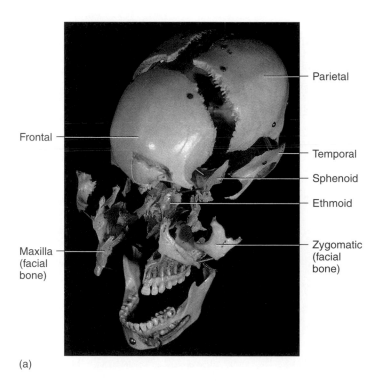

(a)

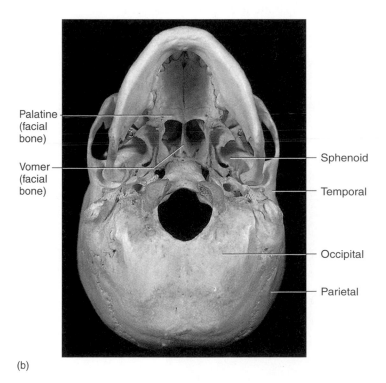

(b)

Figure 11.1   Bones of the cranium. (a) Disarticulated skull, 3/4 view; (b) inferior view. ✶

The bones of the face are listed next. Locate these bones on the skull in the lab as shown in figures 11.1 and 11.2.

Maxilla (2)   Nasal (2)
Mandible (1)   Palatine (2)
Vomer (1)   Zygomatic (2)
Lacrimal (2)   Inferior nasal concha (plural, *conchae*) (2)

As you locate the cranial and facial bones on a skull in the lab, make sure you do not poke pencils, pens, or fingers into the delicate regions of the eyes or nasal cavity. You may break the delicate structures in these regions. Once you have found the major bones of the skull use the following descriptions and illustrations to find the structures of the skull. The skull is described from a series of views, and you should locate the structures listed for those views.

Anterior View

The large bone that makes up the forehead is the **frontal bone.** It is a single bone that makes up the superior portion of the **orbits** (the eye sockets) and has two ridges above the eyes called **supraorbital ridges** (the eyebrows). There is a hole in each of these ridges called the **supraorbital foramina,** which allow for nerves and arteries to reach the face. Most of what you can see inferior to the frontal bone are facial bones. The bony part of the nose is composed of the

paired **nasal bones,** which join with the cartilage that forms the tip of the nose. Posterior to the nasal bones are thin strips of the upper **maxillae,** bones that hold the upper teeth; posterior to the maxillae are the thin **lacrimal bones,** which contain

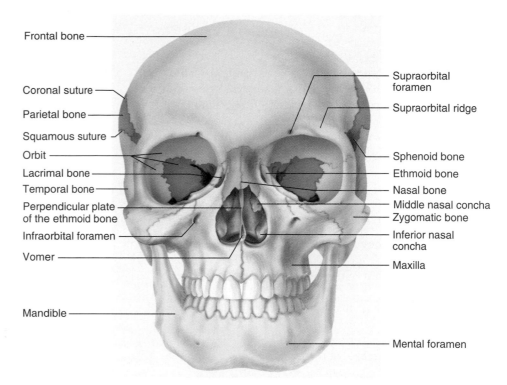

(a)

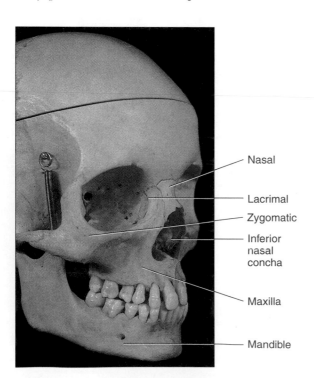

Figure 11.2   Bones of the face. ⚡

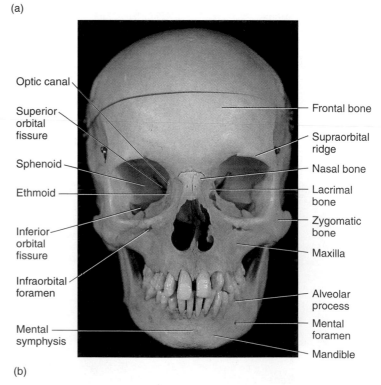

(b)

Figure 11.3   Skull, anterior view. (a) Diagram; (b) photograph. ⚡

the **nasolacrimal duct,** into which tears drain from the eye into the nose. Locate these bones in figure 11.3. Posterior to the lacrimal bones is the **ethmoid bone,** which is very delicate and frequently broken on skulls that have been in the lab for years. Posterior to the ethmoid is the **sphenoid bone,** which forms the posterior wall of the orbit and contains not only the **optic canal** (a passageway for the optic nerve) but also the **superior orbital fissure** and the **inferior orbital fissure.** Note the bones on the lateral side of the orbit. These are the **zygomatic bones.**

The major bone below the orbit is the maxilla, which also makes up the floor of the orbit. The two maxillae have sockets called **alveoli** (singular, *alveolus*) which contain the teeth and extensions of bone between each pair of sockets called **alveolar processes.** The **infraorbital foramen** is a small hole below the eye in the maxilla and serves as a passageway for nerves and blood vessels. The most inferior bone of

the face is the **mandible,** which is commonly known as the lower jaw. The mandible also has alveoli and alveolar processes (figure 11.4). The mandible begins as two bones in utero and fuses at the midline of the chin. This fusion of the **mental symphysis** joins the two mandible bones into one. On the lateral aspects of the mandible are the **mental foramina,** which serve to conduct nerves and blood vessels to the tissue anterior to the jaw.

Superior View

Locate the major suture lines of the skull from this view (figure 11.5). The frontal bone is separated from the pair of parietal bones by the **coronal suture.** The parietal bones are separated from one another by the **sagittal suture.** The parietal bones are separated from the occipital bone by the **lambdoid**

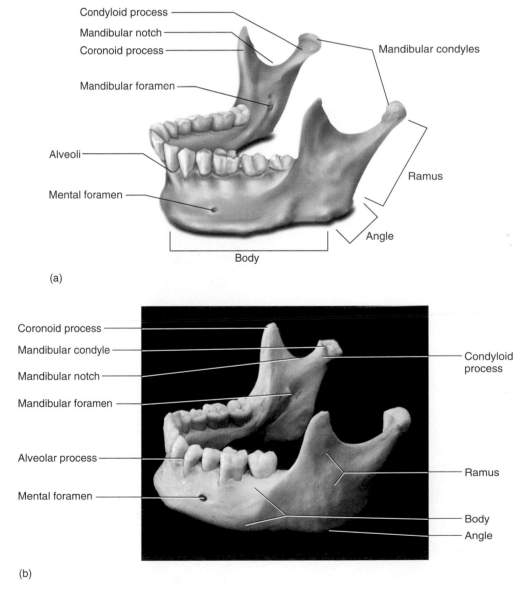

(a)

(b)

Figure 11.4   Mandible, lateral view.

suture. The lambdoid suture is named after the Greek letter lambda, which is Y-shaped. There may be small bones that occur between the occipital bone and the parietal bones (or between other skull bones), and these are known as **sutural,** or **Wormian, bones.**

Lateral View

The coronal and lambdoid sutures also may be seen from the lateral view as well as the **squamous suture,** which separates the **temporal bone** from the **parietal bone.** Locate the cranial bones from a lateral view. These are the **frontal, parietal, occipital, temporal, sphenoid,** and **ethmoid** bones. You may be able to see a bump at the back of the occipital bone from this angle. This is known as the **external occipital protuberance.** The hole on the side of the head where the ears attach is the **external acoustic/auditory meatus** (or external acoustic/auditory canal). The large process posterior and inferior to this opening is the **mastoid process.** A long, thin spine medial to the mastoid process is the **styloid process,** which has muscles that connect it to the hyoid bone. The sphenoid bone can also be seen from this view just anterior to the temporal bone. A bony process is found lateral to the sphenoid bone. This is the **zygomatic arch,** which makes up the upper part of the cheek. The zygomatic arch is composed of part of the temporal bone and part of the **zygomatic bone,** which is commonly known as the cheek bone. It can be seen as forming the lateral wall of the orbit. On the inner wall of the orbit is the ethmoid bone, the lacrimal bone, the maxilla, and the nasal bone. Examine these structures in figure 11.6.

The mandible has a **condyloid process** with a terminal **mandibular condyle,** which articulates with the temporal bone; a **coronoid process,** which lies medial to the zygomatic arch; and the **mandibular notch,** which is a depression between the condyloid process and coronoid process of the mandible. The vertical section of the mandible is known as the **ramus of the mandible** (*ramus* - branch), and the horizon-

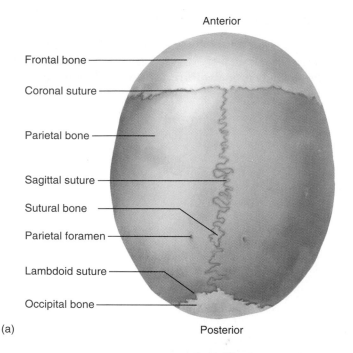

Anterior

Frontal bone

Coronal suture

Parietal bone

Sagittal suture

Sutural bone

Parietal foramen

Lambdoid suture

Occipital bone

(a)

Posterior

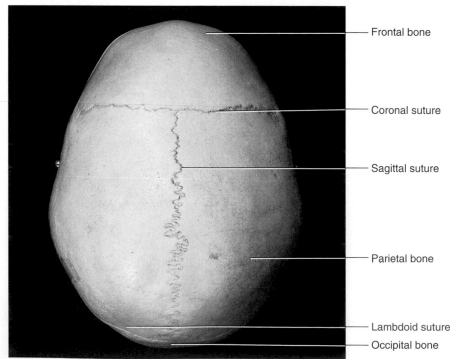

Frontal bone

Coronal suture

Sagittal suture

Parietal bone

Lambdoid suture

Occipital bone

(b)

Figure 11.5 Skull, superior view.

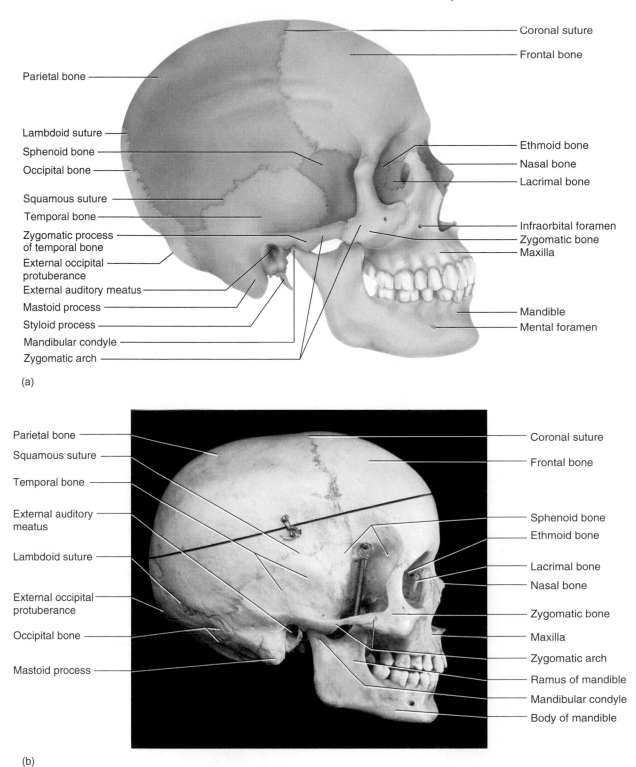

Parietal bone

Lambdoid suture
Sphenoid bone
Occipital bone

Squamous suture
Temporal bone
Zygomatic process
of temporal bone
External occipital
protuberance
External auditory meatus
Mastoid process
Styloid process
Mandibular condyle
Zygomatic arch

Coronal suture
Frontal bone

Ethmoid bone
Nasal bone
Lacrimal bone

Infraorbital foramen
Zygomatic bone
Maxilla

Mandible
Mental foramen

(a)

Parietal bone
Squamous suture

Temporal bone

External auditory
meatus

Lambdoid suture

External occipital
protuberance

Occipital bone

Mastoid process

Coronal suture

Frontal bone

Sphenoid bone
Ethmoid bone

Lacrimal bone
Nasal bone

Zygomatic bone

Maxilla

Zygomatic arch
Ramus of mandible
Mandibular condyle
Body of mandible

(b)

Figure 11.6   Skull, lateral view. 𝕏

tal portion of the mandible is the **body of the mandible.** The **angle** of the mandible is at the posterior of the mandible at the junction of the body and the ramus. On the inside of each ramus of the mandible is the **mandibular foramen,** which is a conduit for a nerve. The parts of the mandible can also be identified in figures 11.4 and 11.6.

Inferior View

Place the skull in front of you with the mandible removed (figure 11.7). The largest hole in the skull, the **foramen magnum,** should be close to you. The foramen magnum is in the occipital bone and is the dividing line between the brain and the spinal cord. Adjacent to the foramen magnum are the

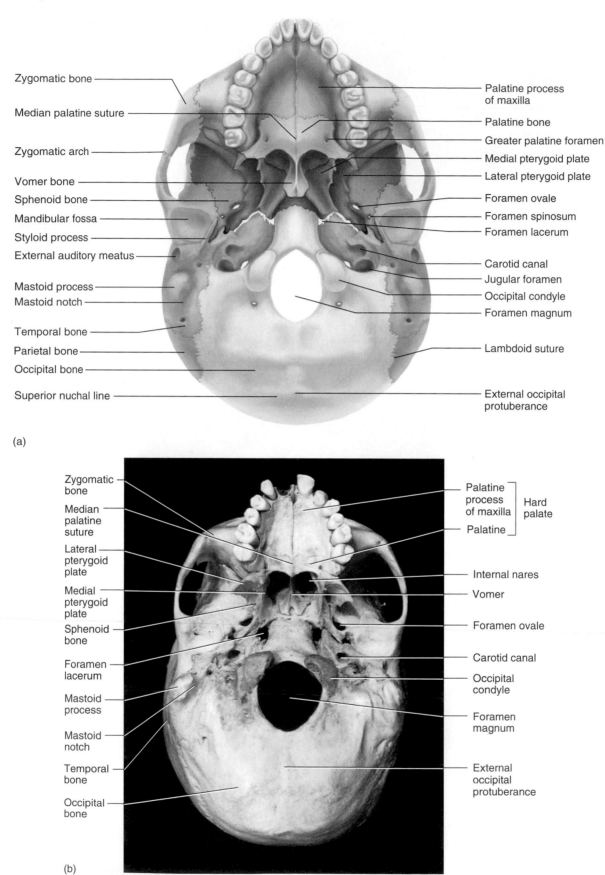

Zygomatic bone

Median palatine suture

Zygomatic arch

Vomer bone

Sphenoid bone

Mandibular fossa

Styloid process

External auditory meatus

Mastoid process

Mastoid notch

Temporal bone

Parietal bone

Occipital bone

Superior nuchal line

Palatine process
of maxilla

Palatine bone

Greater palatine foramen

Medial pterygoid plate

Lateral pterygoid plate

Foramen ovale

Foramen spinosum

Foramen lacerum

Carotid canal

Jugular foramen

Occipital condyle

Foramen magnum

Lambdoid suture

External occipital
protuberance

(a)

Zygomatic
bone

Median
palatine
suture

Lateral
pterygoid
plate

Medial
pterygoid
plate

Sphenoid
bone

Foramen
lacerum

Mastoid
process

Mastoid
notch

Temporal
bone

Occipital
bone

Palatine
process
of maxilla   Hard
                      palate
Palatine

Internal nares

Vomer

Foramen ovale

Carotid canal

Occipital
condyle

Foramen
magnum

External
occipital
protuberance

(b)

Figure 11.7   Skull, inferior view.

occipital condyles, which are processes that articulate with the first cervical vertebra. A small bump at the posterior part of the occipital bone is the external occipital protuberance. At the junction of the occipital bone and the temporal bone is the **jugular foramen,** a hole through which passes the internal jugular vein. If you carefully insert a pipe cleaner into the jugular foramen and turn the skull over, you will see that the foramen leads to the posterior part of the skull.

You can see the mastoid process of the temporal bone and a depression just medial to the process known as the **mastoid notch.** Find the styloid processes in your specimen, though they may be hard to locate since they frequently get broken in lab specimens. The styloid process is an attachment point for muscles that move the tongue, larynx, or hyoid. Medial to the styloid process is the **carotid canal,** through which passes the internal carotid artery to the brain. If you carefully insert a pipe cleaner into the carotid canal of a real skull you will notice that the canal bends at about a 90° angle; if you turn the skull over you will note that the opening occurs in the middle of the skull. You cannot do this with most plastic casts of skulls. At the junction of the temporal bone and the sphenoid bone is the **foramen lacerum,** which joins with the carotid canal as it enters the skull. The temporal bone also has a **mandibular fossa,** which is the articulation site of the mandible. The zygomatic process of the temporal bone can also be seen from this view.

From the inferior view the sphenoid bone can be seen as a bone that runs from one side of the skull to the other. The part of the sphenoid seen from the lateral view is one of the **greater wings** of the sphenoid. Two pairs of flattened process can also be seen, the **lateral pterygoid plate** and the **medial pterygoid plate** (pterygoid means winglike). These serve as attachments for muscles that extend from the sphenoid to the mandible. Just posterior to the pterygoid processes is the **foramen ovale,** which conducts one of the branches of the trigeminal nerve to the mandible.

In the midline of the skull and sometimes looking like a part of the sphenoid is the **vomer.** This is a single bone of the face that forms part of the **nasal septum.** The two holes on each side of the vomer are the **internal nares.** Connected to the vomer and forming part of the **hard palate** is the **palatine bone.** The palatine bones are L-shaped bones with a horizontal plate and a vertical plate. The horizontal plates normally join together at the **median palatine suture.** If this suture does not fuse completely at birth an individual has a cleft palate. Another part of the hard palate is the **palatine process of the maxilla.**

The major openings of the skull are presented in table 11.1. Locate the openings and note the number of structures that pass through these holes.

The Interior of the Cranium

With the top of the skull removed you can see that the cranial cavity is divided into three major regions. These are the **ante-**

Table 11.1 Opening of the Skull

Opening	Function or Structure in Opening
Foramen magnum	Spinal cord, vertebral arteries
Jugular foramen	Internal jugular vein, vagus, and other nerves
Carotid canal	Internal carotid artery
Stylomastoid canal	Facial nerve exits skull
Foramen lacerum	Internal carotid artery
Foramen ovale	Mandibular branch of trigeminal nerve
Foramen spinosum	Meningeal blood vessels
Foramen rotundum	Maxillary branch of trigeminal nerve
Optic canal	Optic nerve
Superior orbital fissure	Nerves to the eye and face
Inferior orbital fissure	Maxillary branch of trigeminal nerve
Mandibular foramen	Mandibular branch of trigeminal nerve
Mental foramen	Mental nerve and blood vessels
Supraorbital foramen	Supraorbital nerve and artery for the face
Infraorbital foramen	Infraorbital nerve and artery for the face
External auditory meatus	Opening for sound transmission
Internal auditory meatus	Vestibulocochlear nerve and facial nerve

rior cranial fossa, a depression anterior to the lesser wings of the sphenoid; the **middle cranial fossae,** which lie between the lesser wings of the sphenoid and the petrous portion of the temporal bone; and the **posterior cranial fossa,** which is posterior to the petrous portion of the temporal bone. Examine figure 11.8 for a view of the interior of the cranium.

Beginning with the anterior cranial fossa you should find the centrally located **ethmoid bone.** A sharp ridge known as the **crista galli** projects from the main portion of this bone. The small horizontal plate of bone with numerous holes lateral to the crista galli is the **cribriform plate.** The holes in this plate are called the **olfactory foramina** and they transmit the sense of smell from the nose to the brain. If the skull was cut close to the orbit, you can see the **frontal sinus** as a hollow space in the anterior portion of the frontal bone.

The dividing line between the anterior and middle cranial fossae is the sphenoid bone. Locate the lesser wings of the sphenoid and the **sella turcica** just posterior to it. The sella turcica is a small depression in which the pituitary gland sits. The posterior, raised part of the sella turcica is known as the **dorsum sella.** The greater wings of the sphenoid are more inferior than the lesser wings, and each one contains the **foramen**

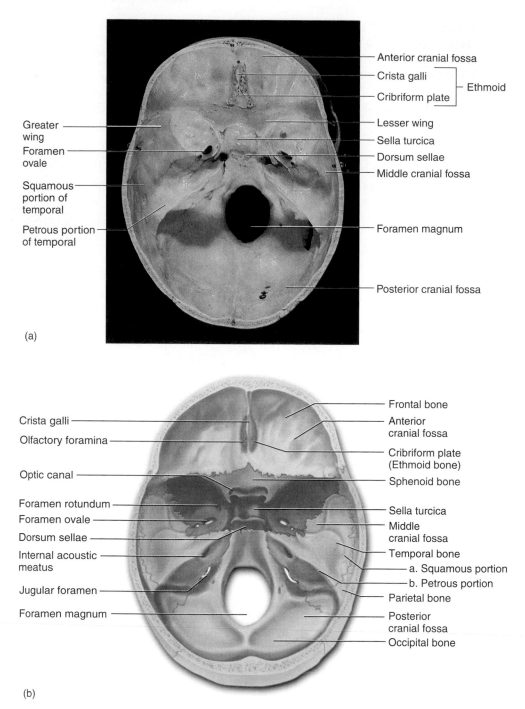

(a)

(b)

Figure 11.8 Interior of the cranium. 𝒯

rotundum, which takes a branch of the trigeminal nerve to the maxilla. The **foramen ovale** can be seen from this view as well.

Behind the sphenoid bone is the temporal bone, which has a flattened lateral section known as the **squamous portion** and a heavier mass of bone known as the **petrous portion.** The petrous portion divides the middle and posterior cranial fossae. The petrous portion also has a hole in the posterior surface, which is the **internal acoustic meatus.** This is a passageway for the nerves that come from the inner ear.

The posterior cranial fossa is located dorsal to the petrous portion of the temporal bone. It contains the foramen mag-num and the jugular foramina. Most of this fossa is formed by the occipital bone.

Midsagittal Section of the Skull

If you have a dissected skull in the lab, examine the structures that can be seen in a midsagittal section. Be extremely careful with the specimen since many of the internal structures are fragile. Use figure 11.9 as a guide to the midsagittal section of the skull whether or not one is present in the lab.

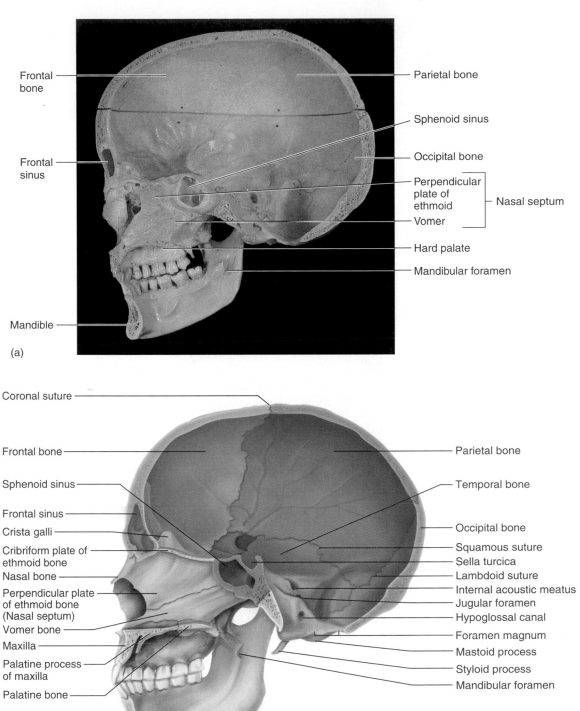

Figure 11.9   **Skull, midsagittal section.**

(a) Labels:
Frontal bone
Frontal sinus
Mandible
Parietal bone
Sphenoid sinus
Occipital bone
Perpendicular plate of ethmoid
Vomer
Nasal septum
Hard palate
Mandibular foramen

(b) Labels:
Coronal suture
Frontal bone
Sphenoid sinus
Frontal sinus
Crista galli
Cribriform plate of ethmoid bone
Nasal bone
Perpendicular plate of ethmoid bone (Nasal septum)
Vomer bone
Maxilla
Palatine process of maxilla
Palatine bone
Mandible
Parietal bone
Temporal bone
Occipital bone
Squamous suture
Sella turcica
Lambdoid suture
Internal acoustic meatus
Jugular foramen
Hypoglossal canal
Foramen magnum
Mastoid process
Styloid process
Mandibular foramen

Locate the **nasal septum,** which is composed of the **vomer,** the **perpendicular plate** of the ethmoid bone, and the **nasal cartilage** (absent in skull preparations). If the nasal septum is removed, you should be able to see the **superior nasal concha** and the **middle nasal concha** of the ethmoid bone. Below these is the **inferior nasal concha,** which is a separate, distinct bone. Look for the junction between the palatine bone and the palatine process of the maxilla. These two bony plates make up the hard palate. If the mandible is present, locate the **mandibular foramen,** which is on the inner aspect of the mandible and transmits branches of the trigeminal nerve and blood vessels to the mandible.

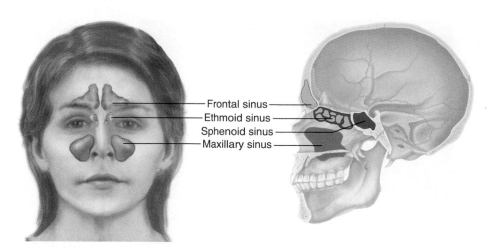

Figure 11.10   Sinuses of the skull.

Sinuses

There are numerous sinuses and air cells in the skull. These sinuses provide shape to the skull while decreasing its weight and provide resonance to the voice. The **paranasal sinuses** occur around the region of the nose and are named for the bones in which they are found. They include the **frontal sinus,** the **maxillary sinus,** the **ethmoid sinus** (or air cells), and the **sphenoid sinus.** These sinuses may fill with fluid during a cold and harbor bacteria in secondary infections. Locate the sinuses in skulls in the lab and compare them to figure 11.10.

Fontanels

The **fontanels** are the "soft spots" of an infant's skull. There are four sets of fontanels, and some of these allow for the passage of the skull through the birth canal by enabling the bones of the cranium to slide over one another. After birth the fontanels allow for further expansion of the skull. The **frontal (anterior) fontanel** is an area found between the frontal bone and the parietal bones. The **occipital (posterior) fontanel** is found between the occipital bone and the parietal bones. The **sphenoid (anterolateral) fontanels** are paired structures on each side of the skull and are located superior to the sphenoid bone, and the **mastoid (posterolateral) fontanels** are also paired structures posterior to the temporal bone. Locate these structures in figure 11.11 and on the material available in the lab.

Select Individual Bones of the Skull

Ethmoid

The **ethmoid bone** is a cranial bone located in the middle of the skull. The **perpendicular plate** of the ethmoid can

be seen in midsagittal view or from the anterior view through the external nares. The **orbital plate** is the part of the ethmoid that lines the medial wall of the orbit. The **middle nasal conchae** can be seen from the nasal cavity as well, but the **superior nasal conchae** are best seen by looking at an inferior view of the skull through the internal nares or at a midsagittal view with the nasal septum removed. Examine isolated ethmoid bones in the lab and find the **crista galli, cribriform plate,** and other structures as shown in figure 11.12.

Sphenoid

The **sphenoid bone** is seen in relation to the entire skull and in isolated views in figure 11.13. Examine an isolated sphenoid bone in the lab and locate the **greater wings,** the **lesser wings,** the **medial** and **lateral pterygoid plates,** the **sella turcica,** the **dorsum sellae,** and other features as seen in figure 11.13.

Temporal

The temporal bone is a paired cranial bone that has a squamous portion which is the lateral part of the bone and forms part of the cranial vault. There is also a medial part of the temporal called the petrous portion. The petrous portion contains the ear ossicles and the opening of the internal auditory meatus as seen in the medial view of the temporal bone in figure 11.14. Examine an isolated temporal bone in lab and locate these features as well as the zygomatic process, which articulates with the zygomatic bone and the mastoid process, which can be palpated (felt) as a bump posterior to the ear.

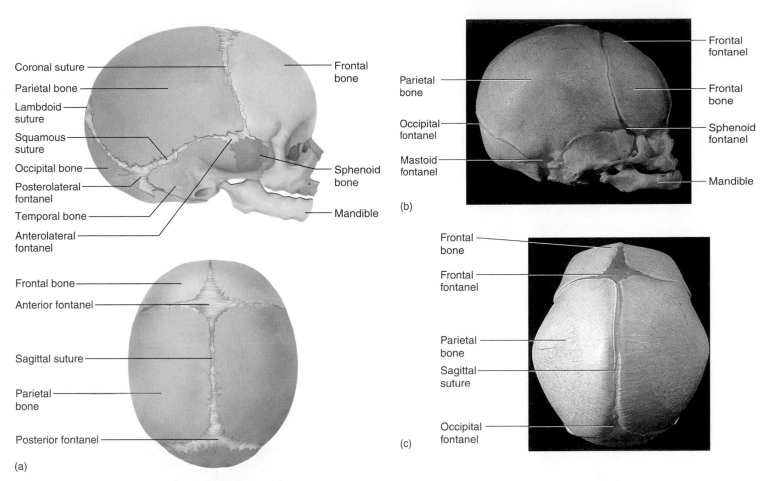

Figure 11.11   Fetal skull and fontanels. (a) Diagram of lateral and superior views; photograph (b) lateral view; (c) superior view.

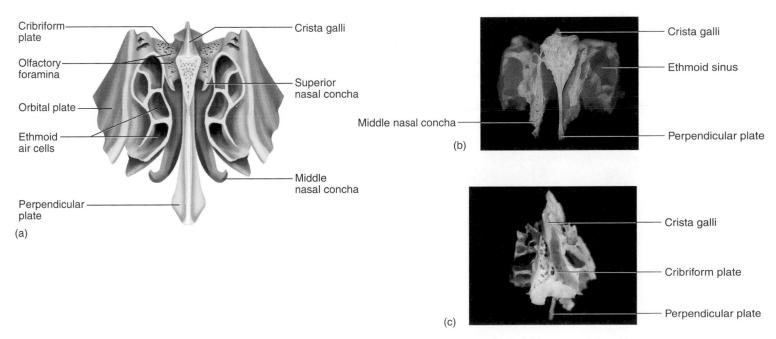

Figure 11.12   Ethmoid bone. Diagram (a) anterior view; photograph (b) anterior view; (c) superior view.

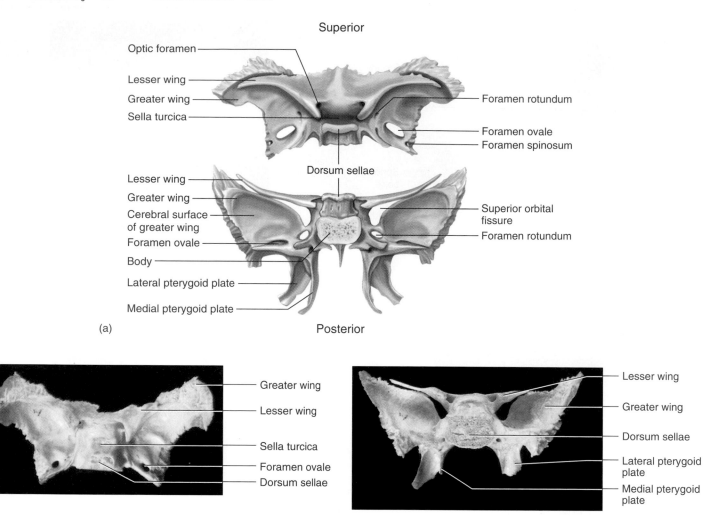

Superior

Optic foramen

Lesser wing

Greater wing

Sella turcica

Foramen rotundum

Foramen ovale
Foramen spinosum

Dorsum sellae

Lesser wing

Greater wing

Cerebral surface
of greater wing

Foramen ovale

Body

Lateral pterygoid plate

Medial pterygoid plate

Superior orbital
fissure

Foramen rotundum

(a)

Posterior

Greater wing

Lesser wing

Sella turcica

Foramen ovale

Dorsum sellae

(b)

Lesser wing

Greater wing

Dorsum sellae

Lateral pterygoid
plate

Medial pterygoid
plate

(c)

Figure 11.13 **Sphenoid bone.** Diagram (a) superior and posterior views; photograph (b) superior view; (c) posterior view.

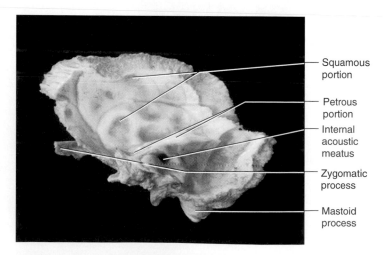

Squamous
portion

Petrous
portion

Internal
acoustic
meatus

Zygomatic
process

Mastoid
process

Figure 11.14 **Temporal bone.** Photograph of interior view of
right temporal bone.

REVIEW

Name _____

1. The eyebrows are over what bone?

2. What is the common name for the zygomatic bone?

3. What is the name of the process that is behind the earlobe?

4. The hard palate is made up of what bones?

5. What are the names of the major paranasal sinuses?

6. The mandible fits into what part of the temporal bone to form the jaw joint?

7. What bone is found just in back of the ethmoid bone in the orbit?

8. The sella turcica is found in what bone?

9. What is the name of the bone that makes up the temple?

10. What are the names of the bones that surround the opening of the nose?

11. The upper teeth are held by what bones?

12. In what bone would you find the foramen magnum?

13. What is the name of the bone that makes up most of the posterior surface of the orbit?

14. What are the two bony structures that make up the nasal septum?

15. The mastoid process is located on which bone?

16. The sagittal suture separates the _____ from the _____.
    a. sphenoid, ethmoid
    b. left parietal, right parietal
    c. frontal, parietal
    d. parietals, occipital

17. Which bone does *not* occur in the orbit?
    a. maxilla   b. zygomatic   c. ethmoid   d. sphenoid   e. temporal

18. Which bone is *not* a paired bone of the skull?
    a. zygomatic   b. temporal   c. lacrimal   d. vomer

19. Label the following illustration.

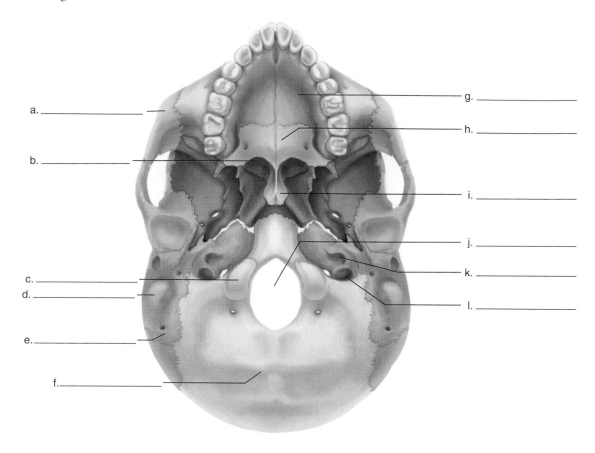

a. _____

b. _____

c. _____

d. _____

e. _____

f. _____

g. _____

h. _____

i. _____

j. _____

k. _____

l. _____

20. Label the following illustration.

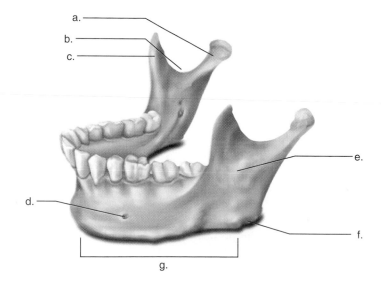

a. _____

b. _____

c. _____

d. _____

e. _____

f. _____

g. _____

Articulations

Introduction

The study of the joints between bones is known as **arthrology.** It is important not only as the point of interplay between the skeletal system and the muscular system but also due to the significant trauma or disease that can occur in joints. Joints are also known as **articulations.** Diseases such as arthritis have an impact on millions of people. Surgery is continually advancing to replace specific joints of the body that can no longer provide either a comfortable range of movement or the stability required for body support.

In this exercise, you study the various types of joints in the body in terms of their structure and function. Joints can be classified according to their physical nature or their range of movement. In terms of structure, joints are classified as fibrous joints, cartilaginous joints, or synovial joints. In terms of range of movement, joints are classified as immovable, semimovable, and freely movable. As you study the joints in the lab refer back to the joints in your body and mimic their actions.

Objectives

At the end of this exercise, you should be able to

1. distinguish between synarthrotic, amphiarthrotic, and diarthrotic joints;
2. explain the structure of the knee, hip, jaw, and shoulder;
3. discuss the nature of a synovial joint;
4. list five different types of synovial joints;
5. classify a joint as fibrous, cartilaginous, or synovial.

Materials

Mammal joint with intact synovial capsule
Dissection tray with scalpel or razor blades, blunt probe, and protective gloves
Waste container
Model or chart of joints including those of the shoulder, knee, hip, and jaw
Articulated skeleton

Procedure

Joints Classified by Movement

There are three major groups of joints classified according to the range of movement they allow. These are synarthrotic, amphiarthrotic, and diarthrotic joints. In synarthrotic joints there is no movement. The bones are tightly bound by connective tissue. In amphiarthrotic joints the bones have some degree of motion, and in diarthrotic joints the bones have significant movement with respect to one another. These three types of joints are discussed next.

Synarthrotic Joints

Synarthrotic joints do not have a joint capsule and allow no movement between bones. A good example of a synarthrotic joint is one that occurs between adjacent bones in the cranium, called a **suture.** In these joints the bones of the cranium are tightly bound together by dense fibrous connective tissue. As a person approaches the age of 35 or so, some of the sutures of the skull begin to fuse from the region closest to the brain towards the superficial surface of the skull. This fusion leads to the union of two bones and is called a **synostosis.** Examine a skull in the lab and locate the various sutures (figure 12.1). Note that they may be serrated, plane, or lap sutures.

Another type of synarthrotic joint is a **gomphosis,** which is represented by teeth in the sockets of the maxilla and the mandible. A gomphosis connects the bone of the jaw to the tooth by fibrous connective tissue called **periodontal ligaments.** Examine a jaw or complete skull in the lab and locate the gomphosis. Compare it to figure 12.2.

There is a developmental joint between the epiphyses and the diaphyses of growing bones. This is a cartilaginous joint called a **synchondrosis** and is considered a synarthrotic joint that eventually fuses to form a single bone (figure 12.3). The cartilage in this joint consists of hyaline cartilage.

Amphiarthrotic Joints

When the bones have a limited range of motion the joint is classified as an **amphiarthrotic joint.** The type of connection between the two bones is variable, and, in some cases,

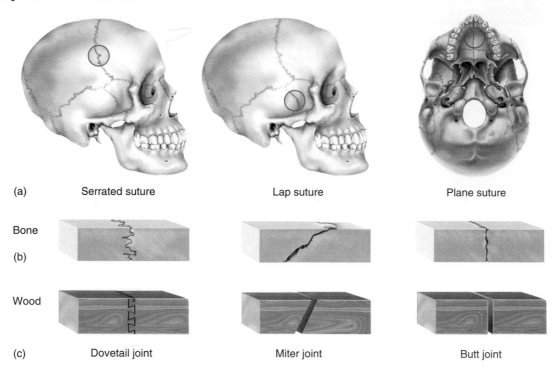

(a) Serrated suture   Lap suture   Plane suture

Bone
(b)

Wood

(c) Dovetail joint   Miter joint   Butt joint

Figure 12.1 Sutures. Serrated, lap, and plane. ⫪

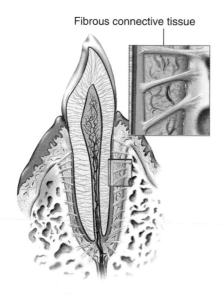

Fibrous connective tissue

Figure 12.2 Gomphosis. ⫪

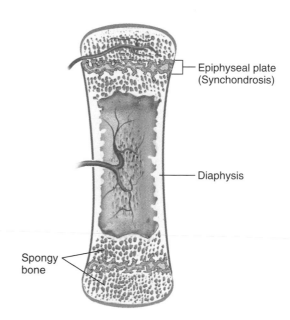

Epiphyseal plate
(Synchondrosis)

Diaphysis

Spongy
bone

Figure 12.3 Synchondrosis of a synarthrosis. ⫪

amphiarthrotic joints can be **fibrous joints,** composed of fibrous material as in the union of the distal radius and ulna. This type of amphiarthrotic joint is called a **syndesmosis,** and the fibrous connective tissue is of greater length in these joints than in sutures. Another example of a syndesmosis is the connection between the distal tibia and fibula. Examine an articulated skeleton in the lab and compare it to figure 12.4.

Another type of amphiarthrotic joint is the **cartilaginous joint.** In this type of joint, hyaline cartilage binds bones together yet lets them move somewhat. This type of joint is an-

other synchondrosis and is found in the joining of the ribs to the sternum by the costal cartilages (figure 12.5).

Another cartilaginous joint is a **symphysis** which is a fibrocartilaginous pad found between the pubic bones (for example, the pubic symphysis) or in the intervertebral discs. Locate the joint in the material in the lab and compare it to figure 12.6. You may have noticed that there are cartilaginous joints that are synarthrotic and some that are amphiarthrotic.

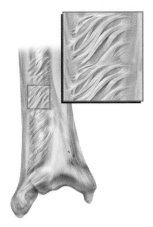

Figure 12.4   Syndesmosis. ✗

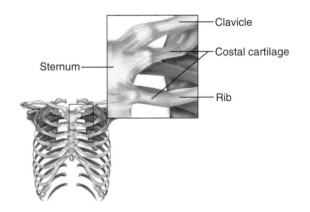

Clavicle

Costal cartilage

Sternum

Rib

Figure 12.5   Synchondrosis of an amphiarthrosis.

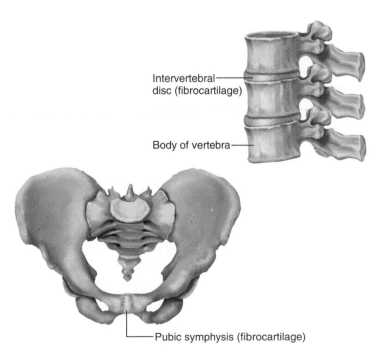

Intervertebral disc (fibrocartilage)

Body of vertebra

Pubic symphysis (fibrocartilage)

Figure 12.6   Symphysis. ✗

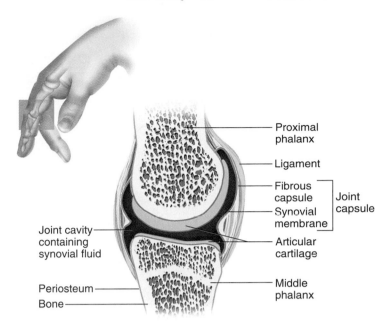

Proximal phalanx

Ligament

Fibrous capsule

Synovial membrane

Joint capsule

Joint cavity containing synovial fluid

Articular cartilage

Periosteum

Bone

Middle phalanx

Figure 12.7   Synovial joint structure. ✗

If the cartilage between the bones is stiff and thin, such as in the epiphyses, the joint does not move and is considered an immovable joint. If the distance between the bones is greater, then the cartilage allows for some movement between the bones and the joint is considered amphiarthrotic.

Diarthrotic Joints

Joints that allow for extensive movement are called **diarthrotic joints.** All diarthrotic joints have the same general structure, which is that of a **synovial joint** consisting of a cavity enclosed by a synovial capsule. Compare the features of the joint in figure 12.7 with models or charts in the lab.

The outer part of the synovial joint is the **joint capsule,** which is made of dense connective tissue. Inside the capsule is the **synovial membrane,** which secretes **synovial fluid,** a lubricating liquid that reduces friction inside the joint. The space inside the joint is called the **synovial cavity,** and each bone of the joint ends in a hyaline cartilage cap called the **articular cartilage.** There are other structures in synovial joints that are characteristic of specific joints, and these are discussed later in this exercise. In general, the more movable a joint is, the less stable it is. The stability of a joint is dependent on the number and types of ligaments, tendons, and muscles and the way the bones fit together.

Dissection of a Mammal Joint

To understand the structure of a synovial joint it is beneficial to examine a fresh or recently thawed mammal joint. If you do not wear protective gloves as you dissect the joint, then wash your hands thoroughly with soap and water after the dissection.

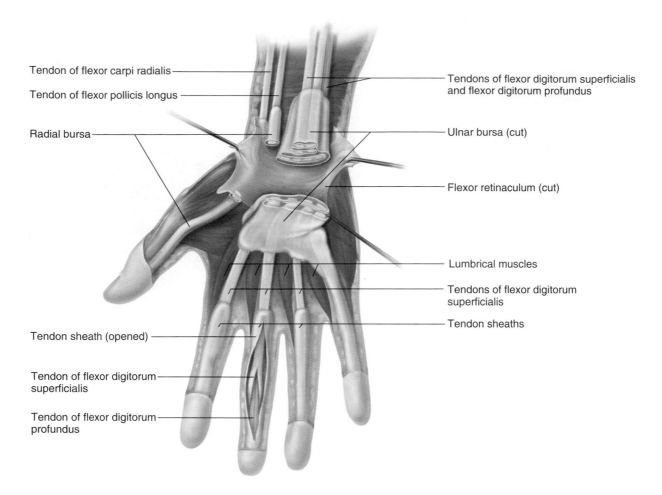

Tendon of flexor carpi radialis

Tendon of flexor pollicis longus

Radial bursa

Tendons of flexor digitorum superficialis and flexor digitorum profundus

Ulnar bursa (cut)

Flexor retinaculum (cut)

Lumbrical muscles

Tendons of flexor digitorum superficialis

Tendon sheaths

Tendon sheath (opened)

Tendon of flexor digitorum superficialis

Tendon of flexor digitorum profundus

Figure 12.8 Modified synovial structures.

Place the joint in front of you on a dissecting tray and cut into the joint capsule with a scalpel or razor blade.

Note the tough, white material that surrounds the joint. This is the **joint capsule,** and it may be fused with **ligaments** that bind the bones of the joint together. Notice the **synovial fluid,** which is a slippery substance that provides a slick feel to the inside of the capsule. Once you cut into the joint examine the **articular cartilage** found on the ends of the bones. Cut into this cartilage with a scalpel or razor blade and notice how the material chips away from the bone. When you have finished with the dissection, make sure to rinse off the dissection equipment and dispose of the mammal joint in the appropriate container.

Modified Synovial Structures

Bursae and tendon sheaths are modified synovial structures. **Bursae** are small synovial sacs between tendons and bones or other structures. The bursae cushion the tendons as they pass over the other structures. **Tendon sheaths** are modified synovial structures that are found encircling the tendons that pass through the palm of the hand. Tendon sheaths serve to lubri-

cate the tendons as they slide past one another. Examine figure 12.8 for these structures.

Diarthrotic Joints Classified by Movement

The diarthrotic joints are classified according to the type of movement they allow between the articulating bones. The joints are listed here in the general order of least movable to most movable.

1. **Gliding joints** allow for movement between two plane surfaces, such as between the superior and inferior facets of adjacent vertebrae or intertarsal or intercarpal joints. Figure 12.9 shows a gliding joint.
2. **Hinge joints** allow for angular movement, such as in the elbow or the knee. You can increase or decrease the angle of the two bones with this joint. Examine figure 12.10 and compare it to material in the lab.
3. **Pivot joints** allow for rotational movement between two bones, such as in the movement of the head to indicate "no." They occur at the proximal radius and ulna and also between the atlas and the axis. Figure 12.11 shows a pivot joint.

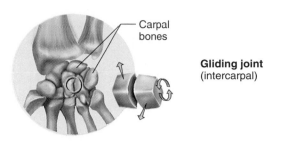

Figure 12.9 Gliding joint. ⚷

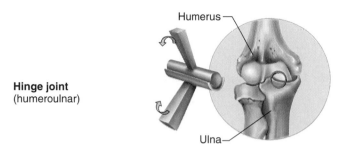

Figure 12.10 Hinge joint. ⚷

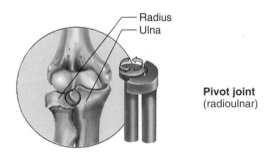

Figure 12.11 Pivot joint. ⚷

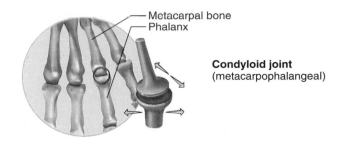

Figure 12.12 Condyloid joint. ⚷

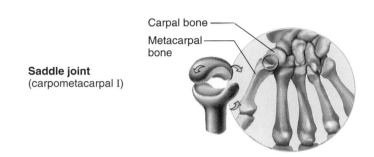

Figure 12.13 Saddle joint. ⚷

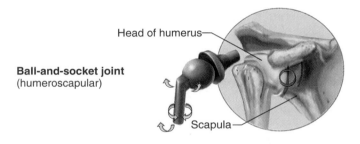

Figure 12.14 Ball-and-socket joint. ⚷

4. **Condyloid joints** allow significant movement in two planes, such as at the wrist. Your hand can easily move anterior to posterior or medial to lateral, yet it does not move well at 45° angles to these planes. Condyloid joints consist of a convex surface articulation paired with a concave surface. The junction between the radius and scaphoid bone is a good example of a condyloid joint. Another type is found between the metacarpals and phalanges. Examine this joint in the lab and compare it to figure 12.12.

5. **Saddle joints** have two concave surfaces which articulate with one another. An example of a saddle joint is between the trapezium and the first metacarpal of the thumb. This provides for more movement in the thumb than the condyloid joint of the wrist. Figure 12.13 shows a saddle joint.

6. **Ball-and-socket joints** consist of a spherical head in a round concavity, such as in the shoulder and the hip. There is extensive movement in these joints, yet they are inherently less stable due to the freedom of movement that they afford. Look at the material in the lab and compare it to figure 12.14.

Monaxial joints move in only one plane. Hinge, pivot, and gliding joints are monaxial joints. Biaxial joints move in two planes. Condyloid and saddle joints are examples of biaxial joints. Multiaxial joints move in many planes. Ball-and-socket joints are multiaxial joints.

Specific Joints of the Body

Temporomandibular

The **temporomandibular joint** is the only diarthrotic joint of the skull. The articular disc is a pad of fibrocartilage that provides a cushion between the condyloid process of the mandible and the temporal bone. Numerous ligaments strengthen this joint (figure 12.15).

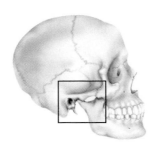

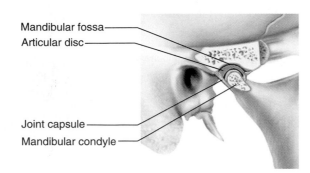

Mandibular fossa
Articular disc

Joint capsule
Mandibular condyle

Figure 12.15 Temporomandibular joint. ⚚

Glenohumeral

The shoulder joint is known as the **glenohumeral joint** as the glenoid cavity articulates with the head of the humerus. The shallow glenoid cavity is deepened by the glenoid labrum, a cartilaginous ring that surrounds the cavity. Numerous bursae, a tough joint capsule, and the rotator cuff muscles also serve to stabilize the joint. Examine figure 12.16 and a model or actual joint in the lab.

Acetabulofemoral

The hip joint is known as the **acetabulofemoral joint.** As with the shoulder joint, the acetabular labrum deepens the hip socket. Numerous ligaments, including the iliofemoral, pubofemoral, and ischiofemoral, bind the femur to the os coxa. The ligamentum teres is a band of dense connective tissue that attaches the acetabulum to the fovea of the femur. Compare the models, charts, or specimens in the lab to figure 12.17.

Tibiofemoral

The **tibiofemoral joint,** also known as the knee joint, is the largest, most complex joint of the body. It consists of several major ligaments, including the **tibial (medial) collateral ligament,** the **fibular (lateral) collateral ligament,** the **anterior cruciate ligament,** and the **posterior cruciate ligament.** Other important structures include the **patellar ligament,** which runs from the quadriceps femoris muscle to the tibial tuberosity, and the **medial** and **lateral menisci,** which are wedge-shaped pads that provide a cushion between the femur and tibia. The knee is primarily a hinge joint with a little lateral movement allowed. Compare specimens in the lab with figure 12.18.

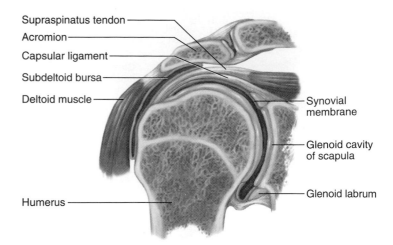

Supraspinatus tendon
Acromion
Capsular ligament
Subdeltoid bursa
Deltoid muscle
Synovial membrane
Glenoid cavity of scapula
Humerus
Glenoid labrum

Figure 12.16 Glenohumeral joint. ⚚

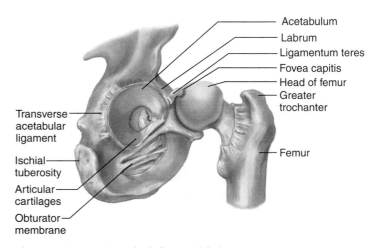

Acetabulum
Labrum
Ligamentum teres
Fovea capitis
Head of femur
Greater trochanter
Transverse acetabular ligament
Femur
Ischial tuberosity
Articular cartilages
Obturator membrane

Figure 12.17 Acetabulofemoral joint. ⚚

Actions

There are many different kinds of movements that occur at joints. These movements, controlled by muscles, are called actions. Actions can decrease a joint angle, increase a joint angle, cause rotation at a joint, and other movements.

Flexion is a decrease in the joint angle from anatomical position. When you reach out to shake a person's hand you are *flexing* your arm and forearm. Flexion of the thigh is in the anterior direction, yet flexion of the leg is in the posterior direction. Bending forward at the waist is flexion of the vertebral column. Looking at your toes is flexion of the head. Examine figure 12.19 for examples of flexion.

Extension is a return of the body to anatomical position after that part of the body is flexed. If you are looking at your toes and lift your head back to anatomical position, you are extending your head. If you straighten your knee after it is bent (flexed), you are extending the leg. Examine figure 12.19 for examples

Anterior

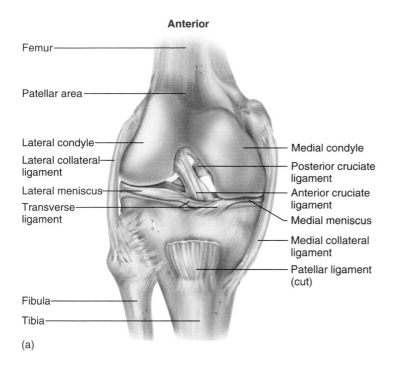

Femur

Patellar area

Lateral condyle

Lateral collateral ligament

Lateral meniscus

Transverse ligament

Medial condyle

Posterior cruciate ligament

Anterior cruciate ligament

Medial meniscus

Medial collateral ligament

Patellar ligament (cut)

Fibula

Tibia

(a)

Posterior

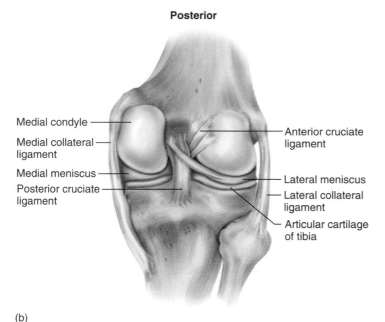

Medial condyle

Medial collateral ligament

Medial meniscus

Posterior cruciate ligament

Anterior cruciate ligament

Lateral meniscus

Lateral collateral ligament

Articular cartilage of tibia

(b)

Figure 12.18 Tibiofemoral joint. (a) Anterior; (b) posterior.

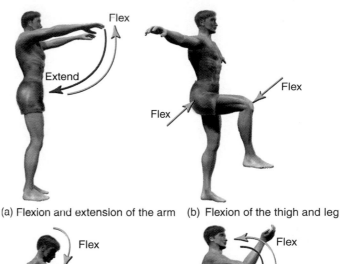

(a) Flexion and extension of the arm

(b) Flexion of the thigh and leg

(c) Flexion of the head

(d) Flexion and extension of the arm

Figure 12.19 Flexion and extension of selected joints. (a) Flexion and extension of the arm; (b) flexion of the thigh and leg; (c) flexion of the head; (d) flexion and extension of the forearm. ✗

of extension. Extension of the part of the body beyond anatomical position is known as hyperextension. When you are about to roll a bowling ball and your arm reaches the very back of the arc, you are hyperextending your arm.

Abduction is movement of the limbs in the coronal plane away from the body (abduct = to take away). Abduction is taking away a part of the body in a lateral direction. This can be seen in figure 12.20.

Adduction is the return of the part of the body to anatomical position after abduction. In doing "jumping jacks" you are abducting and adducting in series. Think of adduction as "adding" a limb back to the body, as seen in figure 12.20.

Rotation is the circular movement of a part of the body. Lateral rotation moves the limb towards the lateral side of the body, and medial rotation turns the limb towards the midline. Lateral rotation of the hand is termed **supination.** The hands are supinated when the body is in anatomical position.

Pronation is medial rotation of the hands. When you turn your palms posteriorly from anatomical position you are pronating your hands.

Circumduction is the movement of a muscle in a conical shape with the point of the cone being at the origin. Examine figure 12.21 for an example of circumduction.

Inversion refers to movement of the feet. Turning the soles of the feet medially so they face each other is known as inversion (figure 12.22).

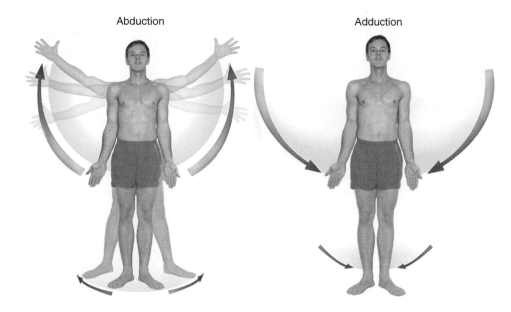

Figure 12.20   Abduction and adduction of the arms and thighs.

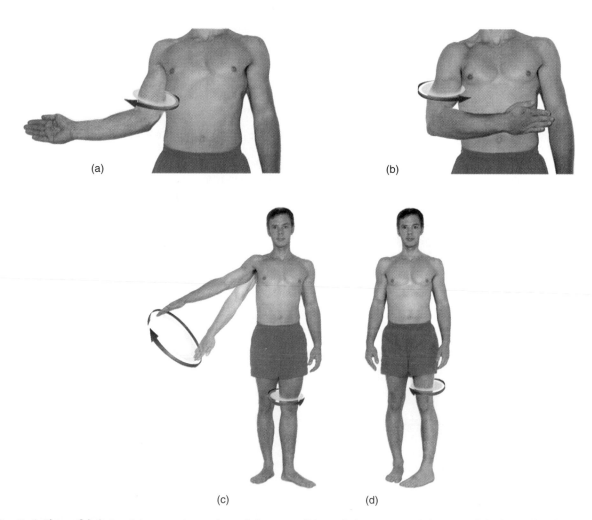

Figure 12.21   Rotation of joints. (a) Lateral rotation of the arm; (b) medial rotation of the arm; (c) circumduction of the arm and lateral rotation of the thigh; (d) medial rotation of the thigh.

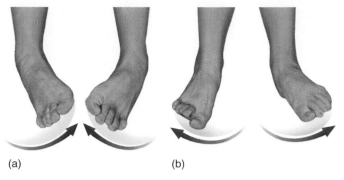

(a)   (b)

Figure 12.22   Inversion and eversion. (a) Inversion of the feet; (b) eversion of the feet.

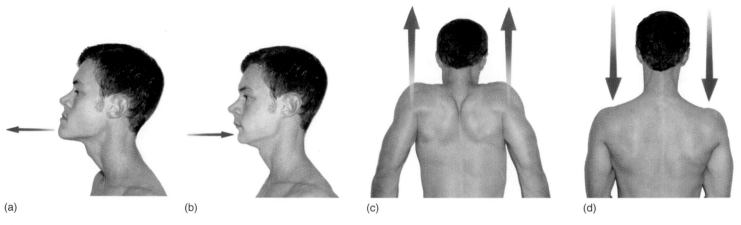

(a)   (b)   (c)   (d)

Figure 12.23   Protraction, retraction, elevation, and depression. (a) Protraction of the jaw; (b) retraction of the jaw; (c) elevation of the scapulae; (d) depression of the scapulae.

Eversion means turning the soles of the feet laterally.

Protraction is a horizontal movement in the anterior direction, as in jutting the chin forward.

Retraction is the reverse of protraction. The jaw that moves from anterior to posterior is retracted. These two actions are illustrated in figure 12.23.

Elevation means to move in a superior direction. Elevation of the shoulders occurs when you shrug your shoulders.

Depression is the opposite of elevation, it is movement in the inferior direction. Elevation and depression are seen in figure 12.23.

Name _____

1. Which one of the following joints has the greatest range of movement?

   a. gomphosis   b. suture   c. synchondrosis   d. hinge

2. In which of these joints would you find a meniscus?

   a. cartilaginous   b. fibrous   c. synovial

3. Label the following illustration.

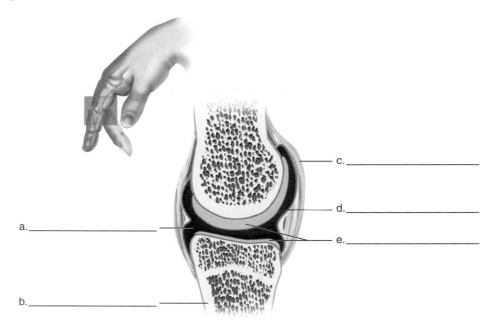

a._____

b._____

c._____

d._____

e._____

4. Match the joint in the left column with the type of joint in the right column.

   _____ acetabulofemoral   a. hinge

   _____ radiocarpal   b. ball-and-socket

   _____ temporomandibular   c. gliding

   _____ vertebrocostal   d. condyloid

5. Rank the following joints in terms of least movable to most movable, with 1 being the least movable and 5 being the most movable.

   gliding   saddle   suture   syndesmosis   ball-and-socket

6. What is the function of the meniscus in the knee?

7. What is the function of the labrum in the glenohumeral joint?

8. Bones that are held together by cartilage are known as _____ joints.

9. A class of joint with great movement is known as a _____ .

10. The teeth are held into the jaw by what kind of joint?

11. What is the name of a joint that is held together by a joint capsule?

12. The joint found between the femur and the tibia is known as what specific type of joint?

13. Synovial fluid is secreted by what structure?

14. A skull suture is what kind of joint, in terms of movement?

15. What kind of joint is found at the wrist (between the radius and the carpal bones)?

16. The joint between the first metacarpal and the proximal phalanx is what kind of joint?

LABORATORY EXERCISE 13

Introduction to the Study of Muscles and Muscles of the Shoulder and Arm

Introduction

The next six exercises focus on skeletal muscles. There are over 600 skeletal muscles in the human body. The muscles are grouped according to their occurrence in particular areas of the body, and most muscles occur as pairs with only a few occurring singly. In the following exercises you will be learning about some of the major muscles of the body. Skeletal muscle is voluntary muscle in that it contracts when consciously stimulated by specific nerves. The muscular system functions in movement, maintenance of posture, generation of heat (shivering), and compression of the abdomen, among other functions.

The origin, insertion, action, and innervation are listed for about 100 muscles. Your instructor may wish to customize the list so you learn specific muscles or specific things about each muscle. The study of muscles can be both fun and challenging. Begin your study of muscles early and not until the night before the lab exam!

Objectives

At the end of this exercise you should be able to

1. locate the muscles of the shoulder and the arm on a torso model, chart, cadaver (if available), or cat (if available);
2. list the origin, insertion, and action of each muscle presented;
3. describe what nerve controls each muscle if your instructor includes innervation with the muscles;
4. list what muscles function as synergists or antagonists to the prime mover;
5. name all of the muscles that have an action on a joint, such as all of the muscles that flex the arm;
6. apply your knowledge of cat musculature to human muscles.

Materials

Human torso model
Human arm models
Human muscle charts
Articulated skeleton
Cadaver (if available)
Cat (if available)

Cat wetting solution
Materials for cat dissection

    Dissection trays
    Scalpel and two to three extra blades
    Gloves (household latex gloves work well for repeated use)
    Blunt (Mall) probe
    String and tags
    Pins
    Forceps and sharp scissors
    First aid kit in lab or prep area
    Sharps container
    Animal waste disposal container

Procedure

The study of human musculature is seen as a daunting task by some students, while other students recall the muscle section of the course as their favorite part of the study of human anatomy and physiology. The satisfaction of studying muscles can be increased if obtainable goals can be reached. This is best accomplished by choosing *small numbers* of muscles to study at a sitting and using flash cards as study aids. It is vital that you know the bones and bony markings before studying muscles. Look over Laboratory Exercises 9 through 11 if you need to review the skeletal markings. You may want to make a quick sketch to help visualize the individual muscles.

A thorough study of muscles involves knowing not only the name of the muscle but also its origin, insertion, action, and the nerves that innervate it. The muscles are listed in this exercise by their name followed by their origin. The **origin** is the attachment point of the muscle that does not move during muscular contraction. The muscles also are listed with their **insertion,** which is the attachment that moves during contraction. Each muscle has one or more **actions,** which represent the effects the muscle has on a part of the body (such as flexion of the arm). Finally, the **innervation** of a muscle is the specific nerve that controls it. Nerves are important in that if the motor portion of a nerve is severed the muscle innervated by that nerve cannot function.

Origins and Insertions

Examine an articulated skeleton as you study the muscles. Try to visualize the muscle as it attaches to the origin or the insertion. As you study the muscles locate them on your body and try to determine which part is the stable origin and which part is the moving insertion.

Actions

Muscles have numerous ways to potentially move joints. To fully understand muscles you must know their actions at joints and associate the actions with the body in anatomical position. When you are learning the action of a muscle, imagine a piece of string tied between the origin and the insertion of the muscle. Think of how the bones would move as you pull the insertion closer to the origin. Mimic the action of the muscle as you learn it. When you perform the action of a particular muscle, you should be able to feel that muscle tighten. You should also review the actions at joints in Exercise 12.

As described in Exercise 12, muscles can have numerous actions at a joint. In addition to moving a joint, muscles can also fix a joint. **Fixing** means to prevent motion in either direction. This is done with antagonist muscles contracting simultaneously. The muscle that has the main force on a joint is called the **prime mover.** Muscles that assist with the prime mover are **synergists,** while those that oppose the muscle are **antagonists.**

Muscle Nomenclature

Another aid to learning muscles is to understand how they are named. Muscles are named by a number of criteria:

Action: the *extensor* digitorum is a muscle that *extends* the fingers.

Origin: the *infraspinatus* muscle originates on the *infraspinous* fossa of the scapula.

Insertion: the flexor *hallucis* longus muscle inserts on the *hallux,* or big toe.

Fiber direction: the *transversus* abdominis muscle has fibers that run horizontally, or in a *transverse* direction.

Number of heads: the *triceps* brachii muscle has *three heads.*

Shape: the *deltoid* muscle is shaped like a *delta,* or triangle.

Examination of Muscles

Look at the charts, models, and cadaver (if available) in the lab and locate the muscles presented next. The muscles of the shoulder and arm are described here if they have some action on the scapula or the humerus. You can find a complete listing of the muscles in table 13.1. Figure 13.1 illustrates the superficial muscles of the back and shoulder. Locate the diamond-shaped **trapezius** muscle, which has an origin in the midline

Figure 13.1   Superficial back muscles, posterior view. ⚲

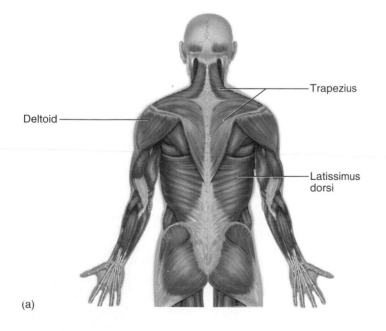

(a)

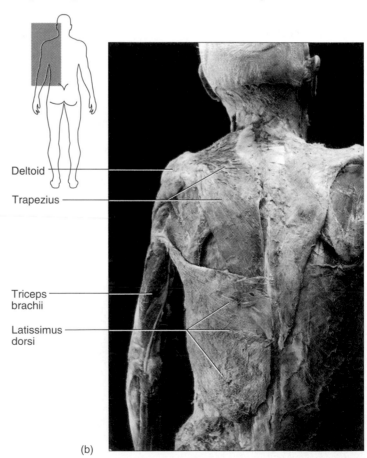

(b)

Table 13.1 Muscles of the Shoulder and Arm

Name	Origin	Insertion	Action	Innervation
Superficial Muscles of the Shoulder				
Trapezius	Posterior occipital bone, ligamentum nuchae, C7–T12	Clavicle, acromion process, and spine of scapula	Extends and abducts head, rotates and adducts scapula, fixes scapula	Accessory nerve (XI), spinal nerves C2–4
Deltoid	Clavicle, acromion process, and spine of scapula	Deltoid tuberosity of the humerus	Abducts arm, flexes, extends, and medially and laterally rotates arm	Axillary nerve
Pectoralis major	Clavicle, sternum, cartilages of ribs 1–7	Crest of greater tubercle of humerus	Flexes, adducts, and medially rotates arm	Medial and lateral pectoral nerves
Pectoralis minor	Ribs 3–5	Coracoid process of scapula	Depresses glenoid cavity, raises ribs 3–5	Medial pectoral nerve
Latissimus dorsi	T7–12, L1–5, S1–5, crest of ilium, ribs 10–12	Intertubercular groove of humerus	Extends, adducts, and medially rotates arm, draws shoulder inferiorly	Thoracodorsal nerve
Deep Muscles of the Shoulder				
Supraspinatus	Supraspinous fossa	Greater tubercle of humerus	Abducts arm, helps stabilize shoulder joint	Suprascapular nerve
Infraspinatus	Infraspinous fossa	Greater tubercle of humerus	Laterally rotates arm, stabilizes shoulder joint	Suprascapular nerve
Subscapularis	Subscapular fossa	Lesser tubercle of humerus	Medially rotates arm, stabilizes shoulder joint	Subscapular nerve
Teres minor	Lateral border of scapula	Greater tubercle of humerus	Laterally rotates and adducts arm, stabilizes shoulder joint	Axillary nerve
Teres major	Inferior angle of scapula	Crest of lesser tubercle of humerus	Extends, adducts, and medially rotates arm	Subscapular nerve
Muscles of the Arm				
Biceps brachii	Long head: superior margin of glenoid fossa Short head: coracoid process of scapula	Radial tuberosity	Flexes arm, flexes forearm, supinates hand	Musculocutaneous nerve
Triceps brachii	Infraglenoid tuberosity of scapula, lateral and posterior surface of humerus	Olecranon process of ulna	Extends and adducts arm, extends forearm	Radial nerve
Coracobrachialis	Coracoid process of scapula	Midmedial shaft of humerus	Flexes and adducts arm	Musculocutaneous nerve
Brachialis	Anterior, distal surface of humerus	Coronoid process of ulna	Flexes forearm	Musculocutaneous nerve
Brachioradialis	Lateral supracondylar ridge of humerus	Styloid process of radius	Flexes forearm	Radial nerve

T = thoracic vertebrae
C = cervical vertebrae
S = sacral vertebrae
L = lumbar vertebrae

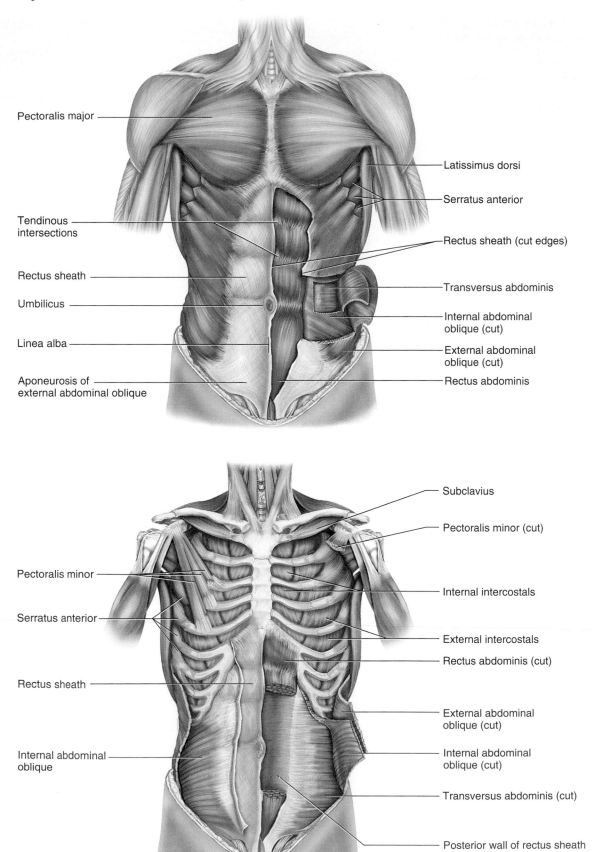

Pectoralis major

Latissimus dorsi

Serratus anterior

Tendinous intersections

Rectus sheath (cut edges)

Transversus abdominis

Rectus sheath

Internal abdominal oblique (cut)

Umbilicus

External abdominal oblique (cut)

Linea alba

Rectus abdominis

Aponeurosis of external abdominal oblique

Subclavius

Pectoralis minor (cut)

Pectoralis minor

Internal intercostals

Serratus anterior

External intercostals

Rectus abdominis (cut)

Rectus sheath

External abdominal oblique (cut)

Internal abdominal oblique

Internal abdominal oblique (cut)

Transversus abdominis (cut)

Posterior wall of rectus sheath (rectus abdominis removed)

(a)

Figure 13.2 Thoracic muscles, anterior view. Diagram (a) superficial and deep muscles; photograph (b) superficial muscles; (c) deep muscles. ✗

Continued

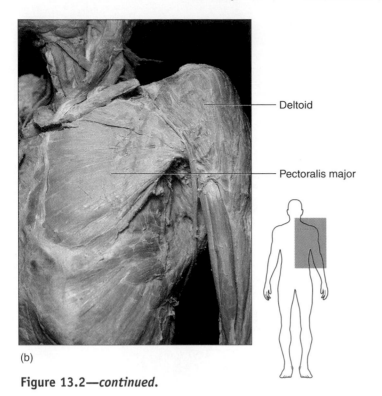

(b)

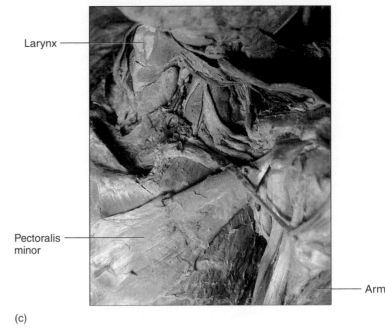

(c)

Figure 13.2—*continued.*

of the vertebral column and head and inserts laterally. If the scapula is fixed the head moves, yet if the vertebral column and head are fixed, then the trapezius has an action on the scapula.

Another superficial muscle of the posterior surface is the **latissimus dorsi** muscle, which arises from the vertebrae by way of a broad, flat lumbodorsal fascia. The latissimus dorsi originates on the back, but the insertion is on the anterior aspect of the humerus. It is a powerful extensor of the arm and is known commonly as the swimmer's muscle (figure 13.1).

The **deltoid** is located on top of the shoulder and has a major action of abducting the arm. Locate the deltoid on material in the lab and compare it to figures 13.1 and 13.2. It is a fan-shaped muscle whose distal attachment partially covers the insertion of the **pectoralis major** muscle seen in figure 13.2. The pectoralis major has fibers that run horizontally across the chest region, and it is a superficial muscle of the chest. Deep to the pectoralis major muscle is the **pectoralis minor** muscle, whose fibers run in a more vertical direction.

Deep to the trapezius and the deltoid are muscles that originate on the scapula proper. The **supraspinatus** is named for its origin on the supraspinous fossa, and it is a synergist to the deltoid. The **infraspinatus** is a muscle originating on the infraspinous fossa, yet it runs laterally and thus laterally rotates the arm. The **subscapularis** is named for its origin on the subscapular fossa, and it medially rotates the arm. The subscapularis is located on the *anterior* surface of the scapula, between the scapula and the ribs, and thus cannot be seen in a

posterior view. Locate these muscles in the lab and compare them to figure 13.3.

The scapula gives rise to two other muscles, the teres minor and the teres major. The **teres minor** appears like a slip of the infraspinatus, while the **teres major** crosses on the medial side of the humerus in a way similar to the latissimus dorsi (figure 13.3).

The **musculotendinous (rotator) cuff** muscles serve to stabilize the shoulder joint. Muscles composing the cuff are the **supraspinatus, infraspinatus, subscapularis,** and **teres minor.** A rotator cuff injury can cause damage to any of these muscles.

The **biceps brachii** muscle is a two-headed muscle of the arm. It has an action on the arm, yet the biceps brachii neither originates nor inserts on the humerus. The biceps brachii has a long tendon that runs between the intertubercular groove of the humerus and a short tendon that originates on the coracoid process. The **triceps brachii** is the only major muscle on the posterior surface of the humerus. The triceps brachii is an antagonist to the biceps brachii.

Alongside the short head of the biceps brachii is a small slip of muscle known as the **coracobrachialis.** This muscle is a short, diagonal muscle of the arm. Underneath the biceps brachii on the anterior surface of the arm is the **brachialis** muscle. The brachialis only crosses the elbow joint and thus serves only to flex the forearm. The **brachio-radialis** is a distal muscle of the arm and is the most lateral muscle of the forearm. Examine figure 13.4 on pages 151 and 152 for these muscles and compare them with descriptions found in table 13.1.

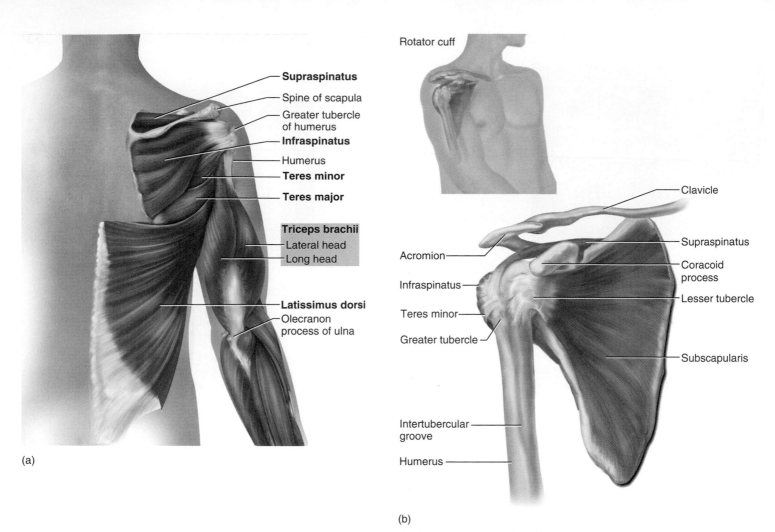

(a)

Supraspinatus

Spine of scapula

Greater tubercle
of humerus

Infraspinatus

Humerus

Teres minor

Teres major

Triceps brachii
- Lateral head
- Long head

Latissimus dorsi

Olecranon
process of ulna

Rotator cuff

Clavicle

Acromion

Supraspinatus

Coracoid
process

Infraspinatus

Lesser tubercle

Teres minor

Greater tubercle

Subscapularis

Intertubercular
groove

Humerus

(b)

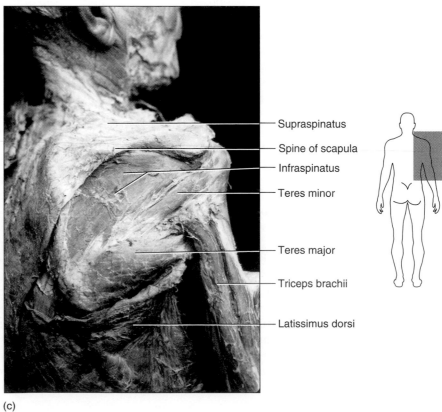

Supraspinatus

Spine of scapula

Infraspinatus

Teres minor

Teres major

Triceps brachii

Latissimus dorsi

(c)

Figure 13.3 Scapular muscles. Diagram
(a) posterior view; (b) anterior view;
photograph (c) posterior-lateral view.

150

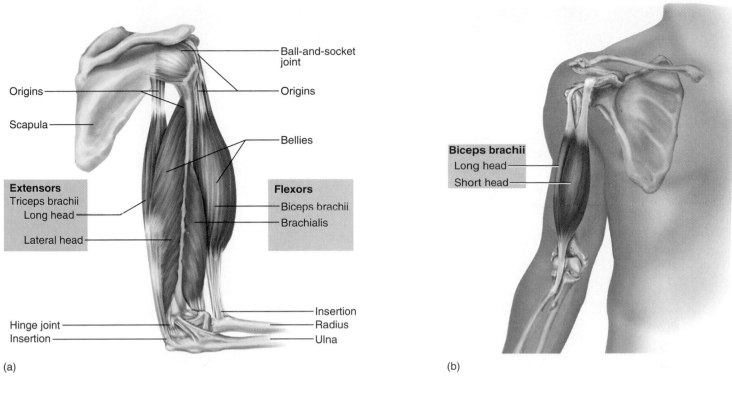

(a)

(b)

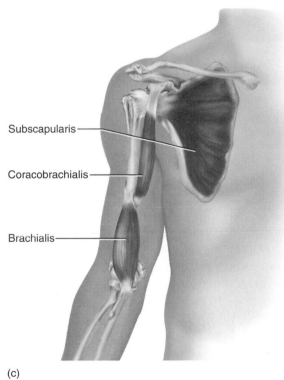

(c)

Figure 13.4   Muscles of the arm. Diagram (a) lateral view; (b) superficial muscles, anterior view; (c) deep muscles, anterior view; photograph (d) anterior view; (e) posterior view.   *Continued*

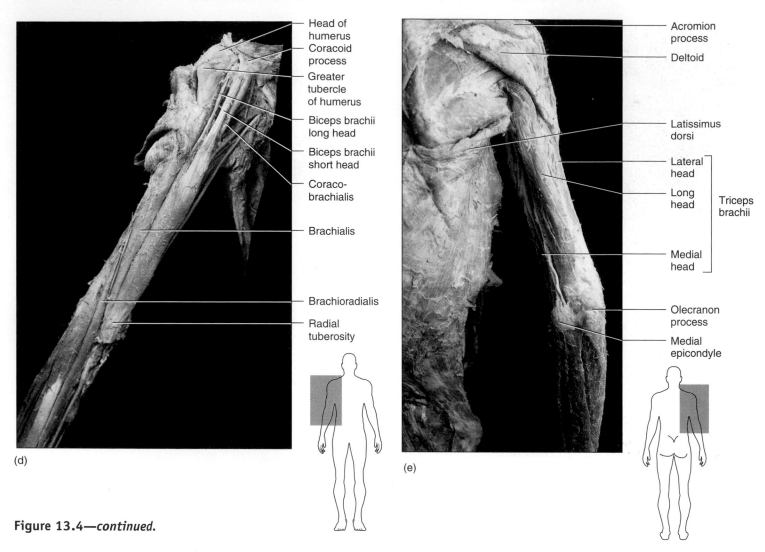

(d)

Head of humerus
Coracoid process
Greater tubercle of humerus
Biceps brachii long head
Biceps brachii short head
Coraco-brachialis
Brachialis
Brachioradialis
Radial tuberosity

(e)

Acromion process
Deltoid
Latissimus dorsi
Lateral head
Long head
Medial head
} Triceps brachii
Olecranon process
Medial epicondyle

Figure 13.4—*continued*.

Cat Dissection

The objective in using a cat for the exercise on musculature is to provide a study specimen for dissection and application to the human system. Differences occur between cat musculature and human musculature, but you should focus on similar structures in order to gain an appreciation of human musculature. Read all of the material in the exercise prior to beginning the dissection. Dissection is a skill in which you try to separate the overlying structures from those underneath while keeping intact as much material as you can.

Dissection Concerns

Wear an apron or an old overshirt as you dissect. Dissection instruments are sharp, and you must be careful when using scalpels. Do not cut *down* into the specimen but rather lift structures gently and try to make incisions so the scalpel blade cuts *laterally*. Cut *away* from yourself and your lab partners. *If you do cut yourself notify the instructor immediately!* Wash the cut with antimicrobial soap and seek medical advice to reduce the chance of infection.

The laboratory should be well ventilated, and if your eyes burn and you develop a headache get some fresh air for a moment. If you dissect without having your face directly over the specimen, that may help. Occasionally students have allergic reactions to formaldehyde. This usually consists of a feeling of restriction of breath. If this happens notify your instructor.

At the end of the exercise wash your dissection equipment with soap and water, taking extra precaution with the scalpel blade. Place all used blades or sharp material in the **sharps container** in lab. Remove all of the excess animal material on the dissection trays and put it in the appropriate animal waste container. *Do not dump excess animal material in the lab sinks!* Wash the dissection trays and place them in the appropriate area to dry.

Cat Care

Keep your cat in a plastic bag. It is recommended that you retain the skin of the cat as a protective wrapping when you are finished with the day's dissection or that you take a small towel (or an old T-shirt) and wrap the specimen in the cloth,

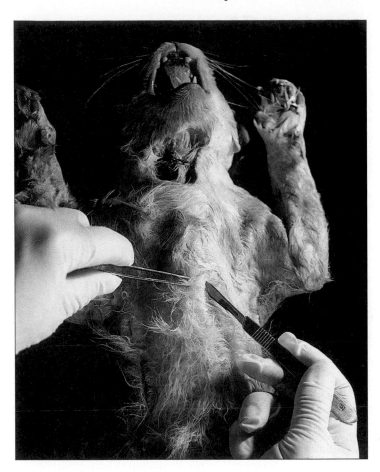

Figure 13.5   Lateral cutting with a scalpel.

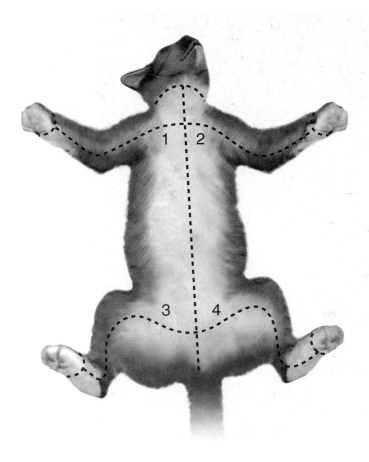

Figure 13.6   Removal of the skin of the cat. After cutting down the midline, make incisions into the limbs (follow the numbers) and gently remove the skin from underlying structures.

soaking it in a cat wetting solution before you place it back in the bag. Usually one lab period is required to skin the cat and prepare the specimen for further study.

External Features

Take a dissection tray and a cat specimen to your table. Remove the cat from the plastic bag and place it on the tray. You may have excess fluid in the plastic bag, and this should be disposed of properly as directed by your instructor. Place the cat on its back and determine whether you have a male specimen or a female specimen. Both sexes have multiple **teats** (nipples), yet the males have a **scrotum** near the base of the tail with a **prepuce** (penile foreskin) ventral to the scrotum. Females have a **urogenital opening** anterior to the **anus** without a scrotum and prepuce. Once you have identified the sex of your specimen, compare yours with others in the class so you can identify the sexes externally. If you have difficulty determining the sex, ask your instructor for help.

Examine the cat and notice the **vibrissae,** or whiskers, in the facial region. Other variations from humans are the presence of **claws** and **friction pads** on the extremities and a **tail.** Review the planes of the body as illustrated in Laboratory Ex-

ercise 2 before you begin the dissection. The terms used in that exercise will be of great importance in the dissection procedures.

Removal of the Skin

Begin your dissection by lifting the skin in the pectoral region with a forceps and making a small cut with a scalpel or sharp scissors in the midline. Work a blunt probe gently into the cut so you free the skin from the underlying fascia and muscle somewhat. Be careful—the muscles are close to the skin. Insert your scalpel, blade side up, and make small incisions in the skin, cutting away from the underlying muscle as illustrated in figure 13.5.

Make a cut that runs up the midline of the sternal region until you reach the neck. Likewise, cut posteriorly until you reach an area craniad to the genital region. Leave the genitals intact and carefully make an incision that runs perpendicular to your first cut. Likewise, make a perpendicular incision along the upper thoracic region (figure 13.6). Continue your caudal incision along the medial aspect of the thigh. Watch out for superficial blood vessels in this area (the great saphenous vein)

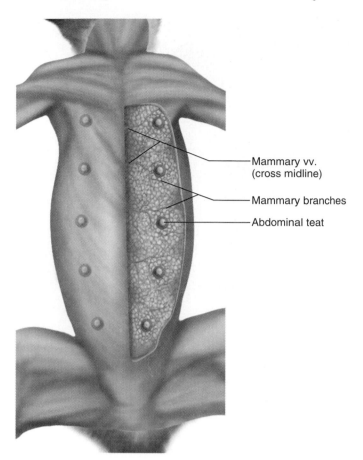

Figure 13.7 **Mammary glands of the cat.**

Mammary vv. (cross midline)

Mammary branches

Abdominal teat

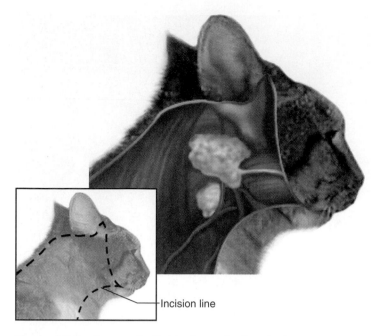

Incision line

Figure 13.8 **Removal of skin from the head of a cat.**

and stop when you reach the knee. Cut the skin around the knee and begin removing it as a layer from the dorsal side of the cat. Cut the skin from the base of the tail and remove the skin from the back.

Once you reach the shoulders, turn the cat back to the ventral side and remove the skin on the medial side of the arm until you reach the elbow. Stop at this level and remove the skin from the lateral aspect of the arm, working back towards the shoulder. Be careful not to damage the superficial veins on the lateral side of the forearm. Continue your removal of the skin from the ventral body region. If your cat is a female, locate the mammary glands, which are elongated, beige, lobular tissue on each side of the midline on the ventral side (figure 13.7).

Make an incision on the ventral side of the neck along the midline. Be careful—there are numerous blood vessels along the neck. Do *not* cut through these blood vessels. Continue up into the face and look for beige lumps of glandular tissue in the region of the mandible. These are the salivary glands. Cut the skin carefully from the face and remove it from the head by cutting around the ears (figure 13.8).

You should be able to remove all of the skin from the cat at this time. Examine the undersurface of the skin and note the superficial muscles that cause the skin to move. These are the **cutaneous maximus** and the **platysma.**

Return to the cat and remove as much fat as you can. Subcutaneous fat is variable from cat to cat, and your specimen may have little or may have significant amounts. The muscle also is covered by fascia, a connective tissue wrapping. Remove the fascia from the muscle so the fiber direction is apparent.

Individual Muscles of the Cat

Begin your **dissection** of the muscles by understanding that the term *dissect* means to *separate.* When you isolate one muscle from another, locate the tendons of that muscle. The **tendon** is the attachment point of the muscle to a bone. Broad, flat tendons are known as **aponeuroses.**

Once you locate the muscle, tug gently on it to locate its **origin** and **insertion.** If you pull too hard, you may rip the muscle. The outer wrapping of the muscle is known as the **fascia,** and it should be removed to find the fiber direction of the muscle. The main part of the muscle is known as the **belly.** You may need to cut a muscle from time to time to locate deeper muscles. This is done by **transecting** the muscle, which is to cut the muscle into two sections perpendicular to the fiber direction. After transecting the muscle you may want to **reflect** it, or pull it towards its attachment site. When separating two muscles you may find a cottonlike material in between the muscles. This is loose connective tissue that forms part of the fascia.

Once you have removed the skin from your cat and removed the superficial fascia, you should identify the major muscles of the cat. Look for the large latissimus dorsi muscle of the back and the external abdominal oblique muscle. You should also find the deltoids and triceps brachii muscles of the shoulder region, and the gluteus and biceps femoris muscles of the hip and thigh region. Compare your cat to figure 13.9.

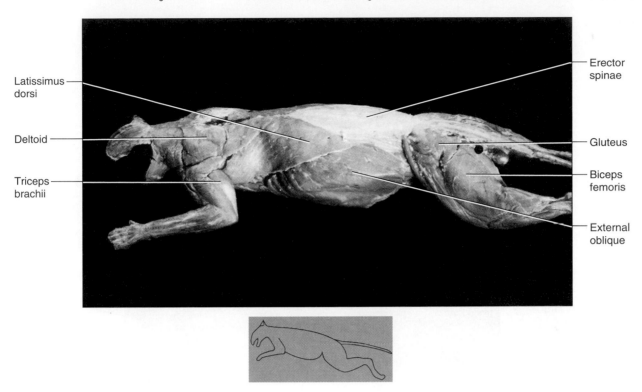

Latissimus dorsi

Deltoid

Triceps brachii

Erector spinae

Gluteus

Biceps femoris

External oblique

Figure 13.9   Major muscles of the cat.

Pectoral Muscles of the Cat

There are four major muscles seen in a superficial view of the pectoral region. From anterior to posterior these are the pectoantebrachialis, pectoralis major, pectoralis minor, and xiphihumeralis. The **pectoantebrachialis** has no corresponding muscle in the human. It originates on the sternum and inserts on the forearm. Transect and reflect the pectoantebrachialis to see the **pectoralis major.** The pectoralis major originates on the sternum and inserts on the upper humerus. The **pectoralis minor** is a large muscle in cats and also inserts on the upper humerus. The **xiphihumeralis** is another cat muscle that has no corresponding human muscle and originates on the sternum and inserts on the proximal humerus along with the pectoralis major and minor. Locate these muscles on your cat and in figure 13.10.

Muscles of the Back

In humans there is a singular trapezius and deltoid muscle on each side of the body. In cats the trapezius consists of three muscles, as does the deltoid. Examine the cat from the dorsal side and locate the **clavotrapezius,** the **acromiotrapezius,** and the **spinotrapezius.** All of these muscles originate on the vertebral column, with the clavotrapezius also originating on the occipital bone. The clavotrapezius inserts on the clavicle, the acromiotrapezius on the acromion process of the scapula, and the spinotrapezius on the spine of the scapula. Compare these muscles to figure 13.11.

The deltoid muscles are also named for their bony attachments. The **clavodeltoid** (clavobrachialis) originates on the clavicle, the **acromiodeltoid** on the acromion process, and the **spinodeltoid** on the spine of the scapula. Insertions of this muscle are on the arm or forelimb. Locate these muscles on the cat and compare them to figure 13.11.

Two other muscles of the region are the **latissimus dorsi** and the **levator scapulae ventralis.** These two muscles are similar to those in the human. Locate these muscles on the cat and compare them to figure 13.11.

The deep muscles of the scapula can be seen by reflecting the overlying muscles. The **supraspinatus, infraspinatus, subscapularis, teres major,** and **teres minor** are roughly equivalent to those same muscles in the human. Locate the supraspinatus, infraspinatus, and teres major and find them in figure 13.12.

Forelimb Muscles

The muscles that have an action on the forelimb of the cat typically either flex the forelimb or extend the forelimb as their primary actions. The **epitrochlearis** is a muscle that does not have a corresponding muscle in humans. The epitrochlearis is on the medial side of the humerus and serves to extend the forelimb. The **biceps brachii** is also a medial muscle, and it serves to flex the forelimb. The **triceps brachii, anconeus, brachioradialis,** and **brachialis** are lateral or posterior muscles. The triceps brachii and the anconeus extend the forearm, while the brachialis flexes the forearm. Find the lateral muscles, using figure 13.13 and 13.14 as a guide.

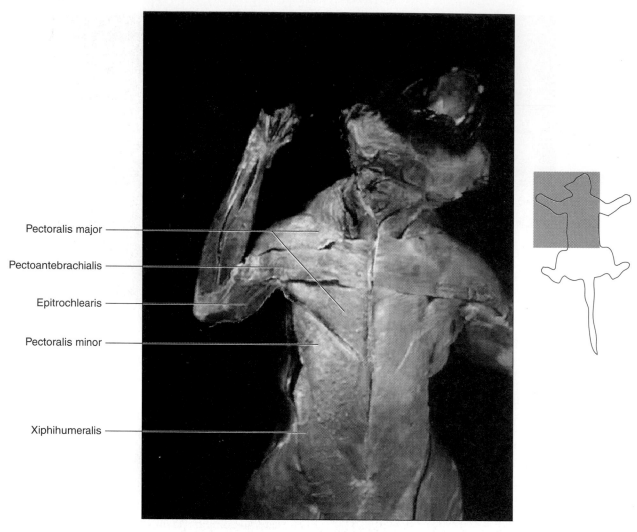

Pectoralis major
Pectoantebrachialis
Epitrochlearis
Pectoralis minor
Xiphihumeralis

Figure 13.10 Muscles of the pectoral region of the cat.

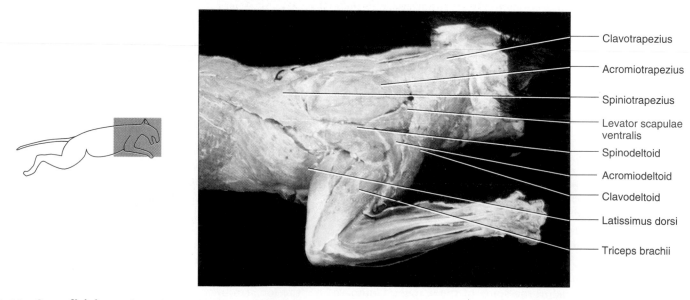

Clavotrapezius
Acromiotrapezius
Spiniotrapezius
Levator scapulae ventralis
Spinodeltoid
Acromiodeltoid
Clavodeltoid
Latissimus dorsi
Triceps brachii

Figure 13.11 Superficial muscles of the shoulder of the cat.

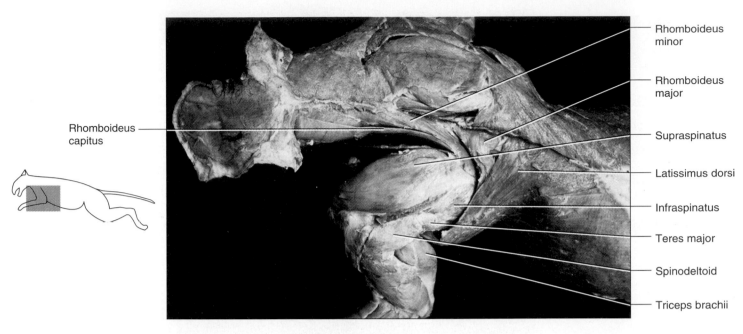

Rhomboideus capitus

Rhomboideus minor

Rhomboideus major

Supraspinatus

Latissimus dorsi

Infraspinatus

Teres major

Spinodeltoid

Triceps brachii

Figure 13.12   Deep muscles of the scapula of the cat.

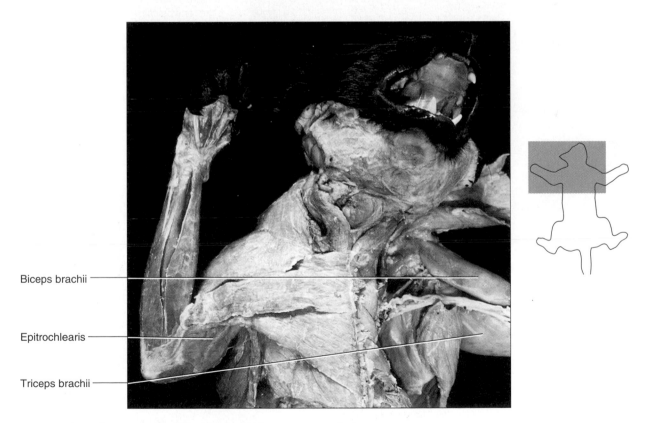

Biceps brachii

Epitrochlearis

Triceps brachii

Figure 13.13   Muscles of the proximal forelimb of the cat, ventral view.

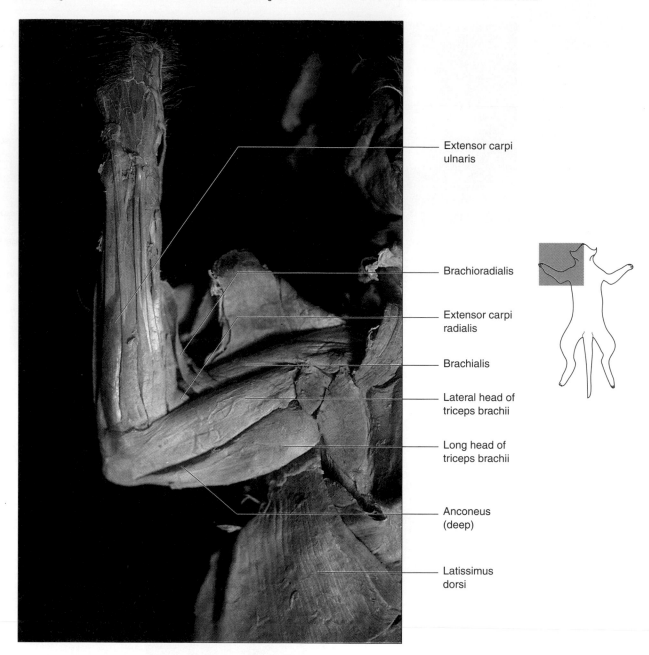

Extensor carpi
ulnaris

Brachioradialis

Extensor carpi
radialis

Brachialis

Lateral head of
triceps brachii

Long head of
triceps brachii

Anconeus
(deep)

Latissimus
dorsi

Figure 13.14   Muscles of the proximal forelimb of the cat, dorsal view.

REVIEW   Name _____

1. In terms of human muscles:

    a. What is the action of the deltoid muscle?

    b. Name the origin of the supraspinatus muscle.

    c. What is the insertion of the trapezius muscle?

    d. Does the biceps brachii muscle originate or insert on the humerus?

    e. What is the insertion of the pectoralis minor?

2. In terms of cat muscles:

    a. What is the action of the epitrochlearis?

    b. Circle the muscle that does *not* correspond to a human muscle: biceps brachii, brachialis, xiphihumeralis, latissimus dorsi

    c. How does the deltoid of the cat differ from the deltoid of the human?

3. Match each term on the left with a description on the right.

    1. dissect   a. what a muscle does
    2. flexion   b. to cut a muscle in half
    3. reflect   c. to stabilize a joint
    4. transect   d. to separate muscles
    5. action   e. to decrease a joint angle
    6. fixing   f. to pull back a muscle

4. Abduction of the arm occurs by what muscles?

5. What is the origin of the trapezius?

6. The pectoralis major has what action?

7. What is the origin of the pectoralis minor?

8. The latissimus dorsi muscle has what action?

9. Name the insertion of the infraspinatus?

10. What is the insertion of the subscapularis?

11. What is the action of the triceps brachii?

12. What is the origin of the brachialis?

13. Name all of the muscles that flex the arm.

14. Which muscles are antagonists of the triceps brachii?

Muscles of the Forearm and Hand

Introduction

In this exercise, we continue the study of skeletal muscles. The muscles of the forearm and hand are very important, particularly in terms of rehabilitating limbs after surgery to relieve carpal tunnel syndrome or after the disabling of the limb. You should learn the origins and insertions as you learn the muscles of the forearm, since many of these muscles look quite similar. By knowing the origins and insertions of a muscle you will not easily mistake it for another muscle. As in the last exercise, you can compare the musculature of the human to that of the cat.

Objectives

At the end of this exercise you should be able to

1. locate the muscles of the forearm and hand on a model, chart, cadaver (if available), or cat (if available);
2. list the origin, insertion, and action of each muscle presented;
3. describe what nerve controls each muscle if your instructor includes innervation with the muscles;
4. name all of the muscles that have an action on a joint, such as all of the muscles that flex the hand;
5. distinguish between the cat musculature and the human musculature and compare them for similarities and differences.

Materials

Human torso model
Human arm models
Human muscle charts
Articulated skeleton
Cadaver (if available)
Cat (if available)
Materials for cat dissection
      Dissection trays and equipment
      Gloves
      String

Procedure

Review the muscle nomenclature and actions of muscles as outlined in Laboratory Exercises 12 and 13. Examine the following muscles on a model, chart, or cadaver and locate the origins and insertions of each. Refer to the following descriptions as you examine the muscles. The specific details of the muscles are provided in table 14.1. Review the bones of the hand as well as the bones of the arm and forearm in Laboratory Exercise 9 to precisely locate the bony origins and insertions.

The muscles of the forearm, due to their similar appearance, provide greater challenges than do the muscles of the shoulder and arm. As you study these muscles, it is important that you locate them and determine their origin and insertion. The origins and insertions will help you determine if you are looking at the right muscle.

The muscles of the forearm and hand are generally named for their action (*pronate* the hand or *flex* the digits) or for their insertion (*carpi* for inserting on carpals or metacarpals, *digitorum* for fingers, and *pollicis* for thumb). In a few instances they are named for the shape of the muscle (*teres* for round or *quadratus* for square).

Notice that much of the muscle mass for moving the fingers is located on the forearm. In this way the fingers move as if on puppet strings. If the muscle mass for the hand was located on the hand proper, then the hands would look like softballs. By having the muscle mass in the forearm, an efficiency of form occurs that allows for a powerful grip yet precise movements of the fingers.

Muscles That Supinate and Pronate the Hand

The first group of muscles in this study are those muscles that insert on the radius. The **supinator** is a muscle that originates on the arm and forearm and functions to supinate the hand. It is a muscle that wraps around the radius and is the deepest, proximal muscle of the forearm. Examine this muscle in the lab and compare it to figure 14.1.

The **pronator teres** is a round muscle that pronates the hand. It originates on the medial side of the arm and forearm and inserts on the lateral side of the radius. The pronator teres is a bit different from the other superficial forearm muscles in that the pronator teres runs at an oblique angle on the forearm while the other muscles run parallel along the length of the forearm.

The **pronator quadratus** is a square muscle that occurs deep to the other forearm muscles on the distal part of the radius and the ulna. It serves to pronate the hand. Compare the two pronator muscles in the lab to figure 14.1.

Table 14.1 Muscles of the Forearm and Hand

Name	Origin	Insertion	Action	Innervation
Supinator	Lateral epicondyle of humerus, anterior ulna	Proximal radius	Supinates hand	Radial nerve
Pronator quadratus	Distal part of anterior ulna	Distal radius	Pronates hand	Median nerve
Pronator teres	Medial epicondyle of humerus, coronoid process of ulna	Lateral, middle shaft of radius	Pronates hand, flexes forearm	Median nerve
Palmaris longus	Medial epicondyle of humerus	Palmar aponeurosis	Flexes hand	Median nerve
Flexor carpi radialis	Medial epicondyle of humerus	Second and third metacarpals	Flexes and abducts hand	Median nerve
Flexor carpi ulnaris	Medial epicondyle of humerus, olecranon process and dorsal border of ulna	Pisiform, hamate, and fifth metacarpal	Flexes and adducts hand	Ulnar nerve
Flexor digitorum superficialis	Medial epicondyle of humerus, proximal ulna, proximal radius	Middle phalanges of second through fifth digits	Flexes proximal and middle phalanges, flexes hand	Median nerve
Flexor digitorum profundus	Anterior, proximal surface of ulna, interosseus membrane	Distal phalanges of second through fifth digits	Flexes phalanges, flexes hand	Median and ulnar nerves
Flexor pollicis longus	Anterior portion of radius and interosseus membrane	Distal phalanx of pollex (thumb)	Flexes thumb	Median nerve
Extensor carpi radialis longus	Lateral supracondylar ridge of humerus	Second metacarpal	Extends and abducts hand	Radial nerve
Extensor carpi radialis brevis	Lateral epicondyle of humerus	Third metacarpal	Extends and abducts hand	Radial nerve
Extensor carpi ulnaris	Lateral epicondyle of humerus, proximal ulna	Fifth metacarpal	Extends and adducts hand	Radial nerve
Extensor digitorum	Lateral epicondyle of humerus	Middle and distal phalanges of second through fifth digits	Extends phalanges, extends hand	Radial nerve
Abductor pollicis longus	Posterior radius and ulna, interosseus membrane	First metacarpal	Abducts thumb	Radial nerve
Extensor pollicis longus and brevis	Posterior radius and ulna, interosseus membrane	Proximal and distal phalanges of pollex (thumb)	Extends thumb	Radial nerve

Flexor Muscles

The next muscles to be studied are grouped by their insertion on the hand. The superficial **palmaris longus** muscle is absent in about 10% of the population. It is centrally located in the middle, anterior forearm and inserts into a broad, flat tendon known as the palmar aponeurosis. This aponeurosis has no bony attachment but rather attaches to the fascia of the underlying muscles.

The **flexor carpi radialis** muscle inserts on the metacarpals on the radial side of the hand. The pulse of the radial artery is frequently taken at the wrist just lateral to the tendon of the flexor carpi radialis muscle. Most of the flexor muscles of the hand originate from the medial epicondyle of the humerus, and the flexor carpi radialis has its origin here as well. The flexor carpi radialis runs underneath a connective tissue band known as the **flexor retinaculum,** which anchors the tendons to the wrist and prevents

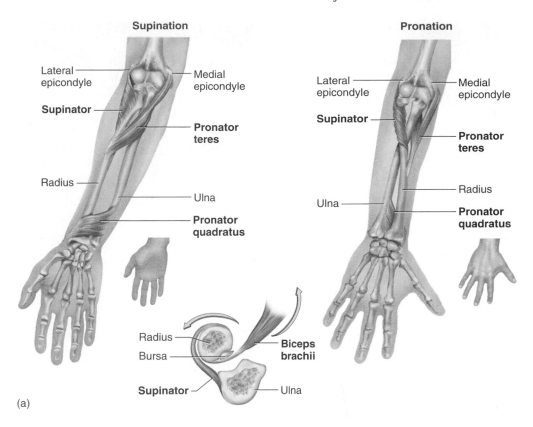

Supination

Lateral epicondyle

Medial epicondyle

Supinator

Pronator teres

Radius

Ulna

Pronator quadratus

Pronation

Lateral epicondyle

Medial epicondyle

Supinator

Pronator teres

Ulna

Radius

Pronator quadratus

Radius

Bursa

Biceps brachii

Supinator

Ulna

(a)

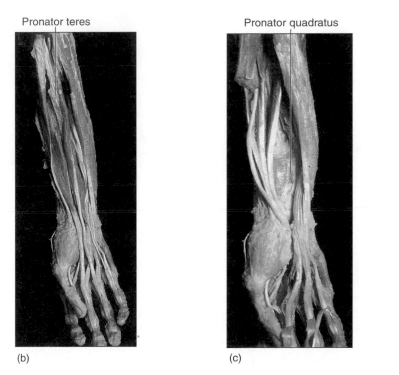

Pronator teres

Pronator quadratus

(b)

(c)

Figure 14.1 Supinator and pronator muscles of the forearm, anterior view. (a) Diagram; (b) superficial muscles; (c) deep muscles. ⚡

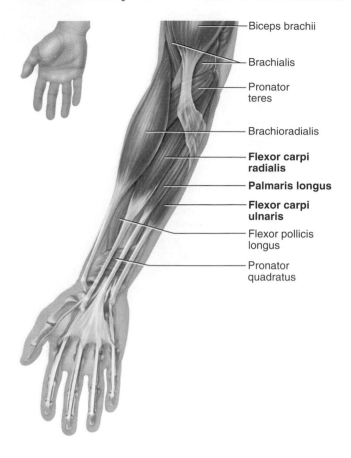

- Biceps brachii
- Brachialis
- Pronator teres
- Brachioradialis
- **Flexor carpi radialis**
- **Palmaris longus**
- **Flexor carpi ulnaris**
- Flexor pollicis longus
- Pronator quadratus

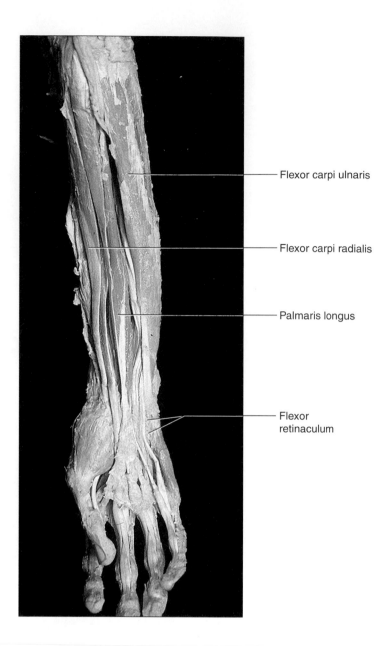

- Flexor carpi ulnaris
- Flexor carpi radialis
- Palmaris longus
- Flexor retinaculum

Figure 14.2   Superficial flexor muscles of the forearm, anterior view.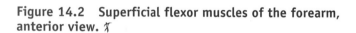

the tendons from pulling away from the wrist when the hand is flexed.

The **flexor carpi ulnaris** muscle also has an origin on the medial epicondyle of the humerus and inserts on the carpals and a metacarpal of the ulnar side of the hand. The flexor carpi ulnaris is a medial muscle of the forearm. Examine these muscles in the lab and in figure 14.2.

The **flexor digitorum superficialis** is a superficial flexor muscle of the digits. It is not the most superficial muscle of the forearm but is actually under the palmaris longus and the flexor carpi radialis and flexor carpi ulnaris. The flexor digitorum superficialis is superficial to the deep digit flexor, which is discussed next. As with the other flexor muscles, the flexor digitorum superficialis has an origin on the medial epicondyle of the humerus and the insertion tendons form a V on the middle phalanges of digits 2 through 5.

The **flexor digitorum profundus** is a deep muscle of the digits. It is an exception to the other hand flexors in that it does *not* originate on the medial epicondyle of the humerus but on the ulna and the membrane between the radius and the ulna (interosseus membrane). It is a deep muscle of the forearm that runs underneath the flexor digitorum superficialis. At the insertion point the tendons of the flexor digitorum profundus actually run through the split tendons of the flexor digitorum superficialis and extend to the distal phalanges of digits 2 through 5.

The **flexor pollicis longus** originates on the radius and the interosseus membrane and runs along the lateral forearm to insert on the pad-side of the thumb. Flexion of the thumb is the curling of the thumb as if you are ready to flip a coin. Examine these muscles and compare them to figure 14.3.

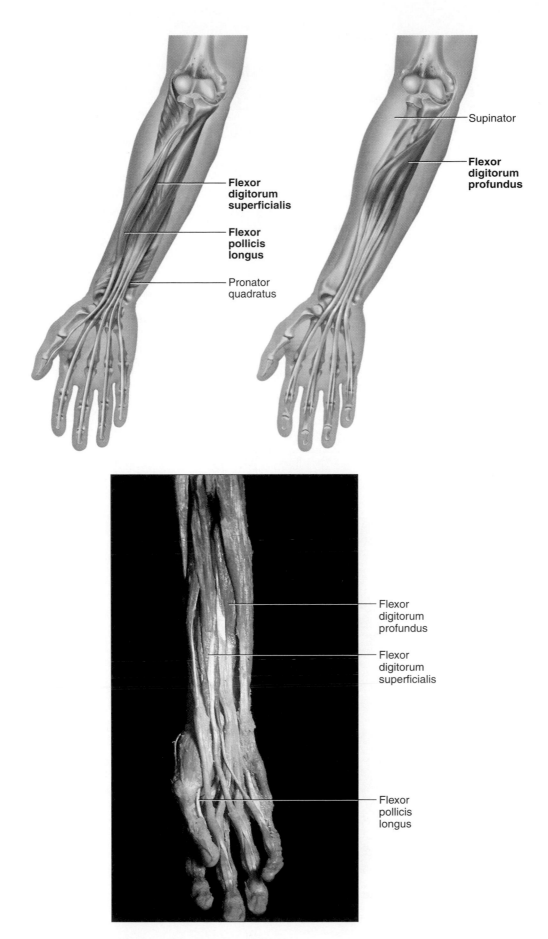

Figure 14.3   Deep flexor muscles of the forearm, anterior view.

In the illustrations:

Flexor
digitorum
superficialis

Flexor
pollicis
longus

Pronator
quadratus

Supinator

**Flexor
digitorum
profundus**

In the photograph:

Flexor
digitorum
profundus

Flexor
digitorum
superficialis

Flexor
pollicis
longus

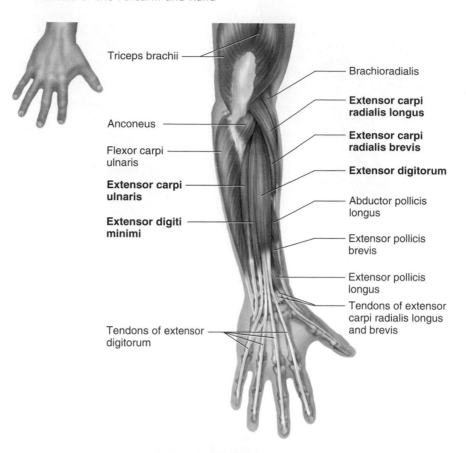

Triceps brachii

Anconeus

Flexor carpi
ulnaris

**Extensor carpi
ulnaris**

**Extensor digiti
minimi**

Tendons of extensor
digitorum

Brachioradialis

**Extensor carpi
radialis longus**

**Extensor carpi
radialis brevis**

Extensor digitorum

Abductor pollicis
longus

Extensor pollicis
brevis

Extensor pollicis
longus

Tendons of extensor
carpi radialis longus
and brevis

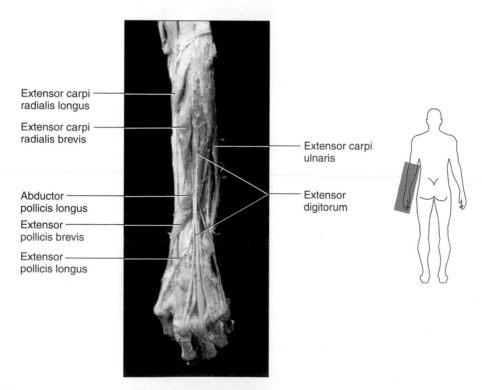

Extensor carpi
radialis longus

Extensor carpi
radialis brevis

Abductor
pollicis longus

Extensor
pollicis brevis

Extensor
pollicis longus

Extensor carpi
ulnaris

Extensor
digitorum

Figure 14.4   Superficial extensor muscles of the forearm, posterior view. ✶

Extensor Muscles

As a general rule, most of the extensors originate on the lateral epicondyle or lateral supracondylar ridge of the humerus. The extensor muscle tendons are held to the posterior surface of the wrist by a connective tissue band known as the **extensor retinaculum.** The **extensor carpi radialis longus** muscle originates on the humerus and inserts on the second metacarpal (on the radial side) of the hand. The **extensor carpi radialis brevis** is deep to the longus, and it inserts on the dorsum of the third metacarpal. Both of these muscles serve to extend and abduct the hand. Examine these muscles in figure 14.4.

The **extensor carpi ulnaris** originates on the lateral epicondyle and inserts on the fifth metacarpal (on the ulnar side) of the hand. As it contracts the hand is extended and adducted. The **extensor digitorum** exists as a singular muscle on the back of the hand (remember there are two flexor digitorum muscles). As an extensor muscle it originates on the lateral epicondyle of the humerus and inserts on the middle and distal phalanges of the second through fifth digits. The tendons of this muscle can be seen on the dorsal surface of the hand and in figure 14.4.

The **abductor pollicis longus** muscle inserts on the metacarpal of the thumb. By pulling your thumb away from the index finger you are abducting the thumb. There are two extensor pollicis muscles, the **extensor pollicis longus** and the **extensor pollicis brevis.** Extension of the thumb is done when you flip a coin. At the end of the flip the thumb is extended. The tendons of the extensor pollicis muscles and the abductor pollicis muscles form a depression in the form of a triangle at the base of the thumb. This depression is known as the anatomical snuff box. These muscles can be seen in figures 14.4 and 14.5.

Cat Dissection

If you did not already remove the skin from the forelimb of the cat, you should do so now. This can be accomplished by making a longitudinal incision along the length of the forelimb. Be very careful not to cut the tendons or the blood vessels or nerves. Pull the skin off as if you were removing a pair of knee socks. As you get to the tips of the digits cut the skin from the digits. Pay particular attention and keep the tendons intact. It is a good procedure to first dissect only one side of the cat at a time. If you make an error on one side, you will have the other side to dissect. As you cut through the pad on the ventral side of the paw of the cat, make sure that you do not cut through the tendons of the flexor digitorum muscles.

As you make the dissection in the cat, you can leave many of the muscles of the forelimb intact. Deeper muscles can generally be seen by moving the more superficial muscles off to the side.

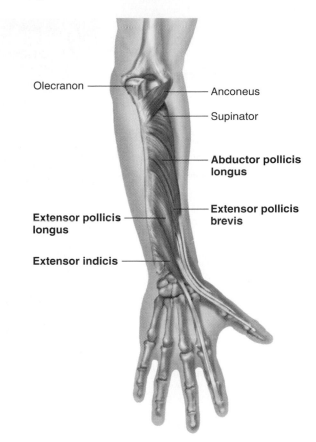

Figure 14.5 Deep extensor muscles of the thumb, posterior view.

Olecranon

Anconeus

Supinator

Abductor pollicis longus

Extensor pollicis longus

Extensor pollicis brevis

Extensor indicis

Superficial Muscles on the Medial Aspect of the Forelimb

Most of the muscles of the forelimb of the cat run parallel to the radius and ulna. An exception to this is the **pronator teres,** which is a small slip of muscle that runs obliquely down the forelimb. It acts to pronate the wrist. The **palmaris longus** is a broad, flat muscle, superficially located on the forelimb with insertions into the digits. In humans the palmaris longus terminates at the palmar aponeurosis.

The **flexor carpi radialis** is a thin muscle inserting on the second and third metacarpals. It is named for its action, its insertion, and its location. The **flexor carpi ulnaris** has an origin on the humerus and ulna and insertion on medial metacarpals and carpals. The **flexor digitorum superficialis** (sublimis) occurs in the forelimb as a middle-level muscle, underneath the palmaris longus. It originates on fascia of other forearm muscles as opposed to originating on bone. Examine the superficial muscles of the medial forelimb and compare them to figure 14.6.

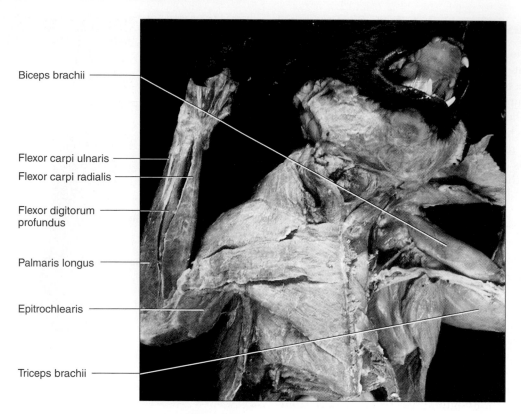

Biceps brachii

Flexor carpi ulnaris

Flexor carpi radialis

Flexor digitorum profundus

Palmaris longus

Epitrochlearis

Triceps brachii

Figure 14.6 **Superficial muscles of the right medial forelimb of the cat.**

Deep Muscles on the Medial Aspect of the Forelimb

Underneath the upper layer of muscles are numerous muscles with varied actions. The **supinator** is a deep muscle that runs diagonally from the lateral epicondyle of the humerus to the proximal radius.

It is the deepest of the proximal forelimb muscles. The **flexor digitorum profundus** is an extensive muscle that in-serts on the first through fifth digits. It replaces the flexor pol-licis longus for the thumb flexion, since this muscle is absent in cats. The **pronator quadratus** is a square muscle located between the radius and ulna deep to the flexor digitorum pro-fundus. Find the deep muscles in the cat and compare them to figure 14.7.

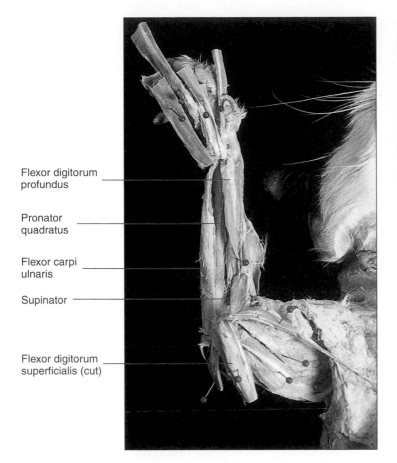

Flexor digitorum profundus

Pronator quadratus

Flexor carpi ulnaris

Supinator

Flexor digitorum superficialis (cut)

Figure 14.7   Deep muscles of the right medial forelimb of the cat.

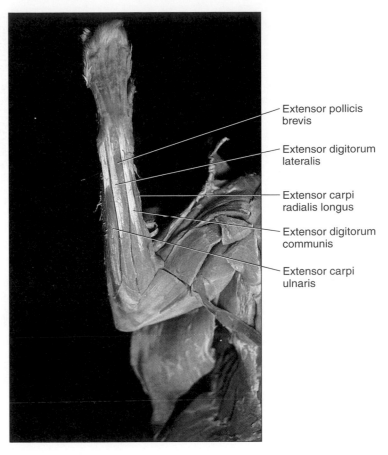

Extensor pollicis brevis

Extensor digitorum lateralis

Extensor carpi radialis longus

Extensor digitorum communis

Extensor carpi ulnaris

Figure 14.8   Lateral muscles of the left forelimb of the cat.

Muscles on the Lateral Aspect of the Forelimb

The lateral muscles of the forelimb are the extensor group, and the **extensor carpi radialis longus** muscle is deep to the brachioradialis and inserts on the second metacarpal. The **extensor carpi radialis brevis** is underneath the extensor carpi radialis longus and inserts on the third metacarpal.

Locate the **extensor carpi ulnaris,** which is next to the extensor digitorum lateralis, inserting on the fifth metacarpal.

The **extensor digitorum communis** can be located on the lateral aspect of forelimb. It is a broad muscle inserting by tendons on the second through fifth digits. Locate these muscles on the cat and in figure 14.8.

The **extensor digitorum lateralis** is specific to the cat and inserts with the tendons of the digitorum communis to the digits. The **extensor pollicis brevis** is a well-developed muscle in the cat, while the abductor pollicis longus is absent in cats.

Name _____

1. What is the origin of the flexor carpi ulnaris in humans?

2. What is an antagonist to the supinator muscle?

3. Where does the flexor digitorum superficialis of the human insert?

4. What is the insertion of the extensor carpi ulnaris muscle?

5. Which muscle is more developed in cats, the flexor digitorum superficialis or the flexor digitorum profundus?

6. What is an antagonist to the extensor pollicis longus and brevis muscles?

7. What is the action of the flexor carpi radialis muscles?

8. What muscle extends both the hand and phalanges?

9. What muscles flex the hand?

10. What muscles extend the thumb?

11. Where are the extensor carpi muscles found, on the anterior or posterior side of the forearm?

12. Label the muscles in the following illustration.

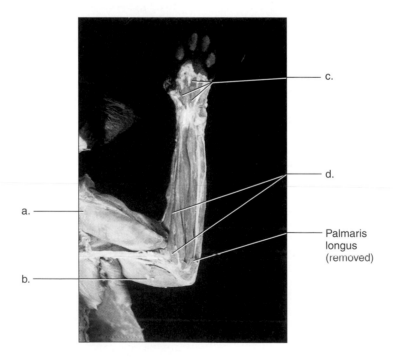

c.

d.

a.

Palmaris
longus
(removed)

b.

Muscles of the Hip and Thigh

Introduction

The muscles of the hip and thigh are primarily muscles of locomotion. They can be grouped generally into muscles that flex or extend the thigh, adduct or abduct the thigh, and flex or extend the leg. Lateral and medial rotations of the thigh are other actions that occur in muscles of this group as well.

The size and shape of the muscles of the hip and thigh in humans are different from other mammals in that humans are bipedal. This two-legged walking habit is much different than a quadruped's locomotion in terms of balance and forward movement. When a quadruped walks, there are typically three legs remaining on the ground. When a biped walks, the inherent instability of standing on one leg brings into focus the importance of stabilizing the support limb. As you study these muscles, think about what happens to the center of balance when you lift a leg to walk or run.

Objectives

At the end of this exercise you should be able to

1. locate the muscles of the hip and thigh on a model, chart, cadaver (if available), or cat (if available);
2. list the origin, insertion, and action of each muscle presented;
3. describe what nerve controls each muscle if your instructor includes innervation with the muscles;
4. list what muscles function as synergists or antagonists to the prime mover;
5. name all of the muscles that have an action on a joint, such as all of the muscles that flex the thigh.

Materials

Human torso model
Human leg models
Human muscle charts
Articulated skeleton
Cadaver (if available)
Cat (if available)

Materials for cat dissection
  Dissection trays
  Scalpel or razor blades
  Gloves
  Blunt probe
  String
  Pins

Procedure

Review the muscle nomenclature and the actions as outlined in Laboratory Exercises 12 and 13. Examine muscle models or charts in the lab and locate the muscles described in the following section and in table 15.1. Correlate the shape of the muscle with the name and begin to visualize the muscles as you study the models or charts. You may want to look at an articulated skeleton as you review the origins and insertions of the muscles so you can better see the muscle attachment points. The descriptions in the text portion of this exercise can help you understand the nature of the muscle, while table 15.1 gives you the particular information about the muscle.

Once you have learned the origin and the insertion, you should be able to understand the action of the muscle. This is done, in part, by imagining how the bony attachments of the origin and insertion would come together if the muscle pulled them closer to one another.

Muscles of the Hip and Thigh

The **iliopsoas** muscle is a major flexor of the thigh. It originates inside the abdominopelvic cavity and crosses over the pelvic brim to insert on the posterior aspect of the femur. The iliopsoas is actually three muscles, the iliacus, the psoas major, and the psoas minor. In this exercise we treat these muscles as one. Examine the material in the lab and compare it to figures 15.1 and 15.2.

The **sartorius** is a slender muscle that runs from a superior, lateral origin and inserts in an inferior, medial location. It allows you to sit cross-legged and is thus named the "tailor's muscle." The sartorius is the most superficial muscle on the anterior thigh. The most superficial muscle on the medial thigh is the **gracilis.**

Table 15.1   Muscles of the Hip and Thigh

Name	Origin	Insertion	Action	Innervation
Iliopsoas	T12, L1-5, sacrum, iliac crest, and iliac fossa	Lesser trochanter of femur	Flexes thigh and lumbar vertebrae	Femoral nerve, spinal nerves L1–3
Sartorius	Anterior superior iliac spine	Proximal, medial tibia	Flexes and laterally rotates thigh, flexes leg	Femoral nerve
Gracilis	Symphysis pubis	Proximal, medial tibia	Adducts thigh, flexes leg	Obturator nerve
Pectineus	Pubis	Femur inferior to lesser trochanter	Adducts, flexes, and laterally rotates thigh	Femoral nerve
Adductor brevis	Pubis	Linea aspera at proximal portion of femur	Adducts, flexes, and laterally rotates thigh	Obturator nerve
Adductor longus	Pubis	Linea aspera at middle portion of femur	Adducts, flexes, and laterally rotates thigh	Obturator nerve
Adductor magnus	Pubis, ischium	Linea aspera, adductor tubercle of distal femur	Adducts, flexes, extends, and laterally rotates thigh	Obturator nerve
Quadriceps Femoris				
Rectus femoris	Anterior inferior iliac spine, margin of acetabulum	Tibial tuberosity by the patellar tendon	Flexes thigh, extends leg	Femoral nerve
Vastus lateralis	Greater trochanter of femur, linea aspera of femur	Tibial tuberosity by the patellar tendon	Extends leg	Femoral nerve
Vastus intermedius	Proximal, anterior femur	Tibial tuberosity by the patellar tendon	Extends leg	Femoral nerve
Vastus medialis	Linea aspera, medial side	Tibial tuberosity by the patellar tendon	Extends leg	Femoral nerve
Tensor fascia lata	Iliac crest and anterior iliac surface	Iliotibial band of fascia lata	Flexes, abducts, and medially rotates thigh	Superior gluteal nerve
Gluteus maximus	Outer iliac blade, iliac crest, sacrum, coccyx	Gluteal tuberosity of femur, iliotibial band of fascia lata	Extends and laterally rotates thigh, braces knee	Inferior gluteal nerve
Gluteus medius	Outer iliac blade	Greater trochanter of femur	Abducts and medially rotates thigh	Superior gluteal nerve
Gluteus minimus	Outer iliac blade	Greater trochanter of femur	Abducts and medially rotates thigh	Superior gluteal nerve
Hamstrings				
Biceps femoris	Ischial tuberosity, linea aspera	Head of fibula, lateral condyle of tibia	Extends thigh, flexes leg	Sciatic nerve
Semitendinosus	Ischial tuberosity	Proximal, medial tibia	Extends thigh, flexes leg	Sciatic nerve
Semimembranosus	Ischial tuberosity	Medial condyle of tibia	Extends thigh, flexes leg	Sciatic nerve

The gracilis is a thin muscle that serves to adduct the thigh. Locate these muscles in the lab and in figure 15.1.

The muscles that occur in the medial aspect of the thigh are adductor muscles. The gracilis is an example of an adductor muscle, as is the **pectineus** and the three **adductor muscles** proper. The pectineus is a small, triangular muscle that is the most proximal of the group. It is found superficial to the adductor brevis muscle and proximal to the adductor longus muscle. Examine the material in the lab and compare the pectineus to figure 15.2.

The **adductor brevis** is the short adductor muscle found deep in the medial aspect of the thigh. It can be seen if the pectineus is reflected. The adductor brevis originates on the pubis and inserts on the proximal third of the thigh.

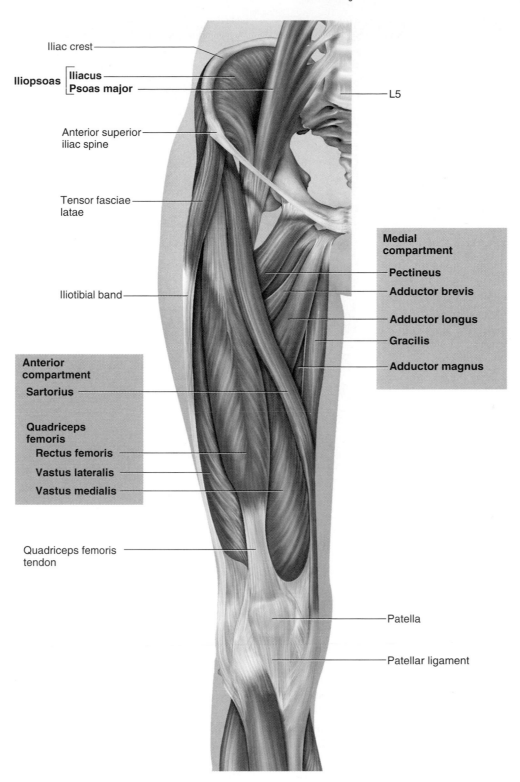

Iliac crest

Iliopsoas [**Iliacus**
Psoas major]

L5

Anterior superior
iliac spine

Tensor fasciae
latae

**Medial
compartment**

Pectineus

Adductor brevis

Adductor longus

Iliotibial band

Gracilis

Adductor magnus

**Anterior
compartment**

Sartorius

**Quadriceps
femoris**

Rectus femoris

Vastus lateralis

Vastus medialis

Quadriceps femoris
tendon

Patella

Patellar ligament

Figure 15.1   Muscles of the thigh, anterior view. ✗

Continued

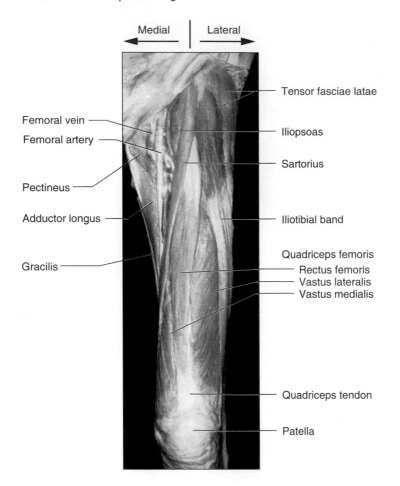

Medial | Lateral

Tensor fasciae latae

Femoral vein

Femoral artery

Iliopsoas

Sartorius

Pectineus

Adductor longus

Iliotibial band

Gracilis

Quadriceps femoris
Rectus femoris
Vastus lateralis
Vastus medialis

Quadriceps tendon

Patella

Figure 15.1—*continued*.

The **adductor longus** also has a pubic origin but its insertion is on the middle third of the thigh. The largest of the adductor muscles is the **adductor magnus,** and it originates along the length of the inferior os coxa (pubis and ischium) and inserts on the distal third of the thigh. Examine these muscles in figure 15.2.

The muscles of the anterior aspect of the thigh belong to the quadriceps group. These muscles all extend the leg, while only one of the group flexes the thigh. The quadriceps muscles all have a common tendon of insertion called the patellar tendon. This tendon inserts on the tibial tuberosity. The most superficial muscle of the group is the **rectus femoris** (figures 15.1 and 15.3). It is a muscle that originates directly inferior to the sartorius. The rectus femoris runs straight down the femur (*rectus* = straight). Because it crosses the hip joint, the rectus femoris is the only quadriceps muscle to flex the thigh.

The three other quadriceps muscles are the vastus muscles. The **vastus lateralis** is so named due to its lateral position. The **vastus intermedius** is deep to the rectus femoris, and the **vastus medialis** is the most medial of the vastus muscles. These muscles originate on the femur, so they do not have an action on the hip. They insert on the tibia and serve

to extend the leg. Locate these muscles in the lab and compare them to figure 15.3.

The **tensor fascia lata** is a lateral muscle of the thigh that attaches to a thick band of connective tissue that runs down the lateral aspect of the thigh like a stripe on a pair of tuxedo pants. This muscle acts to abduct the thigh. Examine the material in the lab and compare it to figure 15.4.

The gluteus muscles in humans are different than in other mammals due to our bipedal stance. The largest of the gluteal muscles is the **gluteus maximus.** It extends the thigh in standing from a sitting position or in actions such as climbing up stairs. The **gluteus medius** and the **gluteus minimus** both aid in keeping balance during walking in that they maintain the center of gravity. The gluteus medius is superficial to the gluteus minimus, but they both originate on the outer surface of the ilium. When you raise one leg to walk, gravity pulls the body in the direction of the lifted leg. The gluteus medius and minimus on the opposite side of the body contract to maintain upright posture. Compare these muscles to figure 15.4.

The last group of hip and thigh muscles covered in this exercise are the hamstring muscles. The hamstring muscles

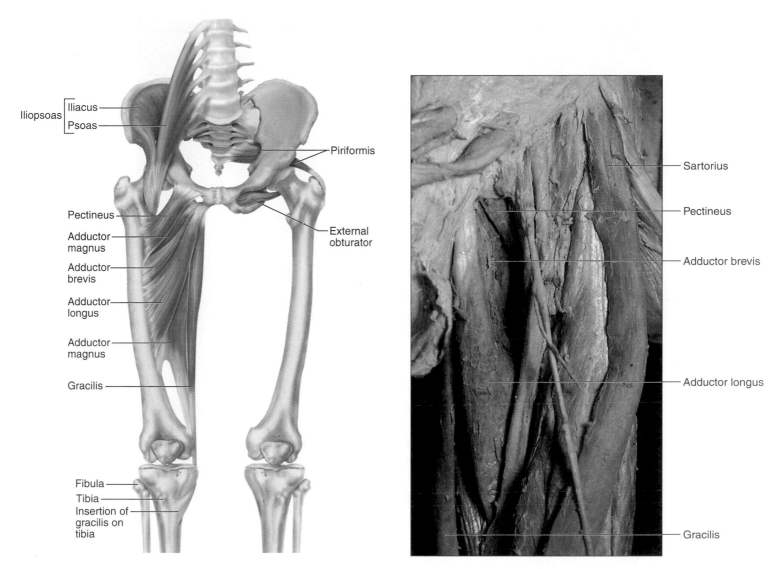

Figure 15.2   Iliopsoas and adductor muscles of the thigh, anterior view. ✗

are so named because these muscles from a pig were once tied together by their tendons (the hamstrings) and hung in a smokehouse to cure. Three muscles make up the hamstrings, the biceps femoris, the semitendinosus, and the semimembranosus. All three of these muscles cross the hip joint and serve as extensor muscles of the thigh in walking. The three muscles also serve as walking muscles in that they flex the leg.

The **biceps femoris** is literally named the "two-headed muscle of the femur." One head originates, with the other hamstring muscles, on the ischial tuberosity, while the second head originates on the femur. The biceps femoris is a muscle that inserts laterally on the proximal leg. The **semitendinosus** also originates on the ischial tuberosity, and it can be distinguished by the long, pencil-like distal tendon. As opposed to the biceps femoris the semitendinosus inserts medially. The **semimembranosus** also originates on the ischial

tuberosity and has a broad, flat membranous tendon on the proximal part of the muscle. The semimembranosus inserts medially on the leg along with the semitendinosus. Examine these muscles in figure 15.5.

Cat Musculature

The cat has many muscles of the hip and thigh that serve as good models for studying human muscles. The quadrupedal nature of the cat lends itself to having a different size or placement of some of the thigh muscles, which are illustrated in this section. Be careful in the dissection of the cat muscles that you *leave the blood vessels and nerves intact* as you dissect your specimen. You will be looking at these structures in later exercises.

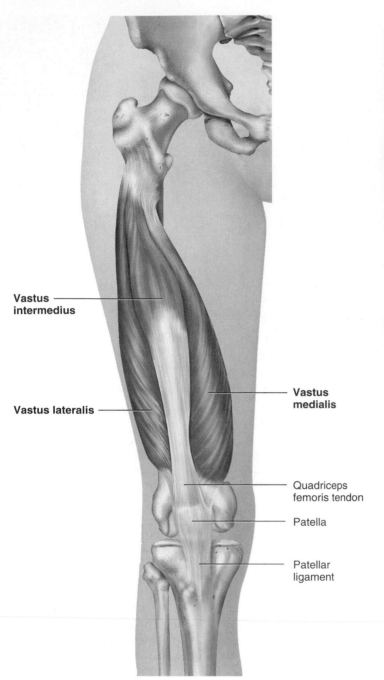

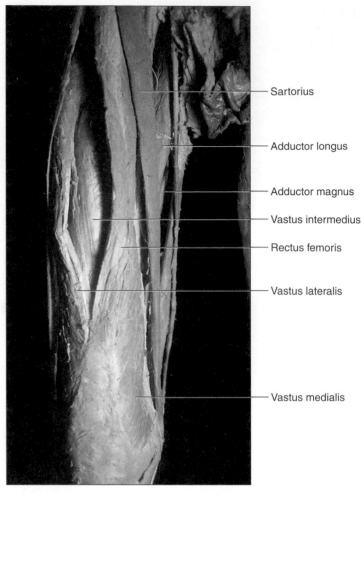

Figure 15.3   Quadriceps muscles of the thigh, anterior view.

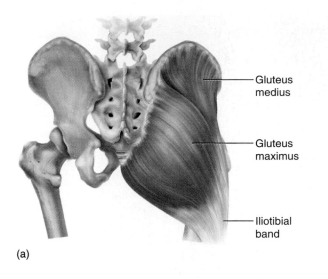

Gluteus medius

Gluteus maximus

Iliotibial band

(a)

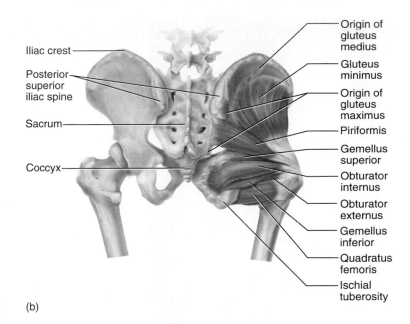

Iliac crest

Posterior superior iliac spine

Sacrum

Coccyx

Origin of gluteus medius

Gluteus minimus

Origin of gluteus maximus

Piriformis

Gemellus superior

Obturator internus

Obturator externus

Gemellus inferior

Quadratus femoris

Ischial tuberosity

(b)

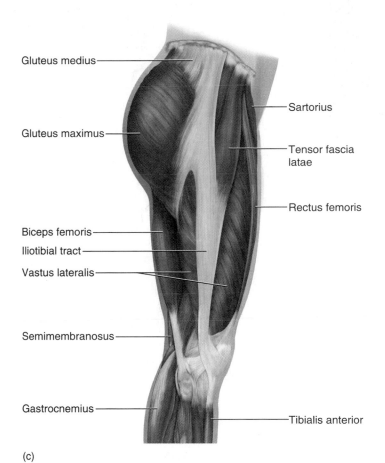

Gluteus medius

Gluteus maximus

Biceps femoris

Iliotibial tract

Vastus lateralis

Semimembranosus

Gastrocnemius

Sartorius

Tensor fascia latae

Rectus femoris

Tibialis anterior

(c)

Figure 15.4   Muscles of the hip and thigh. Diagram (a) superficial muscles, posterior view; (b) deep muscles, posterior view; (c) lateral muscles. Photograph (d) superficial muscles of the hip; (e) deep muscles of the hip. ⚵   *Continued*

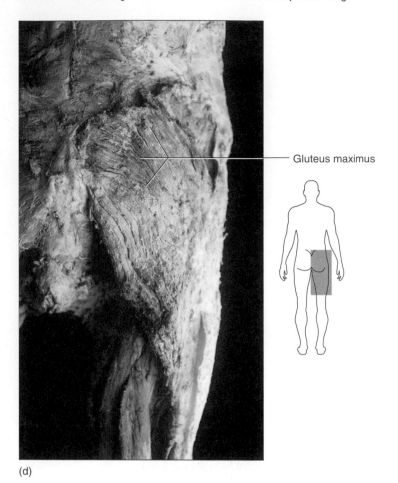

(d)

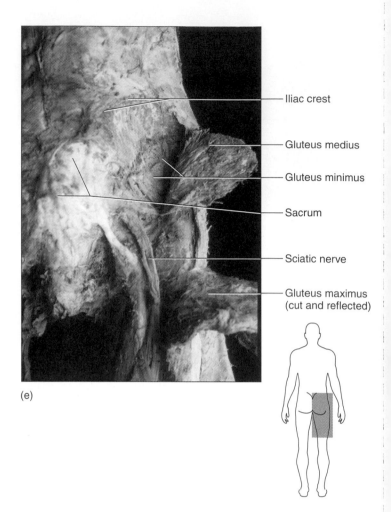

(e)

Figure 15.4—*continued.*

If you have not already done so, remove the skin from the lower limb of the cat. Pay particular attention to the **great saphenous vein,** which runs under the skin and should be preserved. Remove any excess fat and fascia from the muscles as you dissect the material. If you need to cut a muscle, bisect it perpendicular to the fiber direction so that half of it is attached to the origin side and half of it is attached to the insertion side.

Medial Muscles of the Thigh of the Cat

Two major muscles on the medial aspect of the thigh in the cat are the **sartorius** and the **gracilis** muscles. Locate these muscles in figure 15.6 and note that they are much more broad in the cat than in the human.

Cut through the sartorius and gracilis and reflect the ends of these muscles. You should be able to see the deeper muscles of the thigh, including the **vastus medialis,** the **adductor femoris** (a large muscle specific to the cat), and the **semimembranosus.** Locate the external portion of the **iliopsoas** muscle, which is on the medial aspect of the thigh. The distal portion of the **semitendinosus** can also be seen from this

view. Examine figure 15.6 for the deep muscles of the medial thigh and locate the **rectus femoris, vastus medialis,** and **vastus lateralis.** Move the rectus femoris in order to see the **vastus intermedius** muscle. The **adductor longus** is a thin muscle anterior to the **adductor femoris,** and the **pectineus** can also be seen as a small muscle on the medial aspect of the thigh.

Lateral Muscles of the Thigh of the Cat

On the lateral aspect of the thigh are numerous muscles, including the **biceps femoris,** the **tensor fascia lata,** the **gluteus muscles,** and a muscle specific to the cat, the **caudofemoralis.** The biceps femoris is the largest and most lateral muscle of the thigh. Bisect the biceps femoris and reflect the muscle. The **gluteus medius** is larger than the **gluteus maximus** in cats due to the lengthening of the pelvic girdle. The **semimembranosus** is a large muscle in cats (much larger than the semitendinosus), and it can be seen from the lateral side of the cat. The adductor magnus and adductor brevis are not found in the cat. Examine these muscles in figure 15.7.

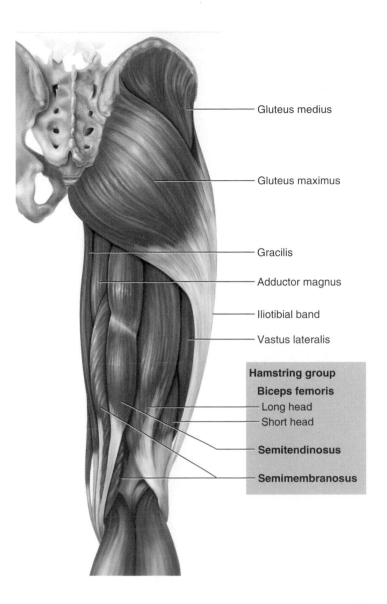

Gluteus medius

Gluteus maximus

Gracilis

Adductor magnus

Iliotibial band

Vastus lateralis

Hamstring group

Biceps femoris
Long head
Short head

Semitendinosus

Semimembranosus

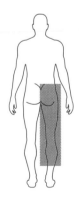

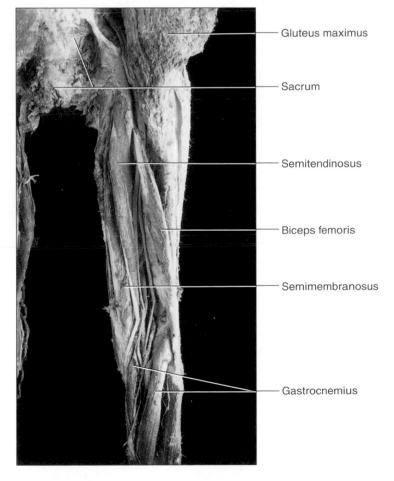

Gluteus maximus

Sacrum

Semitendinosus

Biceps femoris

Semimembranosus

Gastrocnemius

Figure 15.5   Muscles of the thigh, posterior view.

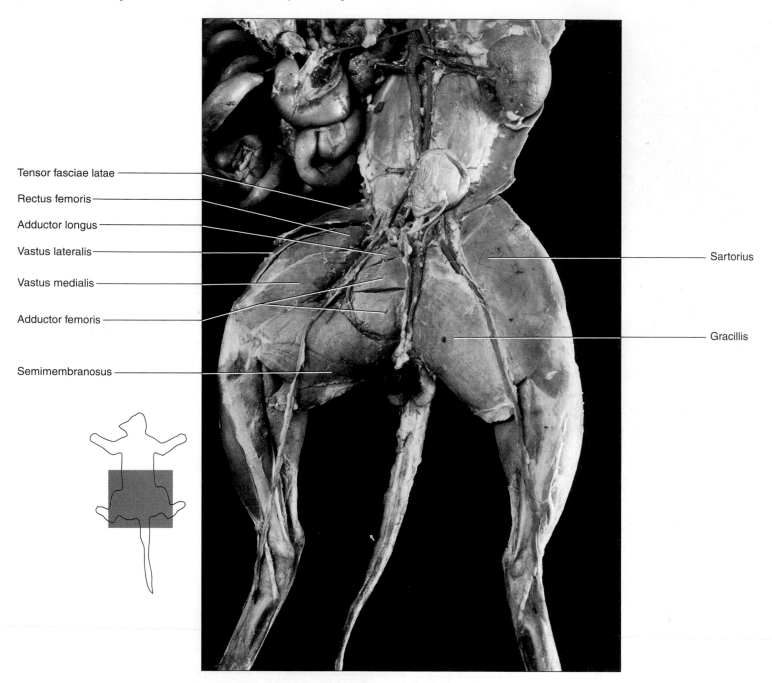

Tensor fasciae latae

Rectus femoris

Adductor longus

Vastus lateralis

Vastus medialis

Adductor femoris

Semimembranosus

Sartorius

Gracillis

Figure 15.6   Thigh muscles of the cat. Superficial muscles (left side of cat); deep muscles (right side of cat). ✕

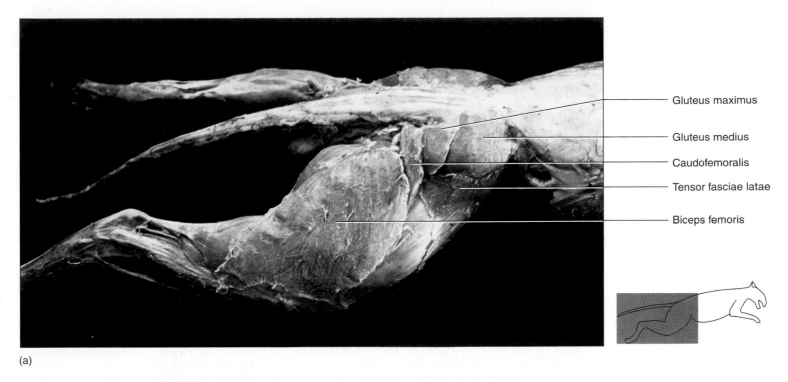

(a)

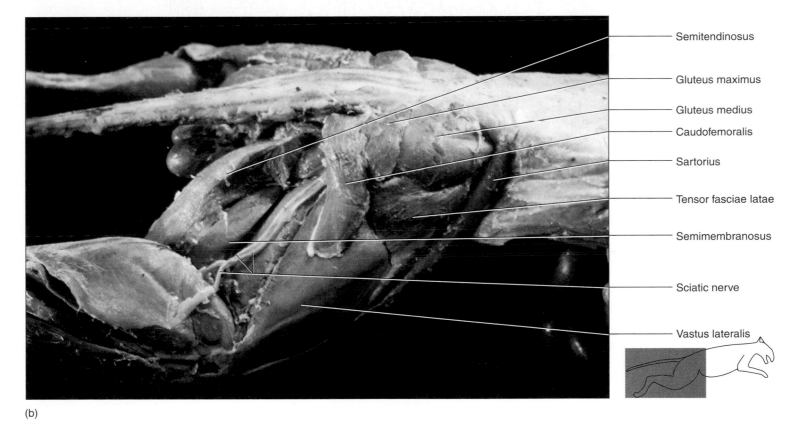

(b)

Figure 15.7   Lateral thigh muscles of the cat. (a) Superficial muscles of the right side; (b) deep muscles of the right side.

1. If you were to ride a horse, what muscles would you use to keep your seat out of the saddle as you ride?

2. How do the gluteus medius and gluteus minimus prevent you from toppling over as you walk?

3. What is a muscle that is an antagonist to the biceps femoris muscle?

4. What are two muscles that are synergists with the biceps femoris muscle?

5. Are all of the hamstring muscles identical in action? What is the action of the hamstring muscles?

6. What is the insertion of all the muscles of the quadriceps group?

7. How does the action of the rectus femoris differ from all the other quadriceps muscles?

8. How many adductor muscles are there?

9. List two muscles in this exercise that are responsible for thigh flexion.

10. Where do the hamstring muscles originate as a group?

11. What is the action of the vastus lateralis?

12. Which muscle group is found on the anterior part of the thigh?

13. Is abduction of the thigh movement away from or toward the midline?

14. What muscle flexes the lumbar vertebrae as part of its action?

15. Label the muscles in the following illustration.

a. _____

b. _____

c. _____

d. _____

e. _____

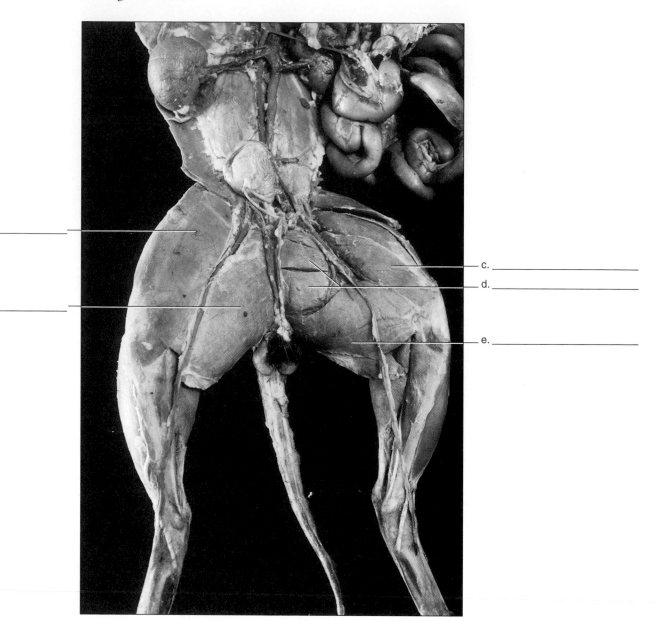

Muscles of the Leg and Foot

Introduction

In this exercise you will study the major muscles of the leg and foot. The muscles of the leg and foot are primarily muscles of locomotion. They originate typically from the femur, tibia, or fibula and insert on the bones of the foot. Actions on the leg are either flexion or extension, whereas actions on the foot are **dorsiflexion** (decreasing the angle between the shin and dorsum or top of the foot), **plantar flexion** (an action that is seen as you stand on your toes to reach something high). Review the other actions of the leg and foot in Exercise 13. Some of the muscles of the feet are analogous to those of the hand, such as the extensor digitorum muscles.

Objectives

At the end of this exercise you should be able to

1. locate the muscles of the leg and foot on a torso model, chart, cadaver (if available), or cat (if available);
2. list the origin, insertion, and action of each muscle presented;
3. describe what nerve controls each muscle if your instructor includes innervation with the muscles;
4. list what muscles function as synergists or antagonists to the prime mover;
5. name all of the muscles that have an action on a joint, such as all of the muscles that flex the leg.

Materials

Human torso model
Human leg models
Human muscle charts
Articulated skeleton
Cadaver (if available)
Cat (if available)
Materials for cat dissection
    Dissection trays
    Scalpel or razor blades
    Gloves
    Blunt probe
    String
    Pins

Procedure

Examine the models and charts in the lab as you study the following descriptions. If you have a cadaver, pay particular attention to the origins and the tendons of insertion of the muscles of the leg and foot. Table 16.1 provides the details of the muscles in this exercise, while the text descriptions serve to point out major features of the muscles. When you are studying the origins and insertions of the muscles, it helps to have an articulated skeleton so you can locate the bony markings as you study the muscle attachments. Review the muscle nomenclature and the actions of muscles as outlined in Laboratory Exercises 12 and 13. The actions of the muscles will be easier when you have identified the origins and the insertions.

Posterior Muscles

The **gastrocnemius** is a thin calf muscle that is the most superficial of the posterior leg group. The gastrocnemius crosses the knee joint and so it flexes the leg. It inserts on the calcaneus by way of the calcaneal tendon, thus serving to plantar flex the foot as well. The **calcaneal tendon** is also known as the Achilles tendon. Deep to the gastrocnemius is the **soleus** muscle. Unlike the gastrocnemius, the soleus does not originate on the femur, therefore it does not cross the knee and has no action on the leg. It inserts on the calcaneus, sharing the calcaneal tendon with the gastrocnemius, and it plantar flexes the foot. The **popliteus** is a small muscle that crosses the knee joint. It serves to unlock the knee joint. These muscles can be seen in figure 16.1.

The **tibialis posterior** is a major muscle deep to the soleus that plantar flexes and inverts the foot. The tendon of the muscle runs along the medial aspect of the ankle. Near the tibialis posterior is the **flexor digitorum longus,** which is a muscle that inserts on the distal phalanges of all of the digits of the foot except for the hallux. The flexor digitorum longus has an action of flexing (curling) the toes in addition to plantar flexing and inverting the foot. The **flexor hallucis longus** flexes the hallux and also aids the flexor digitorum longus in inverting the foot. Locate these muscles in the lab and compare them to figure 16.1.

Anterior Muscles

The **tibialis anterior** is located just lateral to the crest of the tibia on the anterior side of the leg. The tendon of the tibialis

Table 16.1   Muscles of the Leg and Foot

Name	Origin	Insertion	Action	Innervation
Gastrocnemius	Condyles of femur	Calcaneus by the calcaneal tendon	Flexes leg, plantar flexes foot	Tibial nerve
Soleus	Posterior, proximal tibia and fibula	Calcaneus by the calcaneal tendon	Plantar flexes foot	Tibial nerve
Popliteus	Lateral condyle of femur	Proximal tibia	Flexes leg, thus "unlocking" knee	Tibial nerve
Tibialis posterior	Proximal portion of tibia and fibula	Second through fourth metatarsals and some tarsals	Plantar flexes and inverts foot	Tibial nerve
Flexor digitorum longus	Posterior tibia	Distal phalanges of second through fifth digits	Flexes toes, plantar flexes and inverts foot	Tibial nerve
Flexor hallucis longus	Mid shaft of fibula	Distal phalanx of hallux (big toe)	Flexes hallux, inverts foot	Tibial nerve
Tibialis anterior	Lateral condyle and proximal tibia	First metatarsal and first cuneiform	Dorsiflexes and inverts foot	Deep peroneal nerve
Extensor digitorum longus	Lateral condyle of tibia, shaft of fibula	Middle and distal phalanges of second through fifth digits	Extends toes, dorsiflexes foot	Deep peroneal nerve
Extensor hallucis longus	Anterior shaft of fibula	Distal phalanx of hallux (big toe)	Extends hallux, dorsiflexes foot	Deep peroneal nerve
Peroneus longus	Head and shaft of fibula, lateral condyle of tibia	First metatarsal, first cuneiform	Plantar flexes and everts foot	Superficial peroneal nerve
Peroneus brevis	Shaft of fibula	Fifth metatarsal	Plantar flexes and everts foot	Superficial peroneal nerve
Peroneus tertius	Anterior, distal fibula	Dorsum of the fifth metatarsal	Dorsiflexes and everts foot	Deep peroneal nerve

anterior crosses to the medial side of the foot and inserts on the first metatarsal and first cuneiform. As the tibialis anterior contracts, it decreases the angle between the anterior tibial crest and the dorsum of the foot in an action known as dorsiflexion of the foot. Because the insertion of this muscle is medial, it also inverts the foot.

The **extensor digitorum longus** muscle is lateral to the tibialis anterior, and the tendons of this muscle splay out and insert on the middle and distal phalanges of all of the digits of the foot except the hallux. The **extensor hallucis longus** extends the hallux as well as acting as a synergist to the extensor digitorum longus in dorsiflexion of the foot. Examine these muscles in the lab and compare them to figure 16.2.

The peroneus muscles are named for originating on and running the length of the peroneus (fibula). The **peroneus longus** has a tendon that travels from the lateral side of the foot underneath to the medial side, crossing under the arch of the foot. The **peroneus brevis** parallels the peroneus longus except that the tendon stops short and inserts on the lateral side of the foot at the fifth metatarsal.

Both of these muscles have tendons that hook posterior to the lateral malleolus, and they both plantar flex and evert the foot. The final peroneus muscle, the **peroneus tertius**, does not arch behind the lateral malleolus, and it inserts on the dorsum of the fifth metatarsal. When it contracts it dorsiflexes the foot while also everting the foot. Examine these muscles in figure 16.3.

Cat Dissection

If you have not done so already, remove the skin from the lower portion of the hind limb of the cat. Be careful cutting the skin from the distal portions of the leg so you do not cut through the tendons of the foot.

Posterior Muscles

Examine the large **gastrocnemius** on the posterior aspect of the leg. The **soleus** is deeper to the gastrocnemius and inserts communally with the gastrocnemius on the calcaneus. In cats the soleus has only one point of origin, the fibula, while in hu-

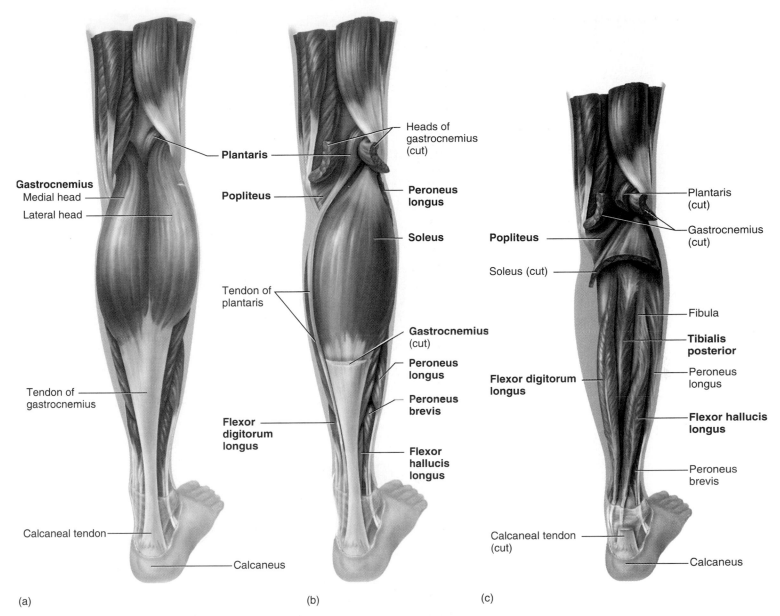

Figure 16.1   Muscles of the leg and foot, posterior view. (a) Superficial muscles of the leg; (b) middle level muscles of the leg; (c) deep muscles of the leg. ✗

mans there are two bones of origin. Examine figure 16.4 with your dissection. Cut through the calcaneal tendon and lift the gastrocnemius and soleus so that you can study the underlying muscles.

The **popliteus** is a small, triangular muscle that crosses the knee joint. Do not damage the nerves and blood vessels that pass over the popliteus because you will study them in the subsequent exercises.

The **flexor digitorum longus** is a muscle that runs along the medial side of the hind limb and flexes the digits of the cat. It joins with the **flexor hallucis longus,** which inserts on all of the digits of the hind limb. The **tibialis posterior** is a narrow muscle that inserts on the tarsal bones of the foot. Locate these muscles in the cat and compare them to figure 16.5.

Anterior Muscles

The **tibialis anterior** is a large muscle of the lower limb and inserts on the dorsum of the foot. The **extensor digitorum longus** originates on the femur in cats and inserts on the distal phalanges in all of the digits in the cat. The extensor hallucis longus is not found in cats. These muscles can be seen in figure 16.6. Examine these muscles and isolate them in your dissection.

The **peroneus longus** extends along the length of the fibula with the **peroneus brevis.** The peroneus longus inserts at the base of the metatarsals, while the peroneus brevis inserts on the fifth metatarsal. The **peroneus tertius** inserts into the tendon of the extensor digitorum muscle. These muscles can be seen in figure 16.7.

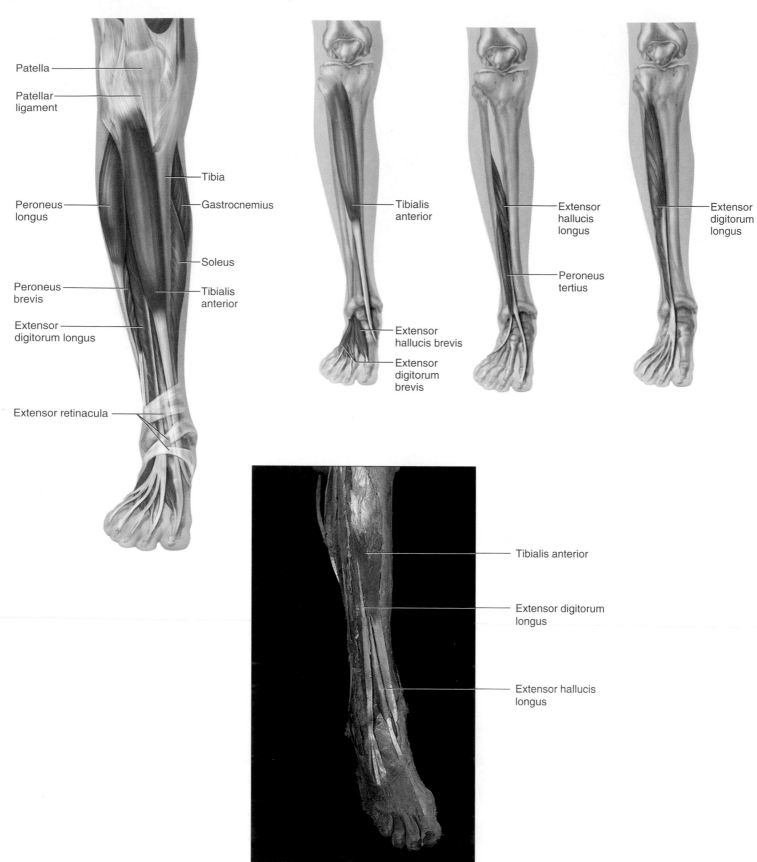

Patella

Patellar ligament

Tibia

Gastrocnemius

Peroneus longus

Soleus

Peroneus brevis

Tibialis anterior

Extensor digitorum longus

Extensor retinacula

Tibialis anterior

Extensor hallucis brevis

Extensor digitorum brevis

Extensor hallucis longus

Peroneus tertius

Extensor digitorum longus

Tibialis anterior

Extensor digitorum longus

Extensor hallucis longus

Figure 16.2   Muscles of the leg and foot, anterior view. ɣ

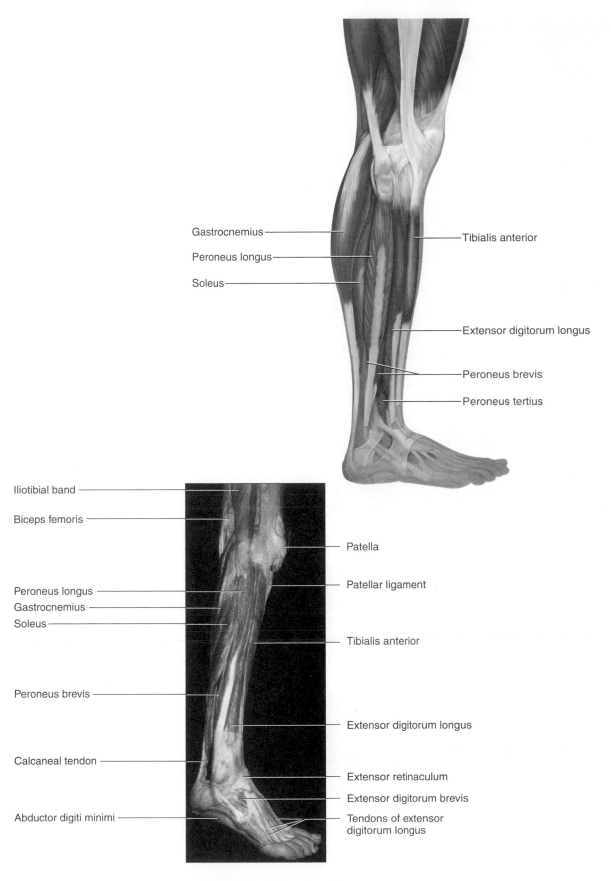

Figure 16.3   Muscles of the leg and foot, lateral view.

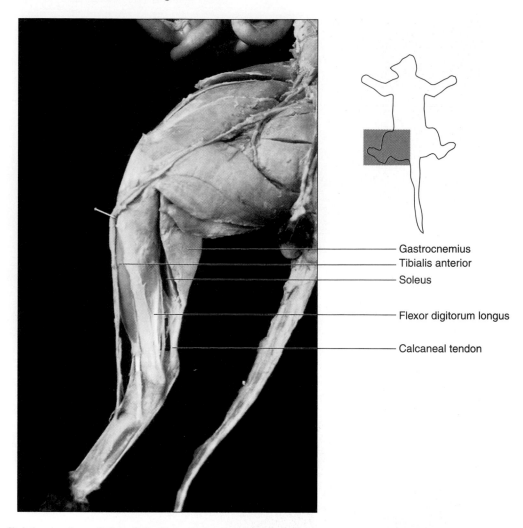

- Gastrocnemius
- Tibialis anterior
- Soleus
- Flexor digitorum longus
- Calcaneal tendon

Figure 16.4   Superficial muscles of the right leg of the cat, medial view.

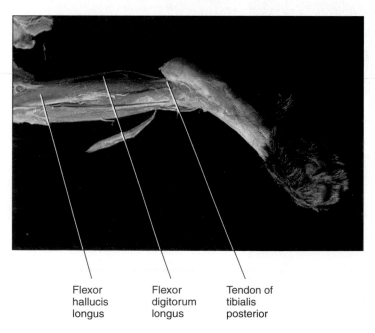

Flexor hallucis longus   Flexor digitorum longus   Tendon of tibialis posterior

Figure 16.5   Deep muscles of the left leg of the cat, lateral view.

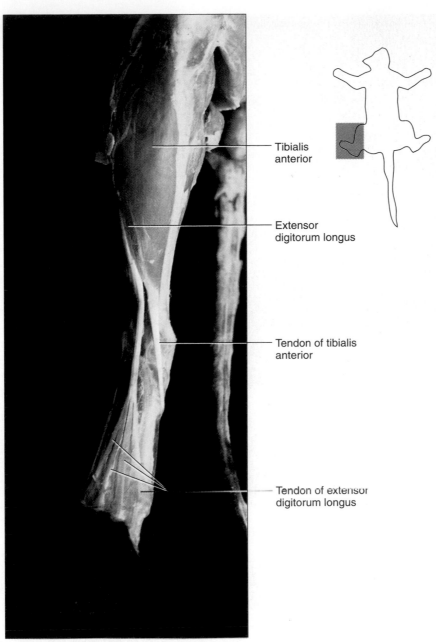

Tibialis
anterior

Extensor
digitorum longus

Tendon of tibialis
anterior

Tendon of extensor
digitorum longus

Figure 16.6   Muscles of the right leg of the cat, anterior view.

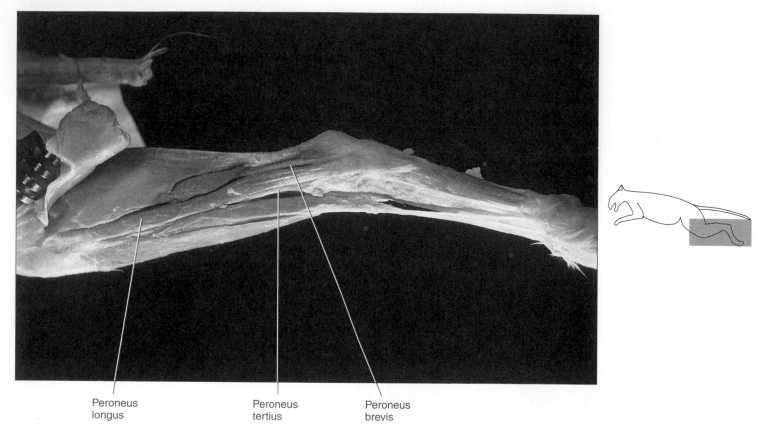

Peroneus
longus

Peroneus
tertius

Peroneus
brevis

Figure 16.7   Muscles of the left leg of the cat, lateral view.

Name _____

1. What is the origin of the gastrocnemius?

2. What is the insertion of the tibialis anterior in humans?

3. How does the action of the peroneus longus in humans differ from that of the peroneus tertius?

4. What is the action of the extensor hallucis longus?

5. Fill in the following illustration.

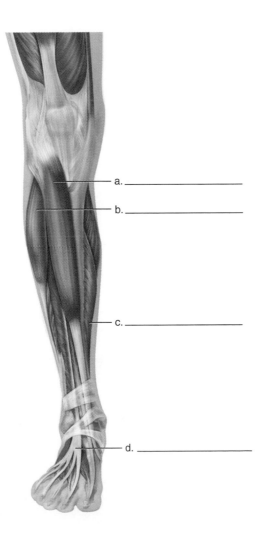

a. _____

b. _____

c. _____

d. _____

6. The calf is made of what two major muscles?

7. Plantar flexion occurs by what two muscles?

8. What muscle has the action of extendng the toes?

9. Name a muscle in this exercise that dorsiflexes the foot.

10. Plantar flexion and eversion of the foot occurs by what muscle?

11. What is the insertion of the soleus?

12. What is the action of the tibialis anterior?

13. What is the insertion of the peroneus tertius muscle?

14. What is the insertion of the flexor digitorum longus?

15. Label the muscles in the following illustration.

a. _____

b. _____

Calcaneal tendon

c. _____

d. _____

e. _____

Muscles of the Head and Neck

Introduction

In this exercise, you learn the major muscles of the head and neck. They are numerous and can be grouped into a few functional classifications. Some of the muscles serve for chewing food or help manipulate the food in the mouth so that chewing or swallowing can occur. Other muscles function in facial expression or in the closing of the eyes or mouth. Still others serve to move the head or neck. Finally, some muscles move the hyoid or the larynx in speech or swallowing.

Objectives

At the end of this exercise you should be able to

1. locate the muscles of the head and neck on a torso model, chart, cadaver (if available), or cat (if available);
2. list the origin, insertion, and action of each muscle presented;
3. describe what nerve controls each muscle if your instructor includes innervation with the muscles;
4. name all of the muscles that have an action on a joint, such as all of the muscles that flex the head;
5. reproduce the actions of select muscles on your own head or neck.

Materials

Human torso model or head and neck model
Human muscle charts
Articulated skeleton
Cadaver (if available)
Cat (if available)
Materials for cat dissection

   Dissection trays
   Scalpel or razor blades
   Gloves
   Blunt probe
   String
   Pins

Procedure

Review the muscle nomenclature and the actions as outlined in Laboratory Exercises 12 and 13. Examine a model or chart of the human musculature and cadaver (if available) as you read the following descriptions. The details of the muscles are listed in table 17.1. Examine a skull or articulated skeleton and review the bony markings as you study the origins and insertions of the muscles in this exercise. Once you know the origins or insertions the actions should be more comprehensible.

Muscles of the Neck

The **levator scapulae** is named for what it does, elevate the scapula. The levator scapulae originates on the lateral side of the neck and inserts on the upper scapula. If the neck is fixed, the levator scapulae raises the scapulae, as in shrugging. If the scapula is fixed it rotates or abducts the neck.

The **scalenus** muscles are found on the lateral side of the neck. They are bounded by the sternocleidomastoid in the front and the levator scapula in the back. They serve to rotate the neck or elevate the ribs. Examine figure 17.1 for these muscles.

The **sternocleidomastoid** muscle rotates the head in a unique way. Place your hand on the right sternocleidomastoid and turn your head to the right. Notice how the muscle does not contract. Now turn your head to the left, and you can feel the muscle contract. The right sternocleidomastoid turns the head to the left and the left sternocleidomastoid turns the head to the right.

The **sternohyoid, sternothyroid,** and **omohyoid** are all named for their origins and insertions. In the case of the sternohyoid and sternothyroid, the sternum anchors the stable part of the muscle (the origin), and the hyoid and thyroid move when the muscle contracts. In the case of the omohyoid, the term *omo* means shoulder, and the scapula anchors the stable part of the muscle. The hyoid is depressed when the omohyoid contracts. Examine figure 17.2 for muscles of the anterior neck.

The **platysma** is a broad, thin muscle that has a soft origin (on the fascia of the pectoral and deltoid muscles). It inserts on the mandible and skin of the lips and can be seen if you elevate your chin and subsequently pout. The thin wings that stick out on the side of your neck are the edges of the platysma muscle. Examine the platysma in figure 17.3 and in the models or on the cadaver in the lab.

Muscles of the Head

The **digastric** is so named because it is a muscle with two bellies. Few muscles open the mandible. The digastric is one that does. In addition to this the digastric has significant action on the hyoid, which is important in tongue movement for speech and swallowing. Deep to the digastric is the **mylohyoid,** a

Table 17.1 Muscles of the Neck and Head

Name	Origin	Insertion	Action	Innervation
Neck				
Levator scapulae	C1–4	Upper vertebral border of scapula	Elevates scapula, abducts and rotates neck	Spinal nerves C3–5 and dorsal scapular nerve
Scalenus (anterior, middle, and posterior)	Transverse process of cervical vertebrae	Ribs 1 and 2	Flexes and rotates neck, elevates ribs 1 and 2	Spinal nerves C4–8
Sternocleidomastoid	Sternum, clavicle	Mastoid process of temporal	Abducts, rotates, and flexes head	Spinal nerves C2–4, accessory nerve (XI)
Sternohyoid	Manubrium of sternum	Hyoid	Depresses hyoid	Spinal nerves C1–3
Sternothyroid	Manubrium of sternum	Thyroid cartilage of larynx	Depresses thyroid cartilage	Spinal nerves C1–3
Omohyoid	Superior surface of scapula	Hyoid	Depresses hyoid	Spinal nerves C1–3
Platysma	Fascia covering pectoralis major and deltoid	Mandible and skin of lower region of face	Depresses lower lip, opens jaw	Facial nerve (VII)
Digastric	Inferior, distal margin of mandible (anterior belly), mastoid notch of temporal (posterior belly)	Hyoid	Elevates, protracts, retracts hyoid, opens mandible	Trigeminal (V) and facial (VII) nerves
Mylohyoid	Inner, inferior margin of mandible	Body of hyoid and median raphe	Elevates hyoid and tongue	Trigeminal nerve (V)
Head				
Frontalis	Galea aponeurotica	Skin superior to orbit	Raises eyebrows, draws scalp anteriorly	Facial nerve (VII)
Occipitalis	Occipital and temporal bone	Galea aponeurotica	Draws scalp posteriorly	Facial nerve (VII)
Temporalis	Temporal fossa	Coronoid process and ramus of mandible	Closes mandible	Trigeminal nerve (V)
Masseter	Zygomatic arch	Angle and ramus of mandible	Closes mandible	Trigeminal nerve (V)
Pterygoids (medial and lateral)	Pterygoid processes of sphenoid bone	Medial ramus of mandible	Medial gliding of mandible (for chewing)	Trigeminal nerve (V)
Orbicularis oculi	Frontal and maxilla on medial margin of orbit	Skin of eyelid	Closes eyelid	Facial nerve (VII)
Orbicularis oris	Fascia of facial muscles near mouth	Skin of lips	Closes lips	Facial nerve (VII)
Corrugator (supercilii)	Medial portion of frontal bone	Skin of eyebrows	Adducts eyebrows (pulls them medially)	Facial nerve (VII)
Risorius	On fascia of masseter	Skin at angle of mouth	Abducts corner of mouth (draws edge of mouth lateral)	Facial nerve (VII)
Mentalis	Anterior mandible	Skin of chin below lower lip	Protrudes lower lip	Facial nerve (VII)
Buccinator	Maxilla and mandible near molar teeth	Orbicularis oris	Compresses cheek	Facial nerve (VII)
Zygomaticus	Zygomatic bone	Muscle and skin at angle of mouth	Elevates corners of mouth (in smiling and laughing)	Facial nerve (VII)
Depressor labii inferioris	Lateral mandible	Muscles and skin of lower lip	Depresses lower lip	Facial nerve (VII)
Levator labii superioris	Maxilla and zygomatic bones	Muscle and skin of lips	Elevates upper lip, flares nostril	Facial nerve (VII)

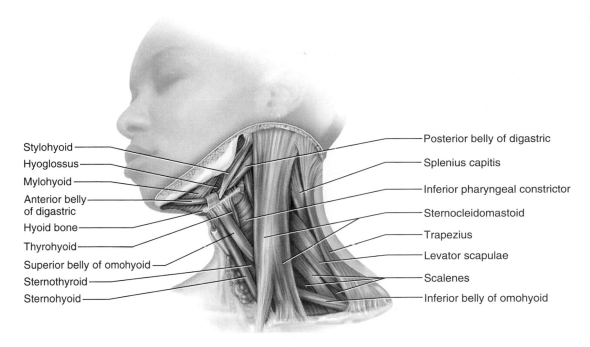

Stylohyoid
Hyoglossus
Mylohyoid
Anterior belly of digastric
Hyoid bone
Thyrohyoid
Superior belly of omohyoid
Sternothyroid
Sternohyoid

Posterior belly of digastric
Splenius capitis
Inferior pharyngeal constrictor
Sternocleidomastoid
Trapezius
Levator scapulae
Scalenes
Inferior belly of omohyoid

Figure 17.1   Levator scapulae and scalenus muscles, lateral view. ✄

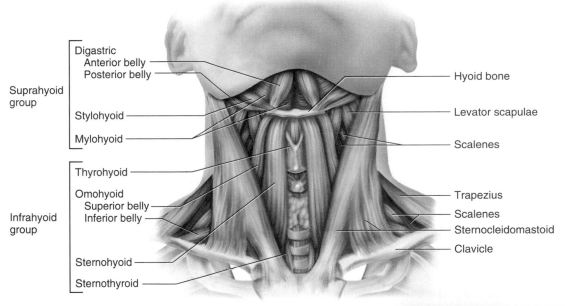

Suprahyoid group

Digastric
 Anterior belly
 Posterior belly

Stylohyoid

Mylohyoid

Infrahyoid group

Thyrohyoid

Omohyoid
 Superior belly
 Inferior belly

Sternohyoid

Sternothyroid

Hyoid bone

Levator scapulae

Scalenes

Trapezius
Scalenes
Sternocleidomastoid
Clavicle

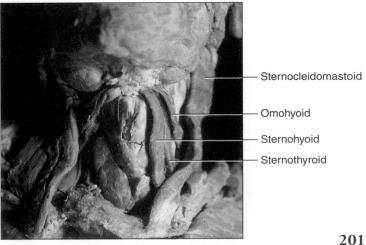

Sternocleidomastoid

Omohyoid

Sternohyoid

Sternothyroid

Figure 17.2   Muscles of the neck, anterior view. ✄

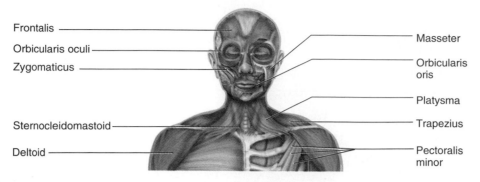

Frontalis

Orbicularis oculi

Zygomaticus

Sternocleidomastoid

Deltoid

Masseter

Orbicularis oris

Platysma

Trapezius

Pectoralis minor

Figure 17.3   Platysma, anterior view. ⚘

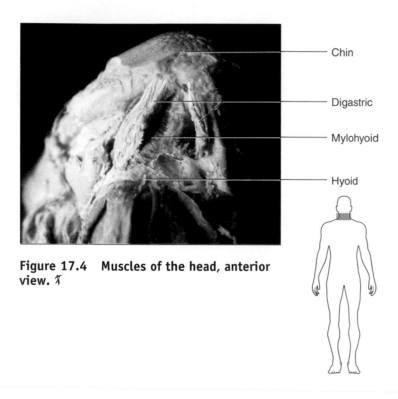

Chin

Digastric

Mylohyoid

Hyoid

Figure 17.4   Muscles of the head, anterior view. ⚘

broad muscle of the floor of the mouth that aids in pushing the tongue superiorly when swallowing. These muscles can be seen in figures 17.2 and 17.4.

The **frontalis** muscle attaches to a broad, flat tendinous sheet on the superior aspect of the skull known as the **galea aponeurotica.** The insertion of the frontalis is on the eyebrow region. If the galea is fixed to the posterior of the skull, then the eyebrows are raised. If the galea is not so anchored, then the scalp is brought forward as in frowning. The **occipitalis** is a functional continuation of the frontalis muscle in that both muscles attach to the galea aponeurotica. The occipitalis pulls the galea aponeurotica posteriorly, and the scalp attached to the galea goes with it. These muscles are sometimes discussed as one muscle, the occipitofrontalis. Examine these muscles in figure 17.5.

The **temporalis** is a powerful muscle that closes the jaw. The temporal fossa is so named because it is a depression medial to the zygomatic arch (though when you examine the skull the

temporal fossa appears slightly domed). The temporalis muscle runs underneath the zygomatic arch and inserts on the coronoid process and on the superior and medial ramus of the mandible.

The **masseter** is a large muscle of the head whose synergist is the temporalis. It acts to powerfully close the jaws. If you place your fingers on the ramus of the mandible and clench your teeth, you can feel the masseter tighten.

The **pterygoid** muscles are deep muscles that originate on the sphenoid bone and insert laterally on the mandible. In this way they act to pull the jaw horizontally, which helps in rotatory chewing. The temporalis and masseter serve to open and close the jaw, while the pterygoids provide the sideways movement so characteristic of a person chewing gum. These muscles can be seen in figures 17.5 and 17.6.

The **orbicularis oculi** and the **orbicularis oris** are sphincter muscles that close orifices. Sphincter muscles act similarly to the strings of a drawstring purse. The orbicularis oculi has a medial, bony origin and an insertion on the eyelid. The muscles ring the eye and close the eyelids. The orbicularis oris originates on fascia and facial muscles near the mouth and inserts on the skin of the lips, thus closing the lips. Examine the facial muscles in figure 17.7.

The **corrugator** muscle has a medial point of origin between the eyebrows and inserts laterally. In this way it serves to furrow the eyebrows.

The **zygomaticus** elevates the corners of the mouth by pulling them superiorly and laterally, as in smiling or laughing. It is named for its origin on the zygomatic bone. The **risorius** is known as the laughing muscle because it pulls the lips laterally. It does not have a bony point of origin but attaches to the fascia of the masseter (figure 17.7).

The **mentalis** originates on the chin (anterior mandible) and is another of the pouting muscles. The **buccinator** muscle of the cheek runs in a horizontal direction. It puckers the cheeks as in trumpet playing and serves to push food towards the molars in chewing.

The **depressor labii inferioris** pulls the lower corners of the mouth inferiorly when pouting. It is named for its action (depressing the lower lip), while the **levator labii superioris** muscle raises the skin of the lips and expands the nostrils, as in the expression of showing extreme disgust. These muscles are seen in figure 17.7.

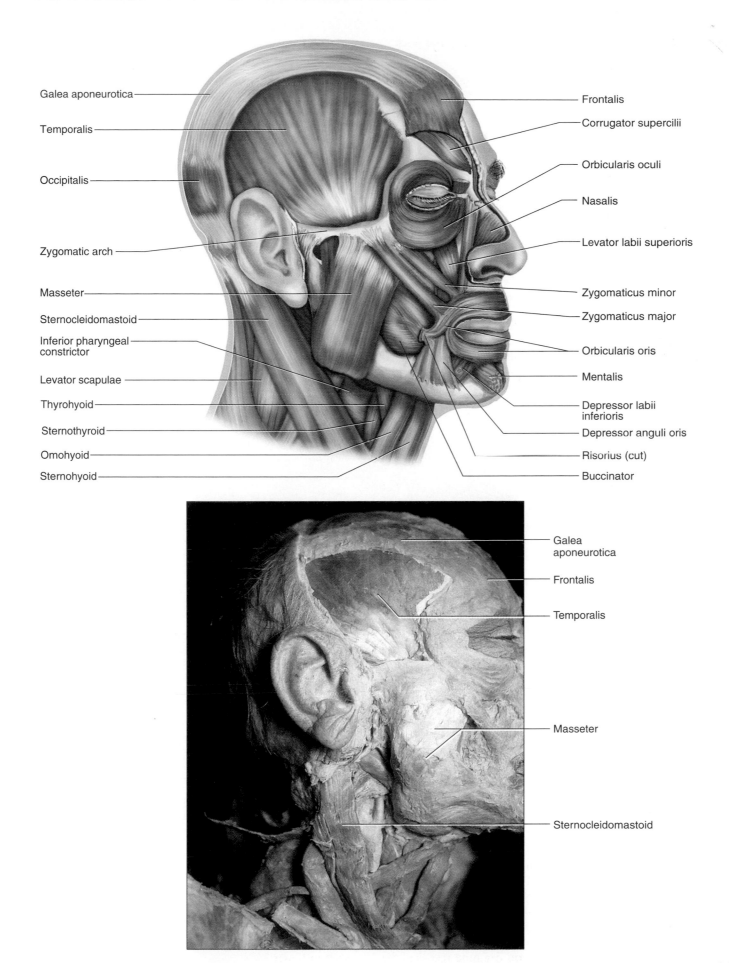

Galea aponeurotica

Temporalis

Occipitalis

Zygomatic arch

Masseter

Sternocleidomastoid

Inferior pharyngeal constrictor

Levator scapulae

Thyrohyoid

Sternothyroid

Omohyoid

Sternohyoid

Frontalis

Corrugator supercilii

Orbicularis oculi

Nasalis

Levator labii superioris

Zygomaticus minor

Zygomaticus major

Orbicularis oris

Mentalis

Depressor labii inferioris

Depressor anguli oris

Risorius (cut)

Buccinator

Galea aponeurotica

Frontalis

Temporalis

Masseter

Sternocleidomastoid

Figure 17.5   Muscles of the head, lateral view. ⚷

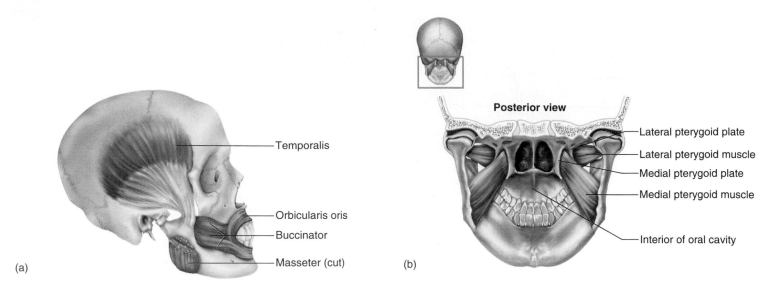

(a)

(b)

Temporalis

Orbicularis oris

Buccinator

Masseter (cut)

Posterior view

Lateral pterygoid plate

Lateral pterygoid muscle

Medial pterygoid plate

Medial pterygoid muscle

Interior of oral cavity

Figure 17.6   Muscles of the head, lateral and posterior views. ⚕

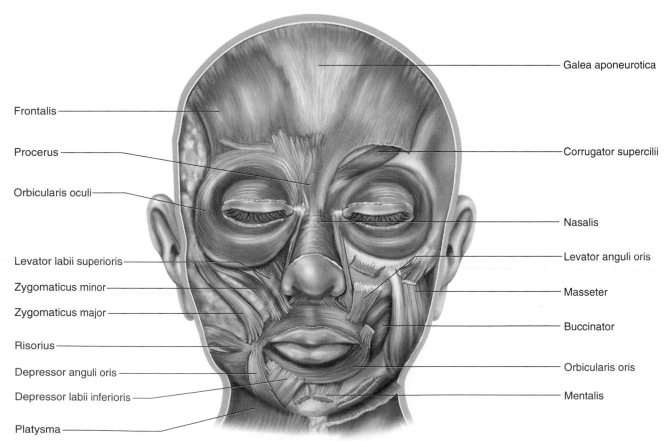

Frontalis

Procerus

Orbicularis oculi

Levator labii superioris

Zygomaticus minor

Zygomaticus major

Risorius

Depressor anguli oris

Depressor labii inferioris

Platysma

Galea aponeurotica

Corrugator supercilii

Nasalis

Levator anguli oris

Masseter

Buccinator

Orbicularis oris

Mentalis

Figure 17.7   Muscles of the face, anterior view. ⚕

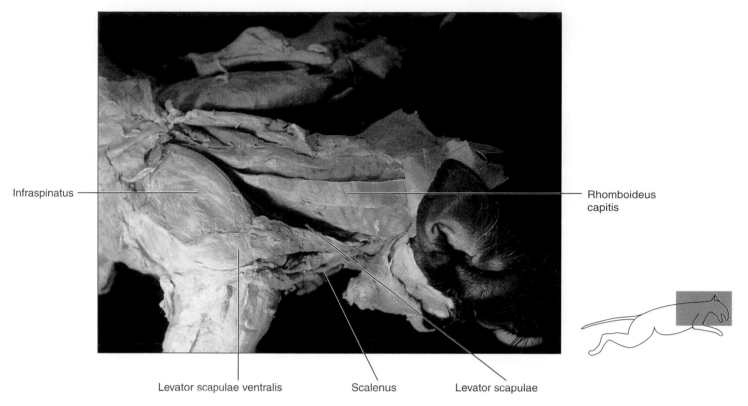

Infraspinatus

Rhomboideus capitis

Levator scapulae ventralis   Scalenus   Levator scapulae

Figure 17.8   Muscles of the neck of the cat, dorsolateral view.

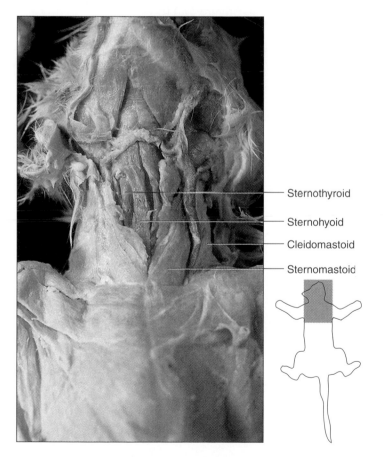

Sternothyroid

Sternohyoid

Cleidomastoid

Sternomastoid

Figure 17.9   Muscles of the neck of the cat, ventral view.

Cat Dissection

The head and neck dissection of the cat is beneficial, but you must take care to not cut through the digestive structures, such as the salivary glands, and the circulatory structures, such as the veins and arteries. You will study these structures in later exercises. A dorsal neck muscle of the cat is the **levator scapulae,** which can be found deep to the trapezius. You will need to cut the clavotrapezius in order to see the levator scapulae. In cats there is an additional muscle called the levator scapulae ventralis, which inserts on the scapular spine. Examine figure 17.8 for the levator scapulae muscles.

The remainder of the muscles you will study in this exercise are seen from a ventral aspect. The **platysma** in the cat was removed during the skinning process, and it will not be seen unless you kept the skin with the cat. The **scalenus** can be dissected by reflecting the pectoralis minor muscle. Notice how the scalenus is composed of separate slips of muscle that run from the ribs to the neck. In humans the muscle is more lateral than in cats. Compare your specimen with figure 17.8.

In the cat the sternocleidomastoid consists of two muscles, the **sternomastoid** and the **cleidomastoid.** The sternomastoid extends from the sternum to the mastoid process of the skull, and the cleidomastoid runs from the clavicle to the mastoid process. Underneath the sternomastoid is the most medial muscle of the neck group, the **sternohyoid.** The **sternothyroid** is deeper and more lateral than the sternohyoid. These muscles can be seen in figure 17.9.

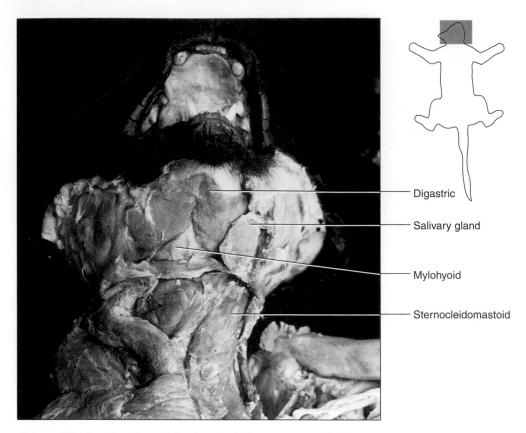

Figure 17.10   Muscles of the head of the cat, ventral view.

— Digastric

— Salivary gland

— Mylohyoid

— Sternocleidomastoid

The **digastric** muscle can be seen running parallel to the lower edge of the mandible and underneath the submandibular gland. Dissect only one side of the head, leaving the structures on the other intact for study of the digestive system. The **mylohyoid** is a broad muscle deep to the digastric. Notice how the muscle fibers run transverse to the direction of the digastric. Compare your dissection to figure 17.10.

The **occipitalis** is a posterior head muscle in the cat that attaches to the **galea aponeurotica** and to the anterior **frontalis** muscle. Lateral to these muscles are the temporalis and masseter muscles. The **temporalis** is located more dorsally than the masseter and can be dissected by removing the skin and fascia anterior to the ear. The **masseter** is a large, well-developed muscle in the cat that originates on the zygomatic arch and inserts on the lateral surface of the mandible. Examine these muscles in figure 17.11.

The **pterygoids** are usually not dissected because you have to cut through the ramus of the mandible to examine them. The muscles of facial expression are generally not studied in the cat. These muscles are small and are frequently removed with the skin. You should study these muscles on human models or a cadaver (if available).

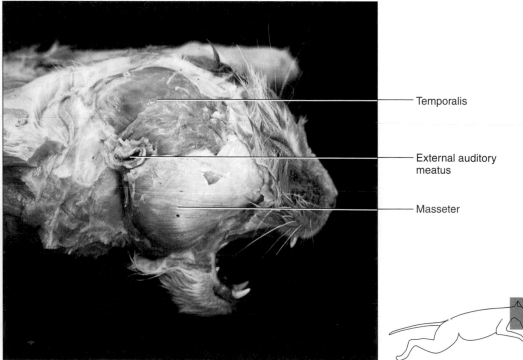

— Temporalis

— External auditory meatus

— Masseter

Figure 17.11   Muscles of the head of the cat, lateral view.

Name _____

1. What is the origin of the masseter muscle?

2. What is a synergist of the masseter muscle?

3. Where is the origin of the levator scapulae muscle?

4. What kind of muscle is the orbicularis oculi or orbicularis oris muscle in terms of function?

5. Fill in the following illustration.

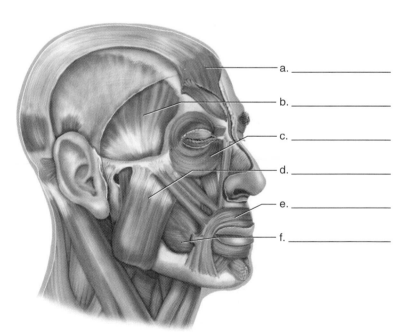

a. _____

b. _____

c. _____

d. _____

e. _____

f. _____

6. What muscle originates on the temporal fossa?

7. Name two muscles that close the jaw.

8. Where does the sternocleidomastoid muscle insert?

9. What muscle closes the lips?

10. Where does the orbicularis oculi insert?

11. What is the insertion of the temporalis?

12. Name a muscle that closes the eye.

13. What is the action of the sternocleidomastoid?

Muscles of the Trunk

Introduction

In this exercise, you continue your study of muscles by studying the muscles of the anterior torso and the deeper muscles of the posterior torso. These muscles can be grouped into a few functional areas. One group is the abdominal muscles, all of which serve to tighten the abdomen. Another group is the respiratory muscles, which assist in breathing. A third group is the postural muscles of the back, and finally there are some muscles that act on the scapula or on the head and neck.

Objectives

At the end of this exercise you should be able to

1. locate the muscles of the trunk on a torso model, chart, cadaver (if available), or cat (if available);
2. list the origin, insertion, and action of each muscle;
3. describe what nerve controls each muscle if your instructor includes innervation with the muscles;
4. list what muscles function as synergists or antagonists to the prime mover;
5. name all of the muscles that have a particular action on the torso, such as all of the muscles that compress the abdomen.

Materials

Human torso model
Human muscle charts
Articulated skeleton
Cadaver (if available)
Cat (if available)
Materials for cat dissection

> Dissection trays
> Scalpel or razor blades
> Gloves
> Blunt probe
> String
> Pins

Procedure

Review the muscle nomenclature and the actions as outlined in Laboratory Exercise 13. Examine a torso model or chart in the lab and locate the muscles described next and in table 18.1. Correlate the shape or fiber direction of the muscle with the name and begin to visualize the muscles as you study the models or charts. You may want to look at an articulated skeleton as you review the origins and insertions of the muscles so you can better see the muscle attachment points. The descriptions in the text portion of this exercise can help you understand the nature of the muscle, while table 18.1 gives you specific information about the muscle.

Once you have learned the origin and the insertion, you should be able to understand the action of the muscle. This is done, in part, by imagining how the bony attachments of the origin and insertion would come together if the muscle pulled them closer to one another.

Anterior Muscles

The muscles of the abdomen all function to compress the viscera, which aids in regurgitation of food and in bowel movements. The **external abdominal oblique** is a broad, superficial muscle of the abdomen with fibers that run from a superior direction to an inferior, medial direction. The **internal abdominal oblique** is deep to the external abdominal oblique and has fiber directions that run perpendicular to the external abdominal oblique. Locate the abdominal muscles in the lab and compare them to figure 18.1.

The deepest of the abdominal muscles is the **transversus abdominis,** which has fibers running in a horizontal direction. The **rectus abdominis** (rectus means straight) muscle runs vertically up the abdomen. The rectus abdominis has small connective tissue bands called **tendinous inscriptions** that run horizontally across the muscle, dividing it into small segments. If the abdominal fat is minimal and the muscles are well developed, the "washboard stomach" or "six-pack" is apparent due to the muscle fibers increasing in girth while the tendinous inscriptions remain undeveloped.

The **serratus anterior** is a broad, fan-shaped muscle that has slips of muscle originating on the upper ribs. These slips of muscles unite and insert on the medial border of the scapula and the inferior angle of the scapula. As the muscle contracts it pulls the scapula towards the front of the ribs.

The **intercostal** muscles and the diaphragm are respiratory muscles. Normally the **diaphragm** is responsible for about 60% of the resting breath volume while the external intercostals contribute to the remaining volume. The

Table 18.1 Muscles of the Trunk

Name	Origin	Insertion	Action	Innervation
Muscles of Thorax, Abdomen, and Pelvis				
External abdominal oblique	Ribs 5–12	Linea alba, iliac crest, pubis	Compresses abdominal wall, laterally rotates trunk	Intercostal nerves from T7–12
Internal abdominal oblique	Inguinal ligament, iliac crest	Linea alba, ribs 10–12	Compresses abdominal wall, laterally rotates trunk	Intercostal nerves from T7–12 and spinal nerve L1
Transversus abdominis	Inguinal ligament, iliac crest, ribs 7–12	Linea alba, crest of pubis	Compresses abdominal wall, laterally rotates trunk	Intercostal nerves from T7–12 and spinal nerve L1
Rectus abdominis	Crest of pubis, symphysis pubis	Cartilages of ribs 5–7, xiphoid process	Flexes vertebral column, compresses abdominal wall	Intercostal nerves from T6–12
Serratus anterior	Ribs 1–8	Vertebral border and inferior angle of scapula	Abducts scapula (moves scapula away from spinal column)	Long thoracic nerve
External intercostals	Inferior border of a rib	Superior border of rib below	Elevates ribs (increases volume in thorax)	Intercostal nerves
Internal intercostals	Inferior border of a rib	Superior border of rib below	Depresses ribs (decreases volume in thorax)	Intercostal nerves
Diaphragm	Xiphoid process, lower ribs, upper lumbar vertebrae	Central tendon	Inspiration (contraction), expiration (relaxation)	Phrenic nerve
Deep Muscles of the Back				

Erector spinae:
    Multifidus
    Iliocostalis
    Longissimus
    Spinalis
	Vertebral column, ilium, ribs	Ribs, vertebral column, occipital and temporal bones	Extends and rotates vertebral column and head	Numerous spinal nerves

Quadratus lumborum	Posterior iliac crest	T12 and L1–4, rib 12	Extends and abducts vertebral column	T12, L1–4
Rhomboideus major	Spines of T2–5	Lower one-third of vertebral border of scapula	Adducts scapula (draws scapulae together)	Dorsal scapular nerve
Rhomboideus minor	Ligamentum nuchae and spines C7–T1	Vertebral border of scapula at scapular spine	Adducts scapula (draws scapulae together)	Dorsal scapular nerve
Splenius	Ligamentum nuchae, C7–T6	C2–4, occipital and temporal bone	Extends and rotates head	Middle and lower cervical nerves
Semispinalis	C7–T12	Occipital bone and T1–4	Extends head and vertebral column, rotates vertebral column	Cervical and thoracic spinal nerves

diaphragm is a domed muscle that has a peripheral origin. The insertion of the diaphragm is central at the base of the mediastinum. If you think of the diaphragm as a trampoline, the outer springs represent the origin while the center (where you jump) represents the insertion. Examine the models in the lab and compare them to figure 18.2.

The intercostal muscles do contribute to the breathing volume at rest, but they also contribute to a greater increase in the movement of the thorax during times of exercise.

There is some debate as to the functions of the intercostals. Some evidence suggests that both the intercostals are involved in inhalation. Other evidence suggests that the **external intercostal** is involved in inhalation while the **internal intercostal** is involved in exhalation. In this exercise, we treat the external intercostals as being responsible for inhalation while the internal intercostals are involved in exhalation. Locate the intercostal muscles and compare them to figure 18.2.

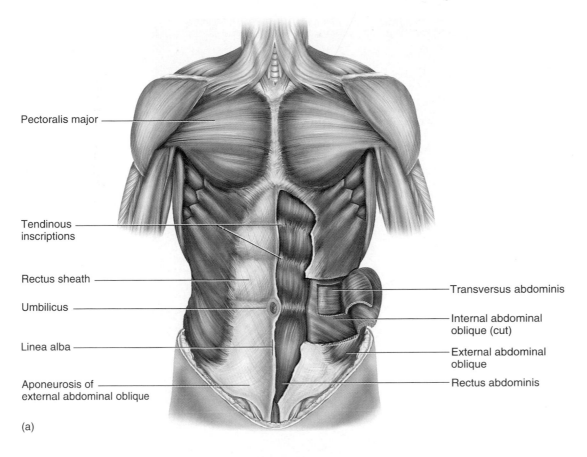

Pectoralis major

Tendinous inscriptions

Rectus sheath

Umbilicus

Linea alba

Aponeurosis of external abdominal oblique

Transversus abdominis

Internal abdominal oblique (cut)

External abdominal oblique

Rectus abdominis

(a)

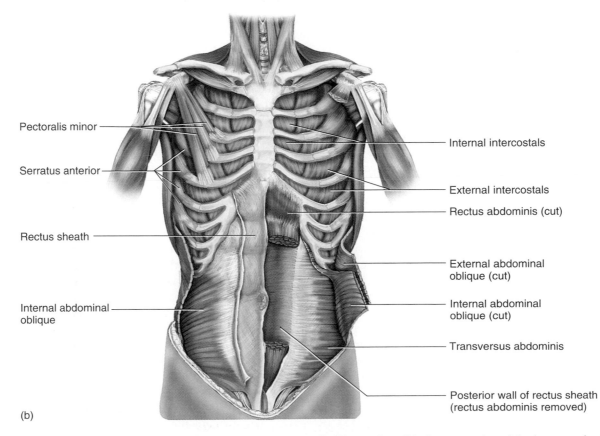

Pectoralis minor

Serratus anterior

Rectus sheath

Internal abdominal oblique

Internal intercostals

External intercostals

Rectus abdominis (cut)

External abdominal oblique (cut)

Internal abdominal oblique (cut)

Transversus abdominis

Posterior wall of rectus sheath (rectus abdominis removed)

(b)

Figure 18.1   Muscles of the abdomen, anterior view. Diagram (a) superficial muscles; (b) deep muscles; (c) photograph.

Continued

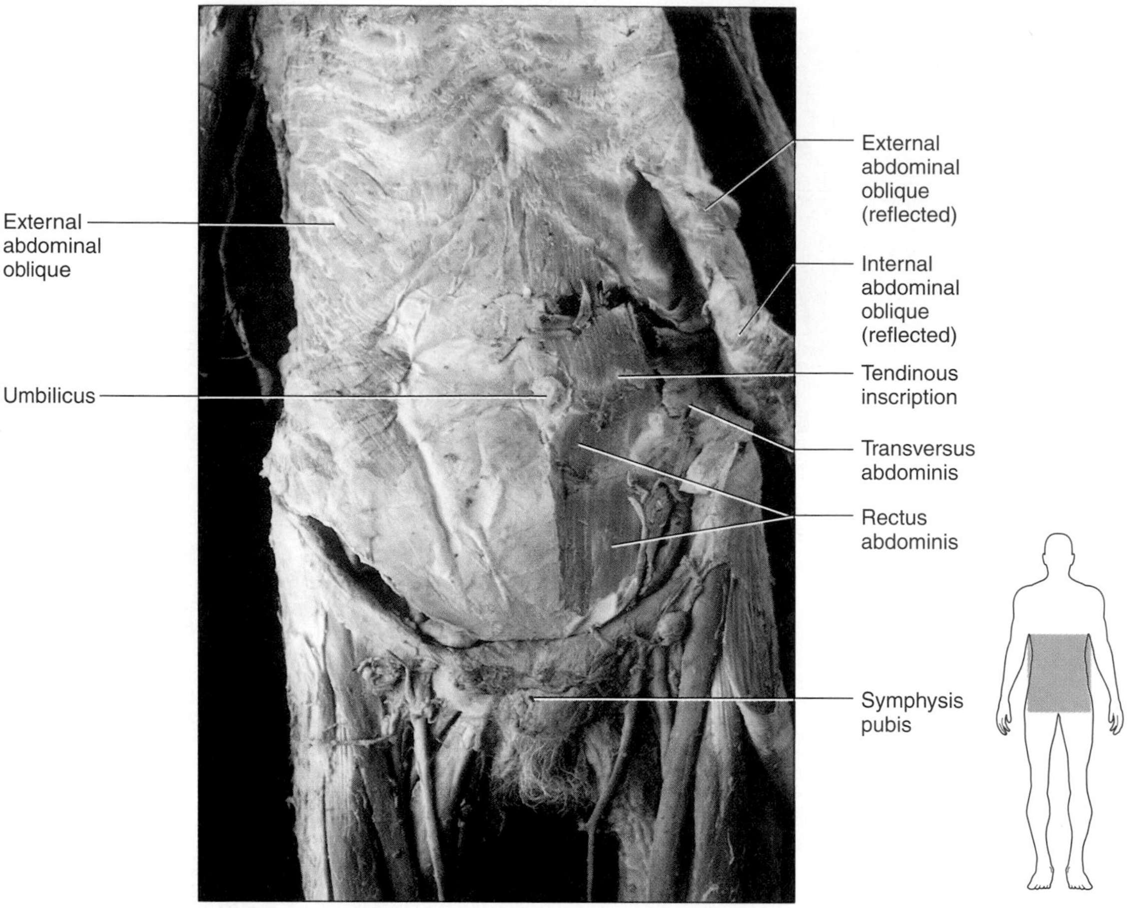

External
abdominal
oblique

Umbilicus

External
abdominal
oblique
(reflected)

Internal
abdominal
oblique
(reflected)

Tendinous
inscription

Transversus
abdominis

Rectus
abdominis

Symphysis
pubis

(c)

Figure 18.1—*continued*

Posterior Muscles

The postural muscles of the back of the torso mostly consist of the **erector spinae** muscles. The erector spinae muscles are actually many muscles that occur between individual vertebrae or between the vertebrae and the ribs. These muscles are grouped conveniently into long strap muscles known collectively as the erector spinae. There are four major groups of erector spinae muscles, the **spinalis,** the **longissimus,** the **iliocostalis,** and the **multifidus** (the *s.l.i.m.* muscles as you move from medial to lateral and then inferior). Another muscle that extends the vertebral column is the **quadratus lumborum,** which is a square muscle that runs from the iliac crest to the lower vertebrae and twelfth rib. These muscles can be seen in figure 18.3.

The **rhomboideus muscles** are those that occur deep to the trapezius. Reflection of the trapezius is necessary to see the rhomboideus muscles. These muscles originate on the vertebral column and insert on the medial border of the scapula. As they contract they pull the scapulae together, thus adducting the scapulae. The deep muscles of the back also consist of the **splenius** and the **semispinalis** muscles. These function to extend and rotate the head and vertebral column. Compare the material in lab to figure 18.4.

Cat Dissection
Anterior Muscles

Place the cat on its back and examine the abdominal muscles. The abdominal muscles in the cat are similar to those in the human in that the **external abdominal oblique** is a broad superficial muscle on the anterior abdomen. Carefully cut through the external oblique to reveal the **internal abdominal oblique** as illustrated in figure 18.5. Deep to this is the **transversus abdominis,** and it can be seen by carefully dissecting the internal abdominal oblique. If you cut too deeply you will enter the abdominal cavity, so be careful in this part of the dissection. The **rectus abdominis** can be seen as a muscle that runs from the pubic region to the sternum.

Move to the thoracic region and examine the muscle of the lateral thorax dorsal to the xiphihumeralis. This is the **serratus anterior** (actually serratus ventralis in the cat), and you should see the scalloped edges of the muscle. Carefully separate this muscle from the others and trace its insertion to the scapula. To see the intercostal muscles you will have to bisect the superficial chest muscles. If you have not done so already, cut through the middle of the belly of the pectoral muscles,

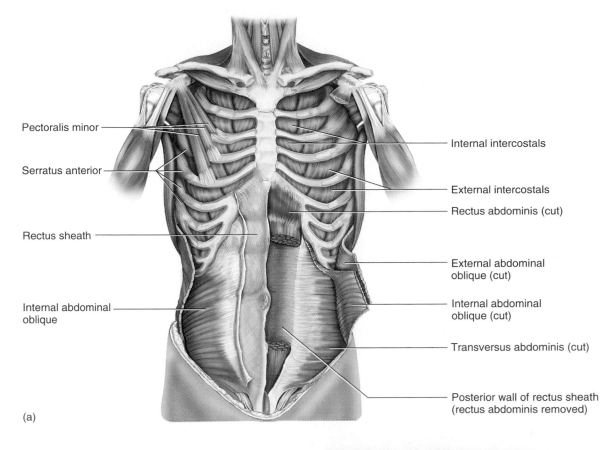

Pectoralis minor

Serratus anterior

Rectus sheath

Internal abdominal
oblique

Internal intercostals

External intercostals

Rectus abdominis (cut)

External abdominal
oblique (cut)

Internal abdominal
oblique (cut)

Transversus abdominis (cut)

Posterior wall of rectus sheath
(rectus abdominis removed)

(a)

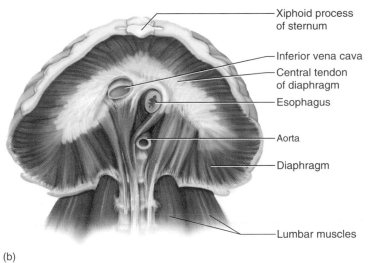

Xiphoid process
of sternum

Inferior vena cava

Central tendon
of diaphragm

Esophagus

Aorta

Diaphragm

Lumbar muscles

(b)

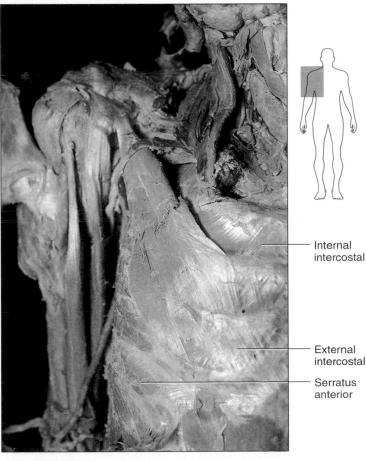

Internal
intercostal

External
intercostal

Serratus
anterior

(c)

Figure 18.2   Muscles of the thorax. Diagram (a) anterior view of intercostal muscles; (b) diaphragm, inferior view. Photograph
(c) anterior view of intercostal muscles. ✸

213

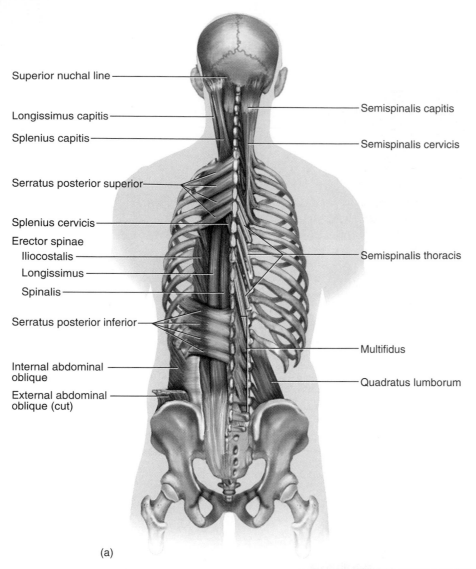

Superior nuchal line

Longissimus capitis

Splenius capitis

Serratus posterior superior

Splenius cervicis

Erector spinae
　Iliocostalis
　Longissimus
　Spinalis

Serratus posterior inferior

Internal abdominal oblique

External abdominal oblique (cut)

Semispinalis capitis

Semispinalis cervicis

Semispinalis thoracis

Multifidus

Quadratus lumborum

(a)

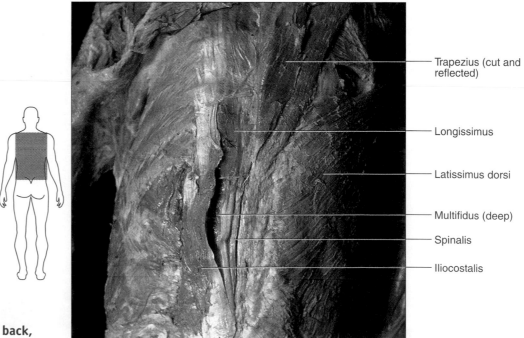

Trapezius (cut and reflected)

Longissimus

Latissimus dorsi

Multifidus (deep)

Spinalis

Iliocostalis

(b)

Figure 18.3 Inferior muscles of the back, posterior view. (a) Diagram; (b) photograph. ⚡

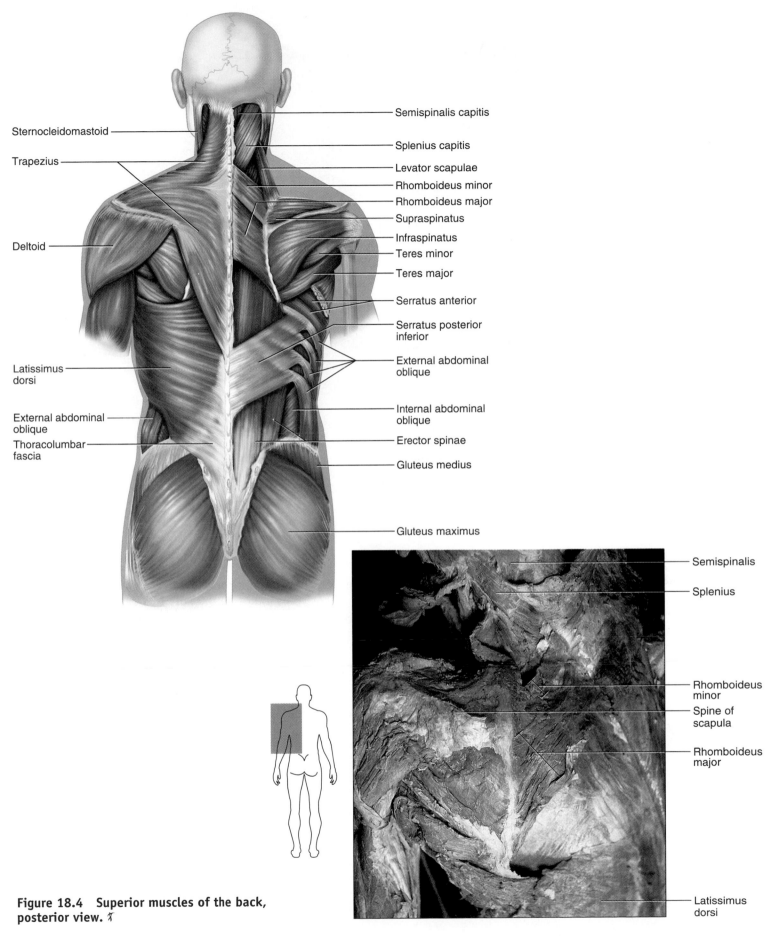

Sternocleidomastoid

Trapezius

Deltoid

Latissimus dorsi

External abdominal oblique

Thoracolumbar fascia

Semispinalis capitis

Splenius capitis

Levator scapulae

Rhomboideus minor

Rhomboideus major

Supraspinatus

Infraspinatus

Teres minor

Teres major

Serratus anterior

Serratus posterior inferior

External abdominal oblique

Internal abdominal oblique

Erector spinae

Gluteus medius

Gluteus maximus

Semispinalis

Splenius

Rhomboideus minor

Spine of scapula

Rhomboideus major

Latissimus dorsi

Figure 18.4   Superior muscles of the back, posterior view.

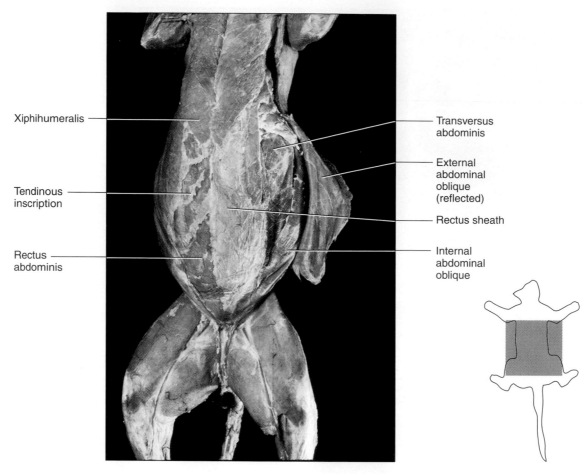

Xiphihumeralis

Tendinous inscription

Rectus abdominis

Transversus abdominis

External abdominal oblique (reflected)

Rectus sheath

Internal abdominal oblique

Figure 18.5   Muscles of the abdomen of the cat, lateral view.

exposing the ribs of the cat. Carefully remove the outer layer of fascia from the muscle between the ribs and locate the **external intercostal** muscle. You should be able to cut part of this muscle away and expose the **internal intercostal** muscle. Note how the fibers run perpendicular to one another. Do not look for the diaphragm at this time. You can see it in Laboratory Exercise 39, as you examine the lungs. Examine these thoracic muscles in figure 18.6.

Dorsal Muscles

Place the cat so you can examine the dorsal surface. You will need to carefully dissect the trapezius to see the **rhomboideus** muscles. These muscles originate on the vertebral column and insert on the scapula. If you examine the muscles of the neck and head, you should see the **splenius** and the **semispinalis** muscles. These are located in figure 18.7.

You will need to bisect the latissimus dorsi and the posterior portion of the external oblique to see the **erector spinae** muscles. The relative position of the cat erector spinae can be seen in figure 18.8. Compare this figure to your dissection. Locate the **multifidus, iliocostalis, longissimus,** and **spinalis** in the cat.

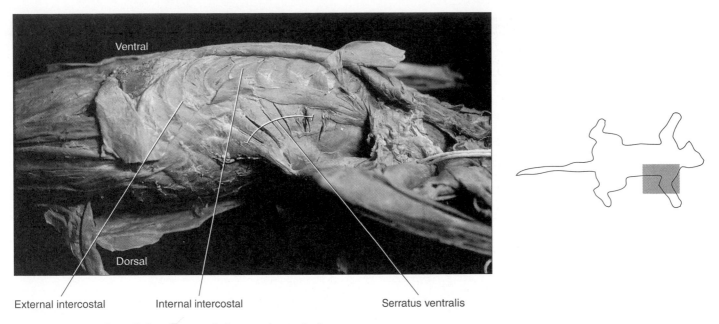

Ventral

Dorsal

External intercostal   Internal intercostal   Serratus ventralis

Figure 18.6   Muscles of the thorax of the cat, lateral view.

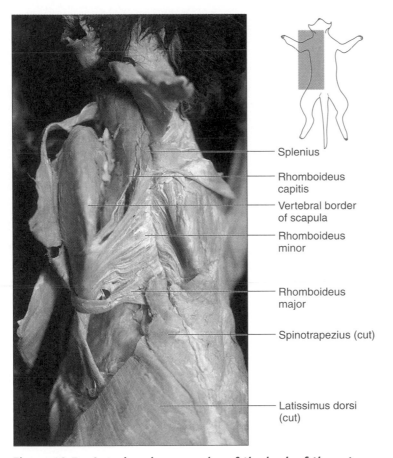

Splenius

Rhomboideus capitis

Vertebral border of scapula

Rhomboideus minor

Rhomboideus major

Spinotrapezius (cut)

Latissimus dorsi (cut)

Figure 18.7   Anterior, deep muscles of the back of the cat, dorsal view.

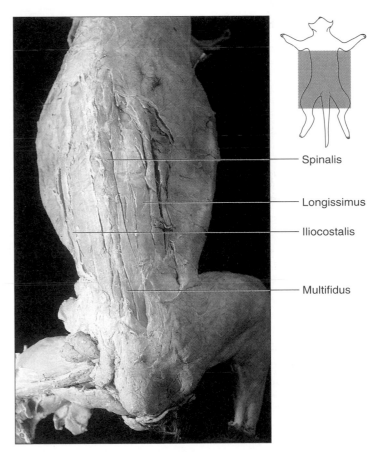

Spinalis

Longissimus

Iliocostalis

Multifidus

Figure 18.8   Posterior, deep muscles of the back of the cat, dorsal view.

REVIEW

Name _____

1. What is the action of the serratus anterior muscle?

2. Name *five* muscles that extend the vertebral column.

3. What is the action of the rhomboideus muscles?

4. How does the serratus anterior function as an antagonist to the rhomboideus muscles?

5. How does the action of the rectus abdominis differ from the other abdominal muscles?

6. What is the physical relationship of the intercostal muscles to each other?

7. Compression of the abdominal wall occurs by what four muscles?

8. Extension and rotation of the vertebral column occurs by what group of muscles?

9. Which muscles adduct the scapulae?

10. What is the action of the intercostal muscles?

11. What muscle inserts on the central tendon?

12. Abduction of the scapula occurs by what muscle?

13. The tendinous inscriptions are found in what muscle?

14. Flexion of the vertebral column occurs by what abdominal muscle?

15. Which is the deepest abdominal muscle?

16. What is the origin of the rhomboideus major muscle?

17. Label the muscles in the following illustration.

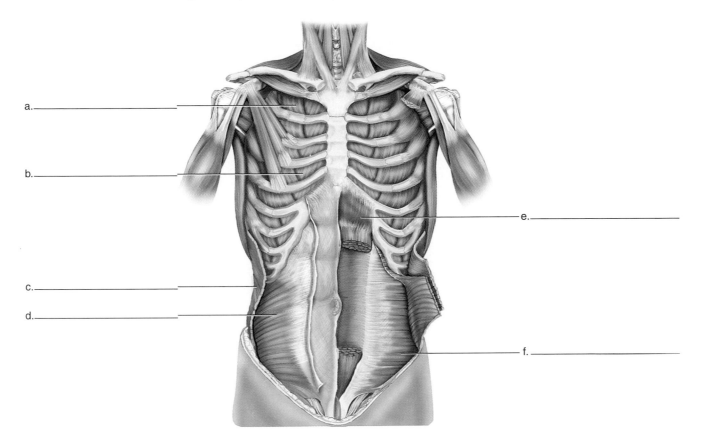

a._____

b._____

c._____

d._____

e._____

f._____

LABORATORY EXERCISE 19

Muscle Physiology

Introduction

Skeletal muscles contract due to stimulation by nerves controlling them. Normally a series of nerve impulses begins to stimulate a muscle and continues, producing the uniform muscle contraction that occurs in our bodies. When the multiple nerve impulses diminish, the muscle relaxes. In this exercise, you explore the nature of skeletal muscle contraction as initiated by external electrical stimulation and apply this information to the functions of the skeletal muscle in your body.

Two events are fundamental to an understanding of muscle physiology. One is an electrochemical event that occurs in muscle membranes, and the other is the physical contraction of the muscle itself. The normal contraction of skeletal muscle occurs when an electrochemical **nerve impulse** travels down an axon and crosses the **synapse** between the nerve and the muscle. This synapse and corresponding neuron and muscle connection are known as the **neuromuscular junction.** Acetylcholine (ACh) is released by the terminal regions of the neuron and stimulates an electrochemical impulse that travels along the length of the muscle fiber.

When the impulse reaches the **T tubules** of the muscle, **calcium** is released and the **actin** and **myosin filaments** join together, producing a power stroke in the muscle cell. As a muscle fiber is **depolarized** the muscle fiber contracts maximally. This particular characteristic represents the **all-or-none law** of muscle fibers. Refer to your text for a more complete description of these events.

Entire muscles in the body do not contract with an all-or-none response, but rather exhibit a **graded response,** where muscles gradually increase from slight to more forceful contractions. This is due to the presence of multiple fibers in a muscle, and the overall contractile strength of the entire muscle is determined by the number of muscle fibers (each of which contracts completely) in that muscle.

When a muscle is stimulated with a single, quick electrical impulse, the muscle undergoes a contraction known as a **twitch.** In a twitch the muscle has three phases; a latent phase, a contraction phase, and a relaxation phase. After the initial **stimulus** the muscle undergoes a **latent phase** (figure 19.1). This is the time when the electrical impulse travels across the muscle cell membrane and calcium ions are released from the sarcoplasmic reticulum. After this short time the muscle goes through the **contraction phase** as the myofilaments slide across one another and the muscle shortens. Finally there is

the **relaxation phase,** which is characterized by the muscle returning to a resting state.

The impulse that causes a muscle to contract must exceed a **threshold** value before any contraction can occur. A subthreshold stimulus will not illicit a response in the muscle. A stimulus above threshold level that occurs too soon after a preliminary stimulus also does not cause the muscle to contract. The time when a stimulus, delivered just after a previous stimulus, produces no contraction is known as the **refractory period.** If a stimulus is applied shortly after the refractory period, a muscle contracts and the contraction strength is more pronounced. This effect is called **wave summation** (figure 19.2).

If rapid, repeated stimuli are sent to a muscle, then the muscle produces a series of contractions called **incomplete tetany** (figure 19.3). If the **frequency** (number of pulses per second) of the stimuli increases, the contractions fuse in a smooth contraction of the muscle known as **complete tetany.** Numerous, sequential stimulations of a muscle produce **temporal summation** in the muscle. Temporal summation results in the smooth, continuous muscle contractions that normally occur in the body.

Complete tetany occurs even in fast muscular movements such as the flicking of a finger. Examine figure 19.3 for recordings of incomplete tetany and complete tetany.

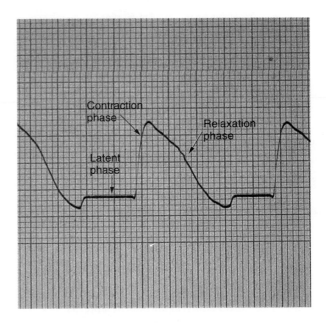

Figure 19.1   Three phases of a muscle twitch.

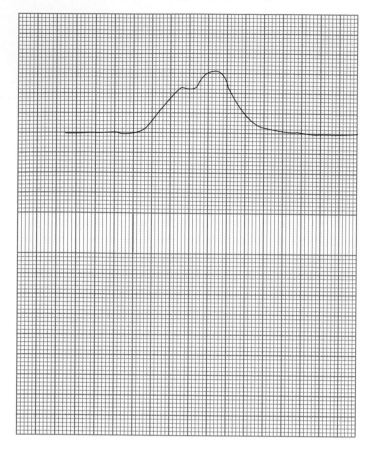

Figure 19.2   Wave summation.

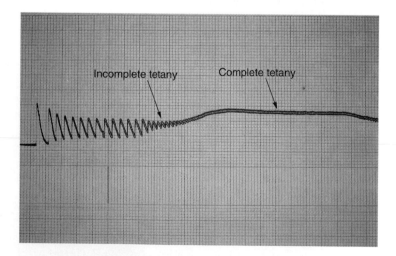

Figure 19.3   Incomplete and complete tetany.

Objectives

At the end of this exercise you should be able to

1. demonstrate the procedure to determine threshold stimulus;
2. differentiate between tetany and treppe;

3. describe incomplete tetany and complete tetany;
4. demonstrate maximum recruitment with lab equipment and a frog;
5. compare the lab experiments on frogs to muscular contractions in humans.

Materials

Large grass frog or bullfrog (one per experiment)
Frog Ringer's solution in dropper bottles (150 ml per experiment)
Thread
Duograph, physiograph, computer
Myograph transducer
Stimulator and cables
Scissors
Clean, live animal dissection pan
Sharp pithing probes
Glass hooks
Scalpel
Pins

Virtual Physiology Lab #3: Frog Muscle

Procedure

You may do this experiment in small groups or your instructor may elect to do a demonstration for the class. If you are doing this experiment as a group, *read the entire exercise first* and then follow the directions.

Frog Preparation

If the frog has not been pithed, follow the directions under number 1. If the frog has been pithed, begin at number 2.

1. The most humane way to conduct frog muscle experiments is to quickly pith the frog by inserting a sharp probe into the braincase.
   a. To do this, firmly grasp the frog and bend the head over your index finger (figure 19.4).
   b. Insert the probe into the braincase and twirl it around in a conical manner, destroying the brain. This is known as a **single pith.** The frog will be killed at this point yet still have reflexes in the lower limbs.
   c. To stop the reflexes, insert the sharp probe into the vertebral canal and run it towards the caudal end of the frog. This is known as a **double pith.** Be careful not to thrust the probe into your hand as you try to locate the vertebral canal. The correct positioning is illustrated in figure 19.4.

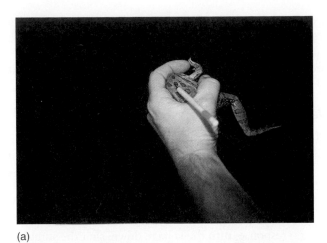

(a)

(b)

Figure 19.4   Pithing a frog. (a) Single pith; (b) double pith.

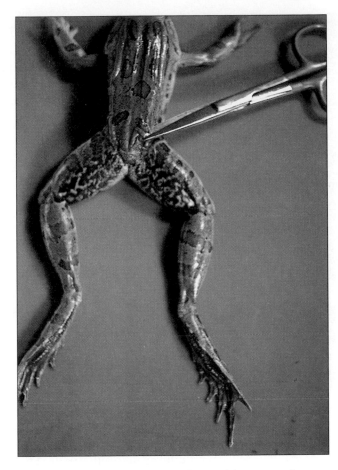

Figure 19.5   Preparation of the frog, skin removal.

2. Once the frog is pithed, carefully snip its skin above the hip joint and peel it back to the foot (figure 19.5).
   a. Locate the **sciatic nerve,** which appears as a thin, white glossy thread that runs along the lateral aspect of the femur. You may have to use a scalpel and tease the muscles away from the sciatic nerve.
   b. Using a glass hook, carefully lift the sciatic nerve away from the thigh muscles, keeping it moist with frog Ringer's solution (figure 19.6).
   c. **Ligate** the nerve by tying a thread around the proximal end of the nerve, near the sacrum, and cut the nerve above the ligature.
   d. Separate the thigh muscles from the femur and remove them leaving the femur exposed. Be careful not to cut or damage the sciatic nerve as you do this.
   e. Locate the **gastrocnemius muscle** of the frog and tie the tendon of the muscle with thread.
   f. Cut the calcaneal tendon distal to the ligature and lift the gastrocnemius away from the other muscles and from the tibiofibula (a fused bone in frogs).
   g. Cut the muscles and the tibiofibula just distal to the knee so you have the femur, the sciatic nerve, the gastrocnemius, and the knee joint intact (figure 19.6).
   h. Anchor the femur to a board or mount it on a tray and lay the sciatic nerve on the gastrocnemius muscle.
   i. Keep the muscle and the nerve moist during the entire experiment. Do not tug on the nerve but gently lay it on the gastrocnemius muscle.
   j. Attach the thread tied to the calcaneal tendon to the end of a myograph transducer leaf (figure 19.6). This should be connected to a recording device such as a physiograph, duograph, kymograph, or physiology computer. If you lightly tap on the leaf of the muscle transducer you should see a response in your recording apparatus.

There are three critical areas of concern in this lab. These are the preparation of the muscle, the stimulation of the muscle, and the recording of the response. Failure in any of these

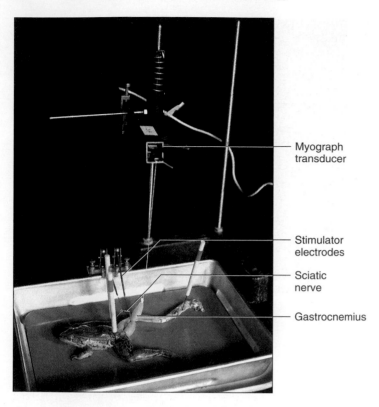

Myograph
transducer

Stimulator
electrodes

Sciatic
nerve

Gastrocnemius

Figure 19.6   Frog hookup to recording apparatus.

three areas will cause poor results or no results at all. The first of these areas is the preparation of the muscle, which has already been outlined. The second and third areas are discussed next.

Stimulator Setup

The preparation of the stimulator first involves determining how many stimuli you deliver in a particular time. Most stimulators can deliver repeating stimuli or single pulses. These

are measured as the **frequency** of the stimulus, and a good frequency to start out with is two pulses per second. Another important factor is the **duration** of the stimulus. This is how long the stimulus is delivered to the sciatic nerve. Durations of 2 to 10 milliseconds usually produce good results.

Determination of Threshold Stimulus

1. You can determine the threshold voltage by keeping the voltage at zero and examining the muscle while the frequency and duration are set as described.
2. Lay the sciatic nerve on the stimulator electrodes and slowly increase the voltage until you see the contraction of the muscle. If you have reached 5 to 8 volts and you still have no response, turn the voltage down, shut the stimulator off, and recheck your connections and settings. Once you see the muscle contract, then the lowest voltage that produces a response is known as the **threshold stimulus.**
3. Record this value in the space provided.

Threshold stimulus: _____

Recording Apparatus Setup

Your lab may be equipped with one or more different physiological recorders. Follow your instructor's directions to set up the apparatus if you are to do the experiment in groups or pay close attention if your instructor demonstrates the experiment. Pay particular attention to the settings of the apparatus.

Spatial Summation (Maximum Recruitment)

The **maximum recruitment** is the lowest voltage stimulus at which all of the muscle fibers are stimulated.

1. To demonstrate this, set the duration of the pulse to 2 milliseconds, keep the frequency at one or two pulses per second, and set the stimulus on repeat.

2. Begin the recording and increase the voltage until you see a response (threshold) in the muscle. Continue to increase the voltage slowly, and you should see the contraction force increase as indicated by an increase in the height of the tracing.

3. As you continue to increase the voltage slowly, the contraction tracings will not get any higher. The minimum voltage it takes to produce the maximum height is known as the maximum recruitment voltage.

4. Record the maximum recruitment voltage in the space provided.

Maximum recruitment voltage: _____

Treppe

1. Set the voltage reading at the maximum recruitment voltage and the frequency at one pulse per second.

2. As you stimulate the muscle, notice that there is a step-wise increase in the peaks of the contractions as the muscle contracts over a period of time. This is thought to be the increased availability of calcium in the muscle fibers and is a possible rationale for "warming up" before exercising.

Phases of Muscle Contraction

1. If you increase the chart speed to 50 mm per second you can obtain tracings of the muscle where the latent phase, contraction phase, and relaxation phase can be seen. If you can record the time when you stimulate the muscle, then you can calculate the latent phase of muscle contraction.

2. Make a tracing of the contraction and compare it to figure 19.7.

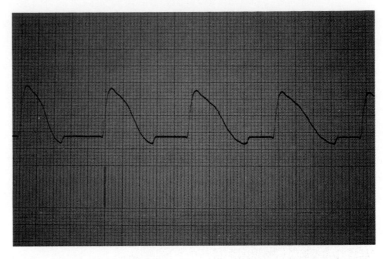

Figure 19.7   Phases of muscle contractions in a single twitch contraction.

If the chart speed is 50 mm per second in your tracing, what is the length of the latent phase? This assumes that you have a mark indicating the time of stimulation.

Latent phase: _____

How long is the contraction phase? Record the length of time.

Contraction phase: _____

How long is the relaxation phase? Record the length of time.

Relaxation phase: _____

Name _____

1. Define subthreshold stimulus.

2. Describe tetany.

3. What is maximum recruitment?

4. How does tetany and twitch as demonstrated in the lab correlate to human muscle contraction? Which one is more reflective of human muscle response? Why?

5. In the following illustration, place an "A" on the latent period, a "B" on the contraction phase, and a "C" on the relaxation phase.

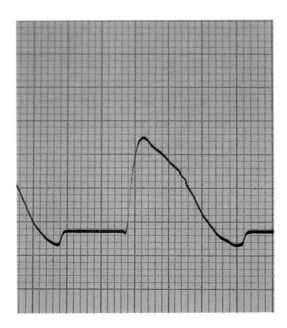

6. What was the threshold stimulus you obtained in lab?

7. What was the voltage at which you first got maximum recruitment?

8. Explain a muscle spasm in terms of recruitment of muscle fibers.

9. A skeletal muscle is stimulated to contract by what structure?

10. What chemical crosses the synapse, causing a muscle to contract?

11. Where is calcium released to cause muscle contraction?

12. What are the two types of filaments found in muscle cells that cause muscle contraction?

13. Does a muscle fiber or an entire muscle contract by an all-or-none response?

14. Name the first phase after a stimulus in a muscle contraction.

15. What happens to the strength of contraction during wave summation?

Introduction to the Nervous System

Introduction

The function of the nervous system, among other things, is communication between the various regions of the body, coordination of body functions (as in digestion or walking), orientation to the environment, and assimilation of information. The functional unit of the nervous system is the **neuron.** It is the cell that carries out the activity of nervous tissue. Neurons are located in the nerves of the body, the spinal cord, and the brain. **Neuroglia** (glial cells) are the supporting cells of the nervous tissue. They function to aid the neurons in terms of increasing the speed of neuron transmission, providing nutrients to the neurons, and protecting the neurons. In this exercise, you learn about the basic structure of the nervous system and the component cells that are part of the nervous system.

Objectives

At the end of this exercise you should be able to

1. describe the three parts of the neuron;
2. list the main divisions of the nervous system;
3. group the organs of the nervous system into the main divisions;
4. describe the functions of the various neuroglia.

Materials

Charts or models of the nervous system
Charts or models of neurons
Microscopes
Prepared slides of:

    Spinal cord smear

    Longitudinal section of nerve

    Neuroglia (if available)

    Cerebrum

    Cerebellum

 Virtual Physiology Lab #1: Action Potential
Virtual Physiology Lab #2: Synaptic Transmission

Procedure

Divisions

The nervous system can be divided into three general divisions based either on location or function. The **central ner-**vous system (CNS) is named for its location and is composed of the **brain** and the **spinal cord.** The **peripheral nervous system (PNS)** is also named for its location and is composed of the **spinal nerves,** dorsal root ganglia, the **somatic nerves** (those that radiate into the extremities and other regions of the body), and the **cranial nerves.** The **autonomic nervous system (ANS)** is named not so much for its location as for its function. The ANS has centers in the central nervous system **(midbrain, pons, medulla oblongata,** and **spinal cord)** and has peripheral branches, autonomic ganglia and nerves that stimulate organs, glands, and smooth or cardiac muscles. The ANS functions independently and provides automatic controls for activities normally under subconscious direction. When you walk up stairs your heart rate increases automatically along with your breathing rate without having to think about controlling these activities.

The ANS can be controlled consciously in some cases. The principle of biofeedback involves the conscious lowering of heart rate or blood pressure. Both of these activities are normally under the control of the ANS.

Examine the models or charts in the lab for the divisions of the nervous system. Compare the material in the lab to figure 20.1.

Histology

The **neuron** is a remarkable cell not only for its functional nature but also for the anatomical extremes it exhibits. The nerves in your thigh and leg are composed of neuron fibers, and the neurons that pick up sensation in your toes continue as *single cells* up the leg and thigh to **synapse** (join) with other neurons in the lower back. When you look at prepared slides of neurons in the microscope in this exercise, remember that these neurons are of great length in some cases.

Neurons consist of three main parts, the **axon,** the **dendrite,** and the **nerve cell body,** or **soma.** Examine figure 20.2 and models or charts in the lab for the structure of neurons. Impulses that reach neurons stimulate dendrites or the nerve cell body. Dendrites are so named because they have branching structures that resemble a tree (*dendros* = tree). Nerve cell bodies consist of the **neuroplasm** (cytoplasm of the neuron), **Nissl bodies** or **chromatophilic substances** (rough endoplasmic reticulum of the neuron), and the **nucleus.** The triangular region of the nerve cell body that is devoid of Nissl bodies is the **axon hillock,** and it leads to the axon that exits the nerve cell body.

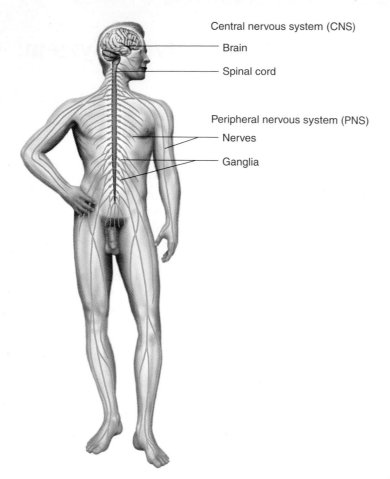

Central nervous system (CNS)
— Brain
— Spinal cord

Peripheral nervous system (PNS)
— Nerves
— Ganglia

Figure 20.1   Divisions of the nervous system.

Axons consist of long strands of neurofibrils wrapped in myelin sheaths. These sheaths are discussed later in the exercise.

Functions of Neurons

There are three types of neurons based on function. **Sensory (afferent) neurons** conduct impulses *to* the central nervous system. They convey information from the external or internal body environment to the spinal cord and/or the brain. **Motor (efferent) neurons** conduct impulses *away from* the central nervous system to organs or glands that carry out an activity. **Interneurons,** or **association neurons,** occur *between* sensory and motor neurons. They function to transmit information to the brain for processing.

Neuron Shapes

Neurons can be classified according to shape. A **multipolar neuron** consists of several dendritic processes, a single nerve cell body, and a single axon. The majority of the neurons of the body are multipolar neurons. Compare material in the lab to figure 20.3. **Bipolar neurons** are so named because the

nerve cell body has two poles. Dendrites receive information and conduct it to one pole of the nerve cell body. At the other pole an axon leaves the nerve cell body and transmits the impulse away from the cell body. Bipolar neurons are found in nerves conducting the senses of smell and vision.

Pseudounipolar, or **unipolar, neurons** have dendrites that lead to an axon that, in turn, takes the impulse to the nerve cell body. The axon enters the nerve cell body at a specific location and another axon leaves the nerve cell body at the same location. Most of the sensory nerves of the body are composed of pseudounipolar neurons.

Histology of the Neuron

Examine a prepared slide of a spinal cord smear under the microscope and locate the purple, star-shaped structures under low power. These are the nerve cell bodies of multipolar neurons. Switch to high power and locate the Nissl bodies, the nucleus, and the axon hillock. If you find the axon hillock you should be able to see the axon leading away from the hillock. All of the other processes that are attached to the nerve cell body are dendrites. The small nuclei that are scattered throughout the smear belong to glial cells in the spinal cord. Compare your slide to figure 20.4.

Synapses

Neurons transmit information **electrochemically** along the length of the axon to the **synaptic knob.** Neurons are not physically attached to one another but communicate by chemical signals that flow across a short space between the neurons. The space is called the **synapse,** and the chemicals that move across the synapse are called **neurotransmitters.** Figure 20.5 illustrates a synapse.

Neuroglia

Numerous cells aid the functioning of the neuron. These cells are called the **neuroglia,** or **glial cells.** The common glial cell of the peripheral nervous system is the **Schwann cell,** or **neurolemmocyte** (figure 20.2). These glial cells wrap around the axon, leaving small gaps between successive cells called the **nodes of Ranvier,** or **neurofibril nodes.** Neurolemmocytes are cells that wrap around the axon much like a thin strip of paper can be wrapped around a pencil. The neurolemmocyte is composed of a **lipoprotein** material called **myelin,** and the series of neurolemmocytes produces a **myelin sheath.** Examine the types of neuroglia in figures 20.2 and 20.6

Myelinated nerve fibers appear white, and so this type of nervous tissue is called **white matter.** The neurolemmocyte nodes allow for the nerve transmission to jump from node to

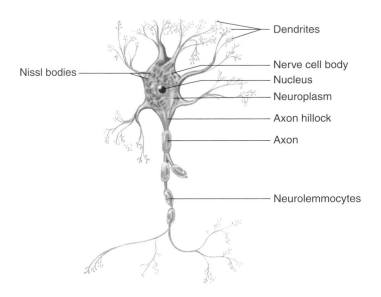

Figure 20.2   Parts of the neuron.

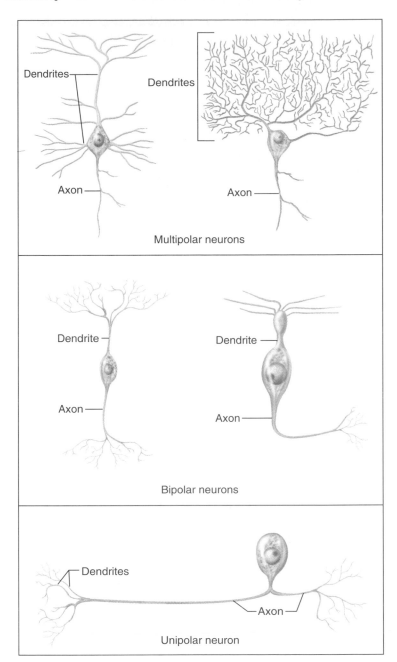

Figure 20.3   Neuron shapes.

node, thus increasing the transmission speed of the neuron. This type of jumping transmission is called **saltatory conduction. Unmyelinated** fibers form the portion of the nervous tissue known as **gray matter.**

Histology of the Neurolemmocyte

Examine a prepared slide of a longitudinal section of nerve under high power. You should be able to see the axon fibers as long dark threads in the microscope. If you scan the slide closely, you should be able to see the junction of two neurolemmocytes. The gap between them is the node of Ranvier. Compare your slide to figure 20.7.

Other types of neuroglia include myelinating fibers in the CNS. Neurolemmocytes occur in the PNS, and **oligodendrocytes** are their counterpart in the CNS. Unlike the neurolemmocytes, oligodendrocytes frequently wrap around several neurons, and the white matter of the spinal cord and brain is due to the lipoprotein of the oligodendrocytes.

Astrocytes are branched glial cells that provide a barrier between the nervous tissue and the blood. Astrocytes and capillary endothelial cells are responsible for the blood-brain barrier that serves to protect the nervous tissue from some blood-borne infections as well as inhibit some medications from reaching the brain.

Microglia are small glial cells that are phagocytic. The microglia function is to digest the foreign particles that invade the nervous tissue. Examine figure 20.6 for examples of microglia and other glial cells.

The last of the glial cells covered in this exercise are the **ependymal cells.** These cells line the ventricles of the brain and serve as a barrier between the fluid in the area (the cerebrospinal fluid) and the nervous tissue.

Specialized Neurons

The cerebral cortex of the brain contains **pyramidal cells** that have extensive dendritic branches. Examine a prepared slide of the cerebrum and locate the pyramidal cells. Compare these to figure 20.8.

In the cerebellum **Purkinje cells** are common. They occur in the gray matter of the cerebellum and also have branching dendrites. Examine a prepared slide of Purkinje cells and compare them to figure 20.8.

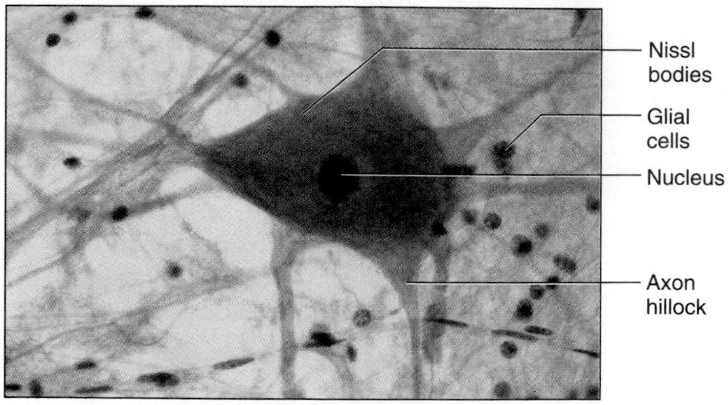

Nissl bodies

Glial cells

Nucleus

Axon hillock

Figure 20.4   Spinal cord smear.

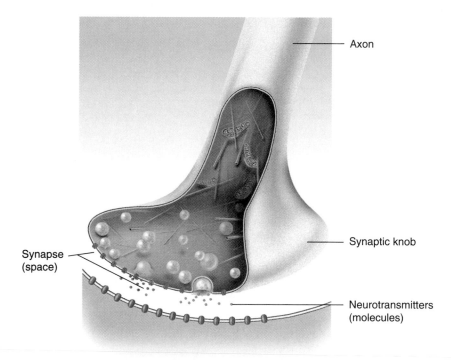

Axon

Synaptic knob

Synapse (space)

Neurotransmitters (molecules)

Figure 20.5   Synapse.

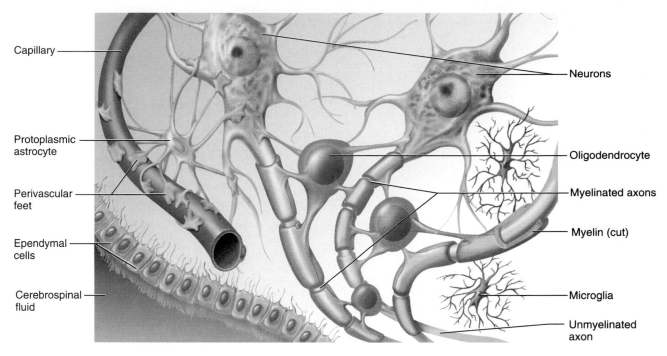

Capillary

Protoplasmic astrocyte

Perivascular feet

Ependymal cells

Cerebrospinal fluid

Neurons

Oligodendrocyte

Myelinated axons

Myelin (cut)

Microglia

Unmyelinated axon

Figure 20.6   Neuroglia.

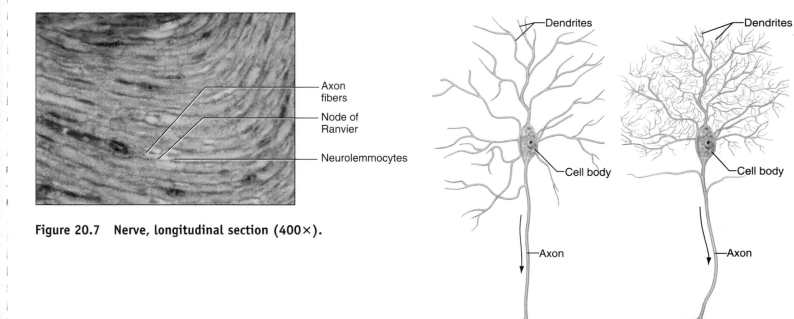

Axon fibers

Node of Ranvier

Neurolemmocytes

Figure 20.7   Nerve, longitudinal section (400×).

Dendrites

Dendrites

Cell body

Cell body

Axon

Axon

(a) Pyramidal cell

(b) Purkinje cell

Figure 20.8   Pyramidal and Purkinje cells.

Name _____

1. Draw a neuron in the space provided and label the axon, dendrite, and nerve cell body along with the Nissl bodies and axon hillock.

2. Draw a pseudounipolar neuron and label the parts.

3. Describe the function of

   a. an astrocyte:

   b. an ependymal cell:

   c. an oligodendrocyte:

4. The brain belongs to what division of the nervous system?

5. A spinal nerve belongs to what division of the nervous system?

6. To what major division of the nervous system does the spinal cord belong?

7. Which division of the nervous system is classified according to function rather than form?

8. What does CNS stand for?

9. What kind of cell performs the main function of the nervous system?

10. Where in a neuron is the nucleus?

11. A neuron has three main parts. What are they?

12. What is another name for an efferent neuron?

13. If a neuron has a soma with a dendrite on one side and an axon on the other, what kind of neuron would this be?

14. Two adjacent neurons are separated by a space. What is this space called?

15. Cells that support a neuron are called what general name?

16. In which one of the three nervous system divisions are neurolemmocytes found?

17. Myelin is made of what kind of material?

Structure and Function of the Brain and Cranial Nerves

Introduction

Two specific traits distinguish humans from other animals. One is our upright posture, and the other is the extensive development of the brain. In this exercise, you examine the anatomy of the human brain and the cranial nerves associated with it. You will also be able to compare the human brain to a sheep brain and identify similarities and differences.

The brain, which is part of the central nervous system, is located in the cranial cavity of the skull and weighs approximately 1.4 kilograms (3 pounds). The brain is derived from three embryonic regions, each of which further develops into more specific areas. The three embryonic regions of the brain are the **prosencephalon,** or forebrain, the **mesencephalon,** or midbrain, and the **rhombencephalon,** or hindbrain. Cranial nerves, though belonging to the peripheral nervous system, can be studied along with the brain since they are closely associated with it. Knowledge of the anatomy of the brain is important in locating the cranial nerves.

Objectives

By the end of this exercise you should be able to

1. name the three meninges of the brain and their location relative to one another;
2. locate the three major regions of the brain;
3. name the main structures that occur in each of the three regions of the brain;
4. describe the function of the thalamus, Broca's area, the occipital lobe, the temporal lobe, the cerebellum, and the medulla oblongata;
5. trace the path of cerebrospinal fluid through the brain;
6. list the major blood vessels that take blood to or from the brain.

Materials

Models and charts of the human brain
Preserved human brains
   (if available)
Cast of the ventricles of the brain
Chart, section, or illustration of the brain in coronal and
   transverse sections
Sheep brains
Dissection trays

Scalpels or razor blades
Gloves
Blunt probes

Procedure

Meninges

There are three layers called **meninges** that surround the brain. The outermost of these is the **dura mater,** a tough, dense connective tissue sheath that encircles the brain and has a series of shelves that extend into the brain. The dura mater is divided into an outer **periosteal layer** and an inner **meningeal layer.** The next deeper layer is the **arachnoid membrane,** which is a thin membrane. Between the dura mater and the arachnoid membrane is the subdural space. Deep to the arachnoid is the **subarachnoid space,** which contains **cerebrospinal fluid (CSF).** The deepest layer is the **pia mater,** which is a membrane directly on the outer surface of the brain. Locate the meninges in preserved brains in the lab (if available) and compare them to figure 21.1.

Blood Supply to the Brain

The blood vessels that supply nutrients and oxygen to the brain do not actually penetrate the brain tissue itself but rather form a meshwork around the brain and into the open spaces in the interior. The main arteries that provide blood to the brain are the **vertebral arteries** and the **internal carotid arteries.** Locate the vertebral arteries as illustrated in figure 21.2. These arteries pass through the transverse foramina of the cervical vertebrae and join to form the **basilar artery** at the base of the brain before branching into the **arterial circle** (circle of Willis) that forms a loop around the pituitary gland. The basilar artery gives rise to the **cerebellar arteries,** which take blood to the cerebellum. A pair of vessels joins the arterial circle at the anterior end, and these are the internal carotid arteries. From the arterial circle numerous **cerebral arteries** take blood superiorly to the surface of the cerebrum and into the interior of the brain. Note the various arteries in figure 21.2 at the base of the brain.

The drainage of the brain occurs as veins take blood from the brain and pass through the subarachnoid membrane to the **venous sinuses** in the subdural spaces. The drainage of blood from the brain flows into the **internal jugular veins** on the return trip to the heart. Examine figure 21.3 for the drainage of blood from the brain.

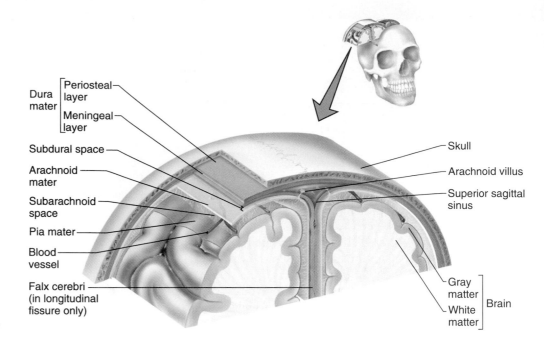

Dura mater
- Periosteal layer
- Meningeal layer

Subdural space

Arachnoid mater

Subarachnoid space

Pia mater

Blood vessel

Falx cerebri (in longitudinal fissure only)

Skull

Arachnoid villus

Superior sagittal sinus

Gray matter } Brain
White matter

Figure 21.1   Meninges of the brain.

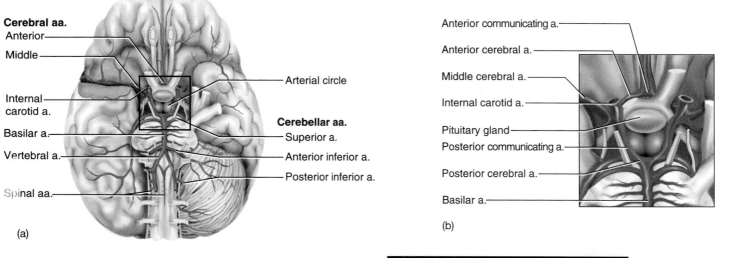

Cerebral aa.
Anterior
Middle

Internal carotid a.

Basilar a.

Vertebral a.

Spinal aa.

Arterial circle

Cerebellar aa.
Superior a.
Anterior inferior a.
Posterior inferior a.

(a)

Anterior communicating a.

Anterior cerebral a.

Middle cerebral a.

Internal carotid a.

Pituitary gland

Posterior communicating a.

Posterior cerebral a.

Basilar a.

(b)

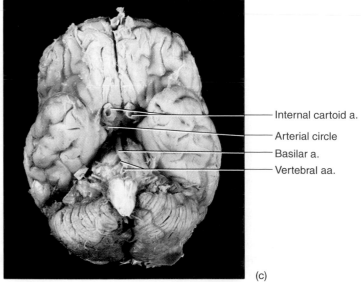

Internal cartoid a.
Arterial circle
Basilar a.
Vertebral aa.

(c)

Figure 21.2   Brain with arteries, inferior view. (a) Overview of arterial supply to brain; (b) close-up of arterial circle; (c) photograph.

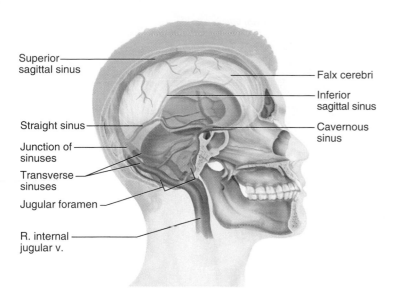

Superior sagittal sinus

Falx cerebri

Inferior sagittal sinus

Straight sinus

Cavernous sinus

Junction of sinuses

Transverse sinuses

Jugular foramen

R. internal jugular v.

Figure 21.3   Major drainage of the brain.

Ventricles of the Brain

The hollow neural tube that develops in the first trimester of pregnancy is seen in the adult as the ventricles of the brain. The two ventricles that occupy the center of each cerebral hemisphere are known as the **lateral ventricles.** These ventricles receive **cerebrospinal fluid** from tufts of capillaries called **choroid plexuses.** You can see the choroid plexuses as small brown areas at the superior portions of the ventricles. Fluid from the lateral ventricles flows through the **interventricular foramina** (foramina of Monro) and into the **third ventricle.** The third ventricle occurs in the thalamus and also receives CSF from choroid plexuses in that area. The third ventricle drains into the **fourth ventricle** by way of the **mesencephalic (cerebral) aqueduct.** If this duct becomes occluded, then CSF accumulates in the lateral and third ventricles. This increases the size of the ventricles producing a condition known as **hydrocephaly.** Normally the mesencephalic aqueduct is open and passes through the region of the midbrain. Posterior to the cerebral aqueduct is the fourth ventricle, which occupies a space inferior to the cerebellum. The fourth ventricle also has a choroid plexus that secretes CSF. Cerebrospinal fluid flows from the fourth ventricle into the subarachnoid space of the spinal cord and brain. CSF is finally absorbed by veins and returns to the cardiovascular system by the **internal jugular veins.** There are approximately 150 mL of CSF in the central nervous system, and it takes about 6 hours to circulate through the system. Locate the ventricles of the brain in figure 21.4.

Surface View of the Brain

Examine a model or chart of the brain and locate its major surface features. Of the major regions of the brain you will be able to easily see the **prosencephalon** (forebrain) and the **rhomben-**cephalon (hindbrain). In the prosencephalon you should examine the large **cerebrum,** which can be seen with folds and ridges known as **convolutions.** The ridges of the convolutions are called **gyri** (singular, *gyrus*), and the depressions are either **sulci** (singular, *sulcus*), or **fissures.** Usually fissures are deeper than the sulci. The lateral view of the brain allows you to see the major lobes of each cerebral hemisphere. These lobes are named for the bones of the skull under which they lie. They are the **frontal, parietal, occipital,** and **temporal lobes.** The **lateral fissure** (sulcus) separates the temporal lobe from the frontal and parietal lobes of the brain. You can locate these features in figure 21.5.

Frontal Lobe

The **frontal lobe** is responsible for many of the higher functions associated with being human. The frontal lobe is involved in intellect, abstract reasoning, creativity, social awareness, and language. An important area responsible for controlling the formation of speech is called **Broca's area** or the **motor speech area.** It is usually located in the left frontal lobe. Locate the frontal lobe in figure 21.5. The posterior border of the frontal lobe is defined by the **central sulcus.**

To find the central sulcus look for two convolutions that run from the superior portion of the cerebrum to the lateral fissure, more or less continuously. The gyrus anterior to the central sulcus is part of the frontal lobe and is known as the **precentral gyrus,** or the **primary motor cortex.** This cortex is important for directing a part of the body to move and has been mapped as in figure 21.6. How much of the precentral gyrus is dedicated to the face? How much of the gyrus is dedicated to the hands? How much to the trunk?

Parietal Lobe

The gyrus posterior to the central sulcus is known as the **postcentral gyrus,** or the **primary somatic sensory cortex.** This is part of the parietal lobe and is involved in receiving sensory information from the body. This area has also been mapped, and you can see the areas associated with various sensations coming from parts of the body in figure 21.6. The primary sensory cortex receives information, yet the material is integrated just posterior to the sensory cortex in the **association areas.** The primary sensory cortex pinpoints the part of the body affected, and the association area interprets the sensation (pain, heat, cold, etc.). Locate the structures of the parietal lobe in figure 21.5.

Wernicke's area is a region located in the parietal lobe and involved in sensory speech. The region is important for language, especially in the forming of coherent sentences.

Occipital Lobe

Posterior to the parietal lobe is the **occipital lobe** of the skull. The occipital lobe is considered the visual area of the brain and damage to this lobe can lead to blindness. Shape, color, and

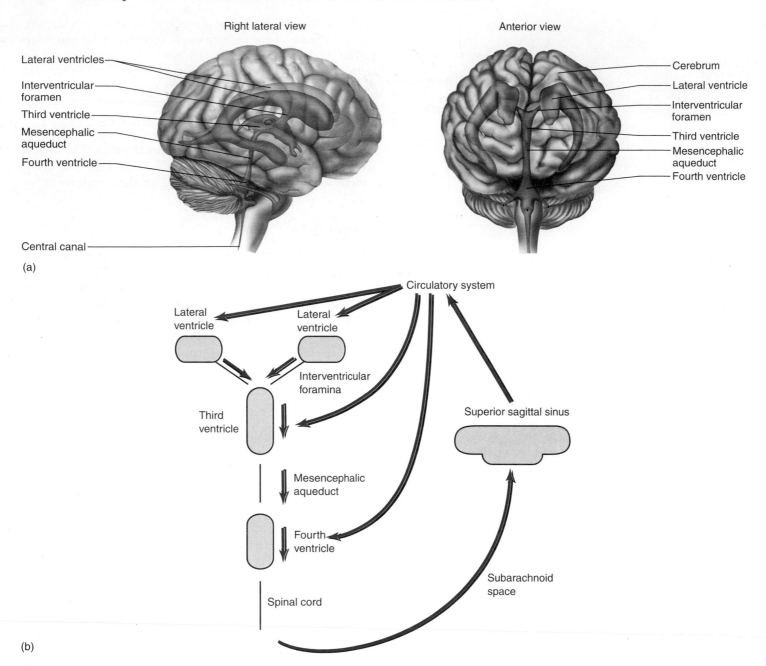

Figure 21.4 Ventricles of the brain and schematic presentation of the flow of cerebrospinal fluid.

distance of an object are perceived here. Recollection of past visual images occurs here as well. As you are reading these words your occipital lobe is receiving the information and transferring it to other regions, which convert the words to thought. Between the occipital lobe and the cerebellum is a **transverse fissure** that separates these two regions of the brain. Locate the occipital lobe and the cerebellum in figure 21.5.

Temporal Lobe

The **temporal lobe** is separated from the frontal and parietal lobes by the **lateral fissure** (sulcus). The temporal lobe con-

tains an area known as the **primary auditory cortex** that interprets hearing impulses sent from the inner ear. This auditory cortex distinguishes the nature of the sound (music, noise, speech) and the location, distance, pitch, and rhythm as well. The primary auditory cortex translates words into thought. The temporal lobe also has centers for the sense of smell (**olfactory centers**) and taste (**gustatory centers).**

Cerebral Hemispheres

In the surface regions of the cerebrum numerous associations are compared and contrasted as the brain integrates informa-

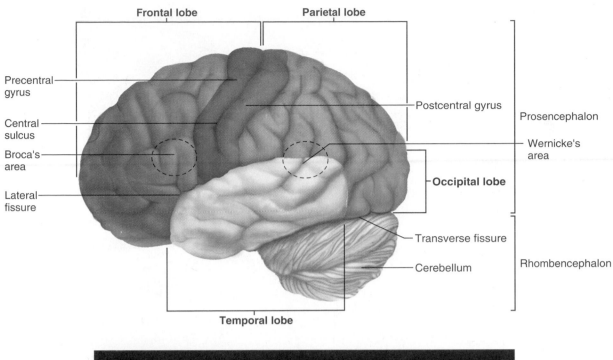

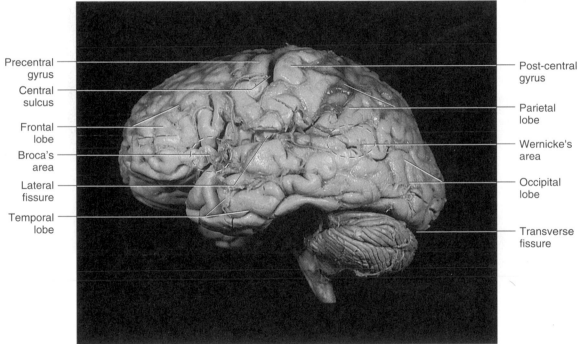

Figure 21.5   Brain, lateral view.

tion about the environment around us. If you rotate the brain so you are looking at it from a superior view, you should be able to see the **longitudinal fissure** that separates the cerebrum into the left and right cerebral hemispheres. The **left cerebral hemisphere** in most people is involved in language and rea-

soning. In most people, Broca's area is on the left side of the brain. The **right cerebral hemisphere** of the brain is involved in space and pattern perceptions, artistic awareness, imagination, and music comprehension. Examine the surface features of the brain in figure 21.7.

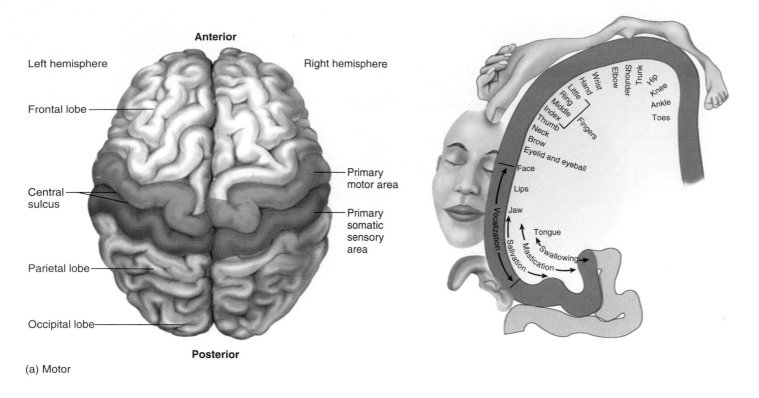

(a) Motor

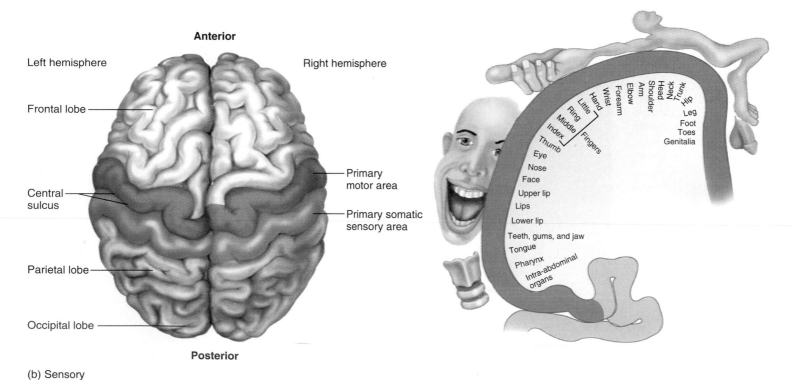

(b) Sensory

Figure 21.6 Primary motor and somatic sensory cortex. (a) Motor cortex (precentral gyrus); (b) somatic sensory cortex (postcentral gyrus).

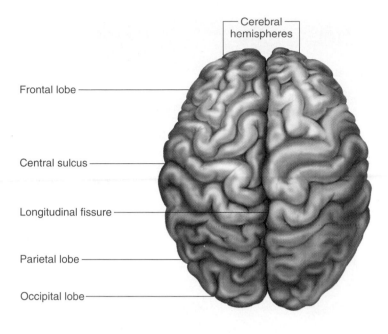

Cerebral hemispheres

Frontal lobe

Central sulcus

Longitudinal fissure

Parietal lobe

Occipital lobe

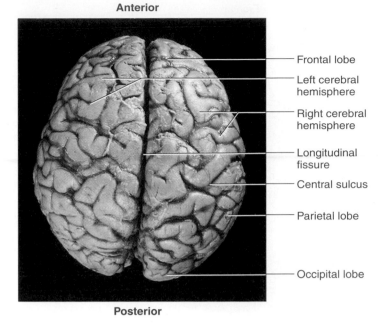

Anterior

Frontal lobe

Left cerebral hemisphere

Right cerebral hemisphere

Longitudinal fissure

Central sulcus

Parietal lobe

Occipital lobe

Posterior

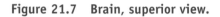

Figure 21.7   Brain, superior view.

Inferior Section of the Brain

Prosencephalon

If you examine the inferior aspect of the brain you can see the frontal lobes of the cerebrum and the temporal lobes as well. You may be able to see the **pituitary gland** if it has not been removed. The **optic chiasma** (*chiasma* = cross) is anterior to the pituitary and serves to transmit visual impulses from the eye to the brain. Two small processes posterior to the pituitary are the **mammillary bodies,** which function in olfactory reflexes.

Rhombencephalon

The inferior view of the brain provides a look at the rhombencephalon, which includes the **pons, medulla oblongata,** and the **cerebellum.** The cerebellum has much finer folds of neural tissue called **folia,** which can be seen in an inferior view of the brain. The medulla oblongata is located inferior to the cerebellum and connects to the spinal cord. The enlarged portion of the brain anterior to the medulla is the pons, which serves as a relay center for information. Examine these structures of the brain in figure 21.8.

Midsagittal Section of the Brain

Prosencephalon

Examine a midsagittal section of a brain as illustrated in figure 21.9. This section is seen by cutting the brain through the longitudinal fissure. Locate the C-shaped **corpus callosum,** which connects the two cerebral hemispheres. The posterior portion

of the corpus callosum is known as the **splenium,** and the anterior portion is the **genu.** Just inferior to the corpus callosum is the **septum pellucidum,** which separates the lateral ventricle of the brain from the third ventricle. If the septum pellucidum is missing, you will be able to look into the lateral ventricle without obstruction. Locate these structures in figure 21.9.

Examine the material in the lab and locate the **diencephalon,** which consists of, in part, the thalamus and the hypothalamus. The **thalamus** forms the lateral wall around the third ventricle and is a relay center that receives information from various tracts in the CNS and sends them to the cerebral cortex.

Below the thalamus is the **hypothalamus,** which has numerous autonomic centers. The hypothalamus in part directs the ANS and is involved with the **pituitary gland** (in the hypothalamopituitary axis) in many endocrine functions. Centers for thirst, water balance, pleasure, rage, sexual desire, hunger, sleep patterns, temperature, and aggression are located in the hypothalamus.

You should also be able to locate the **mammillary bodies** on the inferior portion of the diencephalon and the **optic chiasma** just anterior to it. The **pineal gland** is located posterior to the thalamus and is an endocrine gland that secretes melatonin. Both the pineal gland and the pituitary gland are covered more in depth in Laboratory Exercise 28.

Mesencephalon

The midbrain is a small area posterior to the diencephalon in a midsagittal section. This small area consists of the **cerebral peduncles,** which occupy an area anterior to the pons and on the ventral surface of the brain. The **mesencephalic (cerebral)**

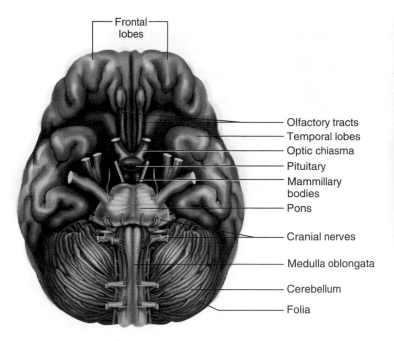

Frontal lobes

Olfactory tracts
Temporal lobes
Optic chiasma
Pituitary
Mammillary bodies
Pons

Cranial nerves

Medulla oblongata

Cerebellum

Folia

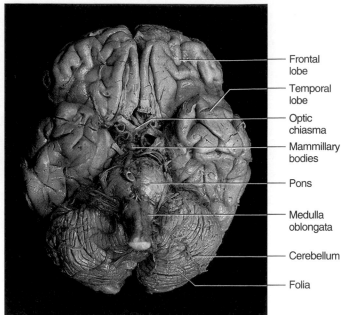

Frontal lobe
Temporal lobe
Optic chiasma
Mammillary bodies
Pons
Medulla oblongata
Cerebellum
Folia

Figure 21.8 Brain, inferior view.

aqueduct passes through the mesencephalon, with the **peduncles** being inferior and the **tectum** as a roof above the aqueduct. Locate these features in the material in the lab and in figure 21.9. Above the tectum are four hemispheric processes known as the **corpora quadrigemina,** which consist of the **superior colliculi** (areas of visual reflexes) and the **inferior colliculi** (areas of auditory reflexes). The midbrain also houses a center known as the **substantia nigra** (not seen in midsagittal sections) that, when not functioning properly, causes Parkinson's disease.

Rhombencephalon

The **rhombencephalon** (hindbrain) consists of an anterior bulge known as the **pons,** a terminal **medulla oblongata,** and the highly convoluted **cerebellum** (figure 21.9). The pons is a relay center shunting information from the inferior regions of the body through the thalamus and to other areas of the brain. The pons has important respiratory centers that are involved in controlling breathing rate.

The medulla oblongata not only has centers for respiratory rate control but also for the control of blood pressure. Some information from the right side of the body crosses over to the left brain in the medulla oblongata. The same occurs for information coming from the left side of the body in that information crosses over to the right side of the brain. The area of crossing over in the medulla is known as the **decussation of the pyramids.** The medulla oblongata terminates at the foramen magnum and becomes the cervical region of the spinal cord.

The cerebellum is a location primarily noted for muscle coordination and maintenance of posture. The cerebellum consists of an outer **cerebellar cortex** and an inner, extensively branched pattern of white matter known as the **arbor vitae.** The **folia** of the cerebellum can be seen in this section. The triangular space anterior to the cerebellum is the fourth ventricle and can be seen in figure 21.9. Examine the features of the rhombencephalon as described here and seen in material in the lab.

Coronal Section of the Brain

The brain consists of **unmyelinated** gray matter and **myelinated** white matter. The **gray matter** of the brain is extensive and forms the **cerebral cortex.** Most of the active, integrative processes of the brain occur in the cerebral cortex (figure 21.10). The cerebral cortex is approximately 4 mm thick and occupies the superficial regions of the brain. Deep to the cortex is the **white matter** of the brain, which consists of **tracts** that take information from deeper regions of the brain to the cerebral cortex for processing. Sensory information coming from the spinal cord moves through the inferior regions of the brain and through the white matter for integration in the cerebral cortex. White matter can also take information from one region of the cerebral cortex to another for integration or from the cerebral cortex back to the spinal cord and to other parts of the body for action.

Gray matter is not restricted to the cerebral cortex, however. Deep islands of gray matter in the brain compose the **basal nuclei,** which can also be seen in this section. Basal

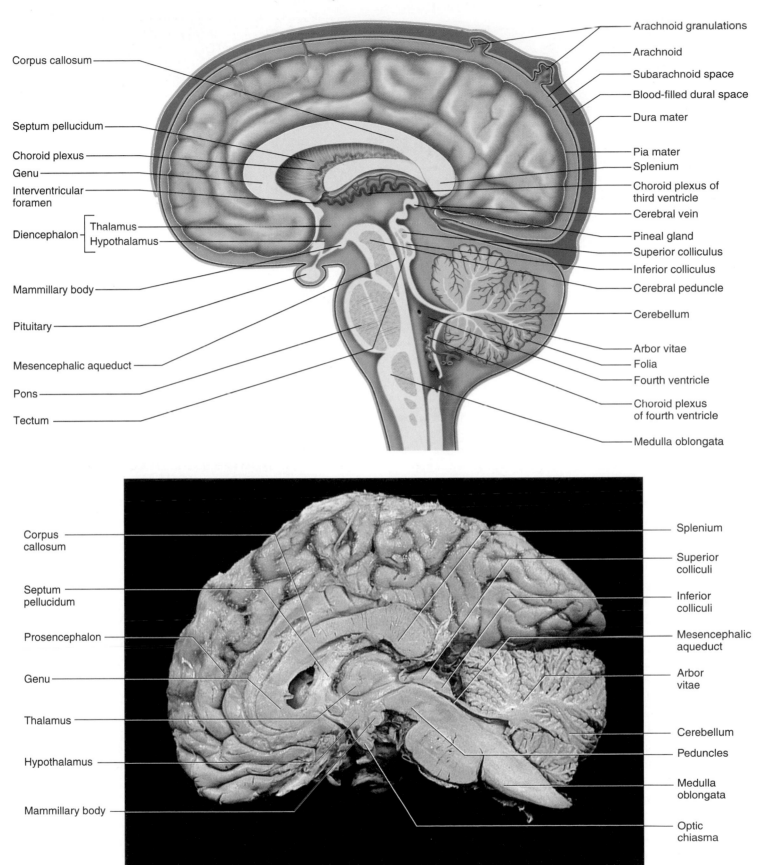

Figure 21.9   Brain, midsagittal section.

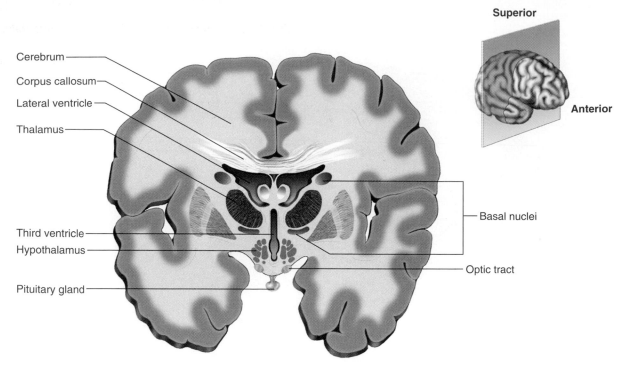

Superior

Anterior

Cerebrum

Corpus callosum

Lateral ventricle

Thalamus

Third ventricle

Hypothalamus

Pituitary gland

Basal nuclei

Optic tract

Figure 21.10   Brain, coronal section.

nuclei serve a number of functions in the brain, many of which involve subconscious processes such as the swinging of arms while walking or the regulation of muscle tone.

Limbic System

Part of the limbic system may be seen in a coronal section. The limbic system is a very complex region of the brain that is involved in mood, emotion, and has centers for feeding, sexual desire, fear, and satisfaction. The inferior portion of the limbic system has neural fibers that come from the olfactory regions of the brain. These are best seen in a model of the limbic system or in a transected brain. Examine figure 21.11 for the major features of the limbic system.

Brainstem

The brainstem consists of the mesencephalon, the pons, and the medulla oblongata. Look at a model or section of brain that has had the cerebrum removed. Locate the corpora quadrigemina (figure 21.12) along with the medulla oblongata and the pons.

Examine table 21.1 for a listing of the specific regions of the brain.

Development of the Central Nervous System

The brain begins development, as does the rest of the nervous system, in the third week of pregnancy as a **neural groove** in the ectoderm. By the fourth week the brain has folded into a **neural tube** that contains the **central canal** (figure 21.13). The posterior portion of the central canal becomes the central canal of the

Table 21.1   Regions of the Brain

Prosencephalon
Telencephalon
   Cerebrum (cerebral hemispheres)
    Cerebral cortex (gray matter)
    Basal nuclei
   Corpus callosum
Diencephalon
   Pineal body
   Thalamus
   Hypothalamus
   Pituitary gland
   Mammillary bodies

Mesencephalon
Peduncles
Tectum
Corpora quadrigemina
   Superior colliculus
   Inferior colliculus

Rhombencephalon
Metencephalon
   Pons
   Cerebellum
Myelencephalon
   Medulla oblongata

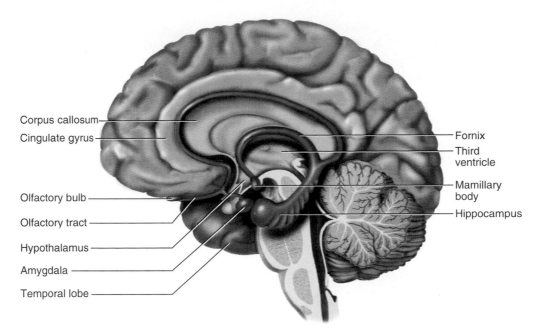

Figure 21.11   Limbic system.

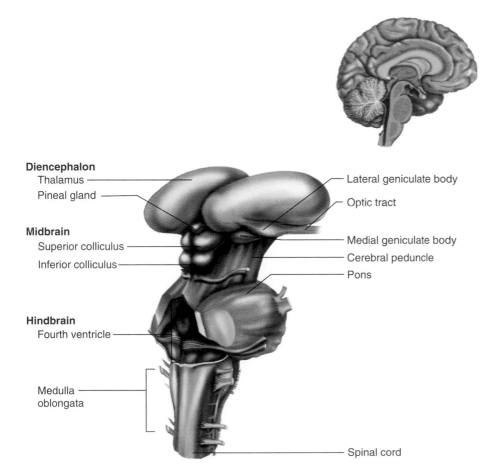

Figure 21.12   Brain stem, superior view.

spinal cord. The anterior portion of the canal becomes the ventricles of the brain in adults. By the sixth week of development the cerebral hemispheres begin to form and continue their development throughout pregnancy. Young adults have perhaps 50 billion neurons in the cerebral cortex.

Cranial Nerves

The paired cranial nerves are part of the PNS but they are frequently studied along with the brain. The cranial nerves are listed by Roman numeral, and you should know the nerve by name and by number. Examine a model of the brain along with figure 21.14 and note that all of the cranial nerves except for nerve XII are in sequence from anterior to posterior. Nerves may be sensory, motor, or mixed (both sensory and motor).

The **olfactory nerve** runs along the anterior base of the brain at the inferior aspect of the frontal lobe. The **optic nerve** comes from the eye to the base of the brain and forms the optic chiasma. Some tracts remain on one side of the brain while others cross to the other side. The **oculomotor nerve** can be found anterior to the pons, more or less in the midline of the brain, while the **trochlear nerve** is found at about a

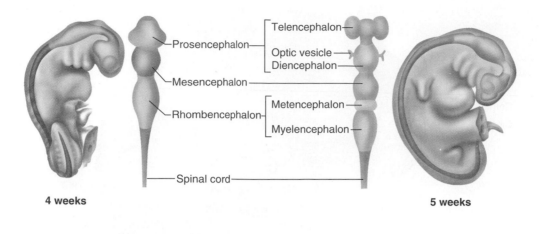

4 weeks   5 weeks

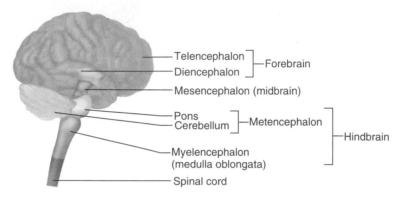

Figure 21.13   Brain development.

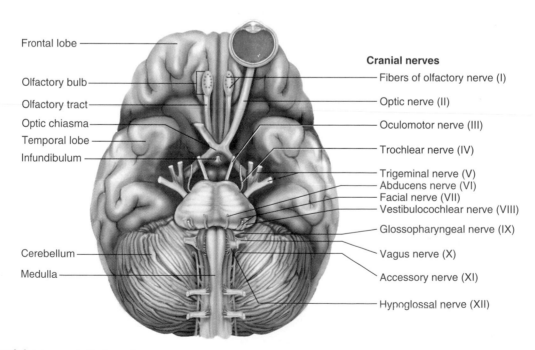

Figure 21.14   Cranial nerves, inferior view.

Continued

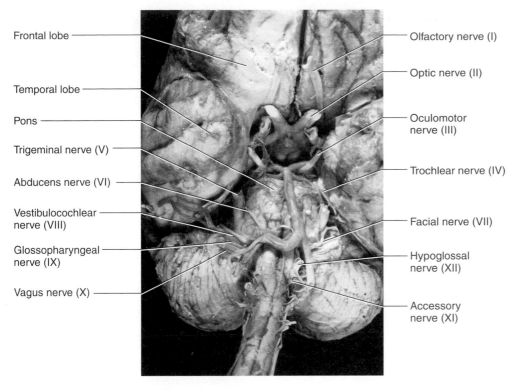

Frontal lobe

Temporal lobe

Pons

Trigeminal nerve (V)

Abducens nerve (VI)

Vestibulocochlear
nerve (VIII)

Glossopharyngeal
nerve (IX)

Vagus nerve (X)

Olfactory nerve (I)

Optic nerve (II)

Oculomotor
nerve (III)

Trochlear nerve (IV)

Facial nerve (VII)

Hypoglossal
nerve (XII)

Accessory
nerve (XI)

Figure 21.14—*continued.*

45° angle from midline on the lateral aspect of the pons. The large **trigeminal nerve** is found at a 90° angle to the pons and is located on the lateral aspect of the pons. The **abducens nerve** is found at the midline junction of the pons and medulla oblongata, while the **facial nerve** is more lateral. Posterior to the facial nerve is the **vestibulocochlear nerve,** and the nerve directly behind that is the **glossopharyngeal nerve.** The **vagus nerve** (*vagus* = wandering) is seen as a large nerve or large cluster of fibers, and on the lateral aspect of the medulla oblongata is the **accessory nerve.** More towards the midline is the **hypoglossal nerve,** which is the last of the cranial nerves. Locate these nerves in figure 21.14 and note their details in tables 21.2 and 21.3.

The cranial nerves are listed in table 21.4 as sensory nerves, motor nerves, or both sensory and motor nerves. A mnemonic device also follows to help you remember the function of the nerve. The first letter of the mnemonic represents the first letter.

Table 21.2   Cranial Nerves—Function

Number	Name	Function
I	Olfactory	Receives sensory information from the nose
II	Optic	Receives sensory information from the eye, transmitting the sense of vision to the brain
III	Oculomotor	Transmits motor information to move the eye muscles particularly to the medial, superior, and inferior rectus muscles and to the inferior oblique muscle
IV	Trochlear	Transmits motor information to move the eye muscles, particularly the superior oblique muscle
V	Trigeminal	A three-branched nerve; transmits both sensory information from, and motor information to, the head
VI	Abducens	A motor nerve to move the eye muscles, particularly the lateral rectus muscle
VII	Facial	A large nerve that receives sensory information from the anterior tongue and takes motor information to the head muscles
VIII	Vestibulocochlear	Receives sensory information from the ear; the vestibular part transmits equilibrium information and the cochlear part transmits acoustic information
IX	Glossopharyngeal	A mixed nerve of the tongue and throat that receives information on taste
X	Vagus	Receives sensory information from abdomen, thorax, neck, and root of tongue; transmits motor information to pharynx, larynx, and controls autonomic functions of heart, digestive organs, spleen, and kidneys
XI	Accessory	A motor nerve to the muscles of the neck that move the head
XII	Hypoglossal	A motor nerve to the tongue

Table 21.3 Cranial Nerves—Location

Number	Name	Location
I	Olfactory	Begin in the upper nasal cavity and pass through the cribriform plate of the ethmoid bone. They synapse in the olfactory bulbs on either side of the longitudinal fissure of the brain. The fibers take information on the sense of smell and pass through the olfactory tracts to be interpreted in the temporal lobe of the brain.
II	Optic	Take sensory information from the retina at the back of the eye and transmit the impulses through the optic canal in the sphenoid bone. They then cross at the optic chiasma and pass via the optic tracts to the occipital lobe, where vision is interpreted.
III	Oculomotor	Emerge from the surface of the brain near the midline and just anterior to the pons. They pass through the superior orbital fissure and innervate the inferior oblique muscle and the medial, superior, and inferior rectus muscles and also carry parasympathetic fibers to the lens and iris.
IV	Trochlear	Seen at the sides of the pons at about a 45° angle from the midline of the brain. They pass through the superior orbital fissure to the superior oblique muscle.
V	Trigeminal	Seen at a 90° angle from the midline at the lateral sides of the pons. The trigeminal has three branches: (1) the ophthalmic branch passes through the superior orbital fissure; (2) the maxillary branch passes through the foramen rotundum of the sphenoid bone; (3) the mandibular branch passes through the foramen ovale of the sphenoid bone and enters the mandible by the mandible foramen and exits by the mental foramen.
VI	Abducens	Begins at the midline junction between the pons and the medulla oblongata and passes through the superior orbital fissure to carry motor information to the lateral rectus muscle of the eye.
VII	Facial	Begins as the first of a cluster of nerves on the anterolateral part of the medulla oblongata. It passes through the internal auditory meatus and through the inner ear to the stylomastoid foramen of the temporal bone to innervate facial muscles and glands. It receives sensory information from the anterior tongue. Sensory information of the tongue is interpreted in the temporal lobe of the brain.
VIII	Vestibulocochlear	Comes from the inner ear and passes through the internal auditory meatus. The conduction passes to the pons, and hearing and balance are interpreted in the temporal lobe.
IX	Glossopharyngeal	Passes through the jugular foramen to innervate muscles of the throat (pharyngeal branches) and the tongue (glossal branches). Motor portions of the nerve control some muscles of swallowing and salivary glands, while sensory nerves receive information from the posterior tongue and from baroreceptors of the carotid artery.
X	Vagus	Passes through the jugular foramen and along the neck to the larynx, heart, and abdominal region. The sensory impulses travel in this nerve from the viscera in the abdomen, the thorax, the neck, and the root of the tongue to the brain.
XI	Accessory	Multiple fibers arise from the lateral sides of the medulla oblongata and pass through the jugular foramen to numerous muscles of the neck.
XII	Hypoglossal	Begins at the anterior surface of the medulla and passes through the hypoglossal canal to innervate the muscles of the tongue.

Dissection of the Sheep Brain

Bring a sheep brain back to your table along with a dissecting tray and appropriate dissection tools. If the brains still have the **dura mater,** examine this tough connective tissue coat that occurs on the outside of the brain. Cut through this layer to examine the meninges that occur underneath it. Deep to the dura mater is a filmy layer of tissue that contains blood vessels. This is known as the **arachnoid membrane.** If you tease some of the membrane away from the brain you will see that it has a cobweblike appearance in the **subarachnoid space.** The subarachnoid space may contain some fluid, which is the CSF. Underneath this layer and adhering directly to the brain convolutions is the **pia mater,** which is the surface lining of the convolutions of the brain. Work in pairs during the dissection of the sheep brain. Examine the features presented in the beginning of this exercise and locate the structures that occur in figure 21.15.

Laboratory 21   Structure and Function of the Brain and Cranial Nerves   **253**

Table 21.4 Type of Cranial Nerves

Number	Name	Type	Mnemonic
I	Olfactory	Sensory	Sally
II	Optic	Sensory	Sells
III	Oculomotor	Motor*	Many
IV	Trochlear	Motor	Monkeys
V	Trigeminal	Both	But
VI	Abducens	Motor	My
VII	Facial	Both	Brother
VIII	Vestibulocochlear	Sensory	Sells
IX	Glossopharyngeal	Both	Bigger
X	Vagus	Both	Better
XI	Accessory	Motor	Mega
XII	Hypoglossal	Motor	Monkeys

* Many of the motor nerves have sensory fibers that come from proprioreceptors in the muscles they innervate. Information about the tension of the muscle is sent back to the brain to make adjustments in contractile rate. Since the main function of these nerves is motor, they are listed as motor nerves even though they do have some sensory capabilities.

Find the major lobes of the brain, the cerebellum, pons and medulla oblongata. Since sheep are quadrupeds, the flexure of the brain does not occur in them as it does in humans. Sheep have a horizontal spinal cord while humans have a vertical one. Sheep also have a reduced cerebrum. Examine the inferior surface of the sheep brain. You should see the olfactory bulbs and tracts, the optic nerve, and optic chiasma easily from this view. The pituitary gland will probably not be attached but you should locate the infundibulum, caudad to the optic chiasma. Locate these structures in figure 21.16. If the sheep brain is intact, you will need to decide which brain will be sectioned in the midsagittal plane and which will be sectioned in the coronal plane. For the midsagittal section, divide the brain along the length of the longitudinal fissure. Your cut should reflect a section illustrated in figure 21.17. Note that the sheep brain has an enlarged corpora quadrigemina compared to humans. Locate the corpus callosum, lateral ventricles, third ventricle, hypothalamus, pineal gland, superior and inferior colliculi, cerebellum, arbor vitae, pons, medulla oblongata, mesencephalic aqueduct, and the fourth ventricle.

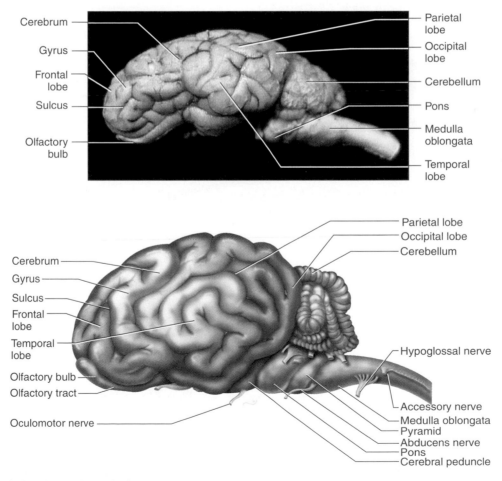

Figure 21.15   Brain of the sheep, lateral view.

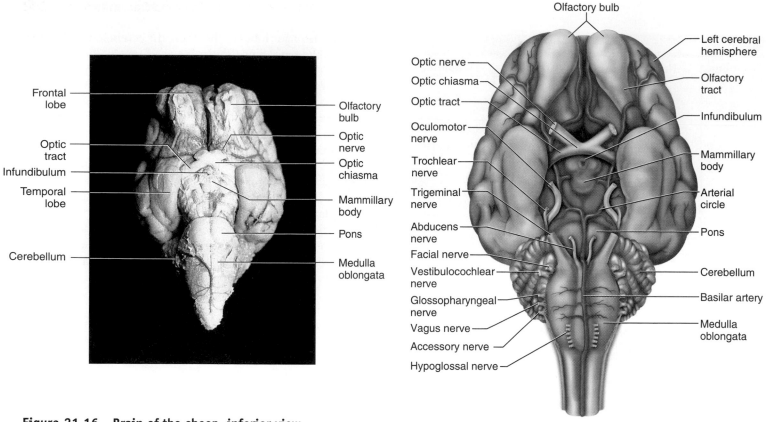

Frontal lobe
Optic tract
Infundibulum
Temporal lobe
Cerebellum

Olfactory bulb
Optic nerve
Optic chiasma
Mammillary body
Pons
Medulla oblongata

Olfactory bulb
Optic nerve
Optic chiasma
Optic tract
Oculomotor nerve
Trochlear nerve
Trigeminal nerve
Abducens nerve
Facial nerve
Vestibulocochlear nerve
Glossopharyngeal nerve
Vagus nerve
Accessory nerve
Hypoglossal nerve

Left cerebral hemisphere
Olfactory tract
Infundibulum
Mammillary body
Arterial circle
Pons
Cerebellum
Basilar artery
Medulla oblongata

Figure 21.16   Brain of the sheep, inferior view.

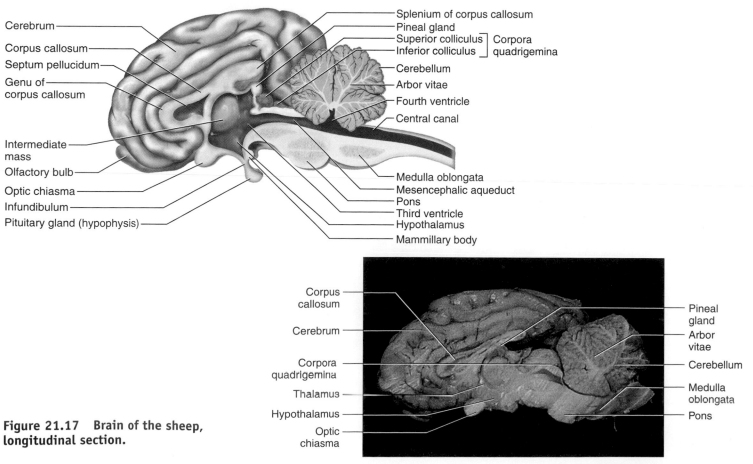

Cerebrum
Corpus callosum
Septum pellucidum
Genu of corpus callosum
Intermediate mass
Olfactory bulb
Optic chiasma
Infundibulum
Pituitary gland (hypophysis)

Splenium of corpus callosum
Pineal gland
Superior colliculus ⎤ Corpora
Inferior colliculus ⎦ quadrigemina
Cerebellum
Arbor vitae
Fourth ventricle
Central canal
Medulla oblongata
Mesencephalic aqueduct
Pons
Third ventricle
Hypothalamus
Mammillary body

Corpus callosum
Cerebrum
Corpora quadrigemina
Thalamus
Hypothalamus
Optic chiasma

Pineal gland
Arbor vitae
Cerebellum
Medulla oblongata
Pons

Figure 21.17   Brain of the sheep, longitudinal section.

The coronal section of a sheep brain is illustrated in figure 21.18. After making a coronal section about midway through the cerebrum, you should locate the cerebral cortex, cerebral medulla, the lateral ventricles, the corpus callosum, the third ventricle, the thalamus, and the hypothalamus.

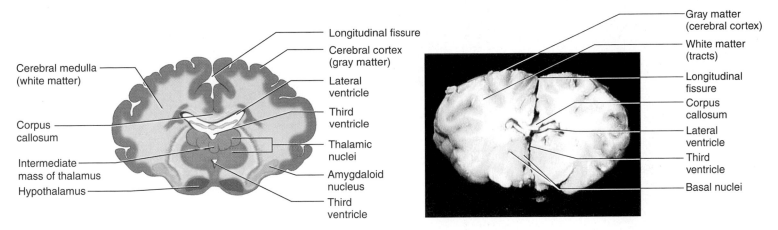

Figure 21.18   Brain of the sheep, coronal section.

REVIEW

Name _____

1. Which of the meninges is between the outer and inner meninges?

2. Name the major veins that take blood from the brain.

3. The basilic artery in the brain receives blood from what two arteries?

4. What fluid is found in the ventricles of the brain?

5. Where does fluid flow from the mesencephalic aqueduct?

6. What shape does a sulcus have compared to a gyrus?

7. What are the lobes of the cerebrum called?

8. What function does the precentral gyrus have?

9. What sense does the temporal lobe interpret?

10. What depression separates the temporal lobe from the parietal lobe?

11. What structure connects the cerebral hemispheres?

12. Name the major regions of the mesencephalon.

13. What function does the cerebellum have?

14. What function does the optic nerve have?

15. What function does the trochlear nerve have?

16. John pulled a "no-brainer" by hitting his forehead against the wall. What possible damage might he do to the function of his brain, particularly those functions associated with the frontal lobe?

17. If a stroke affected all of the sensations interpreted by the brain just concerning the face and the hands, how much of the postcentral gyrus would be affected?

18. Approximately 91% of people are right handed, and the description of cerebral hemispheres is appropriate for them. In left-handed individuals the pattern is typically reversed. One convenient excuse that people often make for their inability to do something is to describe themselves as left-brained or right-brained individuals. Describe what effect the loss of an entire cerebral hemisphere would have on specific functions such as spatial awareness or the ability to speak.

19. Aphasia is loss of speech. There are different types of aphasia that can occur. If Broca's area was affected by a stroke, would the content of the spoken word be affected or would the ability to pronounce the words be affected?

20 Label the following illustration using the terms provided.

Corpus callosum   Arbor vitae
Mesencephalic aqueduct   Hypothalamus
Pons   Medulla oblongata
Pineal gland   Pituitary gland
Thalamus   Fourth ventricle

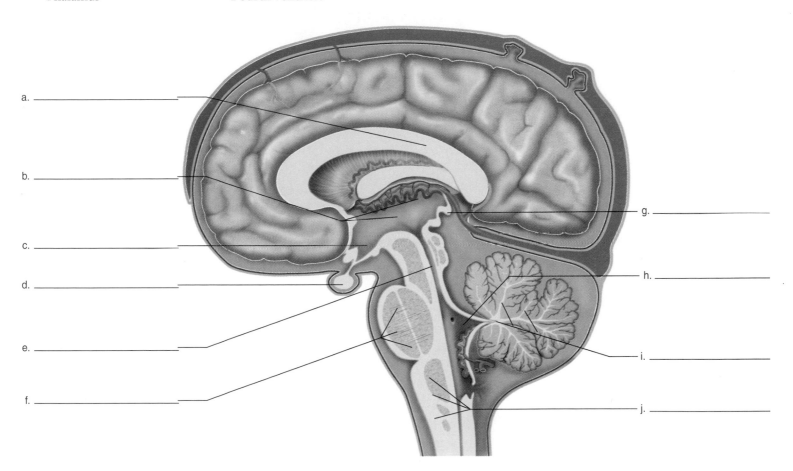

a. _____

b. _____

c. _____

d. _____

e. _____

f. _____

g. _____

h. _____

i. _____

j. _____

Brain, midsagittal section.

21. Label the following illustration using the terms provided.

Olfactory nerve   Pons

Trigeminal nerve   Oculomotor nerve

Vagus nerve   Cerebellum

Medulla oblongata   Optic nerve

Hypoglossal nerves

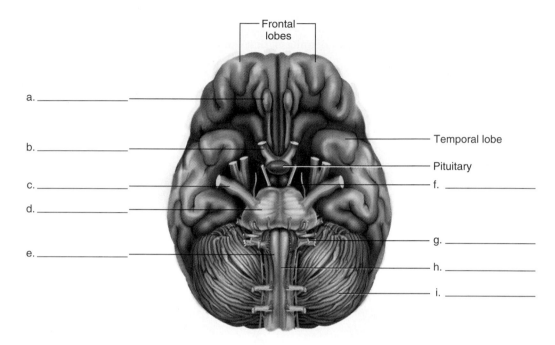

Structure and Function of the Spinal Cord and Nerves

Introduction

The spinal cord is part of the central nervous system, which begins at the foramen magnum of the skull and ends at about vertebra L1 or L2. The spinal cord stops growing early in life, yet the vertebral bodies continue to grow, causing the spinal cord to be shorter than the vertebral canal. The spinal cord receives sensory information from, and transmits motor information to the spinal nerves, which radiate into the body as peripheral nerves. In this exercise, you learn the major features of the spinal cord in longitudinal aspect and in cross section and the major nerves and plexuses of the body.

Objectives

At the end of this exercise you should be able to

1. demonstrate the major regions in a cross section of spinal cord;
2. describe the nature of the longitudinal aspect of the spinal cord;
3. list the major nerves that arise from each plexus;
4. name all of the major nerves of the upper and lower extremities;
5. list the structures that carry impulses to and away from the spinal cord.

Materials

Models or charts of the central and peripheral nervous systems
Prepared slide of a spinal cord in cross section
Model or chart of a spinal cord in cross section and
    longitudinal section

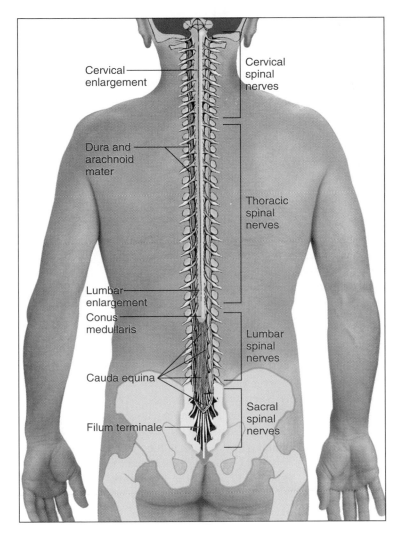

Figure 22.1   Spinal cord, longitudinal view.

Procedure

Longitudinal Aspect of the Spinal Cord

Examine a model or chart in the lab of a longitudinal view of the spinal cord and locate the major features illustrated in figure 22.1. The spinal cord terminates inferiorly as the **conus medullaris** at approximately vertebra L1 and is attached to the coccyx by a continuation of the pia mater known as the **filum terminale.** The neural continuation of the spinal cord exists as

an extension of spinal nerves from L2 through S5. These parallel nerves resemble a horse's tail and are called the **cauda equina.** The **cervical enlargement** occurs at C3 through T2 and represents a bulge in the spinal cord that has increased neural input and output to the upper extremities. The **lumbar enlargement** occurs at about T7 through T11, and this expanse is due to the increased input and output to the lower extremity. Compare the material in the lab to figure 22.1.

261

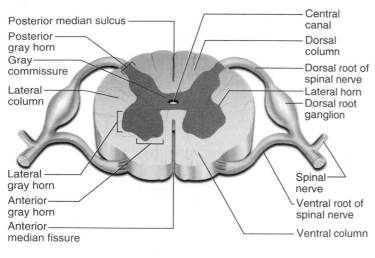

Figure 22.2   Spinal cord, cross section. ✗

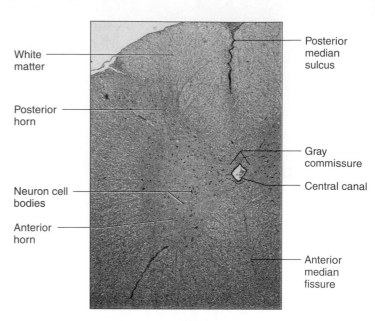

Figure 22.3   Histology of the spinal cord (10×). ✗

Cross Section of Spinal Cord

Examine a model or chart in the lab of a cross section of the spinal cord. Locate the **gray matter,** which appears as an H pattern or a butterfly pattern in the middle of the spinal cord, with the **white matter** located on the periphery of the cord.

In a cross section of the spinal cord you should see the gray matter divided in two narrow horns and two rounded horns. The narrow horns are known as the **posterior gray horns,** and these areas receive sensory information from the spinal nerves. The rounded horns are the **anterior gray horns,** and these send motor signals to the spinal nerves. In some parts of the spinal cord there are additional sections of gray matter known as the **lateral gray horns.**

Each side of the gray matter is connected to the other by a crossbar known as the **gray commissure.** In the middle of the gray commissure is the **central canal,** which runs the length of the spinal cord. The white matter of the cord is divided into **tracts,** or funiculi, which take sensory information to the brain or motor information from the brain. The tracts that take sensory information to the brain are called **ascending tracts,** and those that receive motor information are called **descending tracts.** You should also be able to see a depression in the posterior surface of the spinal cord. This is the **posterior median sulcus,** while the deeper depression on the anterior side is known as the **anterior median fissure.** Examine a model or chart in the lab and compare it to figure 22.2.

Histology of the Spinal Cord

Examine a prepared slide of the spinal cord under low power and locate the **anterior** and **posterior gray horns,** the **gray commissure,** the **central canal,** the **posterior median sulcus,** and the **anterior median fissure.** Examine the anterior gray horn of

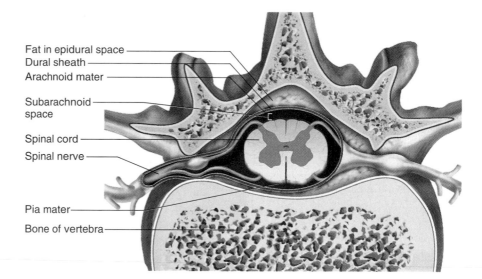

Figure 22.4   Spinal meninges.

the spinal cord and look for the nerve cell bodies of the **multipolar neurons** there. Compare your slide with figure 22.3.

Meninges

The spinal cord is covered by **meninges,** as is the brain. The outer covering of the cord consists of the **dura mater** (dural sheath) the next deeper layer is the **arachnoid mater (membrane).** Deep to the arachnoid membrane is the **pia mater,** which is the innermost of the meninges and a thin cover on the spinal cord proper. Between the pia mater and the arachnoid is the **subarachnoid space,** which contains the **cerebrospinal fluid.** Examine figure 22.4 for an illustration of the meninges of the spinal cord.

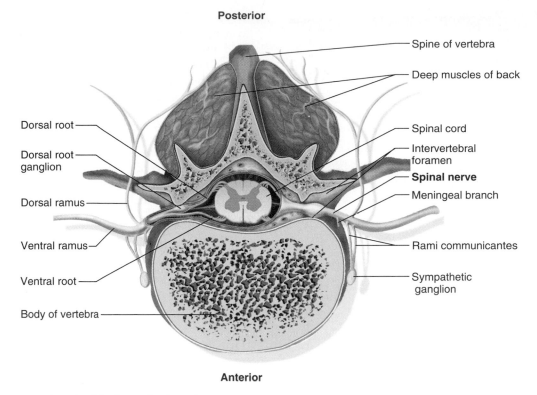

Figure 22.5   Nerves associated with the spinal cord.

Nerves Associated with the Spinal Cord

The **dorsal ramus** is a branch of sensory and motor nerve fibers that, with the **ventral ramus,** unite to form a **spinal nerve.** The spinal nerve enters the **intervertebral foramen** and divides into a **dorsal root** and a **ventral root.** The dorsal root has a dorsal root ganglion that contains the nerve cell bodies of the sensory nerves. The **dorsal root** carries sensory information to the posterior gray horn of the spinal cord. The nerve cell bodies of the motor nerves are located in the anterior gray horn of the spinal cord and exit via the ventral root. Examine the material in the lab and compare it to figure 22.5.

Nerve Structure

Neuron fibers are clustered in parallel arrangements called **nerves** in the peripheral nervous system and **tracts** in the central nervous system. Nerves have a number of connective tissue wrappings that envelope the individual nerve fibers, clusters of fibers, and the entire nerve. The sheath that wraps around single nerve fibers and their myelin sheaths is the **endoneurium,** and the sheath that wraps around groups of nerve fibers (nerve fascicle) is the **perineurium.** The wrapping that covers the entire nerve is called the **epineurium.** Examine figure 22.6 for these layers.

Spinal Nerves and Plexuses

There are 31 pairs of **spinal nerves** that exit from the spinal cord. The spinal nerves are **mixed nerves** carrying both sensory and motor information. The spinal nerves pass through the interver-

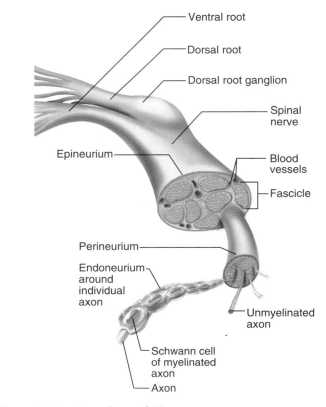

Figure 22.6   Coverings of the nerve.

tebral foramina and are named according to their region of origin. There are **8 cervical nerves, 12 thoracic nerves, 5 lumbar nerves, 5 sacral nerves,** and **1 coccygeal nerve.** Some of these

Table 22.1	Plexuses	
Name	**Spinal Nerves Contributing to Plexus**	**Major Nerves of Plexus**
Cervical	C1–5	Phrenic
Brachial	C5–T1	Radial, median, ulnar, musculocutaneous, axillary
Lumbar	L1–4	Femoral
Sacral	L5–S4	Sciatic (tibial and common peroneal)

nerves exit the spinal cord, and branches of spinal nerves travel to parts of the body individually while others exit the spinal cord and form a branching network with other spinal nerves. These networks are called **plexuses,** and there are generally four recognized plexuses. The composition of the plexuses are outlined in table 22.1 and illustrated in figure 22.7.

Cervical and Brachial Plexus Nerves

An important nerve that exits from each side of the cervical plexus is the **phrenic nerve.** This nerve runs to the diaphragm and is responsible for its contraction in breathing. Examine the nerves of the **cervical plexus** and **brachial plexus** in figure 22.8.

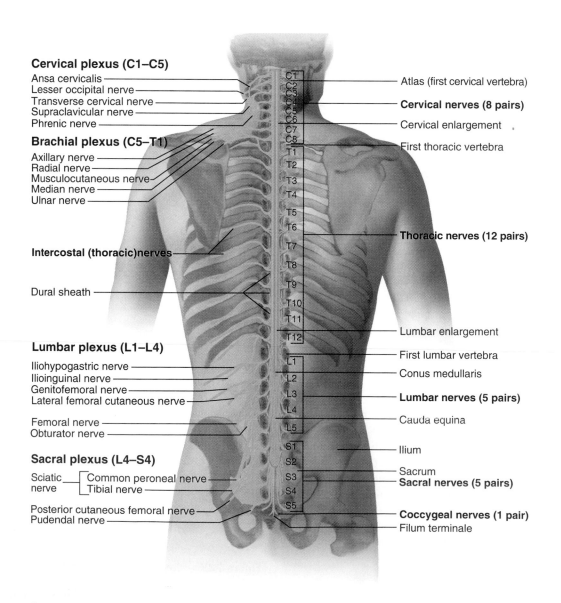

Cervical plexus (C1–C5)
Ansa cervicalis
Lesser occipital nerve
Transverse cervical nerve
Supraclavicular nerve
Phrenic nerve

Brachial plexus (C5–T1)
Axillary nerve
Radial nerve
Musculocutaneous nerve
Median nerve
Ulnar nerve

Intercostal (thoracic) nerves

Dural sheath

Lumbar plexus (L1–L4)
Iliohypogastric nerve
Ilioinguinal nerve
Genitofemoral nerve
Lateral femoral cutaneous nerve
Femoral nerve
Obturator nerve

Sacral plexus (L4–S4)
Sciatic nerve — Common peroneal nerve — Tibial nerve
Posterior cutaneous femoral nerve
Pudendal nerve

C1 C2 C3 C4 C5 C6 C7 C8
T1 T2 T3 T4 T5 T6 T7 T8 T9 T10 T11 T12
L1 L2 L3 L4 L5
S1 S2 S3 S4 S5

Atlas (first cervical vertebra)
Cervical nerves (8 pairs)
Cervical enlargement
First thoracic vertebra
Thoracic nerves (12 pairs)
Lumbar enlargement
First lumbar vertebra
Conus medullaris
Lumbar nerves (5 pairs)
Cauda equina
Ilium
Sacrum
Sacral nerves (5 pairs)
Coccygeal nerves (1 pair)
Filum terminale

Figure 22.7 Nerve plexuses.

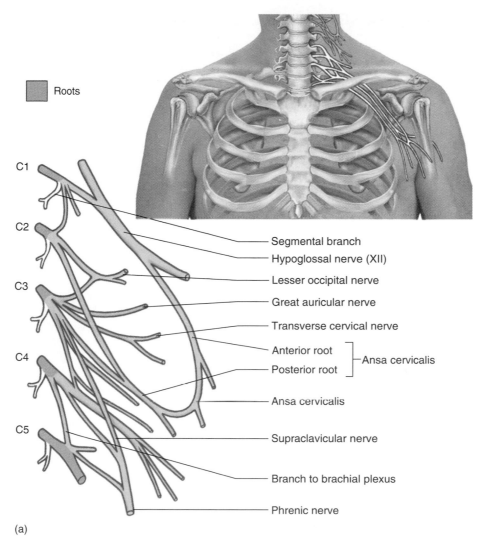

Roots

C1

C2

C3

C4

C5

Segmental branch
Hypoglossal nerve (XII)
Lesser occipital nerve
Great auricular nerve
Transverse cervical nerve
Anterior root ⎤
⎥ Ansa cervicalis
Posterior root ⎦
Ansa cervicalis
Supraclavicular nerve
Branch to brachial plexus
Phrenic nerve

(a)

Figure 22.8   Nerves of the cervical and brachial plexus. (a) Cervical plexus; (b) brachial plexus; (c) peripheral nerves of the upper extremity.   *Continued*

The nerves from the brachial plexus are those that primarily innervate the upper extremities. The muscular aspect of the innervation of these nerves was covered in the exercises on the specific muscles, but the major nerves are the **axillary nerve,** innervating the upper shoulder, the **musculocutaneous nerve,** innervating many of the muscles of the arm, the **median nerve,** running the length of each upper extremity and serving important hand and forearm flexors, the **ulnar nerve,** serving other forearm flexors, and the **radial nerve,** which predominately innervates the triceps brachii and the extensors of the hand. The ulnar nerve crosses behind the medial epicondyle of the humerus and is commonly known as the "funny bone." Examine these nerves in the lab and compare them to figure 22.8.

Lumbar and Sacral Plexus Nerves

The **femoral nerve** is the most significant structure arising from the lumbar plexus. This large nerve passes anteriorly across the inguinal ligament and mostly innervates the muscles of the anterior thigh. There are many nerves that come from the sacral plexus. Many innervate the pelvis and muscles that move the hip, thigh, and leg. Two of the nerves from this plexus, the tibial and common peroneal nerve, unite to form the **sciatic nerve,** a large nerve of the posterior thigh, which innervates the leg and foot. Examine the material in the lab and compare it to figure 22.9.

Thoracic Nerves

There are numerous nerves not associated with a plexus. The **thoracic nerves** are a good example of this. Many of the thoracic nerves exit through the intervertebral foramina of the vertebral column and innervate the ribs, muscles, and other structures of the thoracic wall. These nerves can be seen in figure 22.10.

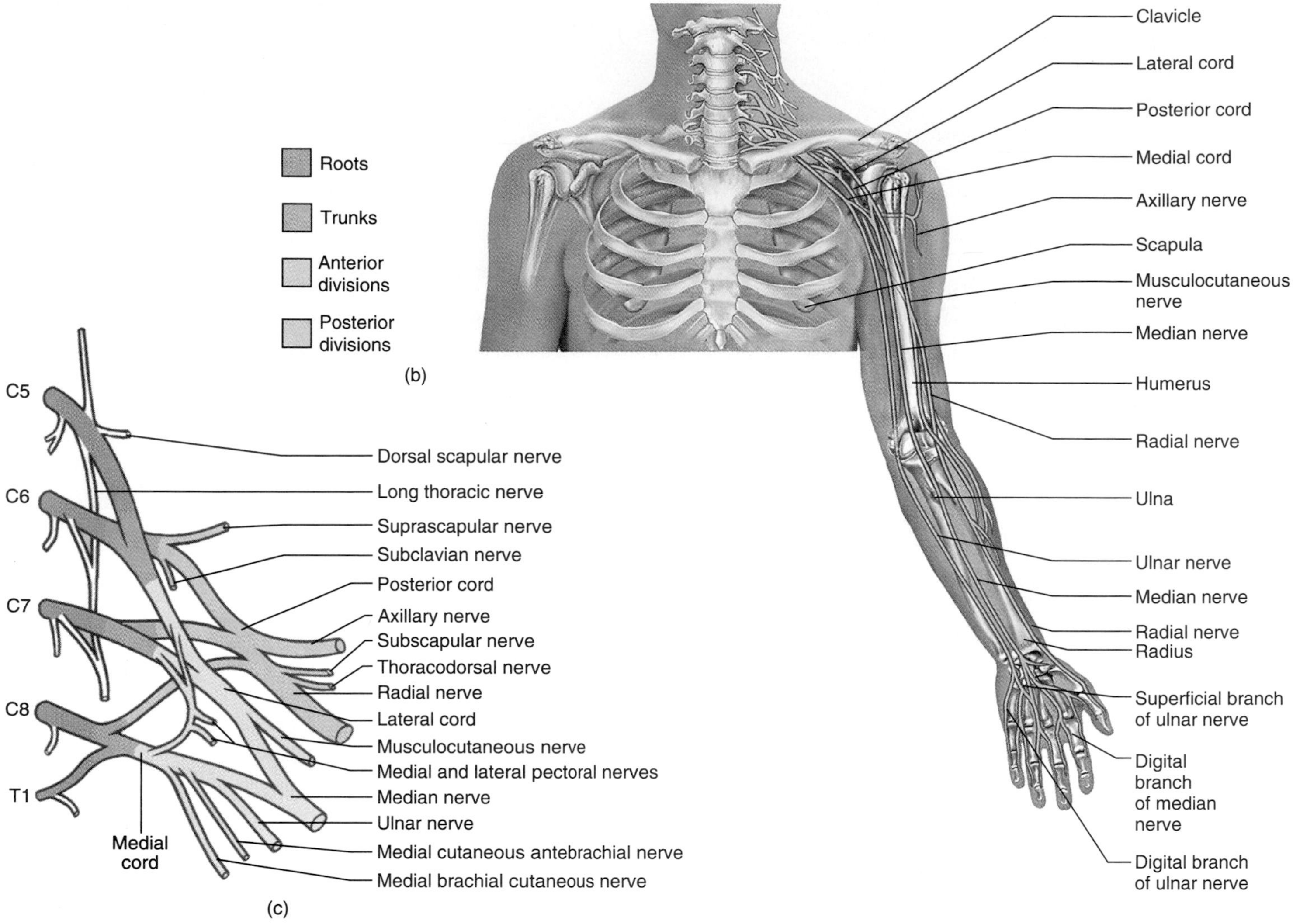

Roots

Trunks

Anterior divisions

Posterior divisions

(b)

C5

C6

C7

C8

T1

Medial cord

(c)

Dorsal scapular nerve
Long thoracic nerve
Suprascapular nerve
Subclavian nerve
Posterior cord
Axillary nerve
Subscapular nerve
Thoracodorsal nerve
Radial nerve
Lateral cord
Musculocutaneous nerve
Medial and lateral pectoral nerves
Median nerve
Ulnar nerve
Medial cutaneous antebrachial nerve
Medial brachial cutaneous nerve

Clavicle
Lateral cord
Posterior cord
Medial cord
Axillary nerve
Scapula
Musculocutaneous nerve
Median nerve
Humerus
Radial nerve
Ulna
Ulnar nerve
Median nerve
Radial nerve
Radius
Superficial branch of ulnar nerve
Digital branch of median nerve
Digital branch of ulnar nerve

Figure 22.8—*continued*.

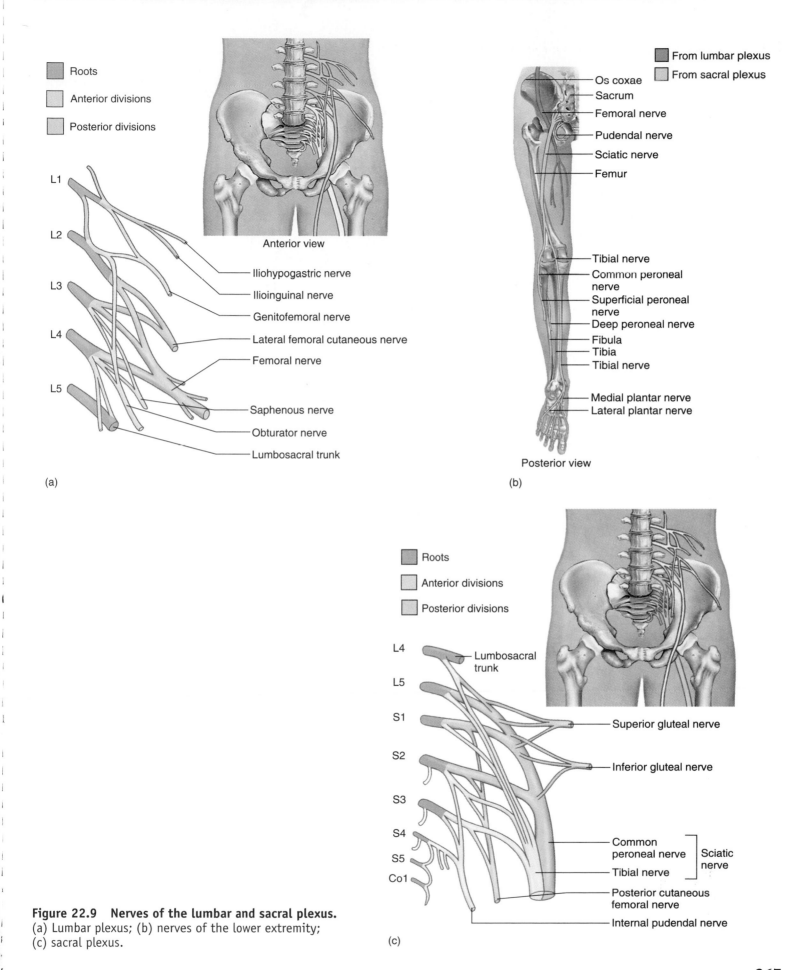

Figure 22.9 Nerves of the lumbar and sacral plexus.
(a) Lumbar plexus; (b) nerves of the lower extremity;
(c) sacral plexus.

Legend (a):
- Roots
- Anterior divisions
- Posterior divisions

Labels (a):
L1
L2
L3
L4
L5

Iliohypogastric nerve
Ilioinguinal nerve
Genitofemoral nerve
Lateral femoral cutaneous nerve
Femoral nerve
Saphenous nerve
Obturator nerve
Lumbosacral trunk

Anterior view

(a)

Legend (b):
- From lumbar plexus
- From sacral plexus

Labels (b):
Os coxae
Sacrum
Femoral nerve
Pudendal nerve
Sciatic nerve
Femur

Tibial nerve
Common peroneal nerve
Superficial peroneal nerve
Deep peroneal nerve
Fibula
Tibia
Tibial nerve
Medial plantar nerve
Lateral plantar nerve

Posterior view

(b)

Legend (c):
- Roots
- Anterior divisions
- Posterior divisions

Labels (c):
L4
L5
S1
S2
S3
S4
S5
Co1

Lumbosacral trunk
Superior gluteal nerve
Inferior gluteal nerve
Common peroneal nerve
Tibial nerve
Sciatic nerve
Posterior cutaneous femoral nerve
Internal pudendal nerve

(c)

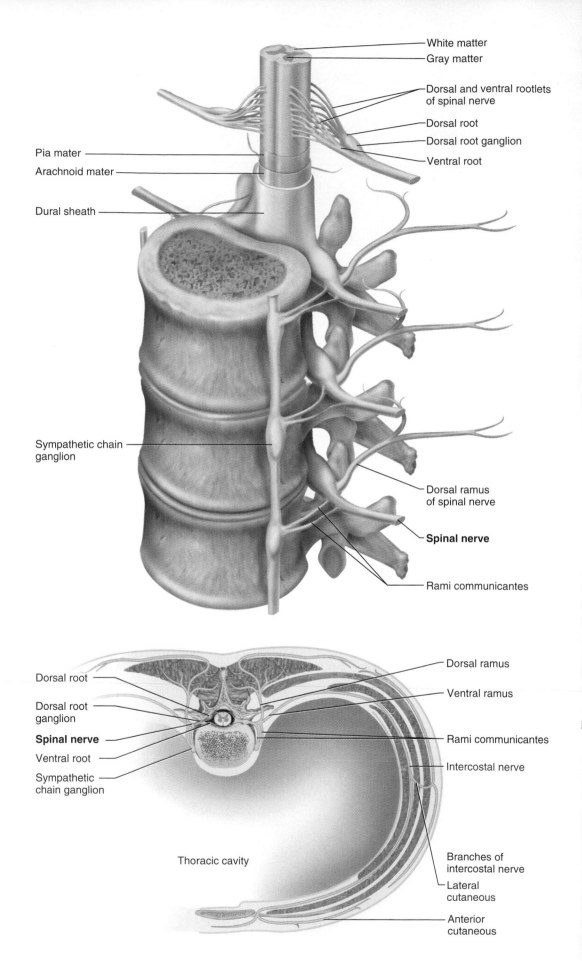

White matter

Gray matter

Dorsal and ventral rootlets of spinal nerve

Dorsal root

Dorsal root ganglion

Ventral root

Pia mater

Arachnoid mater

Dural sheath

Sympathetic chain ganglion

Dorsal ramus of spinal nerve

Spinal nerve

Rami communicantes

Dorsal root

Dorsal root ganglion

Spinal nerve

Ventral root

Sympathetic chain ganglion

Dorsal ramus

Ventral ramus

Rami communicantes

Intercostal nerve

Branches of intercostal nerve

Lateral cutaneous

Anterior cutaneous

Thoracic cavity

Figure 22.10 Thoracic nerves.

Name _____

1. What type of impulse (sensory/motor) travels through the
   a. anterior gray horn?

   b. posterior gray horn?

   c. ascending spinal tracts?

   d. descending spinal tracts?

2. What major nerves arise from the following plexuses?
   a. cervical

   b. brachial

   c. lumbar

   d. sacral

3. How does the dorsal spinal root vary from the ventral spinal root?

4. What causes the cervical enlargement of the spinal cord?

5. Where is the filum terminale found?

6. What is the conus medullaris?

7. What is the cauda equina?

8. What is the endoneurium?

9. In the spinal cord, which is deep to the other, the white matter or the gray matter?

10. What is the area of gray matter that is found between the lateral halves of the spinal cord?

11. The subarachnoid space is filled with what fluid?

12. How do tracts differ from nerves?

13. What is a mixed nerve?

14. The diaphragm contractions are regulated by what nerve?

15. The muscles of the arm such as the biceps brachii have what innervation?

16. The extensor muscles of the hand are controlled by what nerve?

17. The sciatic nerve is composed of two nerves. What are they?

18. A person has sensory perception in the deltoid and biceps brachii region but no feeling in the wrist extensors. Where on the spinal cord has injury occurred?

Nervous System Physiology—
Stimuli and Reflexes

Introduction

Nervous tissue shows two fundamental properties: irritability and conductivity. **Irritability** is the potential of a nerve to respond to some kind of stimulus (chemical, mechanical, electrical), and **conductivity** is the movement of a nerve impulse along the length of the neuron. In this way nerves receive information from a particular area (sense organ, regions in the CNS, etc.) and transmit impulses to either the brain for interpretation or an effector for some kind of action. In this exercise, you study the basic properties of neuronal conduction and their sensitivity to various stimuli. In addition to studying the neurons, you experiment with nerves as they form reflex arcs in several parts of the body.

Objectives

At the end of this exercise you should be able to

1. describe the threshold nature of nerve responses;
2. list three things that cause a nerve to be stimulated;
3. name one substance that stimulates nerves and one that inhibits them;
4. describe reflex arcs;
5. list all of the parts of a monosynaptic and polysynaptic reflex arc;
6. define hyporeflexic and hyperreflexic.

Materials

Nerve Physiology Section
Frog
Latex or plastic gloves
Glass rod with hook at one end
Hot pad or mitt
Bunsen burner
Matches or flint lighter
Frog Ringer's solution in dropper bottles
Microscope slide or small glass plate
Filter paper or paper towel
Dissection equipment for live animals
Scalpel or scissors
Cotton sewing thread
Stimulator apparatus with probe
5% sodium chloride solution

0.1% hydrochloric acid solution (1 ml of concentrated HCl in 1 liter of water)
Procaine hydrochloride solution

Reflex Section
Patellar reflex hammer
Rubber squeeze bulb
Models or charts of spinal cord and nerves

Procedure

Review the structure of the neuron in Laboratory Exercise 20 for descriptions of the dendrites, nerve cell body, and axon, and the anatomy of the nerve in Laboratory Exercise 22. Make sure you read through all of this exercise prior to beginning the experiments. Wear latex gloves as a general precaution when working with fresh specimens such as frogs.

Frog Nerve Conduction

In this exercise, you observe the process of nerve impulse conduction by experimenting on frogs or watching a demonstration, depending on the wishes of your instructor. If you will experiment on frogs, obtain a doubly pithed frog, dissection equipment, frog Ringer's solution, and various test solutions and bring them back to your table. Keep the frog nerve preparation moist with Ringer's solution during the entire experiment. Prepare the frog by cutting the skin away from the hip (figure 23.1). Do not cut, pinch, or otherwise damage the sciatic nerve on the posterior side of the thigh.

Gently remove the nerve from between the muscles with a glass rod (figure 23.1). Do not stretch the nerve; leave it intact alongside the muscles.

Nerve Response to Physical Stimuli

Flush the nerve with Ringer's solution and make sure it remains moist. In this part of the exercise, you examine the effects of physical stimuli on nerve impulse conduction. You will measure the effects of the nerve stimulation by the contraction of the gastrocnemius muscle. If the nerve is stimulated, then the gastrocnemius muscle should contract.

Cut a small (10 cm) section of cotton thread and gently slip it under the sciatic nerve with a pair of fine forceps. Gently move the thread up toward the hip. When you reach the location

Figure 23.1   Removal of the skin and nerve preparation.

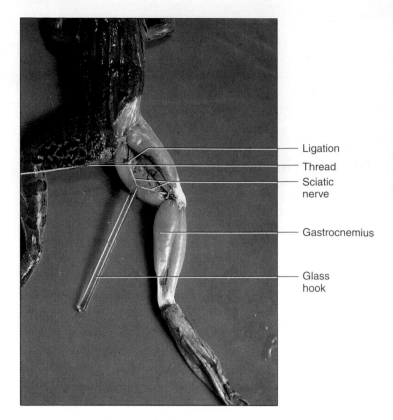

Figure 23.2   Ligation of the sciatic nerve of the frog.

where the nerve descends into the muscle, tie off the nerve with thread and examine the effects. Loop the thread and ligate (tie off) the nerve close to the sacrum (figure 23.2) while watching the gastrocnemius muscle.

As the thread begins to tighten on the nerve, record the response in the space provided.

Response of the nerve to physical stimulation:

After you have tied off the nerve, cut it from the anterior side, leaving the nerve attached to the gastrocnemius muscle. Moisten the nerve with Ringer's solution and prepare for the next experiment.

Nerve Response to Electrical Activity

Place a stimulator probe connected to a stimulator underneath the sciatic nerve, lifting the nerve away from the gastrocnemius muscle. Keep the nerve moist as you determine the minimum voltage (threshold voltage) required for nerve conduction. Set the stimulator to a frequency of two pulses per second and a duration of 10 milliseconds. The voltage should be set at zero (with the knob on 0.1 volt). Slowly increase the voltage until you observe the gastrocnemius twitch. As soon as you see the gastrocnemius twitch at the *lowest voltage*, record this as threshold voltage in the space provided. Turn the voltage to zero, flush the nerve with Ringer's solution, and let it rest for a moment.

Threshold voltage: _____

Continue to increase the voltage until the muscle contracts maximally. Record this as the maximum recruitment voltage. This voltage is obtained when all of the neurons of a particular nerve are stimulated.

Maximum recruitment voltage: _____

Nerve Response to Chemical Stimuli

In the following two experiments, you will test the response of the nerve to different chemical agents. Make sure you observe the nerve as soon as you apply the solution and rinse it as soon as the observation is made.

Acid Solution   Apply a 0.1% hydrochloric acid solution to a cotton applicator stick and touch the applicator gently to the nerve. Record the nerve response.

Response to hydrochloric acid: _____

Flush the nerve with Ringer's solution and let it rest for a moment.

Salt Solution   Now gently apply a 5% sodium chloride solution to the nerve with a new cotton applicator stick. Record the response. Flush the nerve with Ringer's solution and let it rest for a moment.

Response to sodium chloride solution: _____

Nerve Response to Anesthetics

Apply a solution of **procaine hydrochloride** (Novocain) to the nerve by soaking a small square of gauze or cotton with procaine solution and placing it on the nerve for a moment. As the gauze remains on the nerve, set up the stimulator apparatus and place the nerve over the stimulator probes. Remove the gauze and stimulate the nerve with a single pulse stimulus at the voltage that produced a maximum recruitment voltage in the previous experiment. If the nerve responds to the stimulus, leave the procaine hydrochloride on longer. When the nerve does not respond, remove the gauze and stimulate the nerve once every 30 seconds until it recovers from the local anesthetic. Record the recovery time.

Recovery time: _____

Nerve Response to Changes in Temperature

Gently touch the nerve with a glass rod at room temperature. Record the response.

Response to gentle touch: _____

Place the glass rod in an ice bath. As it equilibrates in the ice water place a small chip of ice on the nerve and let it stay there for a moment. Gently touch the nerve with the cold rod and record the response. Flush the nerve with Ringer's solution and let it rest for a moment.

Response to gentle touch with cold stimulation: _____

Take another glass rod in a hot pad or mitts and heat one end of it in a Bunsen burner. Touch the nerve with the hot end of the glass rod. What is the response? Record your result.

Response to gentle touch with hot stimulation: _____

Clean Up   Make sure you clean your station before continuing. Place the specimen in the appropriate container, and use care when cleaning sharp instruments such as scalpels or razor blades.

Reflexes

A reflex is defined as a motor response to a stimulus without conscious thought. Reflexes are involuntary, predictable responses to stimuli. Reflexes occur through reflex arcs, and these arcs have the following structure:

1. Receptor (structure that receives the stimulus)
2. Afferent (sensory) neuron (the neuron taking the stimulus to the CNS)
3. Integrating center (brain or spinal cord)
4. Efferent (motor) neuron (the neuron taking the response from the CNS)
5. Effector (the structure causing an effect)

If the effector is skeletal muscle, we call the reflex a **somatic reflex.** If the effector is a gland, smooth muscle, or cardiac muscle, we call the reflex a **visceral** or **autonomic reflex.**

Most reflexes involve many neurons with many synapses and are called polysynaptic reflexes. A few reflexes involve just two neurons—a sensory neuron and a motor neuron with one synapse between them—and these are called monosynaptic reflex arcs. These are illustrated in figure 23.3.

Note how these reflexes depend on a **stimulus,** or environmental cue, a **receptor** that is sensitive to the stimulus, an **afferent neuron** (or sensory neuron), an **efferent neuron** (or motor neuron), and an **effector.** The polysynaptic reflex has these structures as well as an association neuron, or **interneuron** located between the afferent and efferent neurons.

Testing for reflexes is very important for clinical evaluation of the condition of the nervous system. Decreased response or even exaggerated response to a stimulus may indicate disease or damage to the nervous system. In this experiment, you will test several reflexes and determine if the response is normal, **hyporeflexic** (showing less than average response), or **hyperreflexic** (showing an exaggerated response). In conditions such as hypothyroidism the patient often shows hyporeflexic responses.

Patellar Reflex

The **patellar reflex** tests the conduction of the femoral nerve. It is the most frequent reflex test performed in clinical settings. Sit on the lab table with your leg hanging over the edge and have your lab partner tap you on the patellar tendon with a patellar reflex hammer. The percussion should be placed about 3 to 4 cm below the inferior edge of the patella, and it should

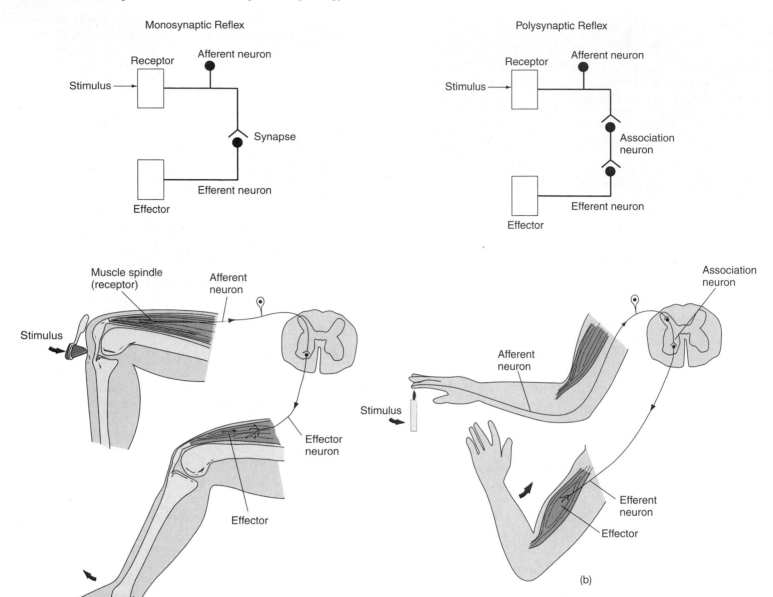

Figure 23.3 Reflex arcs. (a) Monosynaptic; (b) polysynaptic.

be firm but not hard enough to hurt. Look for extension of leg as a response to the patellar reflex (figure 23.4).

The tap stimulates the stretch receptors in the tendon and is representative of a monosynaptic reflex. Record the degree (hyperreflexic, normal, hyporeflexic) of the response.

Patellar reflex: _____

Triceps Brachii Reflex

The **triceps brachii reflex** tests the radial nerve. Lie down on your back on a clean lab table or cot and place your forearm on your abdomen. Have your lab partner tap the distal tendon of the triceps brachii muscle about 2 inches proximal to olecranon process. Look for the triceps muscle to twitch (figure 23.5). Record your result.

Triceps brachii reflex: _____

Biceps Brachii Reflex

The **biceps brachii reflex** tests the activity of the musculocutaneous nerve. Sit comfortably and have your lab partner place his or her fingers on the biceps tendon just proximal to the antecubital fossa (figure 23.6).

Figure 23.4   Patellar reflex.

Figure 23.5   Triceps reflex.

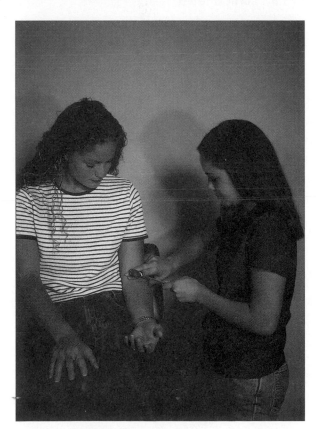

Figure 23.6   Biceps brachii reflex.

Your lab partner should tap his or her fingers with the reflex hammer, while they remain on the tendon, and look for the biceps brachii muscle contraction. Record your results.

Biceps brachii reflex: _____

Calcaneal (Achilles) Tendon Reflex

To test the **calcaneal tendon reflex** kneel on a chair with your foot dangling over the edge (figure 23.7). Have your lab partner tap the calcaneal tendon in order to test the tibial nerve.

As your lab partner taps your calcaneal tendon, look for plantar flexion of the foot. You may see an initial movement of the foot due to the depression of the tendon by the reflex hammer, but there should be a slight pause and then another quick movement of the foot. Record your results.

Calcaneal tendon reflex: _____

Eye Reflexes

The automatic blinking of the eye is important to keep material such as dust away from the outer layer of the eye known as the cornea. In the first part of this experiment, have your lab partner try to make you blink by flicking his or her fingers near your eyes. Can you prevent the blinking response? Record your answer.

Control of blink reflex: _____

Figure 23.7 Calcaneal reflex.

Figure 23.8 Corneal reflex.

Now have your lab partner take a clean rubber squeeze bulb (a large pipette bulb works well) and squirt a sharp blast of air across the surface of the eye (figure 23.8). Use either new bulbs or ones that are free of debris to avoid damage to the eye. Can you inhibit this response? Record your response.

Control of corneal reflexes: _____

Babinski Reflex

Using the *metal end* of the patellar hammer, stroke the foot from the heel along the lateral, inferior surface and then towards the ball of the foot (figure 23.9). The pressure should be firm but not uncomfortable.

Stroking the plantar surface normally results in flexion of the toes in adults. Damage to pyramidal tracts causes extension of the big toe (known as Babinski's response). This is important for determining spinal damage. Adults normally have a negative Babinski response, which is flexion of the foot and toes, while infants under 1 year of age usually have a positive Babinski's response.

Figure 23.9 Babinski reflex.

Name _____

1. What structure receives a stimulus from the external environment and relays it to the afferent neuron?

2. What is another name for an efferent neuron?

3. What is a reflex?

4. In what kind of reflex do you have just two neurons?

5. Polysynaptic reflexes have a neuron specific to them. What is the name of that neuron?

6. In terms of numbers of synapses, what kind of reflex is a patellar reflex?

7. After patients leave the operating room they are transferred to an area called the "recovery room." Correlate the meaning of the word "recovery" in this context with what you have learned about the recovery of nerves in this exercise.

8. Draw a monosynaptic reflex arc in the space provided. Label your illustration with the terms provided.

   Stimulus
   Motor neuron
   Effector
   Sensory neuron
   Synapse

9. What action occurs with a hyperreflexic response? What action happens with a hyporeflexic response?

10. List the positive responses obtained in the frog experiment and correlate this with the specificity of neuronal sensitivity.

11. What was the threshold voltage observed in the nerve response?

LABORATORY EXERCISE 24

Introduction to Sensory Receptors

Introduction

The gateway to understanding our world comes from our ability to sense the environment around us. There are many types of sense organs in the body, and the different types respond to specific stimuli (for example, light, sound, touch). Sense receptors are not uniformly distributed throughout the body but are absent, or few in number, in some areas while densely clustered in other locations. This pattern of uneven distribution is called punctate distribution.

For the sensory system to operate several factors need to be present. The first of these is the presence of a stimulus. There can be no perception without an environmental change. Stimuli are listed according to types of stimuli, or modalities. Examples of modalities are light, heat, sound, pressure, and specific chemicals. Receptors are the receiving unit of the body that respond to stimuli. They transform the stimulus to neural signals that are transmitted by sensory nerves and neural tracts to the brain, which interprets the message. If any link in this sensory chain is broken, the perception of stimuli cannot occur.

Receptors are sensitive to specific stimuli and can be classified according to the stimulus they receive. The human body has photoreceptors, which detect light (for example, the eye); thermoreceptors, located in the skin, which detect changes in temperature; proprioceptors, which detect changes in tension such as those in joints; pain receptors, or nociceptors, present as naked nerve endings in the skin or stomach; mechanoreceptors, which perceive mechanical stimuli (for example, touch receptors or receptors that determine hearing or equilibrium in the ear); baroreceptors, which respond to changes in pressure such as blood pressure; and chemoreceptors, which respond to changes in the chemical environment (for example, taste and smell).

The skin has several different types of receptors and therefore makes a good starting point for understanding sense organs. You may want to review the major sensory structures in the skin such as Meissner's corpuscles, Pacinian (lamellated) corpuscles, and pain receptors in Laboratory Exercise 7 before you begin this exercise.

Objectives

At the end of this exercise you should be able to

1. list the major sense receptors of the body;
2. define adaptation to a stimulus;
3. distinguish between relative and absolute determination of stimuli;
4. define referred pain.

Materials

Blunt metal probes
Dishpan (or large finger bowls) of ice water (2 L)
Dishpan of room temperature water
Dishpan of warm water (45°C)
Towels
Small centimeter ruler
Black, fine-tipped felt markers
Red, fine-tipped felt markers
Blue, fine-tipped felt markers
Two-point discriminators (or a mechanical compass)
Von Frey hairs
Tweezers

Procedure

Two-Point Discrimination Test

The ability to distinguish touch is dependent on the type and number of nerve endings in the skin. You can easily map the relative density of the receptors in the skin by performing a two-point discrimination test. Have your lab partner sit with eyes closed and his or her hand palm up, resting on the lab counter. Using the two-point discriminator or a mechanical compass, touch your lab partner's palm with both points of the instrument and see if he or she can sense one or two points. Determine the *minimum distance* your lab partner senses as two points. Begin the procedure by determining the minimum distance on the palm of the hand. To establish an accurate reading make sure you gently touch both points of the discriminator at the same time. If you sequentially touch the two-point discriminator on the palm, your lab partner may perceive two points in time and not in space. One good way to establish accuracy is to occasionally touch just one of the points on your lab partner's palm. Another way is to vary the spread of the discriminator. You might begin with 3 cm and then adjust it to 1 cm followed by a 2 cm spread. Record your results.

Minimum distance on palm: _____

Now move to the back of the forearm. Establish the minimum distance that is perceived as two points by your lab partner. Record your results.

Distance on back of forearm: _____

Is there a difference between the palm and the back of the forearm?

If there is, what might be the explanation for the difference in terms of the number of nerve endings per unit area?

Now try the fingertip and then the back of the shoulder or neck. Record the results.

Distance on fingertip: _____

Distance on back of shoulder or neck: _____

Warm and Cool Receptors

The skin has receptors that are sensitive to cool or warm temperatures. To determine the difference between the two, take two blunt metal probes and place one in an ice water bath and one in a warm water bath (45°C). Let the probes reach the temperature of each respective bath, which should take a minute or two. Have your lab partner close his or her eyes and extend an arm on the lab counter. Remove one of the probes and quickly wipe it on a clean towel. Test the ability of your lab partner to distinguish between cool and warm by using the handle of the probe on your lab partner's forearm. Repeat the experiment for a total of five cool trials and five warm trials. Frequently return the probes to their respective water baths to maintain the appropriate temperature. Record your results.

Number of accurate cool recordings: _____

Number of accurate warm recordings: _____

Mapping Temperature Receptors

Mark off a square that is 2 cm on a side on the anterior forearm of your lab partner using a felt marker. Now take the pointed end of the blunt probe and place it in cold water. Systematically test areas in the square. When your lab partner perceives cold (not just touch) in a location, mark it with an "X." Retest the area with the warm probe and place an "O" in the location where warm is perceived.

Number of cool receptors in square: _____

Number of warm receptors in square: _____

What is the ratio of "X" to "O" in the square?

Mapping Fine-Touch Receptors

Fine-touch receptors are of two types, Meissner's corpuscles and Merkel discs. You can map these receptors by testing the ability of your lab partner to distinguish fine touch. Draw a square 2 cm

on a side on the anterior surface of the forearm. Using a stiff bristle hair attached to a matchstick, or a Von Frey hair, map the number of areas in the square that can be perceived by your lab partner. Press only until the hair bends a little to stimulate the touch corpuscles. Use a black felt pen to record the location of each positive result. How many positive responses did you get in the square on the forearm? Record your result.

Number of forearm responses: _____

Repeat the experiment on the *lateral* surface of the *arm* using another square 2 cm on a side. How does the number of receptors here compare to those on the forearm? Record your results.

Number of responses on *lateral* side: _____

Adaptation to Touch

There are two types of receptors, tonic receptors and phasic receptors. **Tonic receptors** constantly perceive stimuli while **phasic receptors** adapt to a stimulus. In this experiment you will try to determine if the sense of touch is tonic or phasic. Cut a small piece of paper, about 2 cm on a side and crumple it into a small ball the size of a pea. Have your lab partner place the forearm, anterior side up, comfortably on the lab desk. With a pair of tweezers, place the ball of paper on your lab partner's anterior forearm. Is the paper ball perceived after 1 minute? With your lab partner's eyes closed gently remove the paper ball with the tweezers. Can the perception be felt? Record your results in the following space and determine whether the sense of light touch is tonic or phasic.

Results: _____

Locating Stimulus with Proprioception

In this exercise, you use washable markers of two colors. Have your lab partner close his or her eyes and rest a forearm on the lab counter. Touch your lab partner's forearm with a felt marker and have him or her try to locate the same spot with a felt marker of another color. Test at least five locations on various parts of the forearm, and repeat each location at least twice. Now try the fingertip and palm of the hand and record the result.

Maximum distance error on the forearm: _____

On the palm: _____

On the fingertip: _____

Another method to test proprioreception is to close your eyes and *gently* try to touch the lateral corner of your own eye with your fingertip. Have your lab partner watch you and determine the accuracy of your attempt. Now try to touch the bottom part of your earlobe or the exact tip of your chin. Record the distance for each location.

Corner of eye: _____

Earlobe: _____

Tip of chin: _____

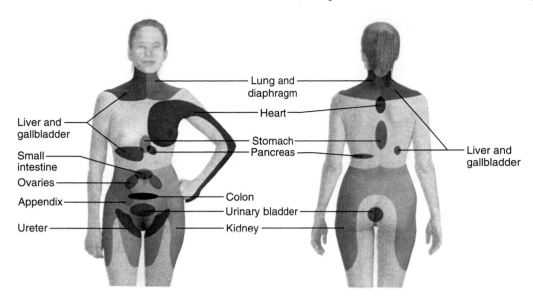

Figure 24.1   Referred pain.

Temperature Judgment

In this exercise, you examine the *adaptation* of thermoreceptors to temperature and the ability to determine temperature by **absolute value** or by **relative value**.

On the lab counter locate three dishpans or large finger bowls full of water. One bowl, located on the right, should be marked "Cold" (it should be about 10°C); another bowl, located on the left, should be marked "Warm" (it should be about 40°–45°C); and the middle bowl should be marked "Room Temperature." Place one hand in the cold dish and the other in the warm dish and let them adjust to the temperature for a few minutes. If your hand begins to ache in the cold water, you may remove it for a short time, but try to keep it in the cold water for as long as possible during the adjustment time. After your hands have equilibrated, place them both in the room temperature water and describe to your lab partner the temperature (cold, warm, hot) of the water as sensed by each hand. How does the hand that was in cold water feel in the room temperature water, and how does the hand that was in warm water feel in the room temperature water? Record your results.

Cold hand perception: _____

Warm hand perception: _____

After completing this part of the exercise, record whether temperature determination is absolute or relative to the skin temperature.

The determination of water temperature is _____.

Referred Pain

Referred pain is the perception of pain in one area of the body when the pain is actually somewhere else. An example of referred pain is the pain felt in the left shoulder and arm when a person is suffering from a heart attack or chest pain (angina pectoris) (figure 24.1).

Referred pain may be due to many pain impulses from different receptors traveling along the same neural tract to the brain. When one area sends a message of pain, the brain may interpret the sensation as coming from more than one location. Place your elbow into a dish of ice water and leave it there for 2 painful minutes. Describe the sensation you feel to your lab partner and the location of the sensation. Record your results.

Description of sensation: _____

Initial location of sensation: _____

Sensation after 2-minute period: _____

Name _____

1. An area with a great number of nerve endings is the upper lip. What can you predict about the ability of the upper lip to distinguish two points?

2. When you extend the temperature beyond the level of cool and warm, the pain receptors in the skin are activated. Cool receptors are activated between 12° and 35°C. Warm receptors are activated between 25° and 45°C. You or your lab partner may have had an experience with very cold conditions such as when cleaning out a freezer or holding dry ice. What perception is sensed?

3. Adaptation is very important to sensory stimulation. We are bombarded with stimuli during most of the day, and much of what we sense is filtered from conscious thought. How is the sense of adaptation used by pickpockets?

4. In terms of receptor density, describe why it is difficult to find the same location on the forearm when your eyes are closed.

5. In regard to sense organs, what is punctate distribution?

6. In reference to the sense organs, what is a modality?

7. What kind of receptors are sensitive to the following modalities?

    a. light

    b. touch

    c. temperature

    d. sound

    e. smell

8. What kind of receptor is responsive to extremely hot sensations?

9. Meissner's corpuscles respond to what kind of sensation?

10. What kind of receptor determines the weight of an object when you pick it up?

11. Which kind of receptor (phasic/tonic) adapts to light in a darkened movie theater?

12. When you drink a liquid that is burning hot, the 'chest pain' felt in the region of the sternum does not really occur there. What is this kind of pain called?

LABORATORY EXERCISE 25

Taste and Smell

Introduction

We take for granted our senses of taste and smell. They are not fully appreciated unless they are lost. People who have lost the sense of smell find food difficult to eat, since they derive no pleasure from the act of eating. In this exercise, you examine the structure and function of these two important senses.

Both taste and smell are examples of chemoreception, where specific chemical compounds are detected by the sense organs and interpreted by various regions of the brain. The sense of taste, or **gustation,** is received predominantly by taste buds in the tongue, although there are also receptors in the soft palate and pharynx. The sense of taste is transmitted by the facial and glossopharyngeal nerves and interpreted in the postcentral gyrus of the parietal lobe of the brain. The sense of smell, or **olfaction,** originates when particles stimulate hair cells in the olfactory epithelium (a specialized neuroepithelium in the upper nasal cavities) and is transmitted by the olfactory nerves. This transmission occurs through the cribriform plate of the ethmoid bone to the olfactory bulb at the base of the frontal lobe of the brain. From here the sense of smell is transmitted to two regions, one in the limbic system and another in the temporal lobe of the brain.

Objectives

At the end of this exercise you should be able to

1. list the two major chemoreceptors located in the region of the head;
2. diagram a taste bud;
3. trace the sense of smell from the nose to the integrative areas of the brain;
4. list the four tastes perceived by humans;
5. compare and contrast the senses of taste and smell.

Materials

Cotton-tipped applicators
Disposable (paper) cups, 4–12 oz cups containing one of the following:

> Solution of salt water (3%) labeled "salty"
> Quinine solution (0.5% quinine sulfate) labeled "bitter"
> Vinegar solution (household vinegar or 5% acetic acid solution) labeled "acidic"
> Sugar solution (3% sucrose) labeled "sweet"

Biohazard bag
Prepared slides of taste buds
Microscopes
Roll of household paper towels
Small bowl of salt crystals (household salt)
Small bowl of sugar crystals (household granulated sugar)
Flat toothpicks
Several small vials (10–20 ml) screw-cap bottles with essential oil labeled "Peppermint," "Almond," "Wintergreen," and "Camphor" (keep vials in separate wide-mouthed jars to prevent cross-contamination of scents)
Four small vials colored red and labeled "Wild Cherry" filled with benzaldehyde solution
One vial of dilute perfume (one part perfume, five parts ethyl alcohol)
Noseclips
Selection of four or five fruit nectars (such as Kern's nectars), two cans each: apricot, coconut/pineapple, strawberry, mango, peach
Small, 3 oz paper cups (89 ml) five per student (or student pair)
Napkins

Procedure

Examination of Taste Buds

Examine the prepared slide of taste buds and compare them to figure 25.1. Note how they are located in the side walls of papilla on the tongue. The taste buds appear lighter than the surrounding tissue (like microscopic onions cut in long sections). Taste buds are composed of neural tissue and epithelial tissue.

Taste Determination of Solid Materials

For something to be tasted it must be in solution. This allows for the fluid to run down the sides of the tongue papilla, where the hair cells of the taste buds are located. Blot your tongue thoroughly with a paper towel. Make sure the surface is relatively dry. Have your lab partner select either the sugar crystals or salt crystals and place a small scoop (with the end of a flat toothpick) on your tongue. Keep your mouth open and do not swirl saliva around. Can you determine what the sample is?

Now close your mouth and see if you can determine the nature of the sample.

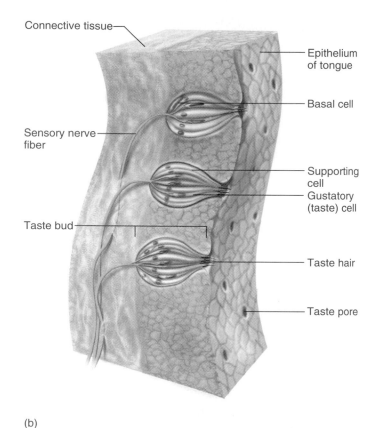

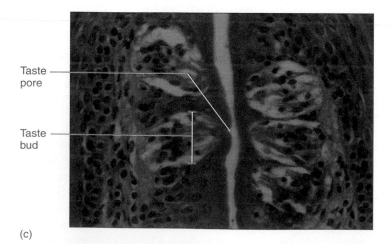

Figure 25.1   Taste buds. (a) Taste buds on sides of a tongue papilla; (b) details of taste buds; (c) photomicrograph of taste buds (100×). ✗

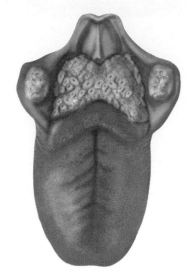

Figure 25.2   Map of taste receptors on tongue. ✗

Mapping the Tongue for Taste Receptors

There are many taste receptors, but most researchers classify taste into four or five primary tastes: sweet, sour, salty, bitter, and recently umami (the taste of glutamate). Individuals have varying degrees of sensitivity to these tastes. Some of us find bitter tastes to be especially objectionable, while others do not seem to mind them as much. As you perform the taste experiments determine for yourself if members of your class are equally sensitive to the same tastes.

Insert a cotton applicator stick into one of the four trial cups that are labeled salty, bitter, acidic, and sweet. Remember to keep track of the kind of substance that is on your applicator stick. Dab the surface of your lab partner's tongue and determine by nod or hand signals when perception of the taste occurs. *Do not* have your lab partner tell you about the taste at first, since the movement of the tongue will wash the solution to other areas of the tongue and you may get a false reading. Map the result of your taste test in figure 25.2. Throw the applicator stick in the biohazard bag. Repeat the test with a *new applicator stick* for each of the four solutions. *Do not contaminate* the solutions by double-dipping the applicator stick into the same or different solutions after it has been in your lab partner's mouth!

Transmission of Sense of Olfaction to Brain

Examine a model, chart, or diagram of a midsagittal section of the head or a model of the brain and locate the olfactory nerves, cribriform plate, olfactory bulb, and olfactory tract. Compare the lab charts or models to figure 25.3.

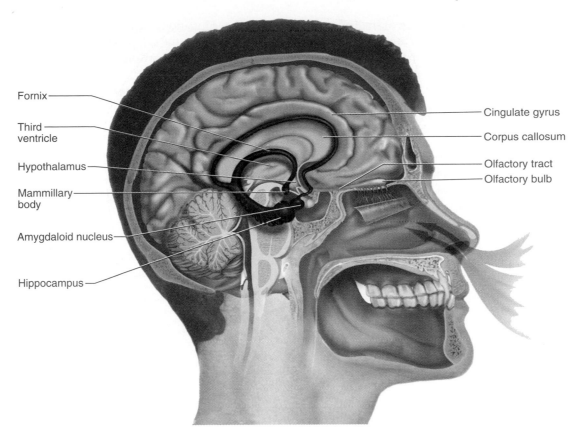

Fornix

Third ventricle

Hypothalamus

Mammillary body

Amygdaloid nucleus

Hippocampus

Cingulate gyrus

Corpus callosum

Olfactory tract

Olfactory bulb

Figure 25.3   Olfactory transmission. ✗

Visual Cues in Smell Interpretation

In this section you will examine the influence of visual cues on the interpretation of smell. You will seek to determine whether the color of a substance has any effect on what you perceive the smell to be. Have your lab partner show you a small vial labeled "Almond" and then smell it. Now examine and smell the small red vial labeled "Wild Cherry." Do you perceive these as two separate smells? Close your eyes and have your lab partner select a vial for you. Can you tell which one it is?

Olfactory Discrimination

Obtain four vials of different scents—peppermint, almond, wintergreen, and camphor. While keeping your eyes closed try to determine the name of each essential oil as your lab partner presents it to you. Record how many of the smells you got correct out of the four. If you are hypersensitive to smells, have your lab partner do this section of the experiment.

Number correct: _____

Adaptation to Smell

Adaptation to smell by the olfactory receptors occurs very rapidly, but the adaptation by the receptors is incomplete. Com-

plete adaptation to smell probably occurs by additional CNS inhibition of the olfactory signals. In this section of the experiment you are trying to determine approximately how long olfactory adaptation takes. Close your eyes and plug one nostril. Inhale the scent from a vial of dilute perfume until you can no longer smell the substance. Have your lab partner record the time when you begin the experiment and how long it takes for the perception of smell to decrease significantly. How long does this take?

Length of time to disappearance of the smell: _____

What might be the evolutionary advantage of adaptation to smell?

Now smell the wintergreen or peppermint vial. Does the adaptation of one smell cause the olfactory receptors to adapt to other smells?

Taste and Olfaction Tests

This experiment demonstrates the dependence of the sense of smell as a component of what we call *taste*. Obtain four or five small cups (3 oz) and label each with the name of the fruit juice or nectar it will contain. Have your lab partner select several types of fruit nectars and pour each into the proper cup.

Caution If you have a particular food allergy notify your instructor. You may wish to omit part or all of this test.

Sit with your eyes and your nose closed (use nose clips or pinch off your nostrils with your finger and thumb) and try to determine the sample presented to you by your lab partner. Your lab partner should place the sample cup (sample unknown to you) in your hand and you should guess what fruit nectar is in the cup. After you have "tasted" the sample try to name it. Test all of the samples with your nose closed. After you make

your determination release your nostrils but still keep your eyes closed and taste the samples again. Record the results. How accurate is your comparison?

Trial Number	Sample	Accuracy or Detection (Yes/No)	
		Nose Closed	Nose Open
1	Apricot		
2	Mango		
3	Coconut/pineapple		
4	Strawberry		
5	Peach		
6	Other		

Name _____

1. What are the primary tastes?

2. What nerves take the sense of smell and transmit it to the brain?

3. What nerves take the sense of taste and transmit it to the brain?

4. Where are the taste buds located?

5. What is the exact region of the nasal cavity that is receptive to smell stimuli?

6. Can you determine the adaptation for having taste buds that determine unpleasant bitter compounds in many plant species?

7. Some individuals with severe sinus infections can lose their senses of smell. How could an infection that spreads from the frontal or maxillary sinus impair the sense of smell?

8. Material must be in solution for it to be tasted. What process would be used (olfaction, gustation) to perceive a lipid-based food?

9. Some smells that we perceive as two separate smells are actually identical. What are the other cues that we use to distinguish these two "smells" as being distinct?

LABORATORY EXERCISE 26

Eye and Vision

Introduction

In most people eyesight accounts for about 80% of accumulated knowledge. The importance of the eye can be inferred in that five of the 12 cranial nerves are dedicated, at least in part, to either receiving visual stimuli or coordinating the movement of the eyes.

Anatomically, the eye consists of an anterior portion visible as we look at the face of an individual and a posterior portion situated in the orbit of the skull. Light from the external environment travels through a number of transparent structures which bend and focus the light on the retina, which is the receptive layer of the eye that converts light energy to neural impulses. Nerve impulses travel from the eyes through the optic nerves to the brain, where they are interpreted as sight in the occipital lobes of the brain. This exercise involves learning the structure of the eye and correlating those structures to the function of vision by performing basic physiology experiments.

Objectives

At the end of this exercise you should be able to

1. identify the major structures of the mammalian eye;
2. determine the visual field for both eyes;
3. demonstrate the Snellen vision tests and those for accommodation and astigmatism;
4. describe the position of the choroid in reference to the retina and the sclera.

Materials

Models and charts of the eye
Snellen charts
Astigmatism charts
Sharp pencil or fine probe
Vision Disk (Hubbard) or large protractor
Microscopes
Prepared microscope slides of eye in sagittal section
Preserved sheep or cow eyes
Dissection trays
Dissection gloves (latex or plastic)
Scalpel or razor blades
3- by 5-inch cards
Ishihara color book or colored yarn
Ruler (approximately 35 cm)

Opthalmoscope and batteries
Penlight
Desk lamp with 60-watt bulb
Paper card with a simple colored image (red circle, blue triangle) printed on it.

Procedure

External Features of the Eye

Examine a model of the eye and note the external features as compared to figure 26.1. The eye has a **sclera,** or "white of the eye," which is composed of dense irregular connective tissue. The sclera is a protective portion of the eye that serves as an attachment point for the muscles of the eye and helps maintain the intraocular pressure (the pressure inside the eye). The pressure maintains the shape of the eye and keeps the retina adhered to the back wall of the eye. Numerous blood vessels traverse the sclera, and if they become dilated anteriorly they give the eye the appearance of being "bloodshot." The sclera is continuous with the transparent cornea in the front of the eye.

Attached to the sclera are the **extrinsic muscles** of the eye. There are six extrinsic muscles that, in coordination, move the eye in quick and precise ways. Locate these muscles on a model and compare them to figure 26.1. These muscles are listed in table 26.1

A structure important in the maintenance of the exterior of the eye is the **lacrimal apparatus.** This consists of the **lacrimal gland** located superior and lateral to the eye (figure 26.2). Lacrimal secretions bathe and protect the eye and clean dust from its surface. The fluid drains through the **nasolacrimal duct** into the nasal cavity.

Table 26.1	Extrinsic Muscles of the Eye	
Muscle Name	**Innervation**	**Direction Eye Turns**
Lateral rectus	VI (abducens)	Laterally
Medial rectus	III (oculomotor)	Medially
Superior rectus	III (oculomotor)	Superiorly
Inferior rectus	III (oculomotor)	Inferiorly
Inferior oblique	III (oculomotor)	Superiorly and laterally
Superior oblique	IV (trochlear)	Inferiorly and laterally

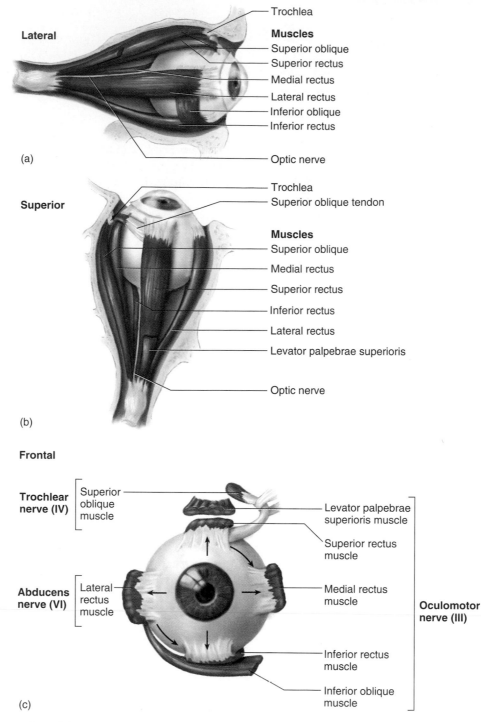

Lateral

- Trochlea
- **Muscles**
- Superior oblique
- Superior rectus
- Medial rectus
- Lateral rectus
- Inferior oblique
- Inferior rectus
- Optic nerve

(a)

Superior

- Trochlea
- Superior oblique tendon
- **Muscles**
- Superior oblique
- Medial rectus
- Superior rectus
- Inferior rectus
- Lateral rectus
- Levator palpebrae superioris
- Optic nerve

(b)

Frontal

Trochlear nerve (IV) — Superior oblique muscle

Abducens nerve (VI) — Lateral rectus muscle

Levator palpebrae superioris muscle

Superior rectus muscle

Medial rectus muscle

Inferior rectus muscle

Inferior oblique muscle

Oculomotor nerve (III)

(c)

Figure 26.1   Right eye, external features.

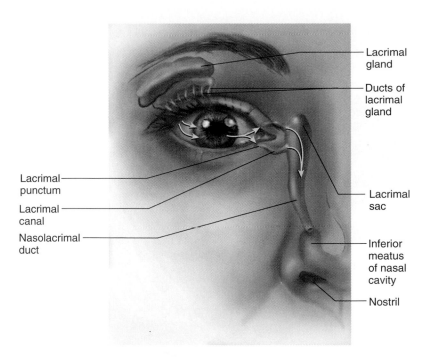

Lacrimal
gland

Ducts of
lacrimal
gland

Lacrimal
punctum

Lacrimal
canal

Nasolacrimal
duct

Lacrimal
sac

Inferior
meatus
of nasal
cavity

Nostril

Figure 26.2   Lacrimal apparatus.

Interior of the Eye

From the front of the eye, the first layer covering the inside of the eyelid and extending across the sclera is the **conjunctiva.** The conjunctiva is composed of epithelial tissue and is an important indicator of a number of clinical conditions (for example, conjunctivitis). In the center of the eye is the transparent **cornea** (figure 26.3). The cornea is the structure of the eye most responsible for the bending of light rays that strike the eye. It is composed of dense connective tissue and is avascular. Why would the presence of blood vessels in the cornea be a visual liability?

Directly behind the cornea is the **anterior cavity,** which is subdivided into the **anterior chamber,** between the cornea and the **iris,** and the **posterior chamber,** between the iris and the **lens.** The anterior cavity is filled with **aqueous humor,** which is produced by the **ciliary body.** Only a few milliliters of aqueous humor are produced each day, and this amount is absorbed by a **venous sinus** (canal of Schlemm).

The iris is what gives us a particular eye color. People with blue and gray eyes are more sensitive to ultraviolet

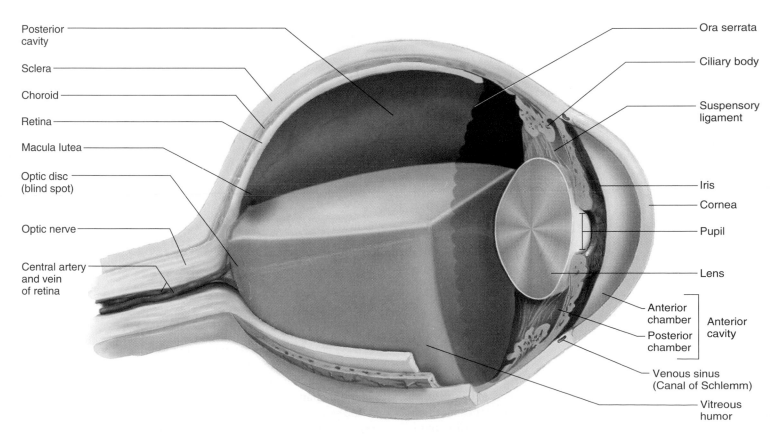

Posterior
cavity

Sclera

Choroid

Retina

Macula lutea

Optic disc
(blind spot)

Optic nerve

Central artery
and vein
of retina

Ora serrata

Ciliary body

Suspensory
ligament

Iris

Cornea

Pupil

Lens

Anterior
chamber

Posterior
chamber

Anterior
cavity

Venous sinus
(Canal of Schlemm)

Vitreous
humor

Figure 26.3   Sagittal section of the eye. ⋌

light than those with brown eyes, due to the protective pigment melanin found in brown eyes. The **circular muscles** of the iris constrict in bright light, reducing the diameter of the **pupil** (the space enclosed by the iris), and the **radial muscles** constrict in dim light, increasing the diameter of the pupil. Behind the pupil is the lens, which is made of a crystalline protein. The lens is more pliable in youth and becomes less elastic as a person ages, which accounts for the need for reading glasses later in life. Suspensory ligaments attach the lens to ciliary muscles in the ciliary body and are involved in pulling the lens for focusing on far-away objects.

Behind the lens is the **posterior cavity,** or vitreous chamber (figure 26.3). This cavity occupies most of the posterior portion or fundus of the eye. The posterior cavity is filled with **vitreous humor,** a clear, jellylike fluid that maintains the shape of the eyeball. Most of the posterior cavity is bounded by three layers, or **tunics,** of the eye. The outermost one, the **sclera,** was already mentioned. Inside of this is the **choroid,** which is a pigmented, vascular layer. The blood vessels found in this layer nourish the eye and the pigmentation prevents light from scattering and blurring vision. The layer closest to the vitreous humor is the **retina.**

The retina is a neural layer that converts light energy into nerve impulses. Light strikes the photoreceptive cells at the posterior portion of the retina, and these cells transmit visual signals to the **bipolar neurons.** The bipolar neurons are in contact with the **ganglionic cells,** thus the visual stimulation that occurs in the posterior portion of the eye is transmitted anteriorly (towards the vitreous humor) to the ganglionic cells. The **ganglionic layer** transmits the neural impulses to the **optic nerve.** From here the neural impulse is transmitted to the occipital region of the brain and integrated further in the temporal lobe (figure 26.4).

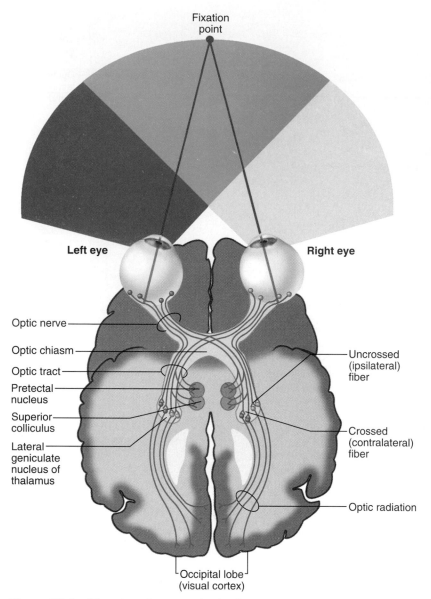

Figure 26.4   Visual pathway to the brain.

Posterior Wall of the Eye

Obtain a microscope and a slide of the retina and compare what you see in the slide to figure 26.5. The retina is composed of three layers, as discussed. Locate the **ganglionic, bipolar,** and **photosensitive layers** of the retina. The photosensitive layer is composed of **rods** and **cones.** Rods are important for determining the motion and general shape of objects and for sight in dim light. Cones are involved in color vision and in visual acuity (determining fine detail).

Examine a model or chart of the eye and locate the **macula lutea** at the posterior region of the eye. Macula lutea means "yellow spot," and in the center of this structure is the **fovea centralis,** a region where the concentration of cones is greatest. In the fovea the cone cells are not covered by the

neural layers as they are in the other parts of the retina. When you focus on an object intently, you are directing the image to the fovea. Locate the fovea in figure 26.6 and in models available in the lab.

Dissection of a Sheep or Cow Eye

Rinse a sheep or cow eye in running water and place it on a dissection tray. Obtain a scalpel, scissors, or new razor blade and a blunt probe. Be careful with the sharp instruments and cut *away from* the hand holding the eye. Wear latex or plastic gloves while you perform the dissection. Using a scalpel, scissors, or razor blade, make a coronal section of the eye behind the cornea (figure 26.7). Do not squeeze the eye with force or thrust the

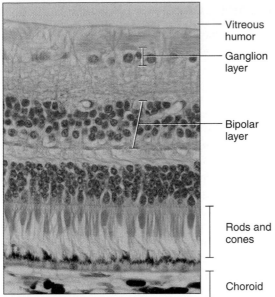

Figure 26.5   **Retina (400×).** 𝒯

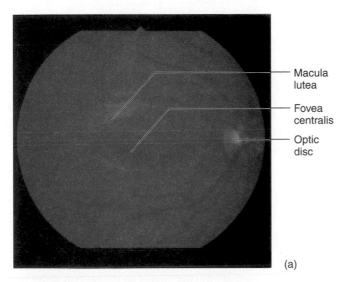

(a)

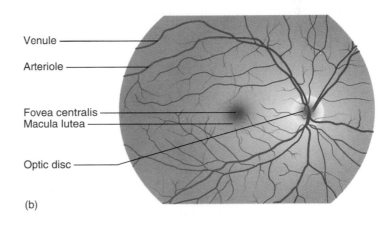

(b)

Figure 26.6   **Eye, posterior view.** (a) photograph; (b) diagram.

blade sharply because you may squirt yourself with vitreous humor. Cut through the eye entirely and note the jellylike material in the posterior cavity. This is the vitreous humor. Look at the posterior portion of the eye. Note the beige retina, which may have pulled away from the darkened choroid. The choroid in humans is very dark, but you may see an iridescent color in your specimen. This is the **tapetum lucidum** and serves to improve night vision in some animals. The tapetum lucidum produces the "eye shine" of nocturnal animals. Also examine the tough, white sclera, which envelopes the choroid.

Now examine the anterior portion of the eye. Is the lens in place? The lens in your specimen probably will not be clear. Normally the lens is transparent and allows for light penetration.

Note that the lens is held to the ciliary body by the suspensory ligaments (see figure 26.3). These ligaments pull on the lens and alter its shape for close or distant vision. Locate the ciliary body at the edge of the suspensory ligaments.

Now turn the eye over. Is there any aqueous humor left in the anterior cavity? Make an incision through the conjunctiva and cornea into the anterior cavity. Can you determine the region of

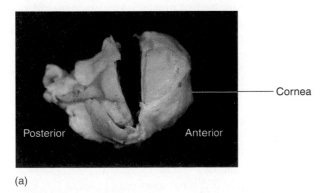

(a)

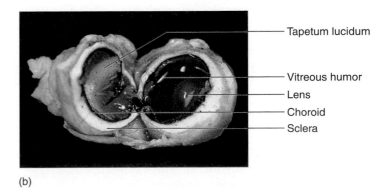

(b)

Figure 26.7   **Dissection of the eye of the sheep.** (a) Coronal cut through the eye; (b) cavity of the eye, posterior view.

the anterior chamber and the posterior chamber that make up the anterior cavity? Locate the iris and the pupil. Dispose of the specimen in a designated waste container and rinse your dissection tools.

Determination of the Near Point

The minimum distance an object can comfortably be held in focus is called the **near point.** The capability of the eye to focus is due to the elasticity of the lens. The pliability of the lens decreases with age. A 10-year-old is able to focus 8 to 10 cm away from the eye, yet a 65-year-old may not be able to focus closer than 80 to 100 cm. The stiffening of the lens becomes noticeable around 40 to 50 years of age, when many people find reading glasses necessary because of the decreasing elasticity of their aging lenses. You can measure your near point by holding a sharp object such as a pencil or fine probe vertically at arm's length in front of you. Close one eye and slowly move the object closer until either you see two objects or it becomes blurry. Have your lab partner measure the distance, in centimeters, from your eye to the object. This is the near point distance. Be careful not to get the sharp point too close to your eye! Measure the near point for both eyes in centimeters and record the data.

Near point of right eye: _____

Near point of left eye: _____

Measurement of the Distribution of Rods and Cones

To determine the distribution of rods and cones, have your lab partner look straight ahead. Slowly move a small colored object (a pen, piece of chalk, comb, etc.), without letting your lab partner see it, from the back of your lab partner's head, around the side, towards the front. Note the approximate angle when your lab partner is able to see the object (the use of rods). Your lab partner should keep his or her eyes directed forward. Continue to slowly move the object forward until recognition of the color of the object occurs. Note the approximate angle from the tip of the nose. You can record the approximate angle by the use of an apparatus called a Vision Disk or by placing a protractor above your lab partner's head while you make your measurements.

Angle where object is perceived: _____

Angle where color is determined: _____

What is the difference in the distribution of rods and cones in the eye?

Rod distribution: _____

Cone distribution: _____

When you are intently focusing on a subject, what cell type (rods or cones) are you using?

Measurement of Binocular Visual Field

Not all animals have the same **visual field.** Some prey species (like deer or sheep) have very extensive visual fields with little binocular vision. On the other hand, many predators, birds, and arboreal animals have a more limited visual field yet they have greater **binocular,** or **stereoscopic, vision.** Binocular vision allows arboreal animals to perceive depth, which is vital when judging how far away the next branch is! You can determine your visual field with the use of a Vision Disk or protractor. If you are using a Vision Disk follow the instructions enclosed. If not, sit down and have your lab partner stand behind you. Close your right eye and look straight ahead with the left. While the right eye is closed, have your lab partner move an object (pen, paper disk, etc.) from behind your head from the left until you can just see the object. Measure the angle from the tip of the nose to where you saw the object (figure 26.8). With the same eye directed ahead, continue moving the object until it is out of view (to the right of the nose somewhere). Determine the angle from the nose to where the object disappears. Add these two values and record the result.

Angle where object is perceived: _____

Angle where object disappears: _____

Now close your left eye and repeat the exercise on the right side. Determine the total visual field for each eye and then for both eyes and the degree of overlap between the eyes (figure 26.9). Record this data.

Visual field for right eye: _____

Visual field for left eye: _____

Complete visual field for both eyes: _____

Degree of overlap: _____

Measurement of Visual Acuity (Snellen Test)

Face the Snellen eye chart from 20 feet away and have your lab partner stand next to the chart. Cover one eye with a 3- by 5-inch card and have your lab partner point to the largest letter on the chart. Do *not* read the chart with both eyes open. Your lab partner should then progressively move down the chart and note the line that has the smallest print in which you made no errors. Your lab partner should record the number at the side of the line. This number refers to your **visual acuity.** Now switch the card to the other eye and repeat the test. Record your results.

Left eye: _____

Right eye: _____

A vision of 20/20 is considered normal. In 20/20 vision you can see the same details at 20 feet that most other people can see at the same distance. If your vision is 20/15, then you can see at 20 feet what most people can see at 15 feet. If your vision is 20/100, then you see at 20 feet what most people see at 100 feet.

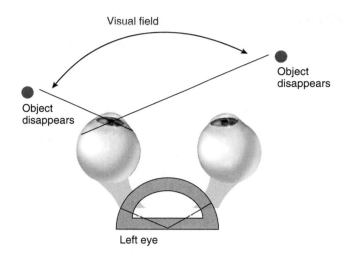

Figure 26.8 Measuring the visual field.

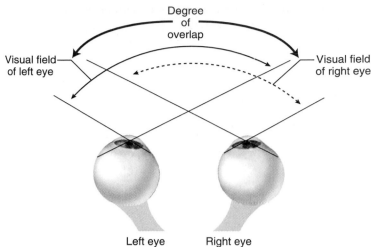

Figure 26.9 Visual field.

Astigmatism Tests

If the cornea or lens of the human eye were perfectly smooth, the incoming image would strike the retina evenly and there would be no blurry areas. In most people the cornea or lens is not perfectly smooth, and various degrees of astigmatism occur. In a normal optical exam the distance correction is made first (to determine visual acuity) and then, with the corrective lenses in place, an astigmatism test is performed.

If you do not normally wear corrective lenses, then cover one eye and examine the astigmatism chart (as represented in figure 26.10) from 20 feet away. The chart consists of a series of parallel lines radiating away from the center. Stare at the center of the chart and determine which of the sets of parallel lines, if any, appears light or blurry. Have your lab partner note the corresponding number on the chart and record the number.

Astigmatism numbers: _____

If you wear corrective lenses, then not only is the condition of nearsightedness or farsightedness corrected but astigmatism is corrected also. To test for astigmatism move about 12 feet from the chart, hold your glasses slightly away from your face, and then rotate them 90°. If your glasses correct for astigmatism, the lines on the chart will be blurry as you rotate your glasses.

Opthalmoscope

Clinical use of the opthalmoscope is important not only for the diagnosis of variances in the eyeball but also as a potential indicator of diseases, such as **diabetes mellitus,** that may affect the eye. Familiarize yourself with the parts of the opthalmoscope (figure 26.11). Locate the ring (rheostat control) at the top of the handle. Depress the button and turn the ring until the light comes on. The head piece of the opthalmoscope consists of a

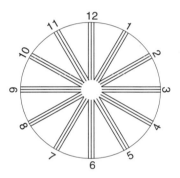

Figure 26.10 Astigmatism chart.

rotating disk of lenses. You can change the diopter (strength) of the lenses by rotating the disk clockwise or counterclockwise. As the numbers get progressively larger, you are looking through more convex lenses. If you rotate the dial in the other direction the lenses become more concave. At the zero reading there are no lenses in place.

Observation with the Opthalmoscope

 Caution Examine the posterior region of the eye only for a short period of time. Extensive use with the opthalmoscope can damage the eye.

Sit facing your lab partner. Select the left eye to examine. Have the opthalmoscope setting at zero and use your left hand to hold the opthalmoscope. Use your left eye to look through the scope and move in close to examine the eye of your lab partner

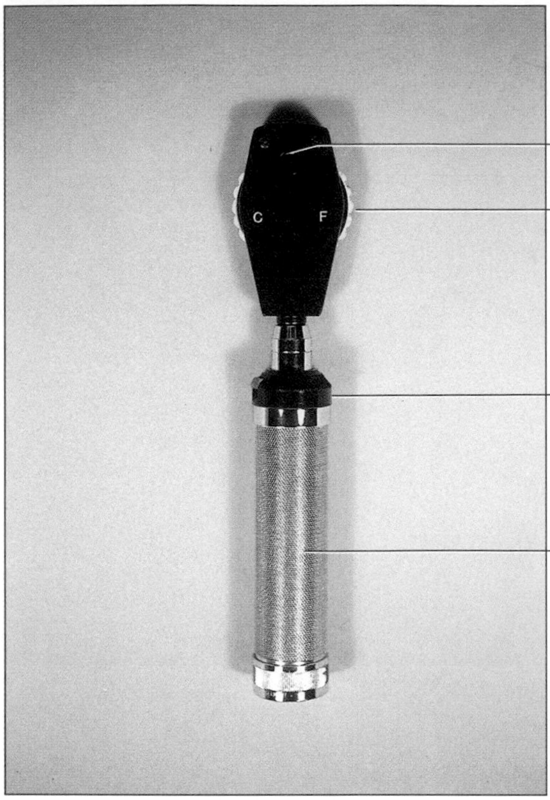

Viewing aperture

Diopter lens selection disk

Rheostat control

Handle

Figure 26.11   Opthalmoscope.

Figure 26.12   Use of opthalmoscope.

(figure 26.12). You may rest your hand on the cheek of your lab partner if you need to steady your hand. Look into the pupil and examine the back of the eye. If the back appears fuzzy, rotate the dial clockwise (positive diopters) and see if it comes into focus. If a positive setting provides a clear view of the eye, then your lab partner has **hyperopia** (hypermetropic vision), or farsightedness. If the image is indistinct, then adjust the dial counterclockwise to obtain a negative diopter reading. If the image is clear with a negative number, then your lab partner has **myopia** (myopic vision), or nearsightedness. This procedure is based on the assumption that your vision, as the examiner, is normal.

Pupillary Reactions

Have your lab partner sit in a dark room for a minute or two. Examine his or her eyes in dim light. Are the pupils dilated or constricted?

Pupil diameter in dim light: _____

Now shine a penlight in the right eye. Record what happens to the pupil diameter.

Right pupil diameter: _____

As you shine the light into the right eye, what occurs in the left pupil? This is called a **consensual reflex.** Record the effect.

Left pupil diameter: _____

Color Blindness

Color vision is dependent on three separate cone cell sensitivities. Cones may be red, green, or blue sensitive. Changes in the genes on the X chromosome are the most common cause of color blindness. Some individuals may be unable to see a particular color at all, while others may have a reduction in their ability to see a particular color. Color blindness is most common in males and relatively rare in females. This can be understood by examining the chromosomal makeup of the two sexes. The male chromosome makeup is XY. The Y chromosome does not carry the gene for color vision. If the X chromosome carries a gene for color blindness, then the male exhibits color blindness. On the other hand, if a female carries the gene for color blindness on the X chromosome, the chances are that she will have a normal gene on the other X chromosome. The normal gene is expressed, cone pigments are produced, and the female has normal color vision.

You can test for color blindness by matching various samples of colored yarn with a presented test sample or you can use Ishihara color charts. If you use the color charts, flip through the book with a lab partner and record which charts were accurately viewed and which plates, if any, were missed. You can calculate the type of color blindness and the degree by following your instructor's directions or those that come with the color chart.

Your degree of color vision: _____

Afterimages

The photosensitive pigment of the eye is called **rhodopsin,** which is composed of the light-sensitive **retinal** (a metabolite of vitamin A) and the protein **opsin.** When light strikes the retina the purple-colored rhodopsin splits into its two component parts and becomes pale (a process known as "bleaching"). You can test the time for separation and reassembly of the photopigments by staring at a colored image on a card under a moderately bright lightbulb (60 watt) for a few moments. Stare long enough to get an image (about 10 to 20 seconds) and then shut your eyes. Have your lab partner record the time. You should see a colored image against a dark background. This is known as a positive afterimage, which is due to the photoreceptors continuously firing. After a few minutes you should see the reverse of the original image (dark against a light background). This is known as a negative afterimage. A negative afterimage reflects the bleaching effects of rhodopsin. Record the time change for the positive afterimage and the negative afterimage in the space following:

Positive afterimage: _____

Negative afterimage: _____

Determination of the Blind Spot

The **optic disc** is a region where the retinal nerve fibers exit from the back of the eye and form the **optic nerve.** The mass exit of the nerve fibers leaves a small circle at the back of the eye devoid of photoreceptors. This region, the optic disc, is also known as the **blind spot.** You can locate the blind spot by holding this lab manual at arm's length. Use your right eye (close your left eye) and stare at the following X. It should be in line with the middle of your nose. *Slowly* move the manual closer to you and stare only at the X. Notice that at a particular distance the dot disappears. Have your lab partner measure the distance between your eye and the manual in centimeters and record the value below.

X   ●

Distance between your right eye and the manual: _____

Now use your left eye and stare at the dot. Make sure it is aligned with your nose (in the midsagittal plane). Record the results for your left eye.

Left eye distance: _____

Name _____

1. Eye shine in nocturnal mammals is different from the "red eye" seen in some flash photographs. Eye shine is the reflection from the tapetum lucidum. What might "red eye" be due to?

2. Fill in the following illustration with the appropriate terms.

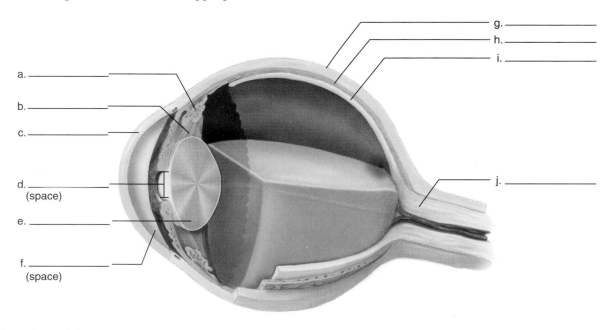

a. _____
b. _____
c. _____
d. _____
   (space)
e. _____
f. _____
   (space)

g. _____
h. _____
i. _____
j. _____

Sagittal section of the eye.

3. Since the lens is made of protein, what effect might the preserving fluid used in lab have on the structure of the lens? How would this affect the clarity?

4. What is the consensual reflex of the pupil?

5. How does the vitreous humor differ from the aqueous humor in terms of location and viscosity?

6. What layer of the eye converts visible light into nerve impulses?

7. What nerve takes the impulse of sight to the brain?

8. What is another name for the sclera?

9. How would you define an extrinsic muscle of the eye?

10. What gland produces tears?

11. What is the name of the transparent layer of the eye in front of the anterior chamber?

12. The iris of the eye has what function?

13. Where is vitreous humor found?

14. What is the middle tunic of the eye called?

15. Is the lens anterior or posterior to the iris?

16. Which retinal cells are responsible for vision in dim light?

17. How would you define the near point of the eye?

18. What do the numbers 20/100 mean for visual acuity?

19. What is an astigmatism?

20. What is at the area of the eye where the blind spot is found?

Ear, Hearing, and Equilibrium

Introduction

The ear is a complex sense organ that performs two major functions, hearing and equilibrium. It consists of three regions, an outer ear, middle ear, and inner ear. Hearing is considered mechanoreception, because the ear receives mechanical vibrations (sound waves) and translates them into nerve impulses. This process begins with the vibrations reaching the outer ear and ends up being interpreted as sound in the temporal lobe of the brain. Equilibrium, on the other hand, involves receptors in the inner ear, along with other sensory perceptions such as visual cues from the eye, and proprioception in joints. There are two types of equilibrium that are sensed by the inner ear. These are **static equilibrium** and dynamic equilibrium. In static equilibrium an individual is able to determine his or her nonmoving position (such as standing upright or lying down). In **dynamic equilibrium** motion is detected. Sudden acceleration, abrupt turning, and spinning are examples of dynamic equilibrium.

Objectives

At the end of this exercise you should be able to

1. explain how mechanical sound vibrations are translated into nerve impulses;
2. list the structures of the outer, middle, and inner ear;
3. describe the structure of the cochlea;
4. perform conduction deafness tests such as the Rinne and Weber tests;
5. compare dynamic and static equilibrium and the structures involved in their perception.

Materials

Models and charts of the ear
Microscope
Slides of the cochlea
Tuning fork (256 Hz)
Rubber reflex hammer
Audiometer, if available
Model of ear ossicles
Meterstick
Ticking stopwatch
Swivel chair

Procedure
Anatomy of the Ear

The ear can be divided into three regions, the **outer, middle, and inner ear.** The outer ear consists of those auditory structures superficial to the **tympanic membrane (eardrum).** The middle ear contains the ear **ossicles (bones)** and **auditory tube,** and the inner ear consists of the **cochlea, vestibule,** and **semicircular canals** (figure 27.1).

Structure of the Outer Ear

The outer ear consists of the **pinna (auricle),** which can further be subdivided into the **helix** and the **earlobe (lobule).** The helix is composed of stratified squamous epithelium overlying elastic cartilage. This cartilage allows the ears to bend significantly. Deep to the pinna the external ear forms the **external auditory canal (tube),** which penetrates into the temporal bone. The tympanic membrane is the border between the outer ear and the middle ear. It is composed of connective tissue covered by epithelial tissue. The membrane is sensitive to sound and vibrates as sound is funneled down the external auditory canal.

Structure of the Middle Ear

The middle ear houses the three ear ossicles and the auditory (eustachian) tube (figure 27.1). The ossicle that is attached to the tympanic membrane is the **malleus** (*malleus* = hammer). As the membrane vibrates the malleus rocks back and forth, amplifying the sound waves. The malleus is attached to the **incus** (*incus* = anvil), which is attached to the **stapes** (*stapes* = stirrup).

Sound consists of pressure waves. As these waves strike the tympanic membrane, it vibrates. This vibration is conducted by the ossicles to the oval window. The process of moving from a large diameter structure (tympanic membrane) to a smaller diameter structure (oval window) magnifies the sound about 20 times. See figure 27.2. Examine the ossicles in figure 27.1 and on models in the lab. In addition to the ossicles the **auditory tube** (eustachian tube) occurs in the middle ear. This tube connects the middle ear to the nasopharynx (see figure 27.1) and provides for equalizing pressure between the middle ear and the external environment when changes of pressure occur (such as during changes of elevation). The auditory tube can be a conduit for microorganisms that travel from the nasopharynx to the middle ear and lead to middle-ear infections.

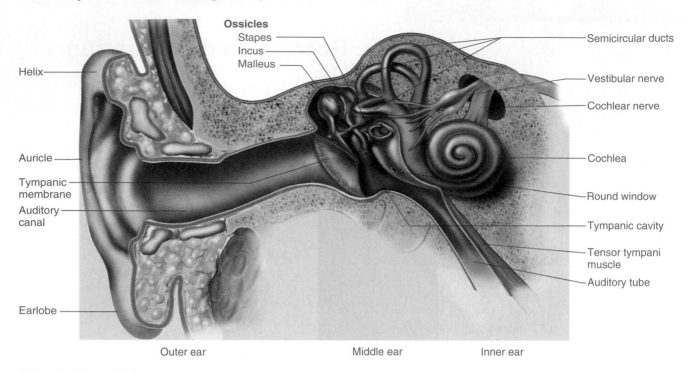

Figure 27.1 **Anatomy of the ear.**

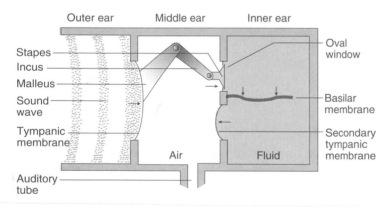

Figure 27.2 **Model of hearing.** Pressure waves of sound vibrate the tympanic membrane, which transfers sound to the oval window, and finally to the basilar membrane where vibration is converted to neural impulses.

Structure of the Inner Ear

The inner ear is a complex structure composed of three separate regions, the cochlea, the vestibule, and the semicircular ducts (figure 27.3). The cochlea is a spiral structure that resembles a seashell (*cochlea* = snail), while the semicircular ducts look like three loops. Between these two structures is the vestibule.

The inner ear is encased in the temporal bone, and the outer bony structure is called the bony labyrinth. Inside the bony labyrinth is perilymph, a clear fluid that is external to the membranous labyrinth. The fluid enclosed by the membranous labyrinth is the endolymph, which is important in both hearing and equilibrium.

Cochlea The cochlea (figures 27.1, 27.4 and 27.5) is involved in hearing. As the sound waves travel down the external auditory canal they cause the tympanic membrane to vibrate. This vibration rocks the ear ossicles, which are connected to the inner ear. As the stapes vibrates it moves back and forth in the **oval window (fenestra ovalis),** and this causes fluid to move back and forth in the cochlea. The cochlea also has a **round window (fenestra rotunda),** which allows the vibration from the ossicles to move fluid back and forth. Sound is measured in cycles per second or hertz (Hz). The human ear is capable of hearing high-pitched sounds up to 20,000 Hz and sounds as low as about 20 Hz. High-pitched vibrations stimulate the cochlea nearest the middle ear while lower pitched vibrations stimulate the cochlea further from the middle ear. The sounds that have the lowest pitch stimulate the ear farthest from the middle ear.

Microscopic Section of Cochlea If you examine a microscopic section of the cochlea in cross section you will see a

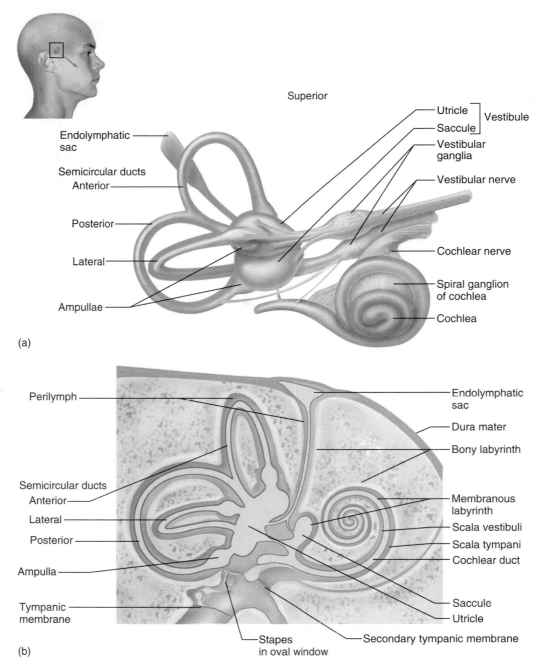

Superior

Endolymphatic sac

Semicircular ducts
  Anterior

Posterior

Lateral

Ampullae

Utricle ⎤ Vestibule
Saccule ⎦

Vestibular ganglia

Vestibular nerve

Cochlear nerve

Spiral ganglion of cochlea

Cochlea

(a)

Perilymph

Semicircular ducts
  Anterior

Lateral

Posterior

Ampulla

Tympanic membrane

Endolymphatic sac

Dura mater

Bony labyrinth

Membranous labyrinth

Scala vestibuli

Scala tympani

Cochlear duct

Saccule

Utricle

Stapes in oval window

Secondary tympanic membrane

(b)

Figure 27.3   Anatomy of the inner ear. (a) Major regions of the inner ear; (b) relationship of membranous labyrinth to bony labyrinth. ⚡

number of chambers. The chambers are clustered in threes. Find the **scala vestibuli, cochlear duct (scala media),** and **scala tympani** on the microscope slide. Compare these to figures 27.4 and 27.5.

Note the **spiral organ (organ of Corti),** which is defined as the area between the **vestibular membrane** and the **basilar membrane.** The spiral organ is seen here in cross section, but remember that it runs the length of the cochlea. The spiral organ is sensitive to sound waves. As a particular region of the spiral organ is stimulated, the basilar membrane and **tectorial membrane** vibrate independently of one another. This produces a tugging on the cilia that connect the hair cells to the tectorial membrane. The hair cells send impulses to the cochlear branch of the **vestibulocochlear nerve.** These impulses travel

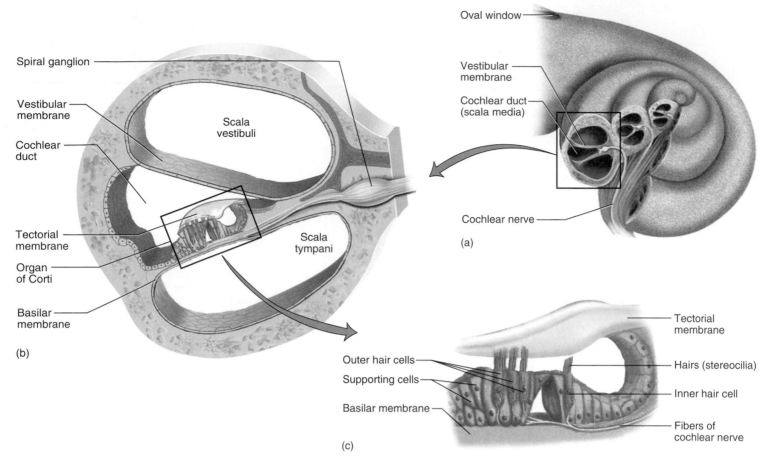

Figure 27.4 Anatomy of the cochlea. (a) Overview of cochlea; (b) details of chambers; (c) spiral organ.

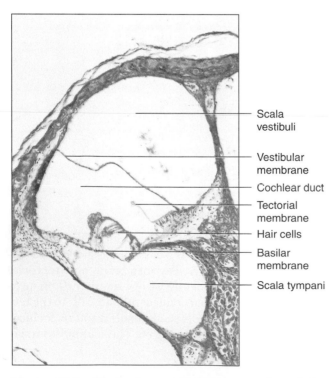

Figure 27.5 Photomicrograph of cross section of the cochlea (100×).

to the **auditory cortex** of the **temporal lobe,** where they are interpreted as sound (figure 27.6).

Vestibule The **vestibule** consists of the **utricle** and **saccule** (figure 27.7). These two chambers are involved in the interpretation of static equilibrium. The utricle and saccule have regions known as maculae, which consist of cilia grouped together with an overlying gelatinous mass and calcium carbonate stones, or **otoliths.** These can be seen in figure 27.7. As the head is accelerated or tipped by gravity, the otoliths cause the cilia to bend, indicating that the position of the head has changed. Static equilibrium is perceived not only from the vestibule but from visual cues as well. When the visual cues and the vestibular cues are not synchronized, then a sense of imbalance or nausea can occur.

Semicircular Ducts The third part of the inner ear consists of the **semicircular ducts,** which are involved in determining dynamic equilibrium. There are three semicircular ducts each at 90° to one another (in the horizontal, sagittal, and coronal planes) (figure 27.8). The semicircular ducts are filled with endolymph and are expanded at their base into an **ampulla.** Inside each ampulla is a **crista ampullaris,** a cluster of cilia with an overlying gelatinous mass called the **cupula.** The endolymph

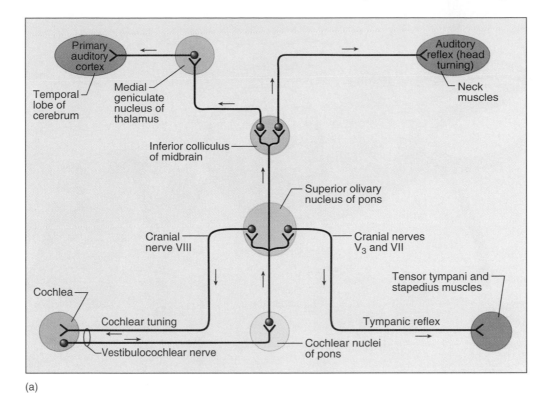

(a)

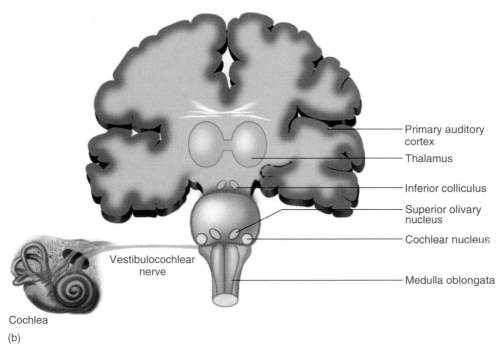

(b)

Figure 27.6   Interpretive pathway of hearing.

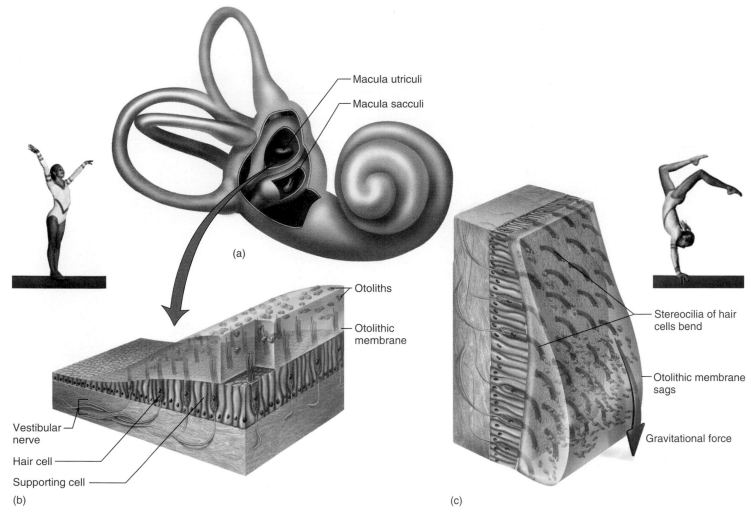

Figure 27.7 **Vestibule.** (a) Inner ear; (b) macula when head is upright; (c) macula when head is tilted.

that is in the membranous labyrinth has inertia; that is, it tends to remain in the same place. As the head is turned the cupula bends against the endolymph, the hair cells (cilia) bend, and angular or rotational acceleration is perceived as shown in figure 27.8*a*. The three semicircular ducts are at right angles to each other. If motion in the forward plane occurs (such as by doing back flips), then the **anterior semicircular ducts** are stimulated. If you were to turn cartwheels, then the **posterior semicircular ducts** are stimulated. If you were to spin around on your heels, then the **lateral semicircular ducts** pick up the information. Motion that occurs in between these areas is picked up by two or more of the ducts and interpreted as movement due to a combination of impulses from the semicircular ducts.

Hearing Tests

Obtain a meterstick and a ticking stopwatch. Have your lab partner sit in a quiet room and slowly move the ticking watch away from one of his or her ears until the sound can no longer be heard. Record the distance in centimeters (when the sound

can no longer be heard) in the space provided. Check the other ear and record the data.

Maximum distance sound perceived for right ear:

_____

Maximum distance sound perceived for left ear:

_____

Audiometer Test (Optional)

If an audiometer is available ask your instructor to demonstrate how to use it to test hearing. Hearing normally decreases somewhat with age, though hearing loss is greatly accelerated by loud music, moderate to extensive stereo headphone use, and exposure to machines that operate at high decibel levels. Another common source of hearing loss is the use of recreational firearms without hearing protection.

To use the audiometer begin with the 1,000 Hz tone. Set the level to 50 decibels (dB). Check frequencies at 125, 250, 500, 1,000, 2,000, 4,000, and 8,000 Hz. Record your ability to hear these frequencies with each ear.

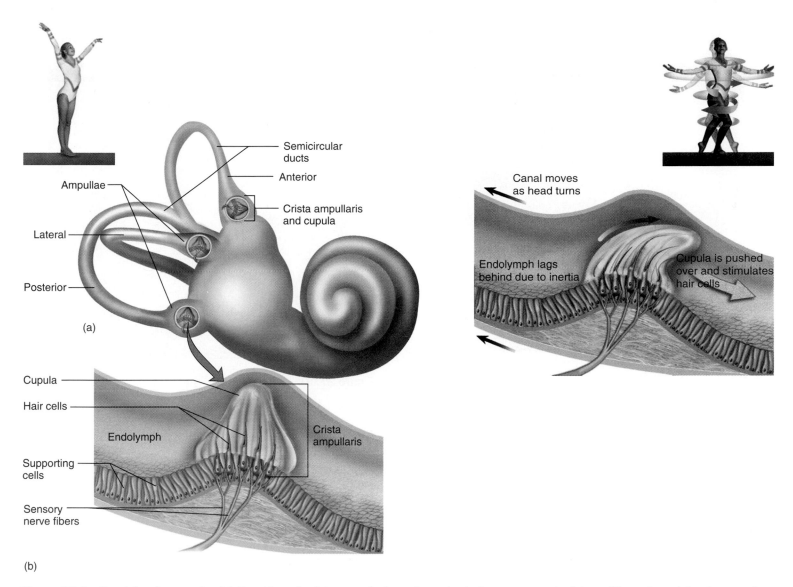

Figure 27.8 Semicircular canals. (a) Function of crista ampullaris and cupula during movement; (b) position of semicircular canals when still. ⚹

	Right Ear	Left Ear
	(yes/no)	(yes/no)
125		
250		
500		
1,000		
2,000		
4,000		
8,000		

The **threshold for hearing** is the lowest level at which your lab partner can hear the tone in at least 50% of the trials. To measure this value set the audiometer to 125 Hz and then lower the level in 10 dB increments from 50 dB to the point where your lab partner can no longer hear the tone. Record the threshold for hearing. You can test other frequencies as well. Hearing loss is significant when the hearing threshold is at 20 dB or more at any two frequencies in an ear or 30 dB at any particular frequency. Record the minimum sound levels.

Right Ear   **Left Ear**

125: _____   _____

250: _____   _____

500: _____   _____

1,000: _____   _____

2,000: _____   _____

4,000: _____   _____

8,000: _____   _____

Figure 27.9 Weber test.

Figure 27.10 Rinne test.

Weber Test

Conduction deafness occurs when damage to the tympanic membrane or ear ossicles is present. You can test for this type of deafness with a tuning fork. Hold onto the handle of a tuning fork at 256 Hz (middle C). Strike the tines (the forked portion) on a solid surface and gently place the end of the handle on your lab partner's chin or forehead (figure 27.9).

If the sound is louder in one ear, then this may be a sign of **conduction deafness.** In this case the affected ear is not picking up sounds from the tympanic membrane and is more sensitive to vibrations coming through the skull bones. The sound will be louder in the ear that has conduction deafness. If the ear has nerve damage, then the sound will be louder in the normal ear. Record your results.

Sound perception (right/left or equal in
both ears): _____

Rinne Test

Strike the tuning fork and place the handle on the mastoid process. When your lab partner indicates that the sound disappears, lift the tuning fork from the mastoid process and bring it to the outside of the pinna (figure 27.10). Can the sound now be heard? If the sound cannot be heard after the tuning fork is placed near the outside of the ear, then there is damage to the tympanic membrane or ear ossicles (conduction deafness). Record your results.

Right ear: _____

Left ear: _____

Sound Location

Have your lab partner sit with eyes closed. Strike the tuning fork with a rubber reflex hammer above his or her head. Have your lab partner describe to you where the sound is located. Strike the tuning fork behind the head, to each side, in front, and below the chin of your lab partner. Record the results.

Location Where Sound Was Struck	Where Sound Was Perceived
Above head	
Behind head	
Right side	
Left side	
In front of head	
Below chin	

Postural Reflex Test

Postural reflexes are important for maintaining the upright position of the body. These reflexes are negative feedback mechanisms. If, for example, you lean slightly to the left your left foot abducts to regain the center of balance.

1. Select an area that is free from any obstacles.
2. While you read this manual, your lab partner should give you a little nudge to the left or right (Not enough to knock you off your feet!) to push you off balance. The

postural reflex is reflected in the movement of the foot on the opposite side of your lab partner. If you are nudged to the left then your left foot should move to the side to correct against the direction of the contact. Did your postural reflex work?

3. Record your result in the following space.

Postural reflex:_____

Barany's Test

Barany's test examines visual responses to changes in dynamic equilibrium. As the head turns, one of the reflexes that occurs is movement of the eyes in the opposite direction of the rotation. Nerve impulses from the semicircular canals innervate the eye muscles and cause the eye movement. When rotating in one direction, the eyes move in the opposite direction so that there is enough time for a visual image to be fixed. The subject for this test should not be subject to dizziness or nausea.

1. Place the subject in a swivel chair with four or five students close by, standing in a circle, in case the subject loses balance and begins to fall. The student chosen for this exercise should grab onto the chair firmly so as not to fall off. The subject should tilt his or her head forward about 30 degrees which will place the lateral semicircular ducts horizontally for maximum stimulation.

2. One member of the group should spin the chair around about ten revolutions while the subject keeps his or her eyes open.

3. Stop the chair and have the subject look forward. The twitching of the eyes is called **nystagmus** and is due to the stimulation of the endolymph flowing in the semicircular ducts. When the chair is stopped the fluid in the endolymph will have overcome the inertia and will continue to flow in the ducts. Did the movement of the eyes occur in the direction of the rotation or in the opposite direction? Record your results below.

Direction of chair rotation relative to subject:_____

Direction of eye rotation:_____

Romberg Test

The Romberg test involves testing the static equilibrium function of the body.

1. Place the subject in front of a blackboard or any surface on which you can see the shadow of the subject.

2. Have the subject stand in that direction for one minute and determine if there are any exaggerated movements to the left or right.

3. Record your results in the following space.

Amount of lateral sway:_____

You should then have your lab partner close his or her eyes and repeat the test. Is the swaying motion increased or decreased? Record your results in the following space.

Amount of lateral sway with eyes closed:_____

Name _____

1. What are the three general areas or regions of the ear?

2. The pinna of the ear consists of what two main parts?

3. The ear is what kind of receptor?

4. The ear performs two major sensory functions. What are they?

5. What separates the outer ear from the middle ear?

6. Name the three ear ossicles.

7. From the choices below, select the function of the cochlea.
   a. static equilibrium   b. taste   c. hearing   d. dynamic equilibrium

8. What area is found between the scala vestibuli and the scala tympani?

9. What is the name of the nerve that takes information of equilibrium and hearing to the brain?

10. What units are used to measure sound energy?

11. Background noise affects hearing tests. In the ticking watch test, what kind of result, in terms of auditory sensitivity, would you have recorded if moderate background noise was present?

12. In the Weber test, the ear that perceives the sound as being louder is the deaf ear. Why is this the case?

13. Fill in the following illustration of the ear.

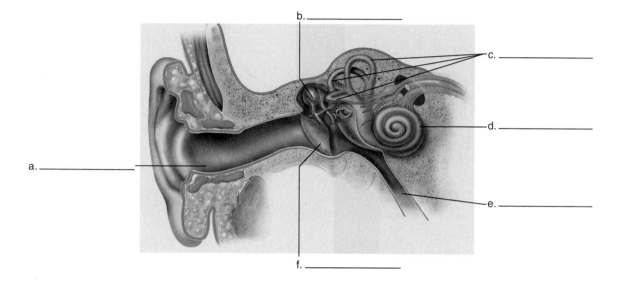

14. Fill in the following illustration of the cross section of cochlea.

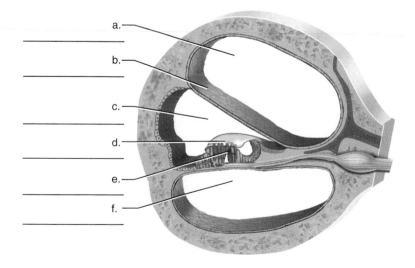

Endocrine System

Introduction

The **endocrine system** consists of glands that produce chemical messengers called **hormones,** which are picked up by blood capillaries. This type of release is called a "ductless secretion," which is characteristic of the endocrine system. Those glands that secrete material in tubules or ducts (such as sweat and salivary glands) or directly to a surface (ovary) are known as **exocrine glands.** The secretions of the endocrine glands enter the interstitial fluid and then travel by blood vessels, which act as highways to carry the hormones throughout the body. Areas that are receptive to hormones are called **target areas,** and these may be organs or tissues. Our traditional concept of the endocrine system is expanding, with research showing that many organs of the body, such as the stomach, heart, and kidney, produce hormones and thus have endocrine functions. This exercise focuses on those glands that have a major endocrine component.

Hormones can have many effects. Some of these actions are growth; changes and development, or maturation; metabolism; sexual development; regulation of the sexual cycle and homeostasis.

Objectives

At the end of this exercise you should be able to

1. list the major endocrine organs of the human body;
2. name the hormones produced by these endocrine organs;
3. describe the detection of hormones by monoclonal antibodies;
4. list the hormones stored in the anterior and posterior pituitary glands;
5. discuss how the secretions of the endocrine glands differ from those of the exocrine glands.

Materials

Models and charts of endocrine glands
Microscopes
Microscope slides of thyroid, pituitary, adrenal gland, pancreas, testis, ovary
Urine samples from a nonpregnant and/or a pregnant individual
Pregnancy test kit
Disposable gloves (latex or plastic) and biohazard container

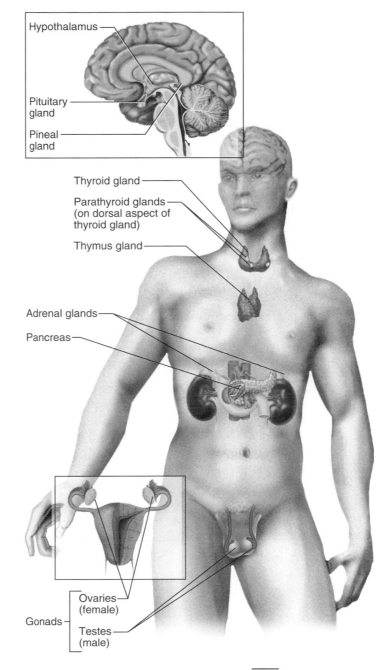

Figure 28.1   Major endocrine glands.

Procedure

Anatomy of the Major Endocrine Organs

Locate the major endocrine glands in figure 28.1, including the pineal gland, pituitary gland, thyroid, parathyroids, thymus, pancreas, adrenals, and gonads (testes or ovaries). Once you have noted their location, you can proceed with a more detailed study.

Pineal Gland

The **pineal gland** develops from the diencephalon of the brain. The gland is named the pineal gland because it looks like a pine nut. Locate the gland in figure 28.1. The pineal gland secretes the hormone **melatonin,** which is an inhibitory hormone and probably regulates circadian rhythms. Melatonin has a psychologically depressing effect in some individuals. It has been named the "hormone of darkness" because it is produced during times of low light levels and its manufacture is inhibited in bright light.

Pituitary Gland

Figures 28.1 and 28.2 illustrate another endocrine gland called the **pituitary gland,** or **hypophysis.** The pituitary is divided into an **anterior lobe,** or **adenohypophysis,** an **intermediate lobe** (part of the anterior lobe), and a **posterior lobe,** or **neurohypophysis.** The pituitary is suspended from the base of the brain by a stalk called the **infundibulum.**

Adenohypophysis

The adenohypophysis originates from the roof of the oral cavity during embryonic development. The result of this development can be seen if you examine a prepared slide of the pituitary gland. In the histologic section, note that the adenohypophysis is composed of simple cuboidal epithelial cells (figure 28.3). The cells of the anterior pituitary have the same embryonic origin yet they have differentiated into several types of specialized cells. These cells produce a number of hormones that have broad effects throughout the body.

Some of the hormones produced in the anterior pituitary and their functions are as follows:

- **Thyroid-stimulating hormone (TSH),** or thyrotropin, stimulates the thyroid gland to produce thyroid hormones.
- **Growth hormone (GH),** or somatotropin, promotes growth of most of the cells and tissues of the body.
- **Prolactin (PRL),** or luteotropin stimulates mammary glands to begin production of milk.
- **Gonadotropins** stimulate ovaries and testes.
  —**Follicle-stimulating hormone (FSH)** stimulates production of sex cells, causes ovarian follicles to mature (develop and produce egg cells), and stimulates testes to produce spermatozoa.
  —**Luteinizing hormone (LH)** stimulates ovaries to produce other hormones (for example, progesterone), causes maturation of Graafian follicles and ovulation, and stimulates testosterone production in testes.

- **Adrenocorticotropin (ACTH)** controls hormone production in the adrenal cortex.

Neurohypophysis

The tissue of the neurohypophysis, or posterior pituitary, is very different from that of the adenohypophysis. The neurohypophysis is composed of nervous tissue that originated from the base of the brain. Compare the tissue of the neurohypophysis in a prepared slide to figure 28.4.

Hormones released from the posterior pituitary are actually secreted by the **hypothalamus** and flow through axons to be stored in the posterior pituitary. These hormones and their actions include:

- **Antidiuretic hormone (ADH)** stimulates reabsorption and retention of water by the kidneys. It is also known as vasopressin because it causes arterioles to constrict, which elevates blood pressure.
- **Oxytocin** stimulates milk release from the breast and causes uterine contractions. External influences on hormone actions can be seen in the release of oxytocin. Look at figure 28.5 and note that the sucking action of an infant on the mother's breast sends impulses to the hypothalamus. The hypothalamus stimulates the posterior pituitary to release oxytocin, which causes a relatively rapid release of milk by causing contraction of the milk secreting glands.

Thyroid

Figure 28.6 illustrates the **thyroid gland,** which is named for its location near the thyroid cartilage of the larynx. The thyroid gland has two lateral **lobes** and a medial **isthmus** that connects them.

The thyroid has a characteristic histologic structure that can be seen by examining a thin section of the organ. Examine a slide of thyroid gland and compare it to figure 28.7. Locate the **follicle cells** that surround the **colloid,** a storage region for thyroid hormones. Colloid is mostly composed of thyroglobulin, which is a large molecular weight compound. Thyroid hormones are formed inside the larger thyroglobulin molecule. Also locate the **parafollicular cells,** which occur in the spaces between the follicles.

The thyroid gland secretes three specific hormones. Two of these are T_3, or **triiodothyronine** (a molecule that includes three iodine atoms) and T_4, or **thyroxine** (containing four iodine atoms per molecule). These hormones are involved with basal metabolic rates and are stored in the colloid. Another hormone of the thyroid gland is **calcitonin.** Calcitonin decreases blood calcium levels by causing excretion of calcium by the kidneys and deposition of calcium in bone by decreasing osteoclast activity and formation. Calcitonin is produced by the parafollicular, or C, cells. Calcitonin is antagonistic to parathyroid hormone (discussed next).

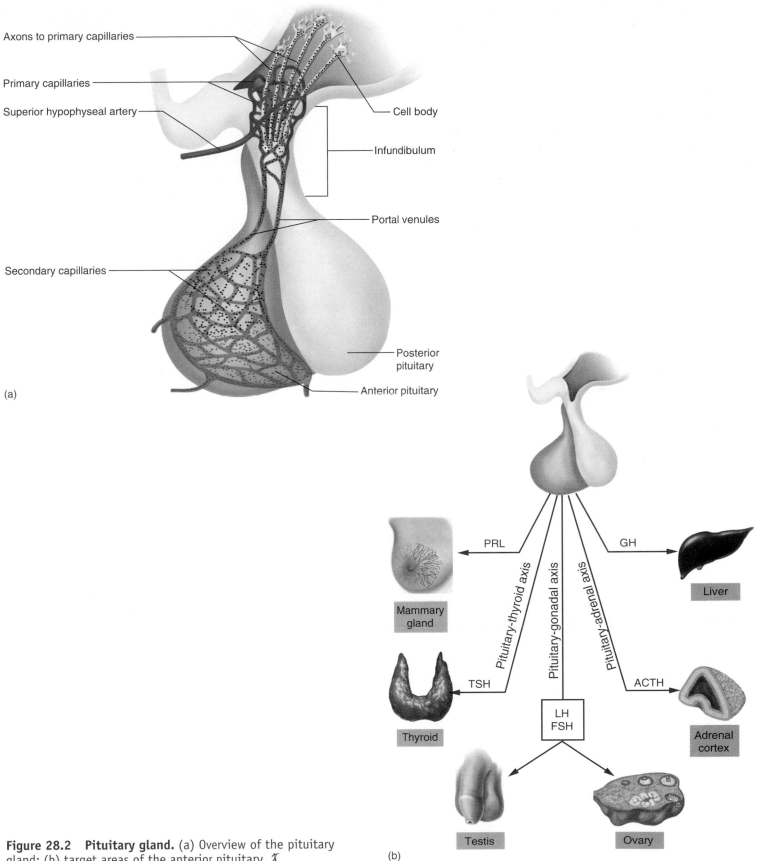

Figure 28.2   Pituitary gland. (a) Overview of the pituitary gland; (b) target areas of the anterior pituitary.

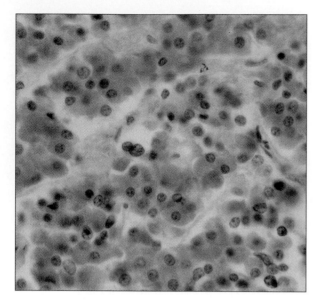

Figure 28.3   Histology of the anterior pituitary gland (400×). ⚘

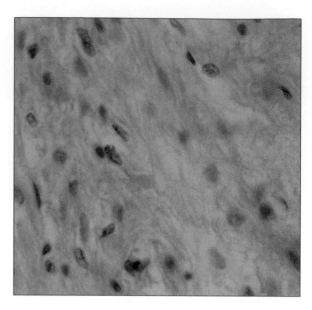

Figure 28.4   Histology of the posterior pituitary (400×). ⚘

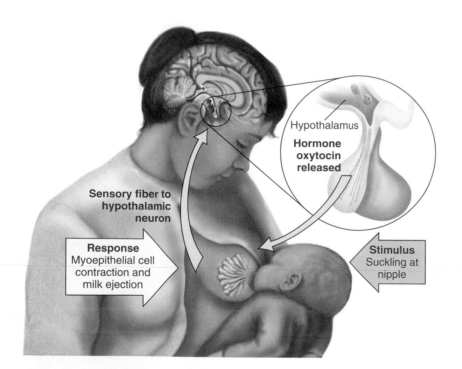

Figure 28.5   External influences on hormonal action.

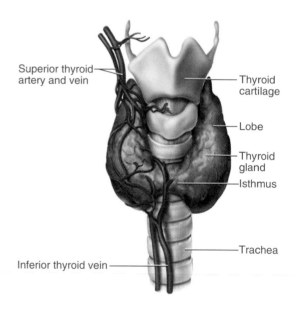

Figure 28.6   Anatomy of the thyroid gland. ⚘

Parathyroid Glands

In the posterior portion of the thyroid are usually two pairs of organs called the **parathyroid glands. Chief cells** in the parathyroid gland secrete **parathyroid hormone (PTH),** or parathormone, which is responsible for increasing calcium levels in the blood. This occurs by increasing calcium uptake in the intestines, increasing kidney reabsorption of calcium, and re-

leasing calcium from bone. Examine the location of the parathyroid glands embedded in the posterior surface of the thyroid gland (figure 28.8).

Thymus

The **thymus** is active in young individuals and plays an important part in immunocompetency. The thymus is located

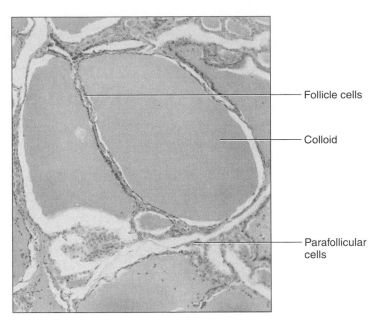

Figure 28.7   Histology of the thyroid gland (100×). ☂

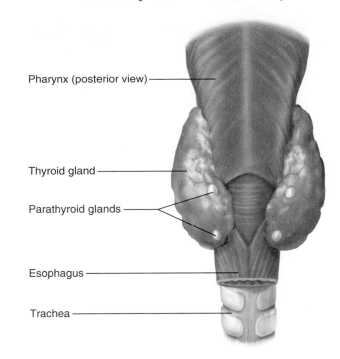

Figure 28.8   Parathyroid glands. ☂

anterior and superior to the heart (figure 28.9). It secretes a hormone, **thymosin,** that causes the maturation of T cells. These cells originate as stem cells in the bone marrow and migrate to the thymus. Under the influence of thymosin the cells mature to provide "cellular immunity" against antigens (foreign material such as bacteria and viruses, which cause immune reactions). The T cells migrate from the thymus predominantly to the lymph nodes and the spleen to carry out their functions.

Pancreas

The **pancreas** is a mixed gland in that it has an exocrine function and an endocrine function. Locate the pancreas in figures 28.1 and 28.10. The exocrine function is digestive in nature because pancreatic juice contains both buffers and digestive enzymes.

The endocrine function of the pancreas consists of the secretion of the hormones insulin and glucagon, which regulate blood glucose levels. When blood glucose levels drop, **glucagon** converts glycogen (a starch storage product) to glucose. Glucagon is produced in specialized cells **(alpha cells)** that occur in clusters called **pancreatic islets** (or the islets of Langerhans) (figure 28.10). Pancreatic islets also produce **insulin,** which lowers the blood glucose level. Insulin, which is produced in **beta cells,** stimulates the conversion of glucose to glycogen. The pancreas also contains delta cells, which secrete somatostatin. Locate the pancreatic islets on a microscope slide and compare them to those in figure 28.10.

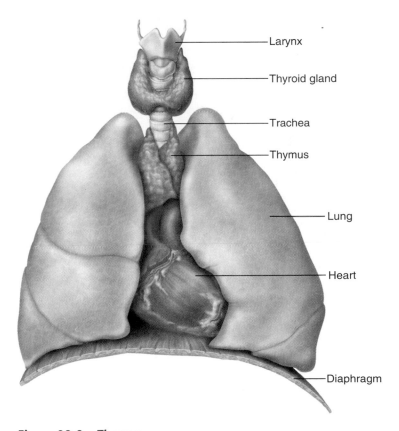

Figure 28.9   Thymus.

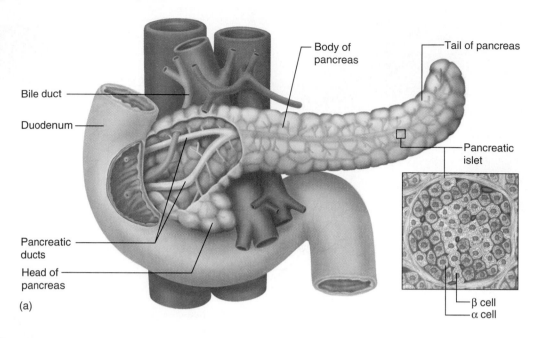

(a)

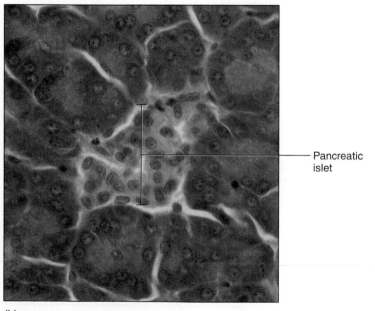

(b)

Figure 28.10   Pancreas. (a) Gross anatomy; (b) histology of the pancreas (400×). ✗

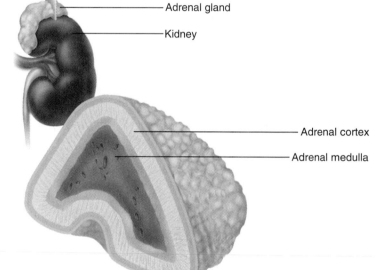

Figure 28.11   Adrenal cortex and medulla. ✗

Adrenal Gland

The **adrenal glands** (*ad* = next to, *renal* = kidney) are superior in position to the kidneys. Each adrenal gland is composed of an outer **cortex** and an inner **medulla.** Examine a microscope slide of an adrenal gland and locate the cortex and the medulla. Compare this to figure 28.11.

The hormones secreted from the adrenal cortex are called **corticosteroid hormones.** They are important in water and

electrolyte balance (Na⁺, K⁺) in the body. They are also important for carbohydrate, protein, and fat metabolism as well as stress management. The cortex can be divided into three regions. The outermost is the **zona glomerulosa** (figure 28.12), which consists of clusters of cells that secrete **mineralocorticoids** (especially aldosterone). Inside this layer (closer to the medulla) is the **zona fasciculata,** which consists of parallel bundles of cells that secrete **glucocorticoids.** The deepest cortical layer is the **zona reticularis,** which consists of a branched pattern of cells that produce both **glucocorticoids** and **sex hormones** (androgens and estrogens). The male sex hormone

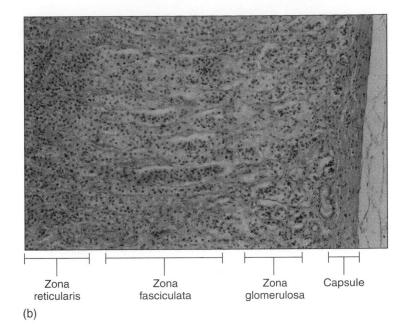

Figure 28.12 **Histology of the adrenal cortex.** (a) Diagram; (b) photomicrograph (100×). ⚷

testosterone is an androgen, yet both males and females produce androgens from the adrenal glands in small amounts. The major source of androgens in males is the testes.

The **hormones epinephrine** and **norepinephrine** are produced in the adrenal medulla. Stimulation by the sympathetic nervous system causes a release of these hormones, which increases heart rate and prepares the body for "fight or flight" reactions.

Gonads

The gonads produce sex hormones and are thus considered endocrine glands. They are stimulated by follicle-stimulating hormone (FSH) from the anterior pituitary, which causes the production or maturation of the sex cells (spermatozoa or oocytes). The gonads are also under the influence of luteinizing hormone (LH), which increases the level of hormone production by the gonads.

Testes

In the male the **testes** produce **testosterone,** which is a hormone responsible for secondary sex characteristics such as the development of facial and body hair, the expansion of the larynx (which produces a deeper voice), and the increased muscle and bone mass seen in males. The testes and mixed glands have both an endocrine and exocrine function. The endocrine function is testosterone production, and the exocrine function is the production of **spermatozoa.** The exocrine function of the testis is explored in Laboratory Exercise 46 on the male reproductive system.

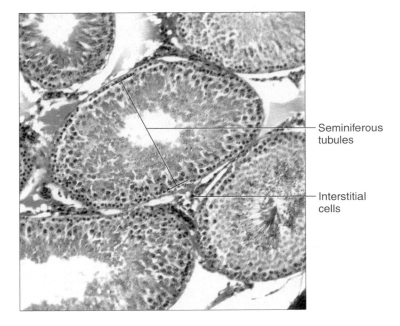

Figure 28.13 **Histology of the testis (100×).** ⚷

Histology of the Testis

Examine a microscope slide of a testis and find the **seminiferous tubules** and **interstitial cells.** Refer to figure 28.13 for assistance. The interstitial cells are found between the tubules and function to produce testosterone. Testosterone not only is responsible for secondary sex characteristics but also influences the development of the male genitalia (penis and scrotum) during embryological and fetal development. Testosterone also aids FSH in the production of spermatozoa.

Ovaries

In females the **ovaries** produce **estrogen** and **progesterone** (figure 28.14). Ovaries are also considered a mixed gland that produces hormones as an endocrine function and **oocytes** (eggs) as an exocrine function. Female hormones are also responsible for secondary sex characteristics in women such as enlarged breasts, development of an additional subcutaneous adipose layer, and a higher voice. Estrogen and progesterone also influence the development of the endometrium, cause maturation of the oocytes, and regulate the menstrual cycle. Estrogen is a generic term for several hormones produced by the female, including **estradiol.**

Endocrine Physiology—Detection of Hormones

In addition to the organs discussed the **embryonic cell mass** can also be considered an endocrine-producing structure. The developing embryo produces **human chorionic gonadotropin (HCG),** which is involved in placental development by preventing the degeneration of the corpus luteum and by increasing production of estrogens and progesterone by the ovary. Working either in pairs or as a larger group, obtain a sample of urine from your instructor from a nonpregnant and/or pregnant individual.

Caution   Remember to treat bodily fluids as if they carry pathogens. Wear protective gloves and eyewear when handling fluids! Dispose of all contaminated disposable materials in the biohazard container. The reusable material should be placed in a 10% bleach solution.

Obtain a pregnancy test kit and, following the instructions provided by your instructor or enclosed with the kit, determine if the urine is from a pregnant individual. Record your results.

Sample #: _____

Pregnant/not pregnant: _____

Home pregnancy tests are available due to advances in **monoclonal antibody** technology, in which cell cultures produce immune reactions against specific molecules. In this case, HCG is injected into mice or other lab animals. The lab animals produce antibodies against HCG. Cells from the animal are fused with tumor cells (tumor cells are easier to grow in culture) and the result is a growth of cells that produce antibodies against particular molecules (in this case HCG). These cloned antibodies are mixed with a dye and applied to an indicator stick. When the antibodies are exposed to their reacting molecule (HCG), the dye becomes visible and gives a positive reaction.

Once you have studied the endocrine glands and material, complete the table below. You should fill in the blank spaces where needed.

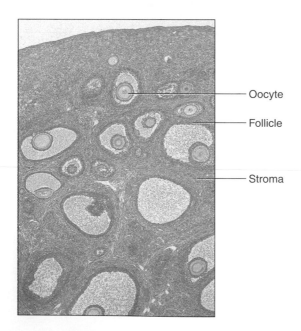

Figure 28.14   Histology of the ovary (40×).

— Oocyte

— Follicle

— Stroma

Organ	Hormones Produced	Effect of Hormones
Thyroid	Antidiuretic hormone	
	Thyroxine	
	Corticosteroid hormones	Regulate electrolyte balance
Ovary		Regulates ovarian cycle
	Melatonin	Regulates sleep cycles
	Luteinizing hormone	
Neurohypophysis	Oxytocin	
Parathyroid		Increases calcium in blood
	Glucagon	Increases blood glucose levels

Name _____

1. What is the general name of the kinds of organs that produce hormones?

2. What name is given to regions that are receptive to hormones?

3. Melatonin is secreted by what gland?

4. Where is ADH produced?

5. What is the effect of TSH and where is it produced?

6. What does glucagon do as a hormone and where is it produced?

7. Which hormones in the adrenal gland function to control water and electrolyte balance?

8. What is the primary gland that secretes epinephrine?

9. Where is growth hormone produced?

10. What is another name for T$_3$?

11. Label the endocrine glands indicated in the following illustration.

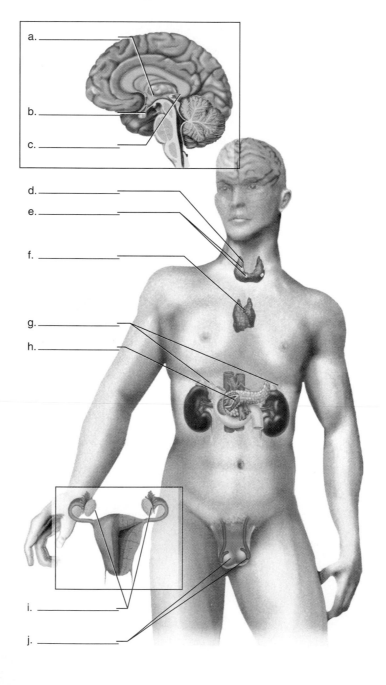

a. _____

b. _____

c. _____

d. _____

e. _____

f. _____

g. _____

h. _____

i. _____

j. _____

12. Identify the parts of the pituitary as seen in the following illustration and label two hormones located in each.

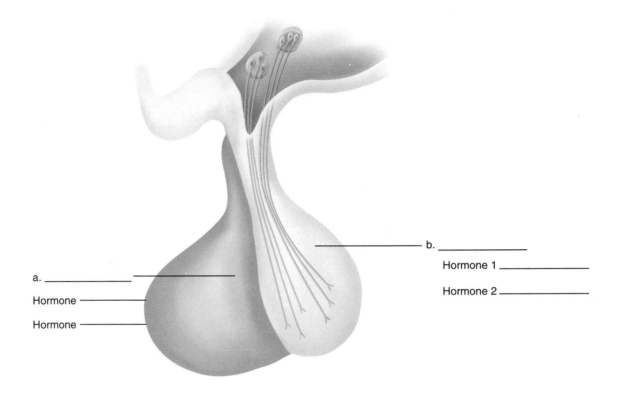

a. _____ _____

Hormone _____

Hormone _____

b. _____

Hormone 1 _____

Hormone 2 _____

13. Identify the three layers of the adrenal cortex as illustrated and list the hormones produced by each layer.

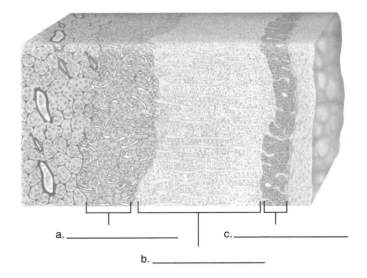

a. _____   c. _____

b. _____

14. Interstitial cells produce which hormone?

15. What structures are responsible for the production of estrogen?

16. What hormone causes an increase in calcium levels in the blood?

LABORATORY EXERCISE 29

Blood Cells

Introduction

Blood is made of two major components, formed elements and plasma. Formed elements make up about 45% of the blood volume and can further be subdivided into erythrocytes (red blood cells), leukocytes (white blood cells), and platelets (thrombocytes).

Plasma constitutes approximately 55% of the blood volume and contains water, lipids, and dissolved substances; colloidal proteins, and clotting factors.

The study of blood cells is important not only as part of the cardiovascular system but because it has significant clinical implications. Changes in the numbers and types of blood cells may be used as indicators of disease. In this exercise, you examine the nature of blood cells.

Caution The risk of blood-borne diseases has been significantly minimized by the use of sterilized human blood and nonhuman mammal blood in this exercise. However, use the same precautions as if you were handling fresh and potentially contaminated human blood. Your instructor will determine whether to use animal blood (from nonhuman mammals), sterilized blood, or synthetic blood. In any of these cases, follow strict procedures for handling potentially pathogenic material.

1. Wear protective gloves during the procedures.
2. Do not eat or drink in lab.
3. If you have an open wound, do not participate in this exercise or make sure the wound is *securely* covered.
4. Your instructor may elect to have you use your own blood. Due to the potential for disease transmission, such as AIDS or hepatitis in fresh blood samples, make sure you keep away from other students' blood and keep them away from your blood.
5. Do not participate in this part of the exercise if you have recently skipped a meal, if you are not well, or if for any other reason you do not feel comfortable with the procedure. Do not hesitate to make your reservations known to the instructor.
6. Place all disposable material in the biohazard bag.
7. Place all used lancets in the sharps container, and all glassware that is to be reused in a 10% bleach solution.
8. After you have finished the exercise, clean and disinfect the countertops with a 10% bleach solution.

Objectives

At the end of this exercise you should be able to

1. distinguish between the various formed elements of blood;
2. determine the percent of each type of leukocyte in a differential white cell count;
3. describe what an elevated eosinophil count might mean;
4. describe hematopoiesis.

Materials

Prepared slides of human blood with Wright's or Geimsa stain
Compound microscopes
Lab charts or illustrations showing the various blood cell types
Dropper bottle of Wright's stain
Large finger bowl or staining tray
Toothpicks
Clean microscope slides
Coverslips
Squeeze bottle of distilled water or phosphate buffer solution
Pasteur pipette and bulbs
Sterile cotton balls
Alcohol swabs
Sterile, disposable lancets
Hand counter
Biohazard bag or container
10% bleach container
Sharps container

Procedure

Plasma

Plasma is the fluid portion of blood and is over 90% water. The remainder mostly consists of proteins such as albumins, globulins, and fibrinogen. Albumins are produced by the liver and make up the majority of the plasma proteins. Some globulins are made by the plasma cells and make up the next largest amount of proteins. Fibrinogen is a clotting protein and this and other clotting factors are produced by the liver. Plasma also contains electrolytes (Na^+, K^+, and Cl^-), nutrients, hormones and wastes.

Examination of Blood Cells

In this part of the exercise you will need to distinguish between erythrocytes, leukocytes, and platelets. If you are using a prepared slide of blood you can move to the section entitled "Microscopic Examination of Blood Cells." If you are to make a smear, then read the following directions.

Preparing a Fresh Blood Smear

Your lab instructor will direct you to use either fresh, nonhuman mammal blood or your own blood. If you are using collected mammal blood, then wear latex gloves and withdraw a small amount of blood with a clean Pasteur pipette. Place a drop of the blood on a clean microscope slide. If you are using your own blood make sure you follow the safety precautions as discussed at the beginning of the exercise.

Withdrawal of Your Own Blood

Obtain the following materials: a clean, sterile lancet, alcohol swab, adhesive bandage, rubber gloves, sterile cotton balls, two clean microscope slides, and a paper towel.

1. Arrange the materials on the paper towel in front of you on the countertop.
2. Clean the end of the donor finger with the alcohol swab and let the hand from which you will withdraw blood hang by your side for a few moments to collect blood in the fingertips. You will be puncturing the pad of your fingertip (where the fingerprints are located).
3. Peel back the covering of the lancet and hold on to the blunt end as you withdraw it from the package. Do not touch the sharp end or lay the lancet on the table before puncturing your finger.
4. Jab your finger quickly and wipe away the first drop of blood that forms with a sterile cotton ball.
5. Place the second drop of blood that forms on the microscope slide about 2 cm away from one of the ends of the slide (figure 29.1) and place another cotton ball on your finger.
6. You can now put an adhesive bandage on your finger.
7. Whether you are using prepared blood or your own blood, use another clean slide to spread the blood by touching the drop with the edge of the slide and *push* it across the slide (figure 29.1). This should produce a smooth, thin smear of blood. Let the blood smear dry completely. Throw the lancet in the sharps container. Never reuse the lancet or set it down on the table!
8. After the slide is dry, place it in a large finger bowl elevated on toothpicks or on a staining tray.
9. Cover the blood smear with several drops of Wright's stain from a dropper bottle. Let the stain remain on the slide for 1 to 2 minutes.

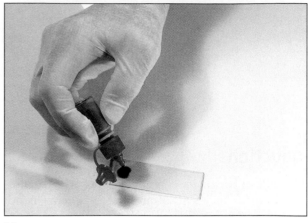

(a)

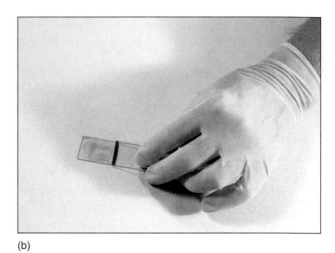

(b)

Figure 29.1 Making a blood smear. (a) Placement of blood on a slide; (b) spreading blood with a glass slide.

10. After this time add water or a prepared phosphate buffer solution to the slide. You can rock the slide gently with gloved hands or blow on it to stir the stain and water. A metallic green material should come to the surface of the slide. Let the slide remain covered with stain and water for 3 to 8 minutes.
11. Wash the slide gently with distilled water until the material is light pink and then stand it on edge to dry. You may also stain blood using an alternate stain (Geimsa stain), following your lab instructor's procedure.

Place all blood-contaminated disposable material in the biohazard container. Place the slide used to spread the blood in the 10% bleach solution and place the lancets in the sharps container if you haven't done so already. Once the slide is completely dry it can be examined under the high-power or oil-immersion lens of your microscope.

Microscopic Examination of Blood Cells

Erythrocytes

Examine a slide of blood stained with either Wright's or Geimsa stain. Erythrocytes are the most common cells you will find on the slide. There are about 5 million erythrocytes per cubic millimeter. They do not have a nucleus but appear as pink, biconcave disks (as if you had placed two dinner plates back to back). This particular shape increases surface area and provides for greater oxygen-carrying capabilities. Blood formation is called hematopoiesis or hemopoiesis. The specific production of red blood cells is called erythropoiesis. Erythropoiesis in the bone marrow is stimulated by erythropoietin (EPO), which is a hormone secreted by the kidneys. Erythrocytes are produced in the sternum, flat bones of the skull, and red marrow of the proximal humerus and femur. In children erythrocytes are also produced in the spleen.

Erythrocytes are about 7.7 μm in diameter, on average, and have a life span of about 120 days, after which time they are broken down by the spleen or liver. The iron portion of the erythrocyte is reabsorbed by the bone marrow, the proteins are broken down to amino acids and reabsorbed for general use by the body, and the heme is metabolized to bilirubin in the liver and secreted as part of the bile. Compare what you see under the microscope to figure 29.2.

Platelets

There are about 150,000 to 350,000 platelets per cubic millimeter of blood. **Platelets,** or **thrombocytes,** are small fragments of **megakaryocytes** that are involved in clotting. Examine the slide for small purple fragments that may be single or clustered and compare them to the platelets in figure 29.2.

Leukocytes

There are far fewer leukocytes in the blood than erythrocytes. The number of leukocytes in a healthy adult is about 7,000 cells per cubic millimeter, with a range of 4,300 to 10,800 cells per cubic milliliter of blood. Leukocytes are formed in bone marrow tissue. The life span of a leukocyte varies from a few hours to several months. Many are capable of ameboid movement as they squeeze between cells (a process termed *diapedesis*), engulfing foreign particles or cellular debris.

Leukocytes, or white blood cells, can be divided into two groups, which are the **granular** and **agranular leukocytes.** Examine the slide under high power or oil immersion and identify the different leukocytes as presented in the following text.

Granular Leukocytes   Granular leukocytes (granulocytes) are so named because they have granules in their cytoplasm. They are also known as polymorphonuclear (PMN) leuko-

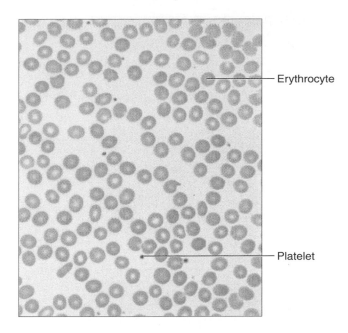

Figure 29.2   Blood smear (400×).

cytes due to the variable shape of their nuclei (which are lobed, not round). The three types of granular leukocytes are neutrophils, eosinophils, and basophils.

Neutrophils are the most common of all leukocytes. Neutrophils typically live for about 6 hours and have ameboid capabilities. They move by diapedesis between blood capillary cells and in the interstitial areas between cells of the body. Neutrophils move toward infection sites and destroy foreign material. They are the major phagocytic leukocyte. The granules of neutrophils absorb very little stain, but mature neutrophils can be distinguished from other leukocytes by their three- to five-lobed nucleus. They are about one and a half times the size of erythrocytes. Examine your slide for neutrophils and compare them to figure 29.3.

Eosinophils typically have a two-lobed nucleus with pink-orange granules in the cytoplasm. The term *eosinophil* actually means "eosin-loving" (eosin is an orange-pink stain and is picked up by the granules). They are about twice the size of erythrocytes. Compare figure 29.3 to your slide as you locate the eosinophils.

Basophils are rare. The granules stain very dark (blue-purple), and sometimes the nucleus is obscured because of the dark-staining granules. The nucleus is S-shaped and the cell is about twice the size of erythrocytes. Their granules contain histamines (vasodilators), and heparin, an anticoagulant which allow for movement of other leukocytes out of the capillaries to infection sites. You may have to look at 200 to 300 leukocytes before finding a basophil. Compare the basophil in your slide to figure 29.3.

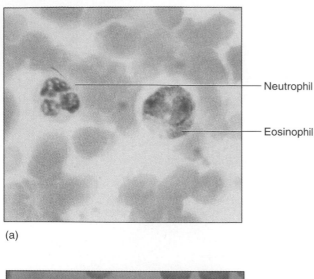

(a)

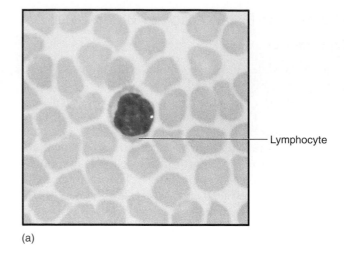

(a)

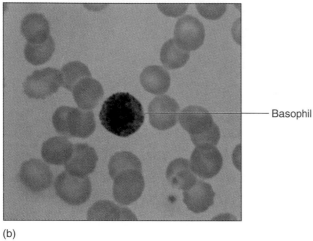

(b)

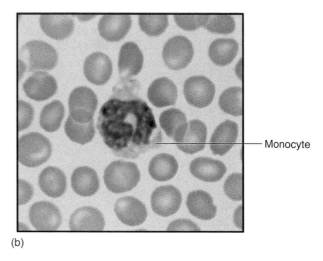

(b)

Figure 29.3   Granular leukocytes (1,000×). (a) Neutrophil and eosinophil; (b) basophil.

Figure 29.4   Agranular leukocytes (1,000×). (a) Lymphocyte; (b) monocyte.

Agranular Leukocytes   Agranular leukocytes are so named because they lack cytoplasmic granules. The nuclei are not lobed but may be dented or kidney-bean-shaped. There are two types of agranular leukocytes, lymphocytes and monocytes.

Lymphocytes have a large unlobed nucleus that usually has a flattened or dented area. The cytoplasm is clear and may appear as a blue halo around the purple nucleus. In terms of function, lymphocytes do not need prior exposure to recognize antigens, and they are remarkable in that separate cells have specific antibodies for specific antigens. There are two groups of lymphocytes, the **B cells** and **T cells.** These cells cannot normally be distinguished from one another in standard histological preparations (for example, Wright's stain) and are considered simply as lymphocytes for the purposes of this exercise. Both B cells and T cells arise from fetal bone marrow. The B cells probably mature in the fetal liver and spleen and T cells

mature in the thymus gland. B cell lymphocytes mature into plasma cells, which make antibodies. Plasma cells provide **humoral immunity** (that is, the plasma cells secrete antibodies that travel in the "humor" or fluid portion of the blood).

T cells, on the other hand, provide **cellular immunity.** In cellular immunity the cells themselves (not antibodies in the blood plasma) move close to and destroy some types of bacteria or virus-infected cells. T cells also attack tumors and transplanted tissues. Most of the T cells are found in the lymph nodes, thymus, and spleen. They enter the bloodstream via the lymphatics. Compare your slide to figure 29.4 for lymphocytes.

Monocytes are very large and have a kidney-bean-shaped or horeshoe-shaped nucleus. They are about three times the size of erythrocytes and are activated by T cells. Locate the large cells on your slide and compare them to figure 29.4.

Differential Leukocyte Count

Once you have identified the various leukocytes on the blood slide, conduct a differential white blood cell count. There is great importance in performing differential leukocyte counts. Changes in the relative percentages of leukocytes may indicate the presence of a particular disease.

1. Use a hand counter and count 100 leukocytes. One lab partner should look into the microscope and methodically call out the names of the different types of leukocytes seen.
2. The other lab partner should record how many of each type are found and keep track of the overall number with the use of a hand counter until 100 cells are counted.
3. Scan the slide in a systematic way so you don't count any cell twice. One method is illustrated in figure 29.5. Tally your results.

   Neutrophils:_____

   Eosinophils:_____

   Basophils:_____

   Lymphocytes:_____

   Monocytes:_____

Neutrophils represent about 60%–70% of all of the leukocytes. They increase in number in appendicitis or acute bacterial infections.

Eosinophils represent about 2%–4% of all leukocytes. Eosinophils increase in number during allergic reactions and parasitic infections (for example, trichinosis).

Basophils represent about 0.5%–1% of all leukocytes. They increase in numbers during allergies and radiation.

Lymphocytes make up 25%–33% of the leukocytes. These cells increase in times of viral infection, antibody-antigen reactions, and infectious mononucleosis.

Monocytes make up about 3%–8% of the leukocytes. They increase in times of chronic infections such as tuberculosis.

Figure 29.5   Counting leukocytes.

Fill in the table below with what you know about the formed elements. You should make sure to read all of the information given before filling in the table.

Characteristics of Formed Elements			
Formed Element	Granules (if present)	Shape of Nucleus (if present)	Cause for Increase
Erythrocyte			
	Not obvious		Mononucleosis
	Orange-staining		Parasitic infections
		3–5 lobed	
	Not obvious	Kidney-bean	
	No granules	No nucleus	
Basophil			

Clean Up Make sure the lab is clean after you finish. Place any slide with fresh blood on it or any material contaminated with bodily fluid in the bleach solution. Place all gloves or contaminated paper towels in the biohazard container. All sharps material (broken slides or coverslips, lancets, etc.) should be placed in the sharps container. Clean the counters with a towel and a 10% bleach solution. If you used immersion oil on the microscope make sure the objective lenses are wiped clean (use clean lens paper only!).

Name _____

1. Formed elements consist of three main components. What are they?

2. What is the most common plasma protein?

3. What is another name for a thrombocyte?

4. Which is the most common blood cell?

5. What is the common name for a leukocyte?

6. What leukocyte is most numerous in a normal blood smear?

7. How many erythrocytes are normally found per cubic millimeter of blood?

8. What is an average number of leukocytes found per cubic millimeter of blood?

9. B cells and T cells belong to what class of agranular leukocytes?

10. What value is there to a change in the percentage of leukocytes to diagnostic medicine?

11. In counting 100 leukocytes you are accurately able to distinguish 15 basophils. Is this a normal number for the white blood cell count, and what possible health implications can you draw from this?

12. What is the function of the thrombocytes of the blood?

13. How does hematopoiesis differ from erythropoiesis?

14. Label the blood cells in the following illustration.

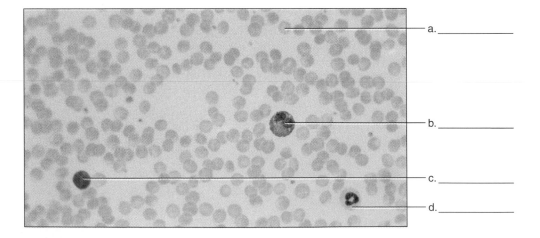

a. _____

b. _____

c. _____

d. _____

LABORATORY EXERCISE 30

Blood Tests and Typing

Introduction

Blood is a complex medium that has multiple functions in the body. Many clinical tests are performed on blood to determine the number of erythrocytes or leukocytes in a known volume of blood, the blood type of an individual, hemoglobin concentration, and other information. In this exercise, you perform selected tests that allow you to evaluate different samples of blood. As in the previous exercise, handle the samples with care. This is important, so we've repeated the caution statement from that exercise here.

Caution The risk of blood-borne diseases has been significantly minimized by the use of sterilized human blood and nonhuman mammal blood in this exercise. However, use the same precautions you would if you were handling fresh and potentially contaminated human blood. Your instructor will determine whether to use animal blood (from nonhuman mammals), sterilized blood, or synthetic blood. In any of these cases, follow strict procedures for handling potentially pathogenic material.

1. Wear protective gloves during the procedures.
2. Do not eat or drink in lab.
3. If you have an open wound do not participate in this exercise or make sure the wound is *securely* covered.
4. Your instructor may elect to have you use your own blood. Due to the potential for disease transmission, such as AIDS or hepatitis in fresh blood samples, make sure you keep away from other students' blood and keep them away from your blood.
5. Do not participate in this part of the exercise if you have recently skipped a meal, if you are not well, or if for any other reason you do not feel comfortable with the procedure. Do not hesitate to make your reservations known to the instructor.
6. Place all disposable material in the biohazard bag.
7. Place all used lancets in the sharps container and all glassware that is to be reused in a 10% bleach solution.
8. After you have finished the exercise, clean and disinfect the countertops with a 10% bleach solution.

Objectives

At the end of this exercise you should be able to

1. perform specific diagnostic tests such as hematocrit and hemacytometer tests;
2. determine the antigens (agglutinogens) present in a particular ABO blood type;
3. list the antibodies (agglutinins) present in a particular ABO blood type;
4. relate Rh-positive or Rh-negative blood to antigens present;
5. correlate hematocrit with erythrocyte counts.

Materials

5 ml of fresh, nonhuman mammal blood (dog, sheep, or cow)
Sterilized human blood (Carolina #k3-70-0120 or other supply company), labeled 1 to 4
Blood typing antisera (anti-A, anti-B, anti-D), test cards, and toothpicks
Rh warming tray
Heparinized blood microcapillary tubes
Capillary tube centrifuge
Hematocrit reader (Criticorp Micro-hematocrit tube reader, Damon Micro-capillary reader, or mm ruler)
Seal-ease
Hemacytometer, coverglass
Microscope
Unopette reservoirs (Becton-Dickinson #5850)
Unopette capillary pipettes (10 microliter capacity)
Latex gloves
Goggles
Contamination bucket with autoclave bag
Container with 10% household bleach solution
Sharps container

Procedure

Blood Typing

An understanding of blood type is critical in clinical work. Proper matching of blood between donor and recipient is a vital process, and you will learn the essentials of this process in this exercise. There are many different typing systems that match blood, but the ABO and Rh systems are the two most common in clinical settings. Blood types are genetically determined. There are other typing systems such as the Kell, Lewis, MNS, and Duffy blood groups. These are antigen/antibody groups that are used to type blood and are frequently used in forensic studies.

Blood cells have surface membrane molecules (glycoproteins) that are antigens **(agglutinogens).** If this blood is

Table 30.1 ABO Blood System		
Blood Type	**Agglutinogens (antigens)**	**Agglutinins (antibodies)**
A	A	Anti-B
B	B	Anti-A
AB	A and B	None
O	None	Anti-A and Anti-B

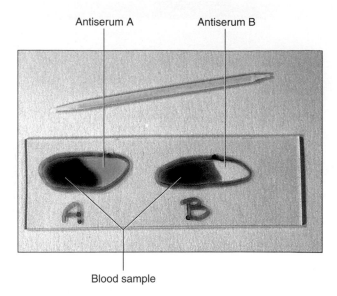

Figure 30.1 Blood testing procedure.

injected into a person with antibodies (**agglutinins**) against that blood, then the injected blood clumps, or **agglutinates.** In the ABO system, no prior exposure to the agglutinogen is needed for granulation or clumping to occur. For example, if a person has blood type A, that individual has agglutinins against blood type B. If the person with blood type A receives a transfusion of blood type B, then the anti-B agglutinins attack the agglutinogens in the blood introduced into the system. This causes a **transfusion reaction,** which is the agglutination and hemolysis of the transfused erythrocytes. If the reaction is severe enough death may occur.

Table 30.1 gives details of the ABO blood system.

In a normal blood transfusion, donated blood is matched to the exact blood type of the recipient (for example, type AB matched with type AB). In emergencies blood type O, with no antigens, can be used for individuals with A, B, or AB blood. Therefore a person with O blood is considered a **universal donor.** A person with which blood type would be considered a **universal recipient?**

Universal recipient: _____

Procedure for Blood Typing

1. Read the entire blood typing procedure before you begin this part of the exercise.
2. Obtain a sample of sterilized blood as directed by your instructor. The vial should be labeled 1, 2, 3, or 4.
3. Record the sample number you are using in the space provided.
4. Place two separate drops of the sample blood on a blood test card or on a very clean glass microscope slide (figure 30.1).
5. Place antiserum-A on one drop and antiserum-B on the other drop.
6. Use separate toothpicks to stir each sample of blood and antiserum.
7. Keep the antisera separate from one another. Examine the sample for clumping or granulation within 2 minutes. Make sure that you dispose of your toothpicks in the biohazard bag. If both blood samples coagulate, you have a sample of type AB blood. If neither sample coagulates,

you have a sample of blood type O. If the blood with antiserum-A coagulates, you have a sample of blood type A, and if blood with antiserum-B coagulates, you have a sample of blood type B.

8. Compare your sample to figure 30.2.

Blood sample number: _____

Blood type: _____

Common ABO blood types for various groups in the United States are listed by percentages in table 30.2.

Determination of Rh Factor

The other blood system of clinical significance is the Rh system. Rh stands for rhesus monkey (the animal in which the Rh system was discovered). The **Rh positive (Rh⁺)** condition represents 85% of the U.S. population; 15% are **Rh negative (Rh⁻).** An Rh⁺ individual carries the Rh antigen (surface membrane marker). The Rh⁻ individual does not carry the Rh antigen but develops antibodies to the antigen after exposure to the Rh antigen.

Determination of the Rh type is a much more subtle test than ABO typing. The Rh antiserum (anti-D) is more fragile in shipping and storage and should not be considered clinically relevant in this exercise. Use the same precautions as outlined at the beginning of the exercise regarding bodily fluids.

1. Place one drop of blood on a slide and place a drop of antiserum-D (Rh antiserum) on the slide.
2. Use a new toothpick and stir the two drops together.
3. Place this mixture on a warming tray and gently rock the slide.
4. Examine the slide for clumping after a minute or two. If clumping occurs, then the sample is Rh⁺ (the Rh antiserum reacted with the Rh antigen in the blood). If no

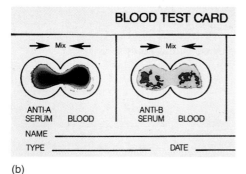

Figure 30.2   Blood types. (a) Type A; (b) type B; (c) type AB. ▭

Table 30.2 Blood Type				
Blood Type	Caucasians	African Americans	Asian Americans	Native Americans
A	41	27	28	16
B	9	21	23	4
AB	3	4	13	1
O	47	48	36	79
Total	**100**	**100**	**100**	**100**

clumping occurs then the sample is Rh⁻. This test is more difficult to determine so look for slight granulation. Rh antiserum should be used fresh.

5. Record your results.

Rh factor determination: _____

Rh determination is very important if a pregnant woman is Rh⁻. If her unborn child is Rh⁺ then she may develop antibodies against the Rh antigen during pregnancy or delivery (a time when fetal and maternal blood frequently mix). If her second child is Rh⁺, then the antibodies that she produced during her first pregnancy may cross the placenta and cause severe reactions in the fetus. This is called **hemolytic disease of the newborn (HDN),** or **erythroblastosis fetalis.** Rho-GAM, an Rh immune globulin, prevents antibody formation in the mother against the Rh antigen and is frequently given to the mother during pregnancy and again after delivery. A woman who is Rh⁺ does not have the antibodies for the Rh factor and therefore will not produce antibodies against the developing fetus, regardless of the Rh factor of the fetus.

Blood Typing Problems

Extensive blood transfusions can lead to problems. Blood that is donated from one person carries antibodies, and these may react with the recipient's blood. There are other blood types in addition to the ABO and Rh systems and sensitivities may develop due to antibodies present in the recipient's blood.

Hematocrit

Hematocrit, or **packed cell volume (PCV),** is the percentage of erythrocytes (or blood cells in general) in the total blood volume. The percentage of erythrocytes, or total cell volume (erythrocytes and leukocytes), can be calculated after centrifuging a sample of blood. The cells end up as a large sediment in the bottom of the microcentrifuge tube, leaving the lighter plasma on top. The hematocrit varies in individuals with values of 37% to 48% being normal in females and 42% to 52% being normal in males. An increase in the hematocrit above normal is known as **polycythemia,** and can exceed 65%.

When blood is lost faster than it is replaced or when the production of erythrocytes is low, **anemia** occurs. In anemia, the hematocrit may drop to 15% or less. Anemia may also be due to low levels of hemoglobin in the blood. Hemoglobin is a complex molecule composed of an iron-containing **heme group** and the protein **globin.** You will next compare the hematocrit of sterilized human blood with that of another mammal.

Procedure for Determining Hematocrit

Refer to the caution statement at the beginning of the exercise concerning working with blood.

1. While wearing latex gloves, fill two capillary tubes with blood. One should have commercially prepared, sterilized human blood and the other should have nonhuman mammal blood.
2. Fill each tube by touching the red end of the capillary tube to the blood sample. Let the blood flow up the tube by capillary action.

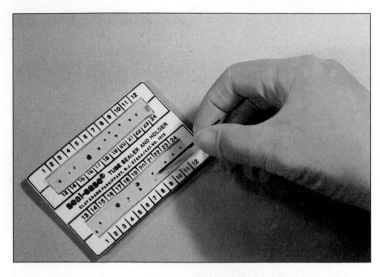

Figure 30.3 Preparation of a hematocrit tube.

3. Once the tube is filled, place your finger on the other end of the tube (the nonred end) to prevent blood from flowing back out. Fill each tube to about two-thirds of the tube length. Both tubes should have about the same volume of blood in them.
4. Seal the red end of the tube with Seal-ease® or modeling clay. Be careful—the capillary tubes are both fragile and sharp! They can break and puncture the skin (figure 30.3).
5. Place the capillary tube in the centrifuge with the clay plug against the outer rubber ring.
6. Place your tubes opposite each other and record the number of each tube. Make sure you keep track of the sample from each tube and don't mix them up!

   Mammal blood tube#: _____
   Sterilized blood tube#: _____

After many tubes have been placed into the centrifuge, make sure that each has the clay seal facing the rubber gasket. If the seals are facing the center of the centrifuge, then the blood will spray out of the tube when the centrifuge begins turning. The tubes should be opposite one another so the centrifuge is balanced. Make sure the metal top is on the centrifuge and screwed in place. Close the cover and spin the capillary tubes for 3 to 5 minutes.

Once the centrifuge has come to a complete stop, remove each tube and calculate the percentage of red blood cells in the tube. This can be done with hematocrit readers or by use of a ruler. Your instructor can assist you with the use of the hematocrit readers. If you use a ruler, you need to measure the total height of the blood column and subsequently the total height of the red blood cells in the column. If you use a ruler to measure the hematocrit, record your results here.

   Sterilized blood:
   Erythrocyte height (mm): _____

Total column height (mm): _____
Mammal blood:
Erythrocyte height (mm): _____
Total column height (mm): _____

The hematocrit can be figured by the following equation:

$$\frac{\text{Millimeters of erythrocyte}}{\text{Millimeters of total blood}} \times 100 = \text{Hematocrit}$$

Record the hematocrit.
Sterilized blood: _____
Mammal blood: _____

How do these values compare?

What commercial procedure might lead to a difference in the hematocrit between these two samples?

Plasma with red coloration indicates hemolysis of the erythrocytes. Normally the plasma should be a straw-colored or light yellow. Do either of your samples show hemolysis?

Note the buff-colored layer between the erythrocyte layer and the plasma layer. This is the leukocyte layer.

Hemoglobin Determination

The number of erythrocytes may be normal in a sample of blood, yet an individual may be anemic due to the lack of hemoglobin in the blood. Hemoglobin can be measured a number of ways, such as by the Tallquist method, by colorimetry, or with the use of a hemoglobinometer. Hemoglobin (Hb) levels are expressed as grams Hb/100 ml of blood. The average value for humans is 12 to 16 grams/100 ml of blood. In males the normal range is 13 to 18 g/100 ml and for females it is 12 to 16 g/100 ml.

To measure the percent hemoglobin follow the directions provided with the hemoglobinometer. Be sure to clean the glass slides with alcohol before and after use. Disinfect the apparatus with bleach afterwards.

   Hemoglobin from sterile blood sample: _____
   Hemoglobin from mammal blood sample: _____

Figure 30.4   Optical cell counter.

Blood Cell Counts

Another way to determine the number of blood cells is to count the number of cells in a known volume and subsequently determine the total number of cells per cubic millimeter (mm^3). Modern evaluation of erythrocyte and leukocyte counts is performed by injecting blood samples into an optical computer system, which automatically calculates the cell counts. Most modern hospitals today use computer-driven optical systems to count cells (figure 30.4).

The process can be demonstrated in the lab by an older method that involves the dilution of blood and counting blood with the use of a glass slide called a **hemacytometer.** This special slide is very expensive because it has etched lines on the upper surface that delineate a precise surface area. When a coverslip is placed over the hemacytometer chamber the grids enclose a specific volume. By counting the cells in the volume and multiplying by the dilution factor and volume factor, you can make a determination of the number of cells per cubic millimeter.

Procedure for Erythrocyte Counts

For males the average count is about 5.4 million erythrocytes per cubic millimeter of blood, with a range of 4.5 to 6.0 million cells per cubic millimeter. For females the average count is about 4.8 million erythrocytes per cubic millimeter of blood, with a range of 4.0 to 5.5 million cells per cubic millimeter.

To make accurate counts, blood must first be diluted on the order of 1 part blood to 200 parts diluent. You will perform this experiment twice: once using commercially prepared, sterilized human blood and once using fresh, nonhuman mammal blood. Obtain the following:

Two Unopette reservoirs
Two Unopette capillaries with pipette shields
Sample of nonhuman mammal blood
Sample of sterilized blood
Hemacytometer and coverslip
Latex gloves
Microscope

1. Puncture the Unopette reservoir with the capillary pipette shield (figure 30.5a).
2. Remove the cover from the capillary tube and draw a sample (the same sterilized blood sample number you used to determine hematocrit) into the tube by capillary action (figure 30.5b). The blood should flow into the capillary tube and fill the entire tube.
3. Squeeze the Unopette reservoir chamber, insert the tip of the capillary into the chamber, and release the chamber, drawing blood from the capillary into the reservoir fluid (figure 30.5c).
4. Withdraw the capillary tube from the reservoir, turning it around and placing it back on the reservoir so it forms a small dropper bottle (figure 30.5d). Gently swirl the reservoir to mix the contents, and invert and discard three to five drops of fluid into a cotton swab to clear the capillary tube. The diluent consists of blood and Hayem's solution, which contains sodium azide, a toxic material. Do not ingest or touch sodium azide. Dispose of any sodium azide solution with plenty of running water.
5. Place the coverslip on the hemacytometer and fill the chamber with the capillary tube (figure 30.6). Do not force excess liquid under the hemacytometer glass. Let the suspension sit 1 to 2 minutes for an even dispersion of the cells.
6. Place the hemacytometer under the microscope and examine the cells under high power (400–450×).
7. Count all of the erythrocytes in each of the five areas marked with an "R" (figure 30.7). Some cells will be on the border of the counting grid. Count only those cells touching the left line and the upper line as part of the sample. Do not count the cells touching the right line or bottom line.

    The blood was diluted 1:200, and the sum of the volume in the five areas represents one-fiftieth of a cubic millimeter. If you multiply the dilution factor (200) and the volume that would occur to fill 1 mm^3 (50), then you would multiply the number of erythrocytes counted by 10,000 to obtain the number of erythrocytes per cubic millimeter.
8. Record your results.

    Number of erythrocytes counted: _____

    Number of erythrocytes per cubic millimeter: _____

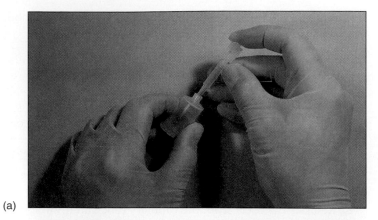

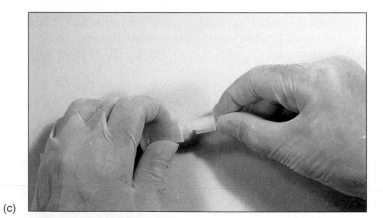

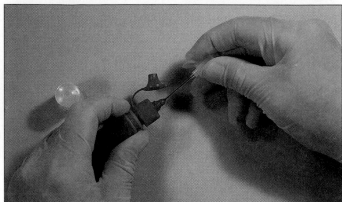

(a)

(b)

(c)

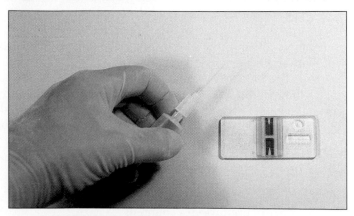

(d)

Figure 30.5   Use of the Unopette system. (a) Puncturing reservoir; (b) filling capillary tube; (c) releasing capillary, filling reservoir; (d) ready to fill hemacytometer.

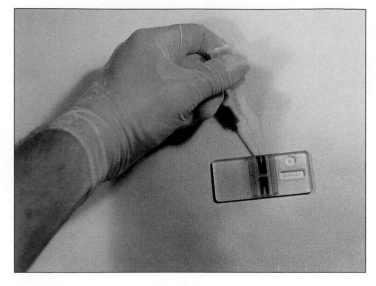

Figure 30.6   Filling the hemacytometer.

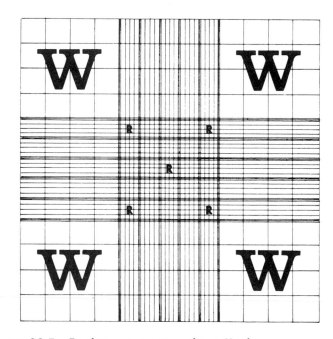

Figure 30.7   Erythrocyte counts using a Neubauer hemacytometer.

Now use the sample of nonhuman mammal blood. Rinse the hemacytometer and coverslip *carefully* with water before following the same procedure. Record the results.

Number of erythrocytes counted: _____

Number of erythrocytes per cubic millimeter: _____

How do the numbers of cells in nonhuman mammal
    blood and sterilized blood compare?

Is there a correlation between the numbers of
erythrocytes counted from these samples and the
hematocrit taken from the same samples?

After you finish your counting, place the hemacytometer
and coverslip *gently* into the bleach solution.

Leukocyte Counts (Optional)

Leukocytes are counted in a similar fashion except that the four
large corner regions of the hemacytometer are used and the
sum of the cell count from these four areas is multiplied by 50.
Your instructor may provide you with additional instructions
for a leukocyte count, using a specific solution for leukocytes.

Number of leukocytes counted: _____

Number of leukocytes per cubic millimeter: _____

Normal blood levels for leukocytes are about 7,000 cells
per cubic millimeter, with a range of 4,300 to 10,800 cells per
cubic millimeter. Leukocytosis is an elevated white blood cell
count typically above 11,000 cells per cubic millimeter. Leukocytosis may be due to infection, poisoning, or metabolic disor-
ders. Leukopenia, on the other hand, is a decreased white blood
cell count with less than 4,000 cells per cubic millimeter. This
may be due to chemotherapy, radiation therapy, HIV infection,
infectious hepatitis, cirrhosis of the liver, typhoid fever, or
chronic infections such as tuberculosis.

Clean Up Make sure the lab is clean before you leave. Wipe
any blood from the microscope objective lenses
with lens cleaner and lens paper. Make sure you
do not leave glass or blood on the lab tables. Use
a 10% bleach solution and a towel to clean off your lab table
before you leave.

Name _____

1. What is the name of a surface membrane molecule that causes an immune reaction?

2. What ABO blood type is found in a person who is a universal donor?

3. What is the average range of hematocrit for people?

4. What percent of the blood volume are formed elements?

5. A person with blood type B has what kind of agglutinins (antibodies)?

6. A person has antibody A and antibody B in his or her blood with no Rh antibody. What specific blood type would this person have?

7. A total of 240 erythrocytes are counted in the hemacytometer chamber. What would be the red blood cell count of this individual in terms of erythrocytes per cubic milliliter?

8. An individual with blood type "B negative" is injected with type A positive blood. What would happen after the injection?

9. What is the normal hematocrit for a healthy female?

10. How might changes in the Unopette technique alter the final determined value of erythrocytes?

11. Using the following illustration, calculate the hematocrit of the individual. Determine if it falls within normal limits.

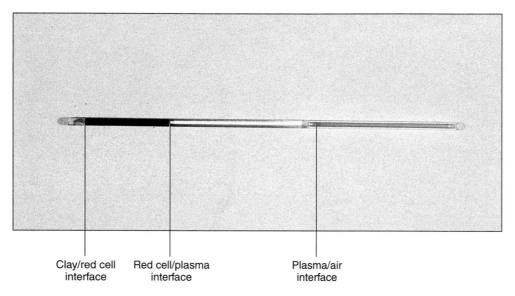

Clay/red cell   Red cell/plasma   Plasma/air
  interface   interface   interface

Hematocrit.

LABORATORY EXERCISE 31

Structure of the Heart

Introduction

In this exercise, you study the structure of the heart. The heart is located deep in the thorax between the lungs in a region known as the **mediastinum.** The mediastinum contains the heart, the coverings of the heart (the **pericardia**), and other structures such as the esophagus and descending aorta. The mediastinum is located between the sternum, lungs, and the thoracic vertebrae. Vessels that return blood to the heart are called **veins,** and those that carry blood from the heart are called **arteries.**

If you were to open the chest cavity, the first structure you would see is the **parietal pericardium.** The parietal pericardium encloses the heart and has two layers; a tough, outer connective tissue sheath called the fibrous layer and an inner layer called the serous layer. Deep to the parietal pericardium is the **pericardial cavity,** which contains a small amount of **serous fluid.** This fluid reduces the friction between the outer surface of the heart and the parietal pericardium. The heart wall itself has an outer layer known as the **visceral pericardium,** or **epicardium.** Locate these pericardial layers in figure 31.1.

In this exercise, you examine models of the heart, preserved sheep and human hearts, if available. As you look at the preserved material try to see how the structure of the heart relates to its function.

Objectives

At the end of this exercise you should be able to

1. describe the general structure of the heart;
2. find and name the anatomical features on models of the heart and in the sheep heart;
3. describe the blood flow through the heart and the function of the internal parts of the heart;
4. discuss the functioning of the atrioventricular valves and the semilunar valves and their role in circulating blood through the heart;
5. describe the position of the heart in the thoracic cavity.

Materials

Models and charts of the heart
Preserved sheep hearts
Preserved human hearts (if available)
Blunt probes (mall probes)
Dissection pans
Razor blades or scalpels
Sharps container
Disposable gloves
Waste container
Microscopes
Prepared slides of cardiac muscle

Procedure

Heart Wall

The heart wall is composed of three major layers. The outermost layer is the **visceral pericardium,** or **epicardium,** which is composed of epithelial and connective tissue. The middle layer is the **myocardium** and is the thickest of the three layers. It is mostly made of **cardiac muscle.** You may wish to review the slides of involuntary cardiac muscle and note the intercalated discs, branching fibers, and fine striations of cardiac tissue (as described in Laboratory Exercise 6). The cardiac muscle is arranged spirally around the heart, and this arrangement provides a more efficient wringing motion to the heart. The inner layer of the heart wall is known as the **endocardium,** which is a serous membrane and consists of endothelium (simple squamous epithelium) and connective tissue.

Caution Be careful when handling preserved materials. Ask your instructor for the proper procedure for working with preserving fluid and for handling and disposal of the specimen. Do not dispose of animal material in the sinks. Place them in an appropriate waste container.

It is best to examine heart models *before* dissecting a sheep heart unless your instructor directs you to do otherwise. Heart models are color coordinated and labeled to make the structures easier to locate.

Examination of the Heart Model

Exterior of the Heart

Examine the heart model and notice that the heart has a pointed end, or **apex,** and a blunt end, or **base.** The apex of the heart is inferior, and the great vessels leaving the heart are located at the base (therefore, in the case of the heart, the base is superior to

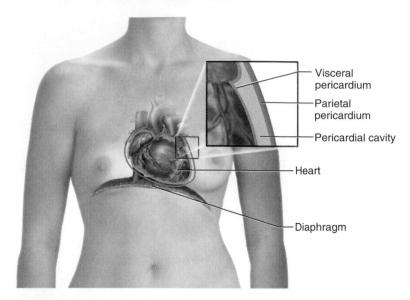

(a)

Visceral pericardium

Parietal pericardium

Pericardial cavity

Heart

Diaphragm

Right lung

Ascending aorta

Left lung

Right atrium

Interventricular groove

Left ventricle

Parietal pericardium (cut)

Visceral pericardium

Pericardial cavity

Apex of heart

Right ventricle

Diaphragm

(b)

Figure 31.1   Heart in thoracic cavity and heart coverings.
(a) Coronal section; (b) photograph of cadaver. ✗

the apex). Compare the model to figure 31.2 and see how the **aorta** curves to the left in an anterior view of the heart and is posterior to the **pulmonary trunk.**

Locate the anterior features of the heart. The heart is composed of two large, inferior ventricles and two smaller and superior atria. The **left ventricle** extends to the apex of the heart and is delineated from the right ventricle by the **interventricular groove,** or sulcus. Running along this interventricular

groove are some of the **coronary arteries** and **cardiac veins,** which are discussed later. If the apex of the heart is positioned directly inferior (pointing straight down), then the right ventricle is more superior and occupies less area than the left ventricle. Note the two earlike flaps that occur on the anterior, superior region of the heart. These structures are the **auricles,** which are part of the atria.

If you examine the heart from the posterior side you will see the atria more clearly. At the junction of the right atrium and the right ventricle is the **atrioventricular sulcus,** or groove. The **coronary sinus,** a large venous chamber that carries blood from the cardiac veins to the right atrium, is located in this sulcus. Locate the **superior vena cava** and the **inferior vena cava,** two vessels that also return blood to the right atrium. Locate the **pulmonary veins,** which carry blood from the lungs to the left atrium. Compare the heart model to figure 31.3.

The major vessels of the heart are illustrated in figures 31.2 and 31.4. Locate the **pulmonary trunk, pulmonary arteries, ligamentum arteriosum** (between the pulmonary trunk and aortic arch), **ascending aorta, pulmonary veins, superior vena cava, inferior vena cava, coronary arteries,** and **cardiac veins.**

The heart tissue is nourished by coronary arteries. The **left coronary artery** arises from the ascending aorta (figure 31.4) and then branches into the **anterior interventricular artery** and the **circumflex artery.** The **right coronary artery** also arises from the ascending aorta and branches to form the **posterior interventricular artery** and the **right marginal artery.** These major arteries of the heart supply blood to the myocardium. On the return flow from the heart muscle the **great cardiac vein** follows the depression of the interventricular groove and the atrioventricular groove to the coronary sinus. On the right side of the heart the **small cardiac vein** leads to the coronary sinus, which empties into the right atrium. Locate these vessels on the external surface of the model of the heart and compare them to figures 31.2, 31.3, and 31.4.

Interior of the Heart

Examine a model of the interior of the heart and locate the right and left ventricles. Note that the **right ventricle** is much thinner walled than the **left ventricle.** The ventricles are separated by the **interventricular septum,** which forms a wall between the two ventricular chambers. Compare the model to figure 31.5.

Examine the **right atrium** and note how thin the wall is compared to the ventricles. Examine the medial wall of the atrium, known as the **interatrial septum,** and locate a thin oval depression in the atrial wall. This depression is the **fossa ovalis.** In fetal hearts this is the site of the **foramen ovale,** but closure

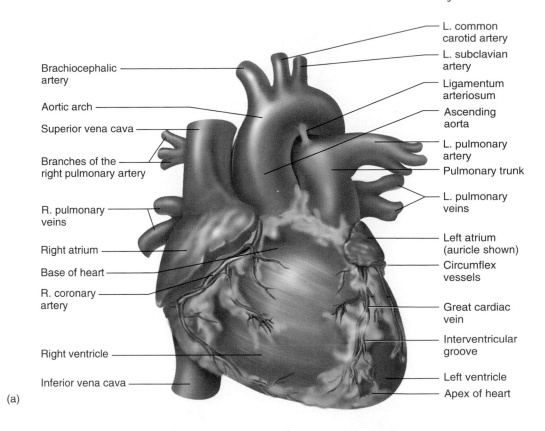

Brachiocephalic artery

Aortic arch

Superior vena cava

Branches of the right pulmonary artery

R. pulmonary veins

Right atrium

Base of heart

R. coronary artery

Right ventricle

Inferior vena cava

(a)

L. common carotid artery

L. subclavian artery

Ligamentum arteriosum

Ascending aorta

L. pulmonary artery

Pulmonary trunk

L. pulmonary veins

Left atrium (auricle shown)

Circumflex vessels

Great cardiac vein

Interventricular groove

Left ventricle

Apex of heart

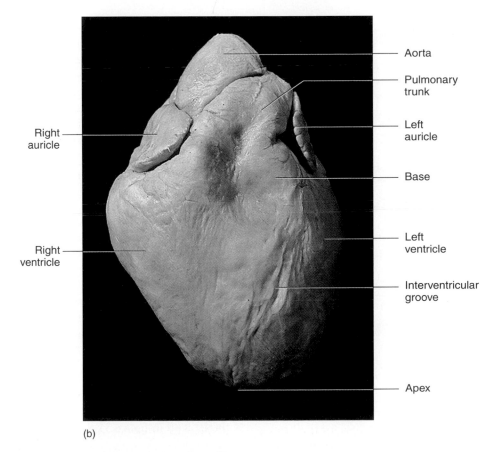

Right auricle

Right ventricle

Aorta

Pulmonary trunk

Left auricle

Base

Left ventricle

Interventricular groove

Apex

(b)

Figure 31.2   Surface anatomy of the heart, anterior view. ✗

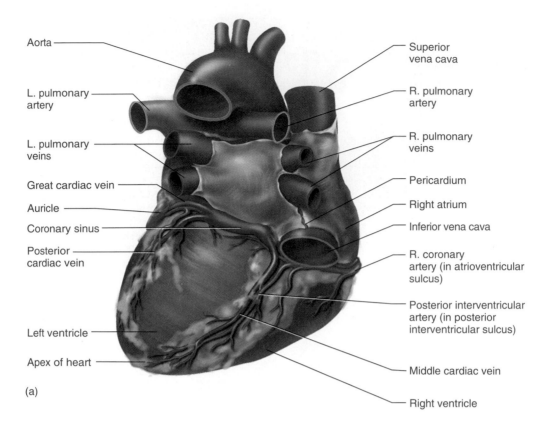

(a)

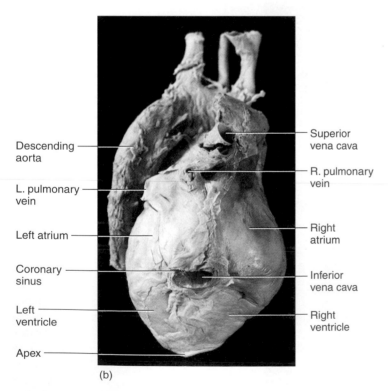

(b)

Figure 31.3   Surface anatomy of the heart, posterior view. ⚲

of the foramen usually occurs around the time of birth. Note the extensive **pectinate muscles** on the wall of the atrium. These provide additional strength to the atrial wall. Blood in the superior vena cava, the inferior vena cava, and the coronary sinus returns to the right atrium. Examine the features of the right atrium in figures 31.5 and 31.6.

The blood from the right atrium flows into the right ventricle. Now examine the valve between the right atrium and right ventricle. This is the **right atrioventricular valve,** or **tricuspid valve,** and it prevents backflow of blood from the right ventricle into the right atrium during ventricular-contraction. Examine the valve for three flat sheets of tissue. These are the three cusps of the tricuspid valve. The tricuspid valve has thin, threadlike attachments called **chordae tendineae.** These tough cords are attached to larger **papillary muscles,** which are extensions from the wall of the ventricle. The right ventricle wall has small extensions called **trabeculae carneae,** which, like the pectinate muscles of the atria, serve to strengthen the ventricle wall. The blood from the right ventricle flows into the pulmonary trunk towards the lungs.

Locate the **pulmonary semilunar valve.** It appears as three small cusps between the right ventricle and the pulmonary trunk and keeps blood from flowing backward from the pulmonary trunk into the right ventricle during ventricular relaxation. Examine the details of the right ventricle in models in the lab and in figure 31.5.

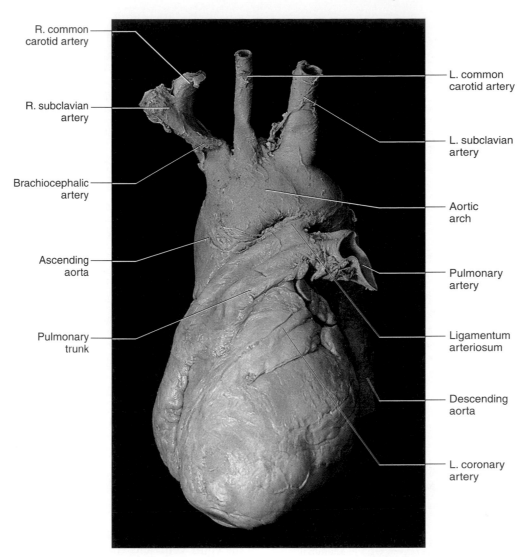

R. common
carotid artery

R. subclavian
artery

Brachiocephalic
artery

Ascending
aorta

Pulmonary
trunk

L. common
carotid artery

L. subclavian
artery

Aortic
arch

Pulmonary
artery

Ligamentum
arteriosum

Descending
aorta

L. coronary
artery

Figure 31.4   Vessels of the heart, anterior view. ✗

Blood from the right ventricle flows through the pulmonary trunk and into the pulmonary arteries prior to entering the lungs. Blood in the lungs releases carbon dioxide and picks up oxygen. The pulmonary veins carry oxygenated blood from the lungs into the left atrium. These vessels are located in the superior, posterior portion of the left atrium. The left atrium contracts and pushes blood into the left ventricle. Locate the two large cusps of the bicuspid valve between the left atrium and left ventricle. The **bicuspid valve** is also known as the **mitral valve,** or **left atrioventricular valve.** It also has attached chordae tendineae and papillary muscles, which you should locate in the models in the lab. Note the thickness of the left ventricle wall compared to the wall of the right ventricle. Compare the left side of the heart to figure 31.5.

The **aortic semilunar valve** is located at the junction of the left ventricle and the ascending aorta. It has the same basic structure and general function as the pulmonary semilunar

valve in that it prevents the backflow of blood from the aorta into the left ventricle. Blood from the left ventricle moves into the aorta and subsequently to the rest of the body.

Dissection of the Sheep Heart

The sheep heart is similar to the human heart and usually is readily available as a dissection specimen. Dissection of anatomical material is valuable in that you can examine structures that are represented more accurately in preserved material than in models. Also, the preserved material has greater flexibility and is easily manipulated. There are some differences between sheep hearts and human hearts, especially in the position of the superior and inferior venae cavae. In sheep these are called the anterior and posterior venae cavae, but we refer to them using the human terminology.

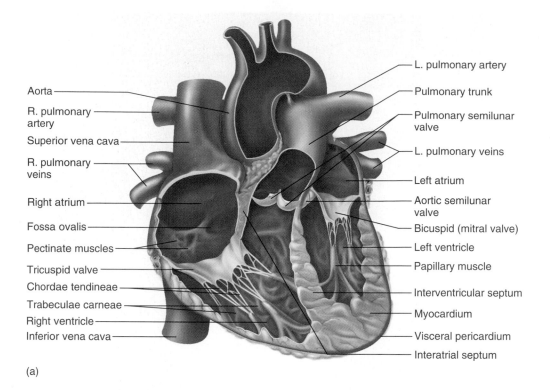

Aorta

R. pulmonary artery

Superior vena cava

R. pulmonary veins

Right atrium

Fossa ovalis

Pectinate muscles

Tricuspid valve

Chordae tendineae

Trabeculae carneae

Right ventricle

Inferior vena cava

L. pulmonary artery

Pulmonary trunk

Pulmonary semilunar valve

L. pulmonary veins

Left atrium

Aortic semilunar valve

Bicuspid (mitral valve)

Left ventricle

Papillary muscle

Interventricular septum

Myocardium

Visceral pericardium

Interatrial septum

(a)

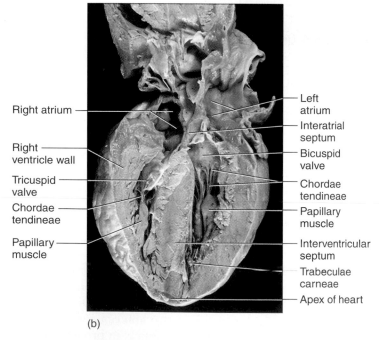

Right atrium

Right ventricle wall

Tricuspid valve

Chordae tendineae

Papillary muscle

Left atrium

Interatrial septum

Bicuspid valve

Chordae tendineae

Papillary muscle

Interventricular septum

Trabeculae carneae

Apex of heart

(b)

Figure 31.5   Heart, frontal section.

If your sheep heart has not been dissected, then you will need to open the heart. If your sheep or other mammalian heart has been previously dissected, then you can skip the next paragraph.

Place the heart under running water for a few moments to rinse off the preserving fluid. Examine the external features of the heart. Note the fat layer on the heart. The amount of fat on the human or sheep heart is variable. Locate the **left ventricle,** the **right ventricle,** the **interventricular groove (sulcus),** the **right atrium,** and the **left atrium.** Note the **auricles** that extend on the anterior surface of the atria.

Using a sharp scalpel or razor blade, make an incision along the right *side* of the heart (lateral side) from the apex of the heart to the lateral side of the right atrium. If you are unsure about how to proceed during any part of the dissection, ask your instructor for directions. Make another long cut from the lateral side of the left atrium through the lateral side of the left ventricle. You will have made a coronal section of the heart if you cut through the **interventricular septum.** Once you have opened the heart, compare the structures of the sheep heart to figure 31.7.

You can locate the vessels of the heart by inserting a blunt metal probe into the vessels and determining which chamber

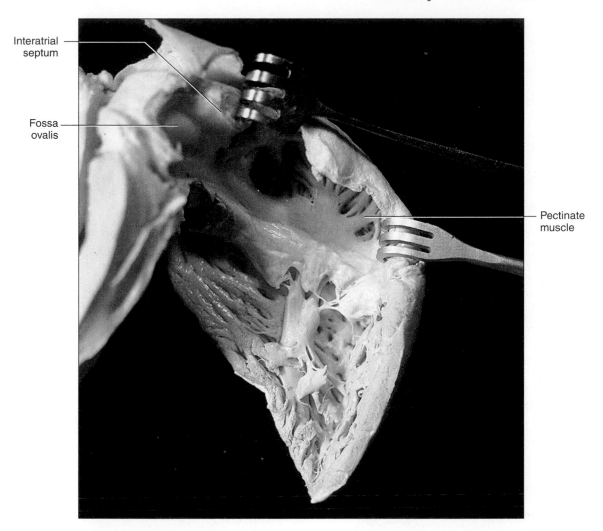

Interatrial septum

Fossa ovalis

Pectinate muscle

Figure 31.6   Details of the right atrium.

the vessel goes to or comes from. Place the heart in anatomical position and insert the probe into the large, anterior vessel that exits toward the specimen's left side. The blunt end of the probe should enter into the right ventricle. This vessel is the **pulmonary trunk.** The pulmonary trunk may still have the **pulmonary arteries** attached. Locate the large vessel directly behind the pulmonary trunk (figure 31.8.) This is the **ascending aorta.** If the vessels are cut farther away from the heart, you can see the **aortic arch.** Insert the probe into this vessel and into the **left ventricle.**

Turn the heart to the posterior surface and locate the **superior (anterior) vena cava** and **inferior (posterior) vena cava.** Insert the probe into the superior and inferior vena cavae,

pushing the probe into the **right atrium.** If you only find one large opening in the atrium, you may have cut through either the superior or inferior vena cava during your initial dissection. The probe can be felt through the wall more easily here than in a ventricle because the atrial walls are thinner than those of the ventricles. On the left side the **pulmonary veins** may either appear as four separate veins or you may just see a large hole on each side of the **left atrium** if the vessels were cut close to the atrial wall. Locate the same structures in the sheep heart as you found on the model and compare them to figure 31.9.

Cut into the right atrium and use your blunt probe to locate the opening of the **coronary sinus** in the posterior, inferior portion of the atrium. It is small and somewhat difficult to find.

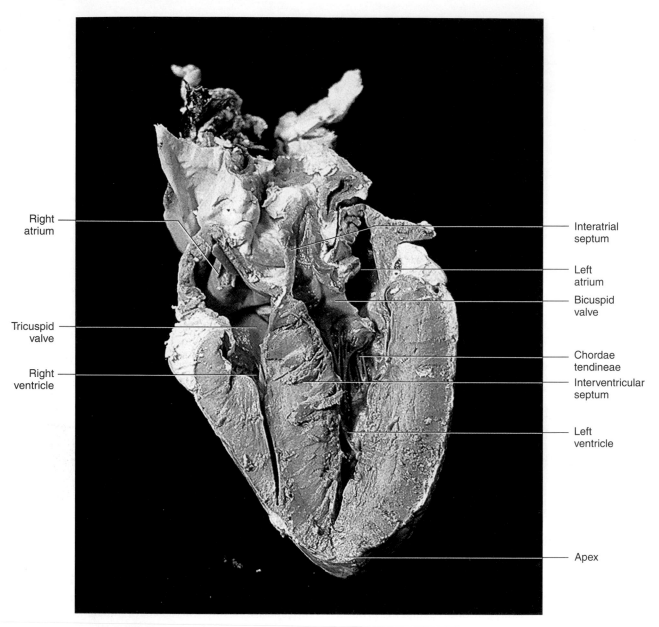

Right atrium

Tricuspid valve

Right ventricle

Interatrial septum

Left atrium

Bicuspid valve

Chordae tendineae

Interventricular septum

Left ventricle

Apex

Figure 31.7   Sheep heart, coronal section.

Examine the opening between the right atrium and the right ventricle to locate the **tricuspid valve.** You may have dissected through one of the cusps as you opened the heart. Locate the major features of the **right ventricle.** Find the **chordae tendineae** and the **papillary muscles.** You can find the **pulmonary trunk** by inserting a blunt probe into the superior portion of the right ventricle. Make an incision in the pulmonary trunk near the right ventricle to expose the three thin cusps of the **pulmonary semilunar valve.** Note how the cusps press against the wall of the pulmonary trunk when the probe is pushed against them in a superior direction. These cusps close when blood begins to flow back into the right ventricle as the ventricle relaxes.

Locate the **left atrium, bicuspid valve,** and **left ventricle.** In the left ventricle you should find the papillary muscles, chordae tendineae, and **trabeculae carneae** as well. You can find the aorta and aortic semilunar valve by inserting a blunt probe towards the superior end of the left ventricle towards the middle of the heart.

Clean Up   When you have finished your study of the sheep heart make sure you clean your dissection equipment with soap and water. Be careful with sharp blades. Place the sheep heart either back in the preserving fluid or in the appropriate waste container as directed by your instructor.

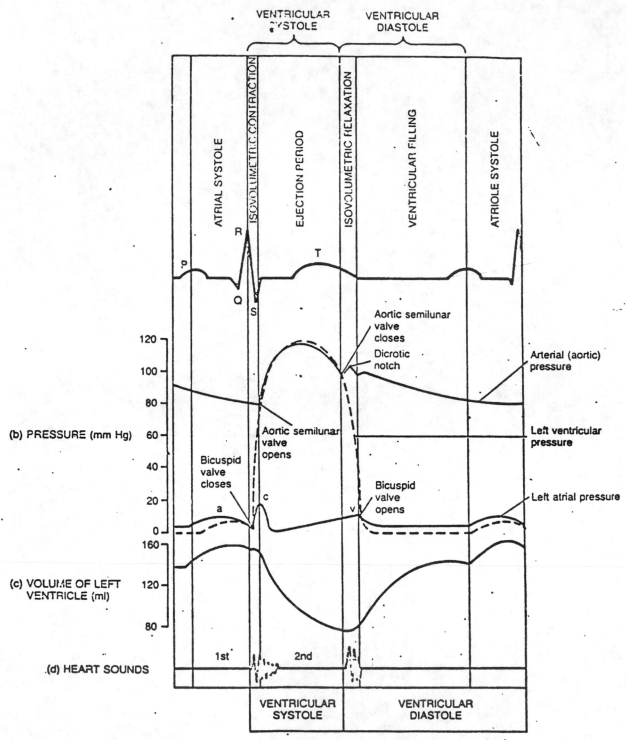

FIGURE 19.5 Cardiac cycle. (a) ECG related to the cardiac cycle. (b) Left atrial, left ventricular, and arterial (aortic) pressure changes along with the opening and closing of valves during the cardiac cycle. The letters a, c, and v represent the elevation waves of the left atrial pressure curve. (c) Left ventricular volume during the cardiac cycle. (d) Heart sounds related to the cardiac cycle.

ANATOMY OF THE HEART

Identify the following structures on the real heart.

✓1. pulmonary trunk (10)
✓2. left auricle (25)
✓3. right auricle (23)
✓4. anterior interventricular sulcas (32)
✓5. pulmonary semilunar valve (63)
✓6. aortic semilunar valve (70)
✓7. superior vena cava (3)
✓8. inferior vena cava (20)
✓9. chordae tendinae (59)
✓10. papillary muscles (58)
✓11. aorta (4)

✓12. pulmonary veins (18&19)
✓13. bicuspid valve (67)
✓14. tricuspid valve (56)
✓15. atrial chambers (right and left) (21)
16. ventricular 22 chambers (right (24) and left) 28
✓17. trabecula carnae 60
✓18. interventricular septum (68)
✓19. coronary sulcas (27)
✓20. posterior → (BACK) interventricular sulcas → (Gap) (33)

in between ←

Identify the following structures on the model heart.

1. interatrial septum
2. fossa ovalis
3. ligmentum arteriosum
4. coronary sinus
5. coronary arteries (right and left)
6. aortic arch (5)
7. pulmonary arteries (right and left)
8. brachiocephalic artery
9. marginal branch of right coronary artery
10. posterior interventricular branch of right coronary artery
11. circumflex branch of left coronary artery
12. anterior interventricular artery
13. greater cardiac vein
14. middle cardiac vein
15. left common carotid
16. left subclavian artery

20

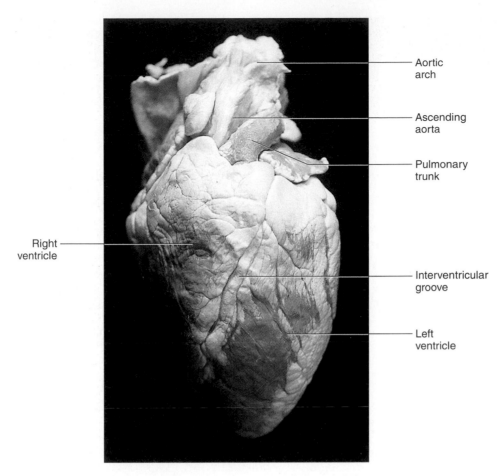

Figure 31.8   Vessels of the sheep heart.

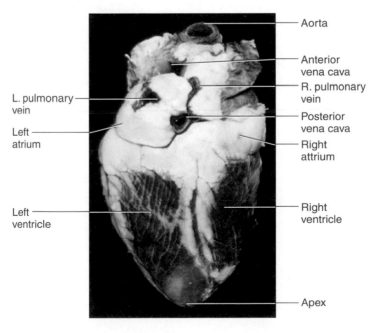

Figure 31.9   Sheep heart, posterior view.

Name _____

1. The heart is located between the lungs in an area known as the:

2. What is the outer layer superficial to the pericardial cavity?

3. What is the innermost layer of the heart wall called?

4. Is the apex of the heart superior or inferior to other parts of the heart?

5. What blood vessels nourish the heart tissue?

6. What separates the left atrium from the right atrium?

7. The bicuspid valve is located between what two chambers of the heart?

8. What is the function of the aortic semilunar valve?

9. What is another name for the tricuspid valve?

357

10. What is the cell type that makes up most of the myocardium?

11. What adaptation do you see in the walls of the left ventricle being thicker than those of the right ventricle?

12. How does cardiac muscle resemble skeletal muscle?

13. In terms of function, how is cardiac muscle different from skeletal muscle?

14. Label the following illustration.

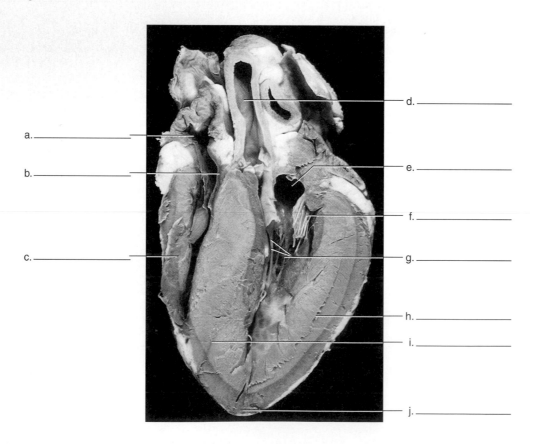

a. _____

b. _____

c. _____

d. _____

e. _____

f. _____

g. _____

h. _____

i. _____

j. _____

LABORATORY EXERCISE 32

Electrical Conductivity of the Heart

Introduction

The living heart is a phenomenal organ. It contracts an average of 72 times per minute for the lifetime of an individual. In this exercise, you observe the electrical activity of the heart and correlate it to the mechanical functioning of the heart. Two concepts are important in understanding the electrical activity of the heart. One is that the heart produces **electrochemical impulses** in a similar way to the production of impulses in the nervous system; the other is that these impulses can travel through the medium of the body and be perceived in areas distant to the heart. The cells of the body are bathed in a saline solution of a little less than 1% salt. This solution is an excellent conducting medium for electrical impulses. As those impulses are generated in the heart they travel through the body including the distal regions of the limbs. In this exercise, you learn the conductive structures of the heart and make a recording of the electrical activity of the heart.

Objectives

At the end of this exercise you should be able to

1. list the structures in the conductive system of the heart;
2. describe the sequence of electrical conductivity of the heart;
3. associate the P wave, QRS complex, and T wave of an ECG with electrical events that occur in the heart;
4. relate how the electrical activity of the heart is transferred to the ECG paper;
5. calculate heart rate, P–Q interval, QRS interval, and Q–T interval from an ECG recording;
6. describe the variance in an ECG from normal if a person has a heart block.

Materials

Electrocardiograph machine, physiograph, or other ECG recording device
Cot or covered lab table
Alcohol swabs
Electrode jelly, paste, or saline pads
Watch with accuracy in seconds
Dissection tray
Squeeze bottle of water
Scissors or razor blade

Fresh or thawed sheep heart
Clamp (hemostat) or string

 Virtual Physiology Lab #5: Electrocardiogram

Procedure

The function of the heart is dependent on the structure of the heart. Review the anatomy of the heart in Laboratory Exercise 31 before beginning this section.

Muscular Function of the Heart

The heart has been described as a muscular pump or, more accurately, two pumps acting in unison. As a pump, the heart has a contraction mode and a relaxation mode. **Systole** is contraction of the heart muscle and can be described more specifically as **atrial systole** and **ventricular systole** (contraction of the atria and ventricles respectively). **Diastole** is the relaxation of the heart muscle. Atrial and ventricular diastole are the relaxation of the atria and ventricles, respectively. Muscular contraction of the heart is preceded by electrical activity that stimulates the muscle to contract.

Electrical Functioning of the Heart

The initiation of the electrical impulse in the heart begins at the **sinoatrial node** (or **SA node**), which is commonly known as the **pacemaker.** Cells of the sinoatrial node use the sodium-potassium pump to generate a difference in electrical voltage across the cells' membranes. The cells are said to be **polarized** because the inside of the cell membrane has more negative ions than the outside of the membrane. This separation of charged particles is reflected in the voltage difference across the membrane. The sinoatrial node spontaneously **depolarizes,** which causes a change in the total amount of charge across the cell membrane. This change in membrane potential occurs approximately 72 times per minute (the average heart rate). Conduction from the sinoatrial node travels across the atria, causing the muscles of the atria to contract. The impulse that spreads out across the atria reaches the **atrioventricular node** (or **AV node**). The impulse has a slight delay (0.1 second) in the node before being conducted further. This delay allows the atrial cardiac muscles to contract prior to ventricular firing (figure 32.1).

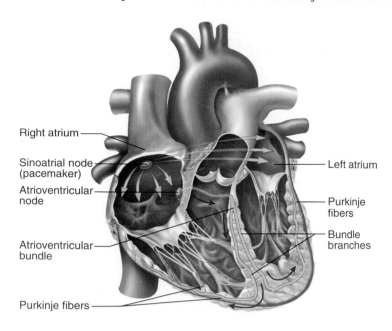

Figure 32.1 Conduction system of the heart. Conduction follows path indicated by arrows. ✗ ▭

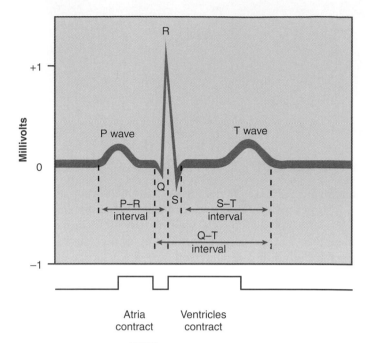

Figure 32.2 ECG. ✗ ▭

The electrical impulse then travels from the AV node to the **atrioventricular bundle (bundle of His),** to the right and left **bundle branches,** and finally to the **Purkinje (conduction) fibers.** The Purkinje fibers stimulate the cardiac muscle of the ventricles to contract. The ventricles are thus stimulated from the apex towards the base, and the contraction proceeds from the inferior end of the ventricles towards the atria. Locate the structures of the heart's conduction system in figure 32.1.

Nature of the ECG

The **electrocardiograph** is a machine that can measure the electrical activity of the heart. It is an instrument that measures slight changes in the voltage related to cardiac activity. The electrocardiograph produces a chart paper recording called an **electrocardiogram (ECG or EKG).** Before you record an ECG, examine the ECG in figure 32.2 and read the accompanying description.

Time is measured along the horizontal axis (each millimeter is equal to 0.04 seconds), and voltage difference is measured along the vertical axis (in millivolts). The ECG typically has three major events. The first event is the **P wave,** which is a small bump that occurs on the chart paper, called a deflection wave. The P wave represents **atrial depolarization.** The **QRS complex** represents the **ventricular depolarization.** Because the ventricles are more massive than the atria, the electrical events produced as the ventricles depolarize are much larger than the P wave generated by the atria. The **T wave** represents the **repolarization** of the ventricles. **Atrial repolarization** is a small electrical event that occurs during ventricular depolarization and is masked by the large QRS complex.

The electrical impulse generated by the atria and ventricles depolarizing and repolarizing can be recorded using an electrocardiograph machine because of the conductivity of the intercellular fluid. Some other bodily functions such as nerve impulse conduction and skeletal muscle contraction also occur by transmitting low-voltage signals. The "average" potential difference of –70 millivolts is a little less than one-tenth of a volt; therefore, the body is operating electrically at slightly less than 0.1 volt. As this electricity is produced, it passes through the saline body fluid and can be picked up by sensors (electrode plates) placed in particular locations around the body. In many college physiology labs the ECG electrode plates are attached to four areas. These are the medial side of the **left** and **right ankles** and the anterior surface of the **left** and **right wrists.** The attachment of the electrode plate to the right ankle serves as an electrical **ground** and is not used for measurement. The ECG is measured as the potential voltage difference between selected electrode plates.

Lead I measures the voltage differences on the horizontal axis of the heart. It measures the conductivity from the electrode plate on the right arm to one on the left arm (figure 32.3).

Lead II measures diagonally as the potential voltage difference across the body as received from the electrode plate on the right arm and one on the left leg.

Lead III indicates the potential voltage differences between the electrode plates on the left arm and left leg.

There are numerous "V" electrodes, which are chest electrodes. You will not use the V electrodes in this exercise unless directed to do so by your instructor.

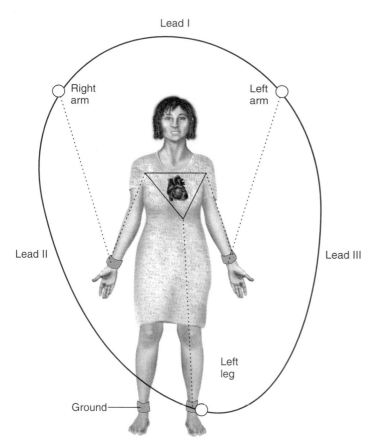

Lead I

Right arm

Left arm

Lead II

Lead III

Left leg

Ground

Figure 32.3 **Leads of a standard ECG.**

Procedure for ECG

Make sure the ECG machine or physiograph is plugged in and turned on. Make sure that the ECG machine is on "standby" when you attach the electrodes. Accidental grounding of the electrodes may cause damage to the equipment. It should warm up for a few minutes prior to running the experiment. Remove metal watches or any jewelry that might interfere with the electrical signal between the heart and the extremities. Have your lab partner scrub the inside of the ankle, about 2 cm above the medial malleolus, with an alcohol swab to remove dust and skin oil. Clean the anterior sides of the wrist in the same way. Saline pads or electrode jelly may be applied to these areas. Attach the electrodes firmly, but not so tight that you cut off blood flow, using the straps provided as follows:

LA attaches to left arm.
RA attaches to right arm.
RL attaches to right leg.
LL attaches to left leg.

Your instructor will explain how to operate your specific electrocardiograph. Have your lab partner rest comfortably on a cot or table for a few moments before monitoring the ECG. Turn the knob on the ECG machine from "Standby" to "Run" and record **lead 1** for about 10 cycles. Turn the knob to

"Standby" after recording each lead. Then record **lead II** and **lead III** for about 10 cycles each. Tear the ECG from the machine and make sure you write your lab partner's name on the paper and note the appropriate lead. It is important that you provide a relaxed atmosphere for your lab partner to produce a good ECG recording. People who are fidgeting or nervous generate extraneous interference and do not produce a level baseline recording. You can demonstrate this electrical "noise" by the following activity.

Interference in an ECG

While your lab partner is still attached to the ECG machine, set the dial to lead II. Run three to four cardiac cycles and then have your lab partner clench his or her fists. Note the electrical interference that occurs as the skeletal muscles depolarize and repolarize. Emotional state also plays a part in electrical activity. If a patient is nervous or excitable, this will show in the ECG. Background electrical noise in the ECG varies from person to person, but you should be able to get a good ECG in the lab.

Analysis of the ECG

ECG recordings are important in assessing the health of the heart. Conduction problems, myocardial infarcts (heart attacks), and heart blocks are a few of the problems that may show up as variances on an ECG. The interpretation of ECGs in this exercise is for educational purposes only. Clinical evaluations of ECGs should be done only by trained health-care workers. Locate the P wave, the QRS complex, and the T wave. The whole cardiac cycle should take, on average, 0.7 to 0.8 seconds.

The standard rate of paper travel in an ECG machine is 25 mm/second. The small squares are millimeters (mm) and the darker lines indicate 5 mm or 0.5 cm. You can calculate the resting heart rate by counting how many millimeters ("Y" mm) occur between the peaks of the R waves. You can calculate the number of beats per minute by using the following formula. Record your results in the space provided.

$$\frac{1 \text{ beat}}{\text{"Y" mm}} \times \frac{25 \text{ mm}}{1 \text{ sec}} \times \frac{60 \text{ sec}}{1 \text{ min}} = \frac{\text{beats}}{\text{min}}$$

Calculated beats per minute: _____

Irregularities in the ECG

An excessively high heart rate is termed **tachycardia.** In young adults a heart rate above 100 beats per minute is considered tachycardia. However, 100 beats per minute in small children may be perfectly normal. An excessively low heart rate is termed **bradycardia.** In young adults a rate below 60 beats per minute is considered bradycardia unless they are highly trained aerobic athletes.

P–Q (P–R) intervals should normally be about 0.16 seconds. The P–Q interval is the time between the beginning of atrial depolarization and the beginning of ventricular depolarization. If the P–Q interval is longer than 0.2 seconds (5 mm on the chart paper), then this could be indicative of a **heart block.** Heart blocks result from reduced conduction between the atria and the ventricles. This may be caused by damage to the AV node or decreased transmission in the AV bundle. In a **complete heart block** the atria do not stimulate the ventricular depolarization, therefore the atria fire independently from the ventricles. In a complete heart block the P waves may be spaced at 0.8 seconds apart, which is the SA node depolarization rate, but the ventricles fire at a much lower rate (1.5 to 2.0 seconds apart). Are the times measured on your recording within normal limits? Record the P–Q interval.

P–Q interval: _____

The **QRS complex** is .08 to 0.10 seconds on average. If the QRS complex spans longer than 0.12 seconds this may indicate a **right** or **left bundle branch block.** In this condition the two ventricles contract at slightly different times, increasing the length of the QRS complex. Measure the length of the QRS complex and record your findings in the space provided.

QRS interval: _____

The **Q–T interval** is 0.3 seconds on average. The interval is shorter as the heart rate increases, and the interval becomes longer as the heart rate slows down. Record your results in the space provided.

Q–T interval: _____

ECG and Exercise

Once you have compared your ECG to normal, you should exercise vigorously for a few minutes. You can do this by running outside of the lab room or doing jumping jacks for a minute or two. Once you feel that your heart is pumping faster and harder, quickly lie down on the cot and remeasure your ECG.

How does your ECG compare to normal in terms of distance between the P waves?

How does the height of the QRS complex compare to the resting ECG?

Heart Valves

The semilunar valves and atrioventricular valves function to prevent the backflow of blood. Your instructor may demonstrate the procedure or you can do it yourself by using a fresh or thawed sheep heart. Flush any remaining blood from the heart before you locate the valves.

Make an incision into the right atrium, exposing the **tricuspid valve.** Pour water into the right ventricle and notice how the water flows past the tricuspid valve and into the right ventricle. Clamp off the pulmonary trunk or tie it with string to prevent blood from flowing out of the pulmonary trunk. *Gently* squeeze the right ventricle and notice how the right atrioventricular valve closes and prevents blood from backing up into the right atrium.

What is the adaptive value for the closing of atrioventricular valves?

Cut the **pulmonary trunk** close to the right ventricle and slowly pour water into it as if you are trying to fill the right ventricle. The **pulmonary semilunar valve** should fill with water and close the entrance of the right ventricle, preventing backflow. This would normally occur as the right ventricle begins diastole. Compare these valve closures to figure 32.4.

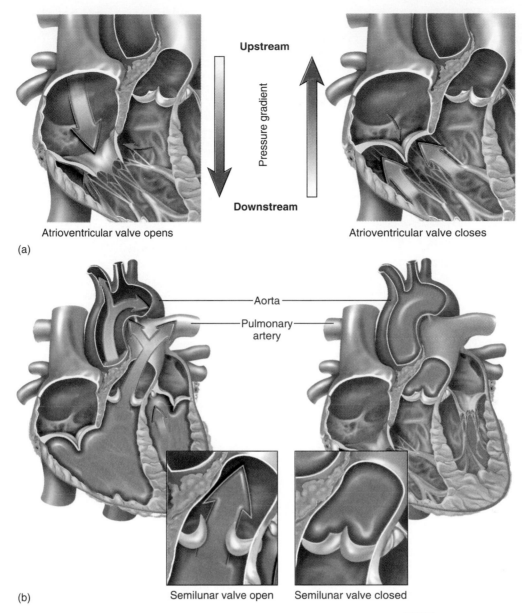

Upstream

Pressure gradient

Downstream

Atrioventricular valve opens

Atrioventricular valve closes

(a)

Aorta

Pulmonary artery

Semilunar valve open

Semilunar valve closed

(b)

Figure 32.4 Closure of the heart valves. (a) Atrioventricular valve; (b) semilunar valve.

Name _____

1. The sinoatrial node has a common name. What is it?

2. Which two chambers of the heart contract last in a normal cardiac cycle?

3. What chambers are stimulated immediately after the SA node depolarizes?

4. After the AV node depolarizes, what three structures directly receive the impulse?

5. What are the main events of an ECG?

6. What electrical event does the QRS complex represent?

7. Ventricular repolarization is represented by what part of an ECG?

8. What ECG event is represented by the atrial depolarization?

9. Why is the ECG event indicating atrial repolarization not seen in an ECG?

10. What is the function of the semilunar valves?

11. Fibrillation is uncoordinated cardiac muscle contraction. Predict what an ECG would look like if there was no uniform conduction of electrical activity in the heart. Draw what it might look like.

12. What consequence does fibrillation have for cardiac muscle contraction and for the pumping efficiency of the heart? Which is more serious—atrial or ventricular fibrillation?

13. What is the "pacemaker" of the heart called?

14. If a myocardial infarct (heart attack) destroyed a portion of the right or left bundle branches, what potential change might you see in an ECG?

15. Tape or paste your ECG in the following space. Label the P wave, the QRS complex, and the T wave.

Functions of the Heart

Introduction

In terms of efficiency and overall endurance, the human heart is a remarkable organ. It pumps blood continuously at the average rate of about 72 beats per minute for the lifetime of the individual. The heart increases the volume of blood it pumps when more oxygen is required by either increasing the rate of contraction or the volume ejected per contraction, or both. In this exercise, you explore some of the functions of the human heart, with particular emphasis on how the heart is regulated in terms of speed and contraction strength.

Specialized muscle cells in the **sinoatrial (SA) node** of the heart wall spontaneously depolarize and send action potentials across the heart wall, causing the cardiac muscle to contract. Normally the **resting heart rate** of mammals is lower than the rate of excised hearts. In human hearts that have been removed for heart transplant surgery, the cardiac rate is about 100 beats per minute. This slowing of the intrinsic heart rate from 100 beats per minute to about 72 beats per minute is due to the **parasympathetic nervous system.** Increase in the heart rate from the resting rate to about 100 beats per minute is due to the progressive inhibition of the parasympathetic nervous system.

The heart rate can be increased beyond 100 beats per minute by stimulation from the **sympathetic nervous system.**

Cardiac Muscle Characteristics

The mechanism by which electrochemical impulses occur in heart muscle depends on three main gates, **fast sodium channels, slow calcium channels,** and **potassium channels.** In the normal **polarization** process of a cardiac muscle cell the **sodium-potassium pump** is activated. The eventual result is sodium and potassium ions occurring in greater concentrations on the outside of the plasma membrane than on the inside. This produces the **resting membrane potential** of a particular cell. Once the **threshold** has been reached the fast sodium channels open and sodium ions rush into the cell. This causes **depolarization** of the membrane. As the sodium channels close the cell begins to slightly **repolarize.** This is the same process that occurs in skeletal muscle. In heart muscle, however, **slow calcium channels** open after the fast sodium channels close, and, instead of the plasma membrane quickly repolarizing, the flow of calcium ions into the cell causes a flattening of the electrical activity of the cell known as a **plateau.** The plateau results as a function of a slowing of repolarization. This can be seen in figure 33.1.

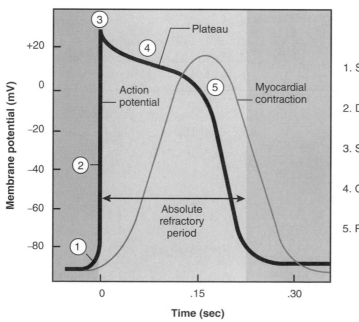

1. Stimulus

2. Depolarization

3. Sodium channels close

4. Calcium channels open, causing a plateau

5. Repolarization

Figure 33.1 Cardiac muscle action potential.

When the calcium channels are open, the potassium channels of the plasma membrane remain closed and potassium ions cannot flow out of the cell. Eventually as the voltage changes, the potassium channels do open and potassium flows out of the cell, thus repolarizing the cell. This **action potential** travels across the cardiac muscle cell, causing further depolarization of the membrane. **Repolarization** occurs relatively quickly by reestablishing the resting membrane potential, and the heart muscle is ready for its next contraction.

Cardiac Pacemaker

The pacemaker cells of the heart spontaneously depolarize. Sodium channels in the plasma membrane of pacemaker cells slowly leak sodium until threshold is reached. Once this occurs, a spontaneous opening of sodium gates leads to depolarization of the pacemaker cells, which subsequently causes a **field effect,** depolarizing adjacent cells near the pacemaker. This depolarization spreads throughout the atria and eventually to the ventricles, stimulating the cardiac muscle to contract.

The native sinoatrial depolarization rate is approximately 100 depolarizations per minute. The parasympathetic nervous system, in a person at rest, slows down this depolarization rate to an average of 72 beats per minute. The **vagus nerve** conveys parasympathetic fibers that innervate the SA node. **Acetylcholine** (ACh) is a neurotransmitter released by this nerve, and it promotes potassium (K$^+$) leakage from the nodal cells to the interstitial fluid, thus **hyperpolarizing** the cells. By increasing the difference in voltage between the inside and the outside of the cells, it takes longer for the cells to reach threshold and depolarize. In this way the heart rate is slowed.

On the other hand, the heart rate can be elevated by increasing the firing rate of the SA node. From the resting heart rate of 72 beats per minute to about 100 beats per minute (which is the native nodal firing rate) the initial control of *increased* firing rate is due to gradual *inhibition* of the vagus nerve. As the nerve secretes less ACh at the node, the firing rate increases. Above 100 beats per minute the heart rate is controlled by the sympathetic nervous system. Nerves from the cervical region of the spinal cord release **norepinephrine,** which binds to **beta-adrenergic receptors.** Calcium channels open somewhat and *decrease* the threshold, thus causing more rapid firing of the SA node.

The electrical activity of the nodes of the heart occurs in specialized *muscle* cells and not in nervous tissue. The process of electrochemical transfer of impulses is known as **myogenic conduction.** The muscle tissue may be *influenced* by the nervous system, but the activity is initiated and generated in specialized muscle fibers. In addition to the influence of the nervous system, certain hormones have an effect on the heart rate.

Because the heart muscle cells are linked together by intercalated discs, they act as one electrical unit. An impulse generated in the nodal tissue spreads to the surrounding muscle, causing it to first depolarize and subsequently contract. Review the conduction pathway of the heart in Laboratory Exercise 32.

In this exercise, you examine the natural rate of heart contraction in a frog and what occurs when various solutions are added to the heart muscle. The heart of a frog is physically somewhat different than the human heart. The frog has two atria and only a single ventricle; however, the physiological response of the frog to various cardiac-influencing substances is similar to that in humans.

Objectives

At the end of this exercise you should be able to

1. correlate the sounds the heart makes with the action of the heart;
2. describe the basic contraction characteristics of the heart;
3. list the effects of various substances such as epinephrine, calcium chloride, and acetylcholine on the heart rate;
4. outline the mechanisms by which these substances change heart rate.

Materials

Clock or watch with second hand
Stethoscope
Alcohol wipes
Frogs
Clean dissection instruments:

    Scissors

    Razor blades or scalpels

    Dissection tray

Small heart hook (fish hook, copper wire, or Z wire)
Pins
Thread
Frog board
Myograph transducer
Physiology recorder (computer, physiograph, or duograph)
Frog Ringer's solution—room temperature, 37°C, and iced
Solutions:

    2% calcium chloride solution

    0.1% acetylcholine chloride solution

    0.1% epinephrine solution

    Saturated caffeine solution

 Virtual Physiology Lab #4: Effects of Drugs on the Frog Heart

Procedure

Pulse Rate and Heart Sounds

Prior to working on the frog heart, measure the pulse rate and listen to heart sounds of your lab partner. Place your fingers on the radial artery or the carotid artery of your lab partner and count the number of beats that occur in 15 seconds. Multiply this number by four and record the resting pulse rate.

Resting pulse rate: _____
beats/minute

Clean the ear pieces of the stethoscope with alcohol swabs and examine the stethoscope. The ear pieces should point in an anterior direction as you place the stethoscope into your ears. Listen for heart sounds generated by your lab partner by placing the bell of the stethoscope inferior to the second rib on the left side of the body. The pulmonary semilunar valves are best heard in this location. On the right side of the body, in the intercostal space between the second and third rib is the location for the aortic semilunar valve. Use figure 33.2 as a guide for listening to heart sounds. The lubb/dupp sounds of the heart reflect the closure of the heart valves. The first heart sound is the **lubb** sound and it occurs due to the closing of the atrioventricular valves. The second heart sound is the **dupp** sound and it is due to the closing of the semilunar valves.

You may hear an additional whooshing sound as you listen through the stethoscope. This is attributed to the imperfect closure of the valves, which is known as a **heart murmur.**

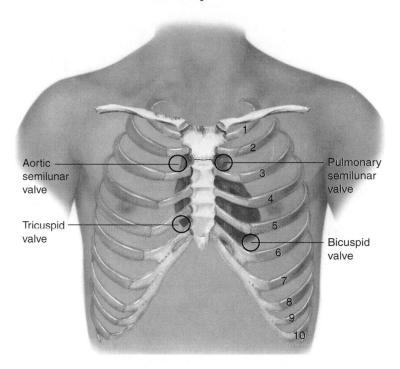

Figure 33.2   Auscultation areas. Circles are locations where specific valvular sounds can best be heard.

Examination of Frog Heart Rate and Contraction Strength

You may either have a frog prepared for you or your instructor may wish for you to double pith a frog for this exercise. If you need to pith a frog follow the instructions outlined in Laboratory Exercise 19 and then follow the procedure described next.

Measurement of Baseline Data

Once the frog is prepared, cut through the pectoral skin with a scissors and through the sternum. The heart should be beating inside the pericardial sac. Pin the frog to a board and cut open the pericardial sac with a pair of scissors or scalpel. Count the number of beats per minute and record this number as the resting heart rate of the frog.

Resting heart rate: _____ beats/minute

Keep the heart moistened with frog Ringer's solution. Unless directed to do otherwise, use the Ringer's solution that is at room temperature.

Carefully cut the heart away from the dorsal surface of the frog and attach a metal hook or wire to the connective tissue attached to the dorsal portion of the heart. You can pierce the apex of the heart with a wire or hook, but you must be very careful not to puncture the ventricle. Attach the hook

to a thread and tie the thread to the myograph transducer, adjusting the tension until you get a deflection in the myograph leaf. Your setup should look similar to figure 33.3.

Set the chart speed to 0.5 cm/second and record the contraction rate and strength of contraction of the heart muscle. The force that is generated by the heart is measured by the height of the stylus on the chart paper. The rate can be determined by the distance between the peaks on the chart paper. Make sure you keep the heart moist with frog Ringer's solution during the experiment.

Effect of Various Chemicals on Heart Rate and Contraction Strength

Addition of Calcium   Add several drops of the 2% calcium chloride solution to the heart muscle. Wait until you get a change in the heart rate. Turn on the recording device and produce a tracing. Calculate the heart rate and record the number.

Heart rate: _____ beats/minute

What happens to the contraction strength?

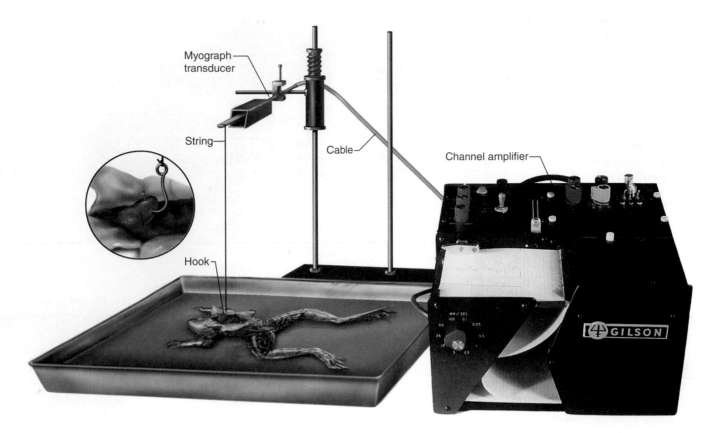

Figure 33.3   Frog set-up.

The addition of calcium chloride can produce pacemakers in regions of the heart other than the SA node. These are called **ectopic foci.** Rinse the heart with room-temperature frog Ringer's solution.

Addition of Acetylcholine   Add several drops of 0.1% acetylcholine solution to the outer surface of the heart. You may have to wait a few minutes to observe any changes in heart rate. Record the heart rate for about 10 seconds on the physiology recorder after waiting 2 minutes. Wait 3 minutes and then record the rate for another 10 seconds. Record the rate in the space. When you have determined a change in the heart rate and recorded your results, flood the heart with frog Ringer's solution.

Heart rate after 2 minutes: _____ beats/minute

Heart rate after 3 minutes: _____ beats/minute

Acetylcholine released by the vagus nerve causes the potassium channels in the plasma membrane of the nodal cells to become more permeable, thus increasing the flow of potassium ions to the outside of the membrane. This increases the voltage difference across the plasma membrane hyperpolarizing the cell. It takes more input to depolarize a cell in this condition, and thus the heart rate slows. Acetylcholine is a parasympathomimetic substance.

Addition of Epinephrine   Add several drops of the 0.1% epinephrine solution to the outer surface of the heart. Examine the rate over the next 2 to 3 minutes, and when you notice a change record the heart rate on the recorder chart paper for about 10 seconds. Write your results in the space provided. After you record your results, flush the heart with room-temperature frog Ringer's solution.

Heart rate after 2 minutes: _____ beats/minute

Heart rate after 3 minutes: _____ beats/minute

Epinephrine increases the heart rate and strength of contraction by making the slow calcium channels *more* permeable, thus lowering the threshold. Calcium (a cation) leaks into the cell, thus decreasing the voltage difference across the membrane. Epinephrine is a sympathomimetic drug.

Addition of Caffeine   Add several drops of the saturated caffeine solution to the outer surface of the heart and note the time when the rate changes. Record the heart rate and strength of contraction of the heart muscle for about 10 seconds on the physiology recorder. Enter the rate.

Heart rate with caffeine: _____ beats/minute

Strength of contraction (increase/decrease): _____

Caffeine is a central nervous system stimulant and also has an effect on heart muscle. It increases heart rate and strength of contraction. High concentrations of caffeine cause arrhythmias.

Changes in Heart Rate with Temperature

Run tests by flooding the heart with iced Ringer's solution. Note the change in heart rate.

Heart rate with iced Ringer's solution: _____ beats/minute

Readjust the rate by adding room-temperature Ringer's solution. Record your results.

Heart rate at room temperature: _____ beats/minute

You can further study the effect of enzymes by flooding the frog heart with frog Ringer's solution at 37°C. Record the contraction rate.

Heart rate at 37°C: _____ beats/minute

The $Q10$ is an enzymatic relationship that states that for every 10 degree rise in temperature (from 0° to 50°C) there is a doubling of enzyme activity. Does this correlate with your findings on the temperature relationships with the heart?

1. Decreasing heart rate is under the control of what nervous division?

2. What is the resting heart rate of the average individual?

3. What particular heart sound is produced by the closure of the atrioventricular valves in the heart?

4. A heart murmur is normally caused by what event?

5. What happens to the net concentration of sodium ions during the resting membrane potential?

6. What happens to sodium ions when a membrane depolarizes?

7. What region in the heart depolarizes spontaneously?

8. What occurs to the heart when an action potential is generated in the SA node?

9. Movement of electrochemical impulses in the myocardium is called _____ conduction.

10. Beta-adrenergic blockers are those that bind to norepinephrine sites, preventing these neurotransmitters from having an effect. What effect would the use of "beta blockers" have on heart rate?

11. When would a murmur occur in the lubb/dupp cycle if the AV valves were not properly closing?

12. What causes the plateau phase of cardiac muscle contraction?

13. How much of a change in the heart rate of the frog did you see after the addition of calcium chloride?

    What was the change in the contraction strength?

    What process might account for this in terms of cardiac muscle interactions with calcium?

Introduction to Blood Vessels and Arteries of the Upper Body

Introduction

The heart serves as the mechanical pump of the cardiovascular system, and the blood vessels are the conduits that carry oxygen, nutrients, and other materials to the cells as well as remove wastes from the tissue fluid near the cells. These vascular conduits consist of numerous vessels, such as large **arteries** that become progressively smaller **arterioles.** Other vessels are **capillaries,** which serve as the sites of exchange with the cells; **venules,** which return blood to the **veins,** which carry blood back to the heart.

Arteries are defined as blood vessels that carry blood away from the heart. Most arteries carry oxygenated blood, but there are a few exceptions to this. Arteries have thicker walls than veins, and the thickness of arterial walls reflects the higher blood pressure found in them.

Arteries are frequently named for the region of the body they pass through. The brachial artery is found in the region of the arm, while the femoral artery is found in the thigh region. In this exercise, you learn the basic structure of blood vessels and locate the arteries of the upper body.

Objectives

At the end of this exercise you should be able to

1. draw a cross section of the wall of a generalized blood vessel showing the three layers;
2. distinguish between arteries and veins;
3. describe the sequence of major arteries that branch from the aortic arch;
4. list the arteries of the upper extremity;
5. trace the arteries that supply blood to the head;
6. describe the major organs that receive blood from the upper arteries.

Materials

Microscopes
Prepared slides of arteries and veins
Models of the blood vessels of the body
Charts and illustrations of the arterial system
Cats

Dissection equipment:
    Dissection tray or pan
    Scalpels
    Pins
    Blunt probes
    Gloves
    Scissors or bone cutter
    String
    Forceps
    Plastic bag with label

Procedure

Cross Sections of Arteries and Veins

Obtain a prepared microscope slide of a cross section of artery and vein. Both arteries and veins have walls that consist of three layers. The outer layer is known as the **tunica externa (tunica adventitia)** and consists of a connective tissue sheath. The middle layer or **tunica media** is composed of smooth muscle in both arteries and veins. The tunica media is much thicker in arteries and there are pronounced elastic fibers in the wall of the tunica media in arteries. The innermost layer (near the blood) is the **tunica interna** and consists of a thin layer of connective tissue and a thin layer of simple squamous epithelium, known as **endothelium.** In arteries there is an **inner elastic membrane** which is a thin layer of elastic fibers. Valves are features of veins not found in arteries. You can easily distinguish veins from arteries in a prepared slide where the vessels are cut in cross section. Veins have a large lumen (though the vein is frequently collapsed in prepared sections) and a thin tunica media relative to its overall size. You may also find nerves in the prepared slide. These are also circular structures, but they are solid, not hollow. Examine a prepared slide of an artery and vein and locate these layers. Compare your slide to figure 34.1.

Arteries

Examine the models, charts, and illustrations of the cardiovascular system in the lab and locate the major arteries of the upper body. Read the following descriptions of the arteries. Name

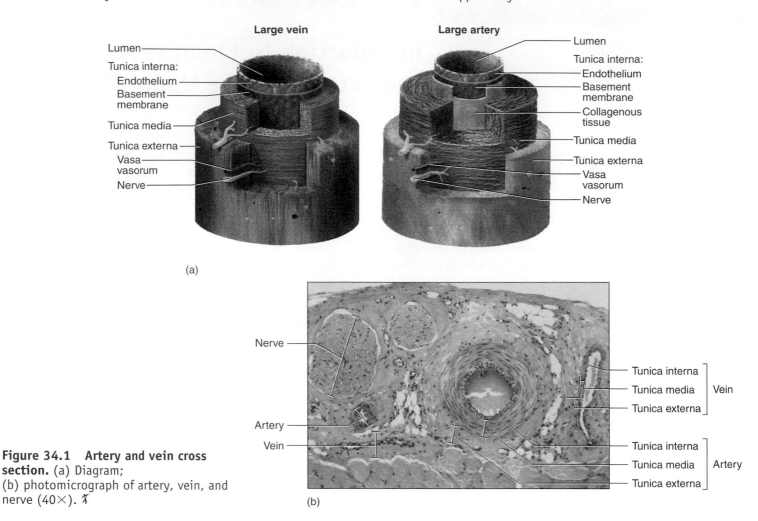

Figure 34.1   Artery and vein cross section. (a) Diagram; (b) photomicrograph of artery, vein, and nerve (40×). ⚘

the vessels that take blood to an artery and those that receive blood from the artery in question. An overview of the major arteries of the body is shown in figure 34.2.

Aortic Arch Arteries

Locate the heart and find the large **aorta** that exits from the left ventricle. This is the **ascending aorta,** and it is a large vessel about the size of a garden hose. The ascending aorta is relatively thick-walled due to the high pressure of the blood coming from the heart. The first two arteries that arise from the ascending aorta are the **coronary arteries,** which were covered in Laboratory Exercise 31. The ascending aorta curves to the left side of the body and forms the **aortic arch.** In humans, there are three

main arteries that receive blood from the aortic arch. The first major artery to receive blood is on the right side of the body and is called the **brachiocephalic artery (trunk).** This artery shortly divides into arteries that feed the right side of the head (the **right common carotid artery**) and the right upper extremity (the **right subclavian artery**) (figure 34.3). The aortic arch has two other arteries. On the left side is the **left common carotid artery,** which takes blood to the left side of the head, and the **left subclavian artery,** which takes blood to the left upper extremity.

As the aortic arch turns inferiorly behind the posterior part of the heart, it becomes the **descending aorta,** which is composed of two segments. Above the diaphragm the descending aorta is known as the **thoracic aorta** and below the diaphragm it is known as the **abdominal aorta** (figure 34.4). The thoracic

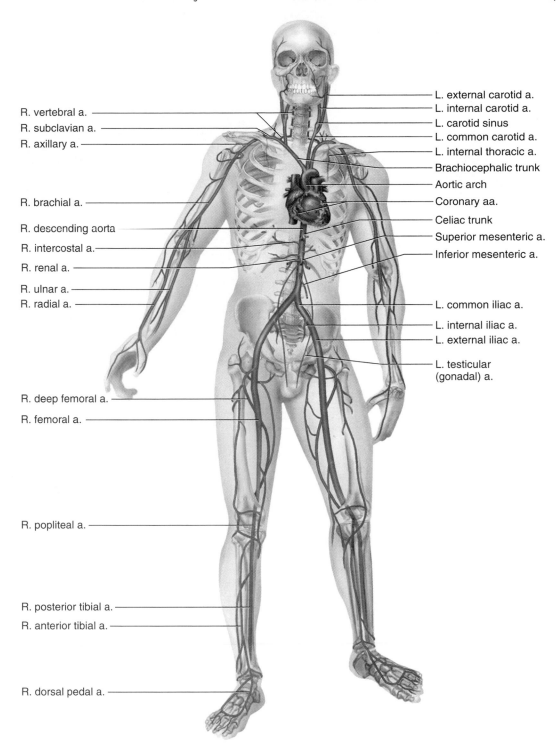

R. vertebral a.
R. subclavian a.
R. axillary a.

R. brachial a.

R. descending aorta
R. intercostal a.
R. renal a.
R. ulnar a.
R. radial a.

R. deep femoral a.

R. femoral a.

R. popliteal a.

R. posterior tibial a.
R. anterior tibial a.

R. dorsal pedal a.

L. external carotid a.
L. internal carotid a.
L. carotid sinus
L. common carotid a.
L. internal thoracic a.
Brachiocephalic trunk
Aortic arch
Coronary aa.
Celiac trunk
Superior mesenteric a.
Inferior mesenteric a.

L. common iliac a.

L. internal iliac a.
L. external iliac a.

L. testicular (gonadal) a.

Figure 34.2 Major arteries of the body.

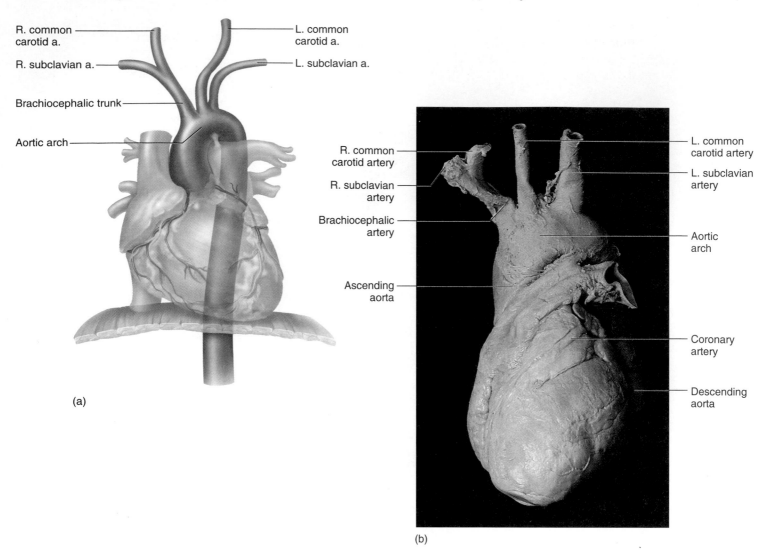

(a)

(b)

Figure 34.3   Arteries of the aortic arch.

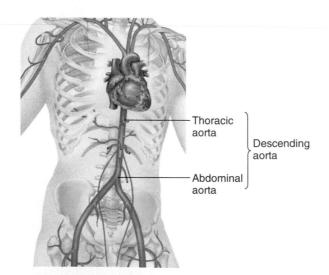

Figure 34.4   Thoracic and abdominal aorta.

aorta has numerous branches that run between the ribs called **intercostal arteries** (figure 34.5).

Arteries That Feed the Upper Extremities

If you return to the brachiocephalic artery, you can see that it is a short tube that takes blood from the aortic arch and then branches into the right common carotid artery and the right subclavian artery. The right subclavian artery has one branch that takes blood to the brain, while another branch becomes the **axillary artery.** The left subclavian artery also has a branch that takes blood to the brain and another branch that becomes the left axillary artery (figures 34.5 and 34.6).

The axillary artery on each side of the body turns into the **brachial artery,** which has a pulse that can be palpated by finding a groove on the medial, distal region of the arm between the biceps brachii and brachialis muscle. This location is the site for the placement of the bell of a stethoscope during blood pressure

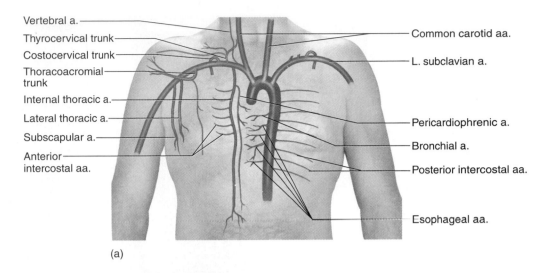

Vertebral a.

Thyrocervical trunk

Costocervical trunk

Thoracoacromial trunk

Internal thoracic a.

Lateral thoracic a.

Subscapular a.

Anterior intercostal aa.

Common carotid aa.

L. subclavian a.

Pericardiophrenic a.

Bronchial a.

Posterior intercostal aa.

Esophageal aa.

(a)

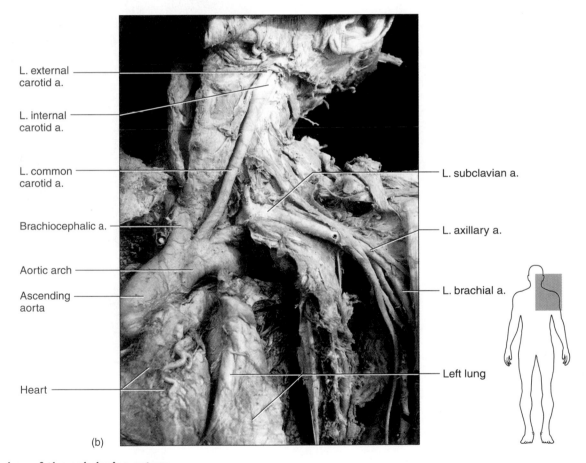

L. external carotid a.

L. internal carotid a.

L. common carotid a.

Brachiocephalic a.

Aortic arch

Ascending aorta

Heart

L. subclavian a.

L. axillary a.

L. brachial a.

Left lung

(b)

Figure 34.5   Branches of the subclavian artery.

measurement. The brachial artery supplies blood to the triceps brachii muscle and the humerus. The brachial artery bifurcates (splits in two) to form the **radial artery** on the lateral side of the forearm and the **ulnar artery** on the medial side of the forearm. These arteries supply blood to the forearm and part of the hand. The radial artery can be palpated just lateral to the flexor carpi radialis muscle at the wrist, which is a common site for the measurement of the pulse. The radial artery and ulnar artery become united again as the **palmar arch arteries.** There is a superficial palmar arch artery and a deep palmar arch artery. They join, or anastomose, and send out the small **digital arteries** that supply blood to the fingers (figure 34.6).

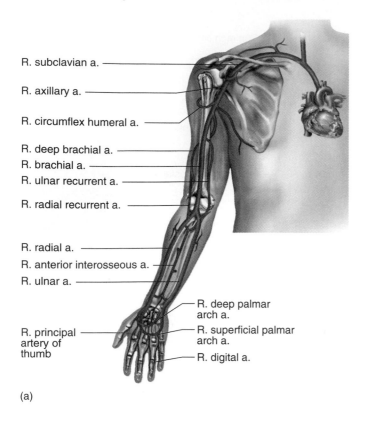

R. subclavian a.

R. axillary a.

R. circumflex humeral a.

R. deep brachial a.

R. brachial a.

R. ulnar recurrent a.

R. radial recurrent a.

R. radial a.

R. anterior interosseous a.

R. ulnar a.

R. deep palmar arch a.

R. principal artery of thumb

R. superficial palmar arch a.

R. digital a.

(a)

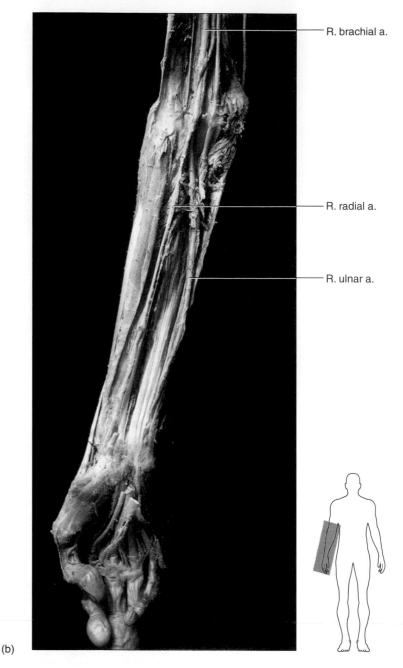

R. brachial a.

R. radial a.

R. ulnar a.

(b)

Figure 34.6   Arteries of the upper extremity.

Arteries of the Head and Neck

The **vertebral arteries** receive blood from the subclavian arteries and take it to the brain by traveling through the transverse foramina of the cervical vertebrae and then into the foramen magnum of the skull. The other main arteries that take blood to the brain are the **common carotid arteries,** which occur on the anterior sides of the neck. Each common carotid artery branches just below the angle of the mandible to form the **external carotid artery,** which takes blood to the face, and the **internal carotid artery,** which passes through the carotid canal and takes blood to the brain. The external carotid artery on each side has several branches, including the **facial artery,** the **temporal artery,** the **maxillary artery,** and the **occipital artery** (figure 34.7).

The brain receives blood from four major arteries. These are the left and right **vertebral arteries** and the left and right **internal carotid arteries.** The vertebral arteries pass through the foramen magnum and unite to form the **basilar artery** at the base of the brain. These were discussed in Laboratory Exercise 21 and illustrated in figure 21.2.

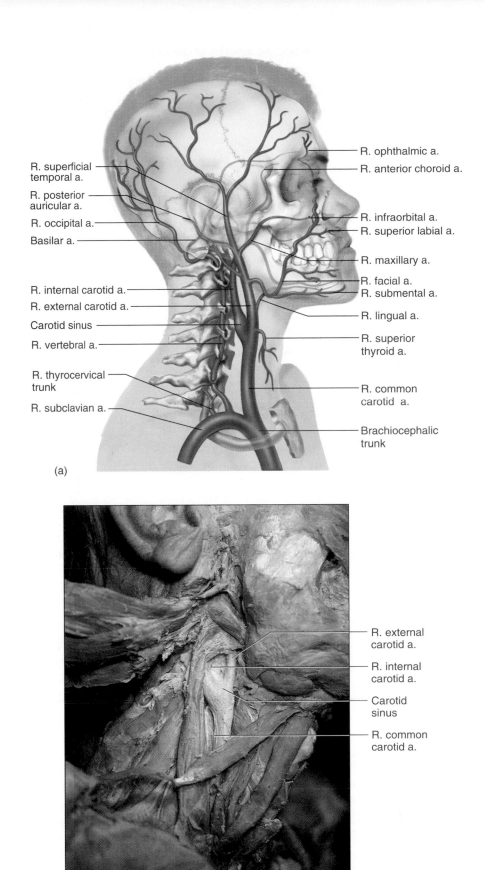

R. ophthalmic a.

R. superficial temporal a.

R. anterior choroid a.

R. posterior auricular a.

R. occipital a.

R. infraorbital a.

R. superior labial a.

Basilar a.

R. maxillary a.

R. internal carotid a.

R. facial a.

R. external carotid a.

R. submental a.

Carotid sinus

R. lingual a.

R. vertebral a.

R. superior thyroid a.

R. thyrocervical trunk

R. common carotid a.

R. subclavian a.

Brachiocephalic trunk

(a)

R. external carotid a.

R. internal carotid a.

Carotid sinus

R. common carotid a.

(b)

Figure 34.7   Arteries of the head.

Cat Dissection

Prepare for the cat dissection by obtaining a dissection tray, scalpel, scissors or bone cutter, string, forceps, and plastic bag with a label. Remember to place all excess tissue in the appropriate waste container and not in a standard wastebasket or down the sink!

Be careful as you make an incision into the body cavity of the cat. Cut with the scalpel blade facing to one side or, even better, use a lifting motion and cut upwards through the tissue. If you use a downwards sawing motion you will damage the internal organs of the cat.

Open the thoracic and abdominal regions of the cat if you have not done so already. Make an incision into the body wall of the cat in the belly, slightly lateral to midline. Cut into the muscle and then grasp the tissue with a forceps and lift the body wall away from the viscera. Continue cutting anteriorly through the belly. Stop at the level of the diaphragm, and then continue cutting a little off center as you cut through the costal cartilages. Use a scalpel or scissors to cut through the thoracic region.

You can now make a transverse incision in the lower region of the belly so you have an inverted T-shaped incision (figure 34.8). Lift the tissue near the ribs and expose the diaphragm. Gently and carefully snip or cut the diaphragm away from the ventral region by cutting as close to the ribs as possible.

Next make an incision by cutting transversely across the upper part of the chest. Be careful not to cut the blood vessels of the neck that were exposed during the original skinning process. Open the body cavity by pulling the two flaps laterally, exposing the thoracic and abdominal cavities. You may want to cut through the ribs at their most lateral point to facilitate the opening of the thoracic cavity.

Carefully expose the thoracic region of the cat and locate the two lungs and the heart. Note that the heart is enclosed in the pericardial sac. Look for the fatty material called the greater omentum that covers the large liver, stomach, and intestines in the abdominal region. As you dissect the arteries of the cat pay careful attention to the veins and nerves that travel along with the arteries. *Do not dissect other organs.* You will study these systems later (for example, respiratory, digestive, and urogenital systems).

As you open the abdominal cavity you will see the space that contains the internal organs, known as the coelom, or body cavity. The lining that occurs on the inside of the body wall is the parietal peritoneum, and the lining that wraps around the outside of the intestines is the visceral peritoneum.

Arteries of the Cat

The blood vessels in the cat have been injected with colored latex. If the cat is doubly injected the arteries are red and the veins are blue. If the cat has been triply injected the hepatic portal vein is typically injected with yellow latex. Be careful locating the arteries in the cat. You can tease away some of the connective tissue

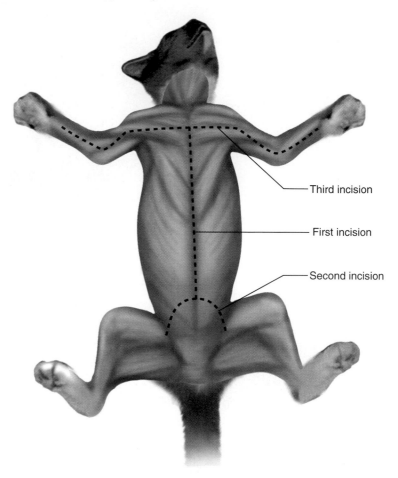

Third incision

First incision

Second incision

Figure 34.8   Incisions to open the thoracic and abdominal cavities.

with a dissection needle to see the vessel more clearly. Note the difference in the aortic arch arteries in the cat compared to that of the human. Look for the **coronary arteries,** which take blood from the ascending aorta to the heart tissue.

You can begin your study of the arteries of the cat by examining the ascending aorta. Note that the **ascending aorta** bends and forms the **aortic arch** and then moves posteriorly to form the **descending aorta.** The descending aorta is composed of the **thoracic aorta** above the diaphragm and the **abdominal aorta** below the diaphragm. The pattern of arteries that leave the aortic arch is somewhat different in cats than in humans. In the cat typically only two large vessels leave the aortic arch, the **large brachiocephalic artery** and the **left subclavian artery.** The brachiocephalic artery further divides into the **left common carotid artery,** the **right common carotid artery,** and the **right subclavian artery.** Examine figure 34.9 for the pattern in the cat. How is the pattern in the cat different from that in the human?

Several arteries arise from the **subclavian artery.** The **vertebral artery** takes blood to the head, and the **subscapular artery** takes blood to the pectoral girdle muscles, along with the

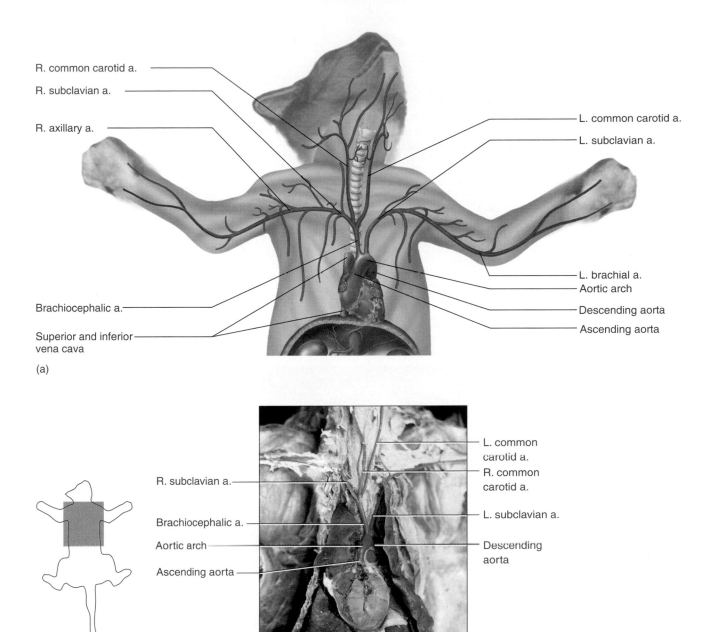

R. common carotid a.

R. subclavian a.

R. axillary a.

L. common carotid a.

L. subclavian a.

L. brachial a.

Aortic arch

Descending aorta

Ascending aorta

Brachiocephalic a.

Superior and inferior vena cava

(a)

R. subclavian a.

Brachiocephalic a.

Aortic arch

Ascending aorta

L. common carotid a.

R. common carotid a.

L. subclavian a.

Descending aorta

(b)

Figure 34.9 Aortic arch arteries of the cat.

ventral thoracic artery and **long thoracic artery.** The ventral thoracic and long thoracic arteries supply the latissimus dorsi and pectoralis muscles with blood. The left and right subclavian arteries become the left and right axillary arteries, which in turn take blood to the **brachial arteries.** The brachial artery takes blood to the **radial** and **ulnar arteries.** Additional arteries that branch from the subclavian artery are the **thyrocervical arteries,** which take blood to the region of the shoulder and neck (figure 34.10).

The **common carotid artery** takes blood along the ventral surface of the neck to feed the small **internal carotid artery,**

which supplies the brain. The common carotid artery also empties into the **external carotid artery,** which takes blood to the external portions of the head. The **occipital artery** receives blood from the common carotid artery and supplies the muscles of the neck.

The thoracic aorta not only takes blood to the abdominal aorta but also supplies the rib cage with blood from the **intercostal arteries. Esophageal arteries** also branch from the thoracic aorta, supplying the esophagus with blood. The thoracic aorta continues as the abdominal aorta as it passes through the diaphragm.

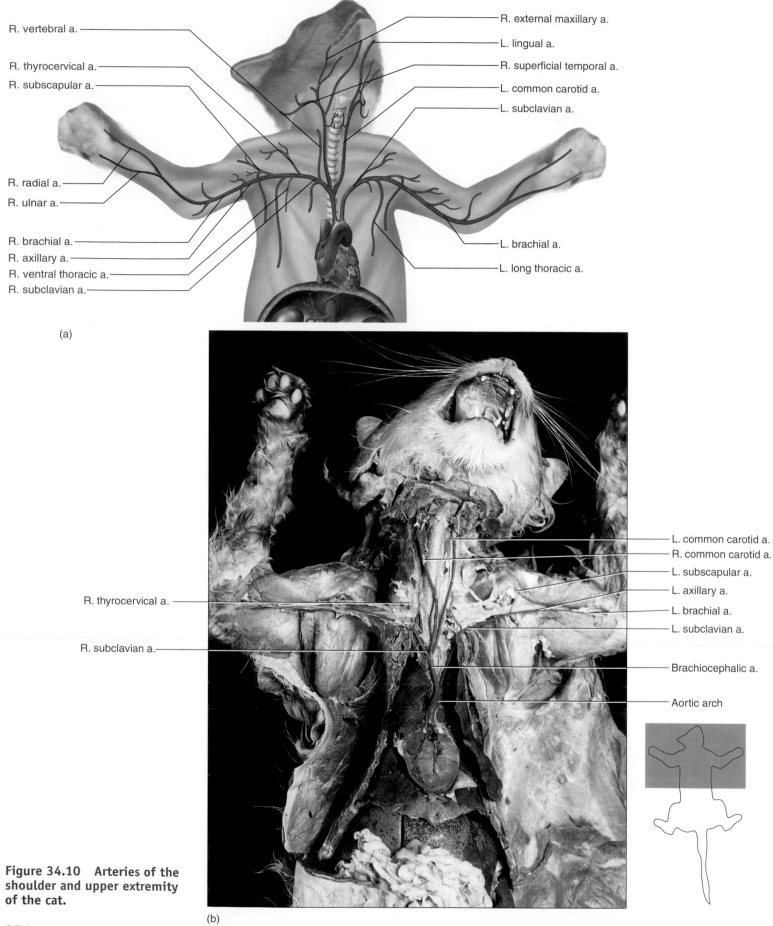

R. vertebral a.

R. thyrocervical a.

R. subscapular a.

R. radial a.

R. ulnar a.

R. brachial a.

R. axillary a.

R. ventral thoracic a.

R. subclavian a.

R. external maxillary a.

L. lingual a.

R. superficial temporal a.

L. common carotid a.

L. subclavian a.

L. brachial a.

L. long thoracic a.

(a)

R. thyrocervical a.

R. subclavian a.

L. common carotid a.

R. common carotid a.

L. subscapular a.

L. axillary a.

L. brachial a.

L. subclavian a.

Brachiocephalic a.

Aortic arch

Figure 34.10 Arteries of the shoulder and upper extremity of the cat.

(b)

Name _____

1. Label the following illustration with the major arteries of the body. Try to complete the illustration first and then review the
    material in this exercise to determine your accuracy.

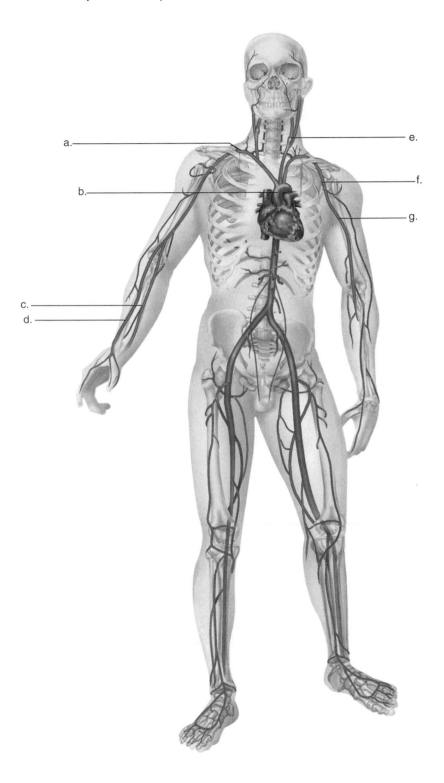

2. Working in pairs, have your lab partner select an artery for you to name. Quiz each other on the material learned in this exercise.

3. Blood from the left subclavian artery would flow into what vessels as it moves towards the left arm?

4. Blood in the radial artery comes from what blood vessel?

5. An aneurysm is a weakened, expanded portion of an artery. Ruptured aneurysms can lead to rapid blood loss. Describe the significance of an aortic aneurysm versus a digital artery aneurysm.

6. The pulmonary arteries carry deoxygenated blood from the heart to the lungs. Umbilical arteries carry a mixture of oxygenated and deoxygenated blood. Why are these blood vessels called arteries?

7. What is the name of the outermost layer of a blood vessel?

8. What kind of blood vessels have valves?

9. Blood from the common carotid artery next travels to what two vessels?

10. Blood from the right brachial artery travels to what two vessels?

11. Where does blood in the right subclavian artery come from?

12. The internal carotid artery takes blood to what organ?

13. From where does the descending aorta get blood?

14. What is the general name of a large vessel that takes blood away from the heart?

15. Blood in the left common carotid artery receives blood from what vessel?

16. Name three blood vessels that exit from the aortic arch.

Arteries of the Lower Body

Introduction

The arteries of the lower body receive blood directly or indirectly from the descending aorta. As with the arteries of the upper extremities these vessels are frequently named for their location, such as the iliac artery, the femoral artery, and the mesenteric arteries, or for the organs they serve, such as the common hepatic artery and the splenic artery. In this exercise, you continue your study of arteries. It is very important that you are careful when dissecting the arteries of the abdominal region of the cat. Do as little damage as possible to the other structures as you examine the arteries in this region.

Objectives

At the end of this exercise you should be able to

1. describe the sequence of major arteries that originate from the abdominal aorta or its derivatives;
2. list the arteries of the lower extremities;
3. describe the major organs that receive blood from the lower arteries;
4. name the vessels that take blood to a particular artery and the vessels that receive blood from a particular artery.

Materials

Microscope
Prepared slide of arteriosclerosis
Models of the blood vessels of the body
Charts and illustrations of arterial system
Cadaver (if available)
Dissection equipment
Cats

Procedure

Examine the models and charts in the lab and the accompanying illustrations to locate the major arteries in the abdomen and lower body. Read the following descriptions of the vessels and locate the individual arteries.

Abdominal Arteries

The **abdominal aorta** is the portion of the descending aorta inferior to the diaphragm. The first major branch of the abdominal aorta is the **celiac artery** (also known as the **celiac**

trunk). The celiac trunk splits into three separate arteries: the **splenic artery,** which takes blood to the spleen, pancreas, and part of the stomach; the **left gastric artery,** which takes blood to the stomach and esophagus; and the **common hepatic artery,** a vessel that takes blood to the liver. Locate the celiac artery and the branches of the celiac in figure 35.1.

Just below the celiac artery is the **superior mesenteric artery.** This vessel takes blood from the abdominal aorta and continues through the mesentery until it reaches the small intestine and proximal portions of the large intestine, including the cecum, ascending colon, and part of the transverse colon. The major branches of the superior mesenteric artery are the **intestinal arteries,** the **ileocolic arteries,** and the right and middle **colic arteries.** The next vessels to branch from the aorta are the paired **suprarenal arteries,** which take blood to the adrenal glands. Below the suprarenal arteries are the left and right **renal arteries.** These arteries take blood to the kidneys. Locate these vessels in the lab and in figures 35.2 and 35.3.

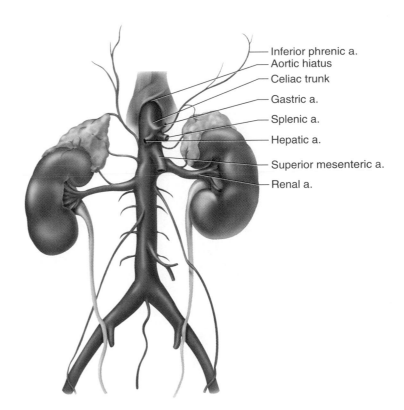

Inferior phrenic a.
Aortic hiatus
Celiac trunk
Gastric a.
Splenic a.
Hepatic a.
Superior mesenteric a.
Renal a.

Figure 35.1   Arteries of the anterior abdominal region.

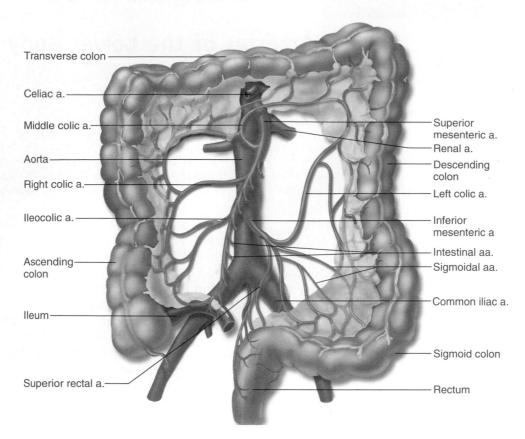

Figure 35.2   Middle abdominal arteries.

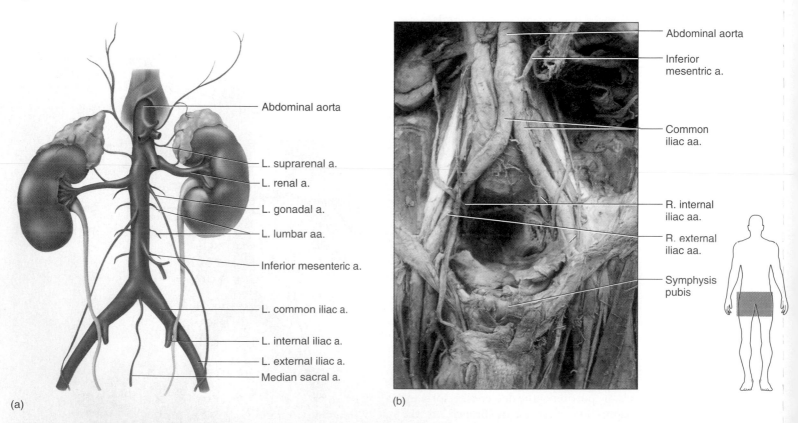

(a)

(b)

Figure 35.3   Lower abdominal and iliac arteries. (a) Diagram; (b) photograph.

The **gonadal arteries** branch inferior to the renal arteries and descend to either the testes or ovaries. The **inferior mesenteric artery** is the next vessel to branch from the aorta, and it takes blood to the lower portion of the large intestine, including part of the transverse colon, the descending colon, and the rectum. These can be located in figure 35.3.

The abdominal aorta terminates by dividing into the two **common iliac arteries.** The common iliac arteries take blood to the **external** and **internal iliac arteries.** The internal iliac artery takes blood to the pelvic region, including branches that feed the rectum, pelvic floor, external genitalia, groin muscles, hip muscles, and the uterus, ovary, and vagina in females. These arteries are illustrated in figures 35.3 and 35.4.

Each external iliac artery branches from a common iliac artery and terminates as it exits the body wall near the inguinal canal. Once the external iliac artery leaves the body cavity, it continues into the thigh as the **femoral artery.** The femoral artery has a superficial branch that feeds the thigh and external genitalia and a deeper branch, called the **deep femoral artery,** that takes blood to the muscles of the thigh, the knee, and the femur. The femoral artery divides into the **posterior tibial artery** and **anterior tibial artery,** which take blood to the knee and leg, and the **fibular (or peroneal) artery,** which supplies the muscles on the lateral side of the leg. The foot is supplied with blood from several arteries, including the **plantar arteries,** the **dorsal pedal artery,** and the **digital arteries.** Locate these arteries in figure 35.4

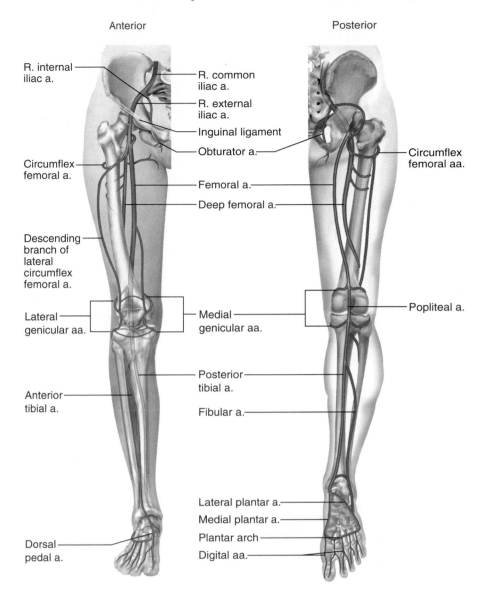

Anterior Posterior

R. internal iliac a. — R. common iliac a.

R. external iliac a.

Inguinal ligament

Obturator a.

Circumflex femoral a.

Circumflex femoral aa.

Femoral a.

Deep femoral a.

Descending branch of lateral circumflex femoral a.

Lateral genicular aa.

Medial genicular aa.

Popliteal a.

Posterior tibial a.

Anterior tibial a.

Fibular a.

Lateral plantar a.

Medial plantar a.

Plantar arch

Digital aa.

Dorsal pedal a.

Figure 35.4 Arteries of the lower extremity.

Arteriosclerosis

Arteriosclerosis is a condition known commonly as hardening of the arteries. Examine a microscope slide or illustration of arteriosclerosis and note the development of cholesterol plaque under the endothelial layer. Draw what you see in the space provided.

Drawing of an artery with arteriosclerosis:

Cat Dissection

Prepare for the cat dissection by obtaining a dissection tray, scalpel, scissors or bone cutter, string, forceps, and plastic bag with label. Remember to place all excess tissue in the appropriate waste container and not in a standard classroom wastebasket or down the sink!

Follow the dissection procedures discussed in the previous exercise. Make sure you do not cut the blood vessels away from the organs that will be studied later.

Abdominal Arteries

The determination of the abdominal arteries is perhaps best done by looking for both the exit of these arteries from the aorta and their entrance into the organs supplied by these vessels. The

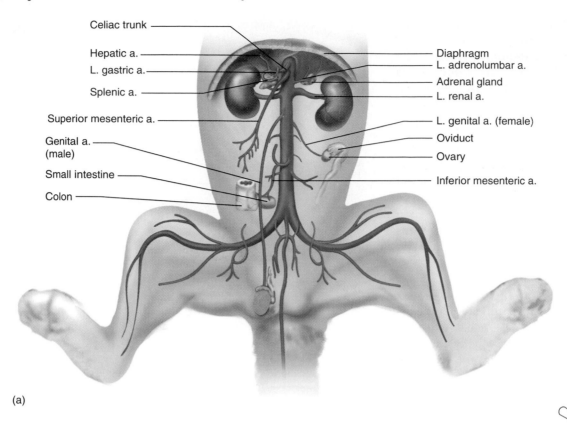

(a)

abdominal cavity of the cat should already be opened. If not, make an incision in the lower abdominal wall and carefully cut towards the sternum. Do not cut deeply into the body cavity or you will puncture or cut into the viscera. Locate the major organs of the abdominal region, such as the stomach, spleen, liver, and small and large intestines. Examine the specimen from the *left side*, gently lifting the stomach, spleen, and intestines towards the ventral right side as you look for the abdominal arteries. The **descending aorta** is located on the left side of the body, while the **posterior vena cava (inferior vena cava** in humans) is located on the right side. Locate the short **celiac trunk (celiac artery),** which quickly divides into three major vessels, the **splenic artery,** the **left gastric artery,** and the **hepatic artery.** The splenic artery takes blood to the spleen, the hepatic artery reaches the liver, and the left gastric artery supplies the stomach. Compare your dissection to figure 35.5.

The **anterior (superior) mesenteric artery** is the next major vessel to take blood from the abdominal aorta.

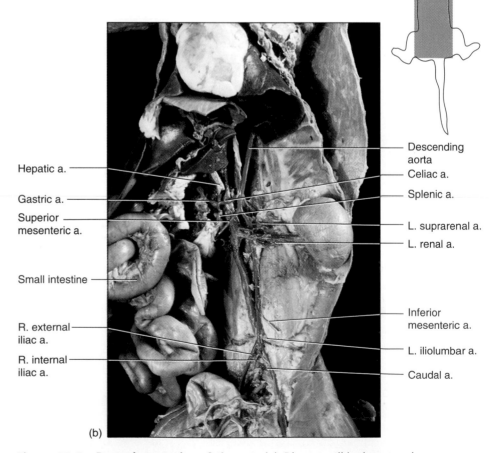

(b)

Figure 35.5   Posterior arteries of the cat. (a) Diagram; (b) photograph.

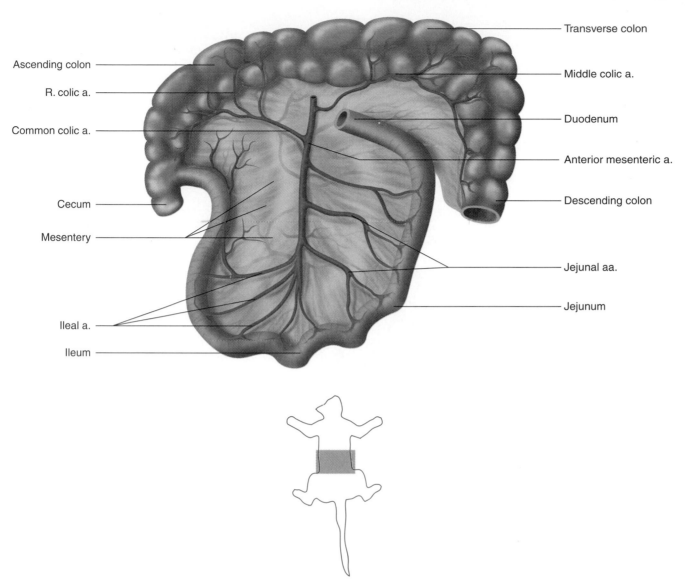

Transverse colon

Ascending colon

R. colic a.

Common colic a.

Middle colic a.

Duodenum

Anterior mesenteric a.

Descending colon

Cecum

Mesentery

Jejunal aa.

Jejunum

Ileal a.

Ileum

Figure 35.6   Branches of the superior mesenteric artery of the cat.

The anterior mesenteric artery travels to the small intestine and the proximal portion of the large intestine. It branches into the **jejunal arteries,** the **ileal arteries,** and the **colic arteries.** Compare the anterior mesenteric artery illustrated in figures 35.5 and 35.6 with your specimen.

The paired **adrenolumbar arteries** branch on each side of the abdominal aorta and take blood to the adrenal glands, parts of the body wall, and the diaphragm. Just caudal to the adrenolumbar arteries are the paired **renal arteries,** which take blood to the kidneys. The **genital arteries** (gonadal arteries in humans) are the next set of paired arteries, and these take blood to the testes in male cats, in which case they are called the **internal spermatic arteries.** If the genital arteries occur in fe-

males, they are called the **ovarian arteries** because they take blood to the ovaries (figure 35.5).

A single **inferior mesenteric artery** takes blood from the abdominal aorta to the terminal portion of the large intestine. The inferior mesenteric artery can be found caudal to the genital arteries. Locate the lower abdominal arteries in your specimen and compare them to figure 35.5.

The pelvic arteries are somewhat different in cats than in humans. The abdominal aorta splits caudally into the **external iliac arteries,** and a short section of the aorta continues on and then divides to form the two **internal iliac arteries** and the **caudal artery.** These are illustrated in figures 35.7 and 35.8. There is no common iliac artery in cats as there is in humans.

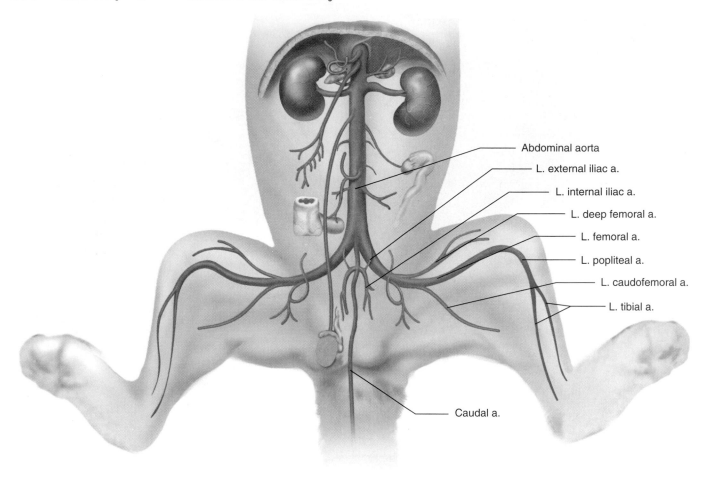

Figure 35.7   Arteries of the pelvis and lower extremity of the cat.

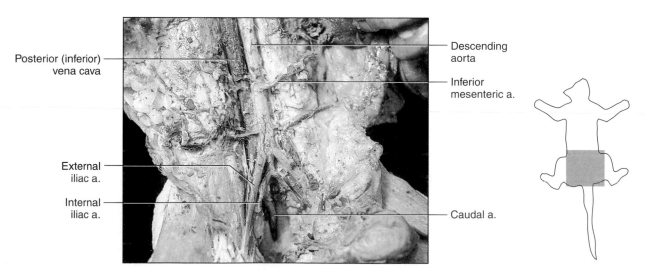

Figure 35.8   Pelvic arteries of the cat.

In cats the caudal artery takes blood to the tail. The internal iliac artery takes blood to the urinary bladder, rectum, external genital organs, uterus, and some thigh muscles. As the external iliac artery exits from the pelvic region and enters the thigh, it becomes the **femoral artery,** which supplies blood to the thigh, leg, and foot. The femoral artery becomes the **popliteal artery** behind the knee, which further divides to form the **tibial arteries** (figures 35.7 and 35.9).

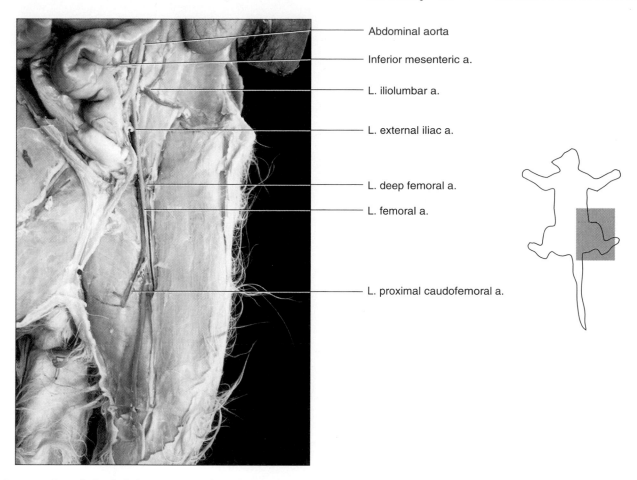

Abdominal aorta

Inferior mesenteric a.

L. iliolumbar a.

L. external iliac a.

L. deep femoral a.

L. femoral a.

L. proximal caudofemoral a.

Figure 35.9   Arteries of the left lower extremity of the cat.

Name _____

1. Blood from the popliteal artery comes directly from what artery?

2. Blood from the celiac artery flows into three different blood vessels. What are these vessels?

3. Blood from the superior mesenteric artery feeds what major abdominal organs?

4. In what part of the arterial wall does cholesterol plaque develop?

5. How do the lower pelvic arteries in humans differ from those in cats?

6. Label the following illustration.

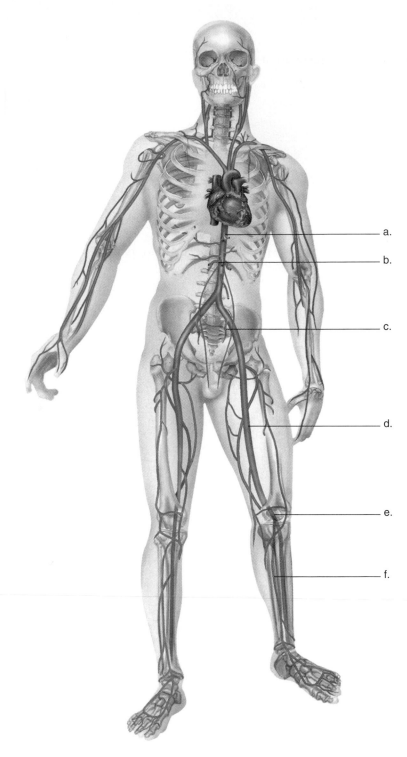

a.

b.

c.

d.

e.

f.

Inferior arteries of the human.

7. Name the section of descending aorta superior to the diaphragm.

8. What artery takes blood directly to the femoral artery?

9. In humans, where does blood in the external iliac artery come from?

10. Blood in the inferior mesenteric artery travels to what organs?

11. What is arteriosclerosis?

12. Name the vessel that takes blood to the adrenal glands in the cat.

13. What vessels take blood to the kidneys?

14. The ovaries or testes receive blood from which arteries?

LABORATORY EXERCISE 36

Veins and Fetal Circulation

Introduction

Veins are defined as blood vessels that carry blood *toward* the heart. They can be superficial (underneath the skin) or deep. The deep veins of the body frequently travel alongside the major arteries and take on the arterial names, while superficial veins have names specific to each vessel. Examples of veins that carry the arterial names are the femoral vein, brachial vein, and subclavian vein. Veins are thinner walled than arteries and contain valves for a one-way flow of blood to the heart. A description of the differences in the walls of arteries and veins was covered in Laboratory Exercise 34. In this exercise you begin the study of veins by examining charts or models of the entire venous system and then studying the veins of specific regions of the body.

Objectives

At the end of this exercise you should be able to

1. describe the sequence of major veins that drain the upper extremities;
2. list the veins of the lower extremities;
3. trace the blood flow from the brain to the heart;
4. distinguish a portal system from normal venous return flow;
5. describe the major digestive organs that supply blood to the hepatic portal system.

Materials

Models of the blood vessels of the body
Charts and illustrations of venous system
Cadaver (if available)
Dissection equipment
Cats

Procedure

Examine the models, charts, and illustrations in the lab and locate the major veins of the body. Read the following descriptions of the veins and find them as they are represented in lab. As you locate a specific vein, name the vessel that takes blood to the vein and those that receive blood from the vein. Veins in this exercise are studied in the direction of their flow from the

cells of the body to the heart. In this way they resemble tributaries of rivers as smaller veins flow into larger veins.

An overview of the major veins of the body is shown in figure 36.1. Compare that figure as well as the material in lab to the following list of some of the major veins of the body.

_____ Radial vein
_____ Ulnar vein
_____ Brachial vein
_____ Axillary vein
_____ Subclavian vein
_____ Brachiocephalic vein
_____ Superior vena cava
_____ Inferior vena cava
_____ Internal jugular vein
_____ Femoral vein
_____ Great saphenous vein
_____ Common iliac vein
_____ External iliac vein
_____ Internal iliac vein

Veins of the Upper Extremities

Examine the models and charts in the lab and locate the veins of the upper extremities. The fingers are drained by the small **digital veins,** which lead to the **palmar arch veins.** The major superficial veins of each upper extremity are the **basilic vein,** which is on the anterior, medial side of the forearm and arm, and the **cephalic vein,** which is on the anterior, lateral side of the forearm and arm. The two vessels have many anastomosing branches (cross-connections) between them. One of the significant anastomosing veins is the **median cubital vein,** which crosses the anterior cubital fossa and is a common site for the withdrawal of blood. Locate these superficial veins in figure 36.2.

The deep veins of the forearm are the **radial vein** and the **ulnar vein,** each of which can be found traveling near the artery of the same name. The deep vein of the arm is the **brachial vein,** which is next to the brachial artery in the proximal portion of the arm. The brachial vein is formed by the union of the radial and ulnar veins and merges superiorly with the **basilic vein** to form the short **axillary vein.** The axillary vein connects with the **cephalic vein** to form the **subclavian vein.** The drainage of the upper extremities is carried to the heart by the left and right

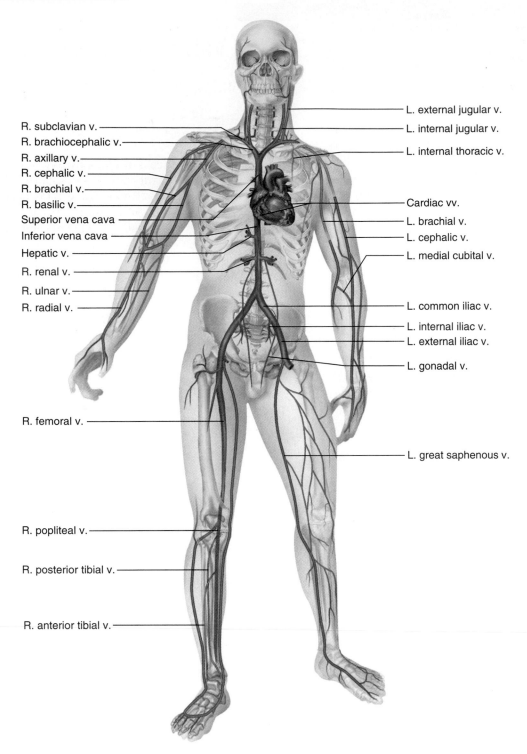

R. subclavian v.
R. brachiocephalic v.
R. axillary v.
R. cephalic v.
R. brachial v.
R. basilic v.
Superior vena cava
Inferior vena cava
Hepatic v.
R. renal v.
R. ulnar v.
R. radial v.

L. external jugular v.
L. internal jugular v.
L. internal thoracic v.

Cardiac vv.
L. brachial v.
L. cephalic v.
L. medial cubital v.

L. common iliac v.
L. internal iliac v.
L. external iliac v.

L. gonadal v.

R. femoral v.

L. great saphenous v.

R. popliteal v.

R. posterior tibial v.

R. anterior tibial v.

Figure 36.1   Major systemic veins.

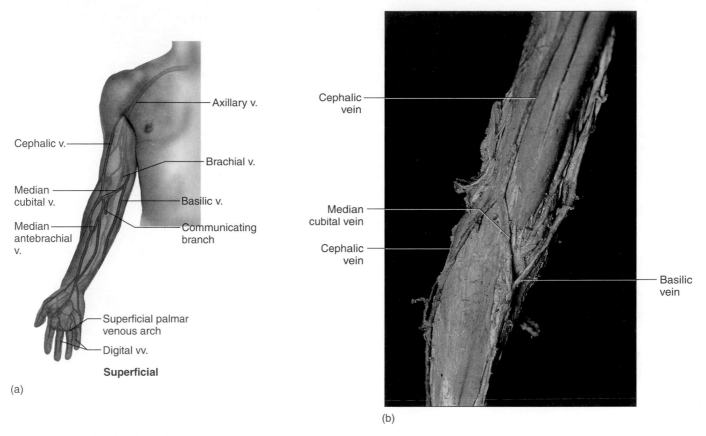

Figure 36.2   Superficial veins of the forearm.

subclavian veins, which take the blood via the **brachiocephalic veins** to the **superior vena cava** and finally to the **right atrium** of the heart. Locate the deep veins of the upper extremities in figure 36.3.

Veins of the Head and Neck

Examine the models and charts in the lab and locate the various veins that drain blood from the head. Blood from the head returns to the heart by a number of routes. Each **internal jugular vein** receives blood from the brain and passes through a jugular foramen of the skull. It then passes along the lateral aspect of the neck as it moves towards the **brachiocephalic vein.** The brachiocephalic vein is formed by the union of the **internal jugular vein** and the **subclavian vein.** The superficial regions of the head (musculature and skin of the scalp and face) are drained by the **external jugular vein.** The external jugular veins join up with the subclavian veins prior to reaching the brachiocephalic veins. The left and right brachiocephalic veins take blood to the superior vena cava. Another vessel that takes blood from the brain is the **vertebral vein,** which, like the vertebral arteries, travels through the transverse foramina of the cervical vertebrae. The vertebral veins take blood to the **subclavian veins,** which, in turn, flow to the **brachiocephalic veins.** Locate these major vessels of the head and neck in figure 36.4.

Veins of the Lower Extremity

The blood vessels that drain the lower extremities operate under relatively low pressure and must bring blood back to the heart against gravity. The longest vessel in the human body is the **great saphenous vein,** which can be found just underneath the skin on the medial aspect of the lower extremity beginning near the medial malleolus and traversing the lower extremity to the proximal thigh. This vessel frequently is imbedded in adipose tissue below the skin in humans, yet it is still considered a superficial vein. Two other vessels that occur in the thigh are the **femoral vein** and the **deep femoral vein.** The femoral vein travels alongside the femoral artery. As the great saphenous vein reaches the inguinal region it joins with the femoral vein, which takes blood from the thigh region and passes under the inguinal ligament. Once the **femoral vein** crosses under the ligament it becomes the **external iliac vein.** Examine these vessels in the lab and compare them to figure 36.5.

Veins of the Abdomen and Pelvis

The **external iliac vein** joins with the **internal iliac vein,** which drains the region of the pelvis, and they form the **common iliac vein.** The common iliac veins unite and form the **inferior vena cava,** which travels superiorly along the right side

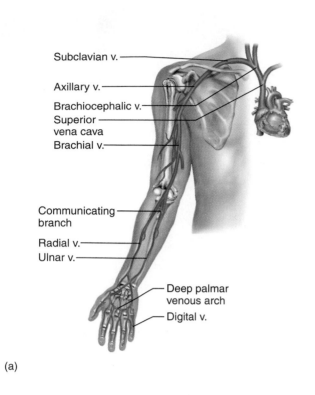

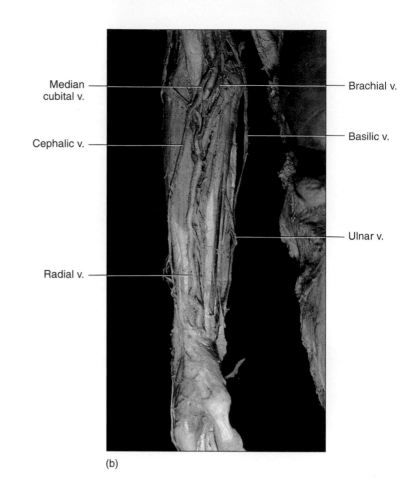

(a)

(b)

Figure 36.3   Deep veins of the upper extremity.

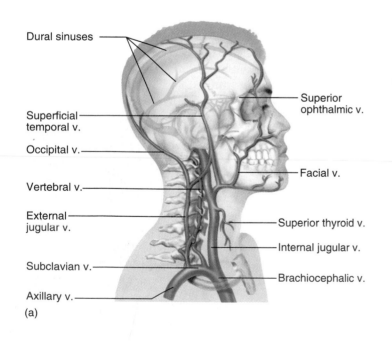

(a)

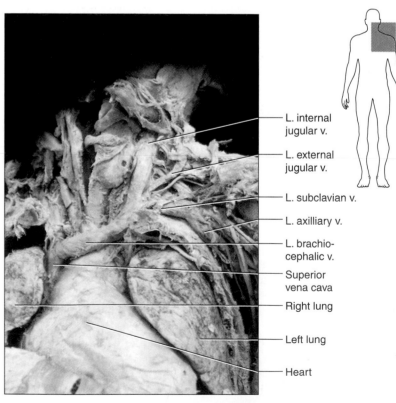

(b)

Figure 36.4   Veins of the head and neck. (a) Diagram of right side; (b) photograph of left veins.

404

Anterior   Posterior

Superficial
epigastric v.

Superficial
circumflex iliac v.

Femoral v.

Great
saphenous v.

Popliteal v.

Anterior tibial v.

Small
saphenous v.   Posterior tibial v.

Great
saphenous v.   Fibular
                                   (peroneal) v.

                                   Small
                                   saphenous v.

Anterior
tibial v.

Dorsal pedal v.

                                   Plantar arch
Dorsal
venous arch

Metatarsal vv.   Digital vv.

(a)

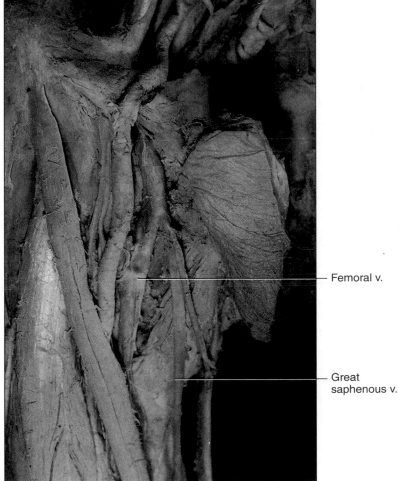

Femoral v.

Great
saphenous v.

(b)

Figure 36.5   Veins of the lower extremity. (a) Diagram of entire lower extremity; (b) photograph of upper right thigh.

of the vertebrae, taking blood to the right atrium of the heart. Locate these veins and compare them to figure 36.6.

Some abdominal veins take blood directly to the inferior vena cava and some pass through the liver before reaching the inferior vena cava. Those that take blood directly to the inferior vena cava are the **renal veins,** the **suprarenal veins,** and the **lumbar veins.** The **right gonadal vein** takes blood directly to the inferior vena cava, but the **left gonadal vein** takes blood to the renal vein prior to flowing into the inferior vena cava. These vessels can be seen in figure 36.6.

The veins that flow into the liver before returning to the heart are known as being part of the **hepatic portal system.** These vessels drain the major organs of the digestive tract and other abdominal organs such as the spleen. Frequently these veins are named for the organs from which they receive blood (for example, the splenic vein takes blood from the spleen). The hepatic portal system is different from the normal venous blood flow. Generally blood from an organ begins the return flow to the heart in the capillary bed of that organ. The blood then travels through venules to veins and finally to the heart.

In the hepatic portal system, a series of vessels takes blood from the *capillary* beds of the abdominal organs through a series of veins and then to another *capillary* bed in the liver. The liver receives the blood from the digestive organs and processes it before sending it through the **hepatic vein** to the inferior vena cava and towards the heart. Examine figure 36.7 for the main vessels of the hepatic portal system and identify the following veins:

_____ Inferior mesenteric vein
_____ Superior mesenteric vein
_____ Splenic vein
_____ Gastroepiploic vein
_____ Hepatic portal vein

Thoracic veins

The major veins of the thoracic region consist of the **inter-costal veins,** which drain the intercostal muscles, and the **azygos vein** and **hemiazygos vein,** which drain blood from the thoracic region. Locate these vessels in the lab and compare them to figure 36.8.

Cat Dissection

Prepare for the cat dissection by obtaining a dissection tray, scalpel, scissors or bone cutter, string, forceps, and plastic bag with a label. Remember to place all excess tissue in the appropriate waste container and not in a standard classroom wastebasket or down the sink!

Begin your dissection by reviewing the arterial system previously studied in Laboratory Exercises 34 and 35. As stated in previous exercises, the blood vessels in the cat have been injected with colored latex. If the cat has been doubly in-jected the arteries are red and the veins are blue. If the cat has been triply injected the hepatic portal vein is also injected, typically with yellow latex.

Veins of the Upper Extremity

Begin your dissection by examining the veins of the upper extremity. The basilic vein is not found in cats, but you should be able to find the **ulnar vein** and note that it joins the **radial vein** to form the **brachial vein.** The brachial vein is next to the brachial artery. Locate the **median cubital vein** and the **cephalic vein.** The **axillary vein** and **subscapular vein** (from the shoulder) join to form the **subclavian vein,** which takes blood to the **brachiocephalic vein.** The cephalic vein continues into the transverse scapular vein that takes blood to the external jugular vein. Locate these veins in figure 36.9.

Veins of the Head and Neck

The veins in this region of the cat have a few variations from those in the human. In the cat the **external jugular vein** is larger than the **internal jugular vein.** The reverse is true in humans. What anatomical difference between the cat and human might explain the difference in the volume of blood carried by these two vessels?

The **transverse jugular vein,** which is present in cats and absent in humans, connects the two external jugular veins. The jugular veins, along with the **costocervical veins** and the **subclavian veins,** unite to form the **brachiocephalic veins** (figure 36.9). The vertebral veins take blood from the head of the cat and empty into the subclavian veins.

Thorax

Anterior to the diaphragm are the veins of the thorax. The **internal mammary vein** (internal thoracic vein) may be difficult to locate since it is frequently cut while opening up the thoracic cavity. The internal mammary vein receives blood from the ventral body wall. Another vein of the thoracic region is the **azygos vein.** This vessel is found on the right side of the body and takes blood from the esophageal, intercostal, and bronchial veins. Compare the veins in your cat to figure 36.9.

Veins of the Lower Extremity

Locate the superficial **great saphenous vein** in the cat on the medial side of the lower extremity. The great saphenous vein joins the femoral vein just above the knee. Note the long **tibial veins** of the leg that join and form the **popliteal veins.** Find the **femoral vein,** which is found with the femoral artery. The vessels of the distal part of the lower extremity take blood to the femoral vein, which turns into the **external iliac vein** as it passes into the body cavity at the level of the inguinal ligament. Find these veins in the cat and compare them to figure 36.10.

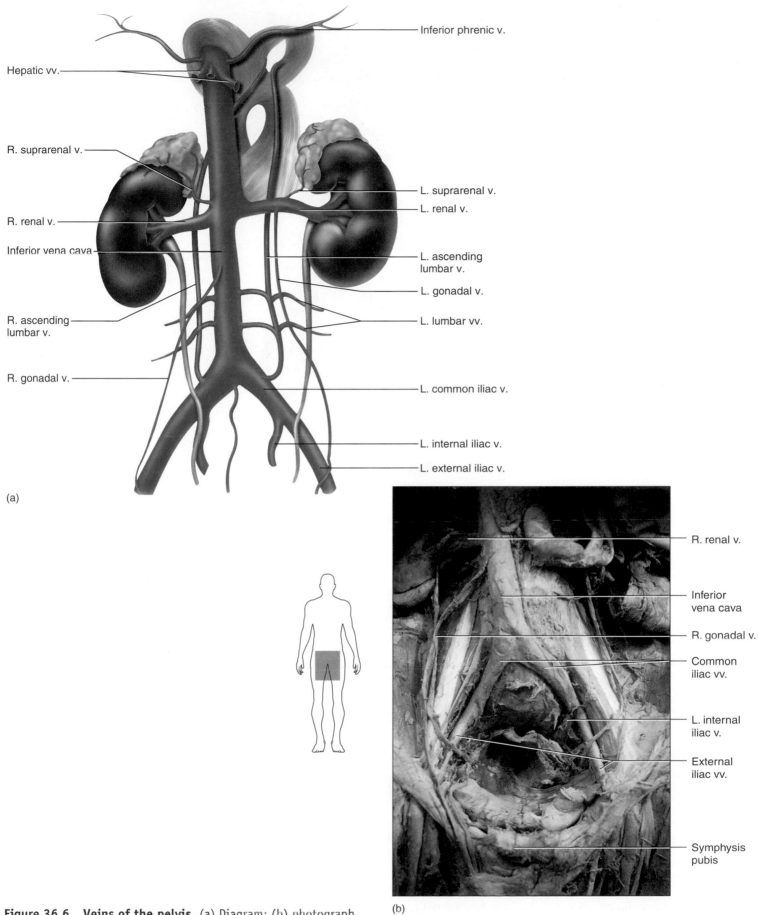

Inferior phrenic v.

Hepatic vv.

R. suprarenal v.

R. renal v.

Inferior vena cava

R. ascending lumbar v.

R. gonadal v.

L. suprarenal v.

L. renal v.

L. ascending lumbar v.

L. gonadal v.

L. lumbar vv.

L. common iliac v.

L. internal iliac v.

L. external iliac v.

(a)

R. renal v.

Inferior vena cava

R. gonadal v.

Common iliac vv.

L. internal iliac v.

External iliac vv.

Symphysis pubis

(b)

Figure 36.6   Veins of the pelvis. (a) Diagram; (b) photograph.

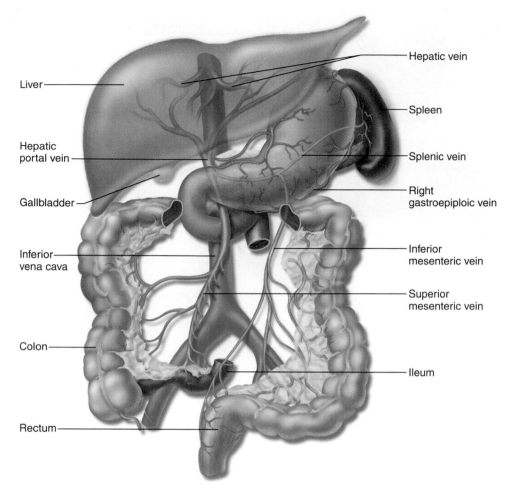

Figure 36.7 Hepatic portal system.

Liver

Hepatic vein

Spleen

Hepatic portal vein

Splenic vein

Gallbladder

Right gastroepiploic vein

Inferior vena cava

Inferior mesenteric vein

Superior mesenteric vein

Colon

Ileum

Rectum

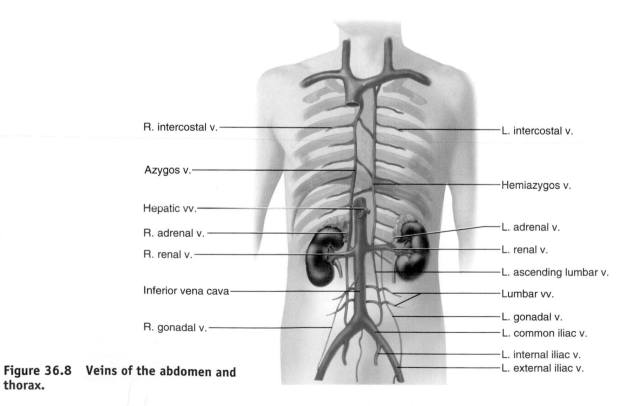

Figure 36.8 Veins of the abdomen and thorax.

R. intercostal v.

L. intercostal v.

Azygos v.

Hemiazygos v.

Hepatic vv.

R. adrenal v.

L. adrenal v.

R. renal v.

L. renal v.

L. ascending lumbar v.

Inferior vena cava

Lumbar vv.

L. gonadal v.

R. gonadal v.

L. common iliac v.

L. internal iliac v.

L. external iliac v.

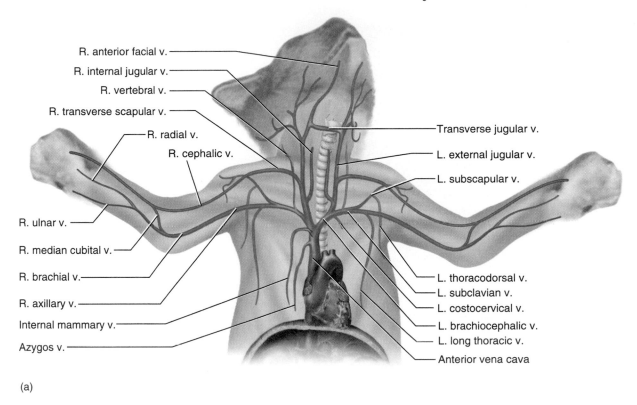

R. anterior facial v.
R. internal jugular v.
R. vertebral v.
R. transverse scapular v.
R. radial v.
R. cephalic v.
R. ulnar v.
R. median cubital v.
R. brachial v.
R. axillary v.
Internal mammary v.
Azygos v.

Transverse jugular v.
L. external jugular v.
L. subscapular v.
L. thoracodorsal v.
L. subclavian v.
L. costocervical v.
L. brachiocephalic v.
L. long thoracic v.
Anterior vena cava

(a)

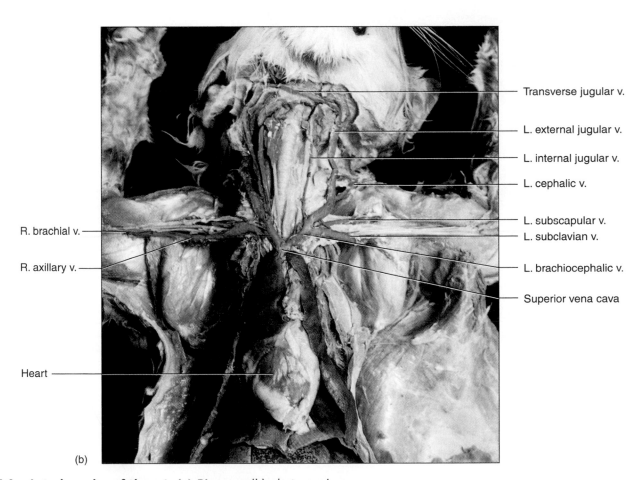

Transverse jugular v.
L. external jugular v.
L. internal jugular v.
L. cephalic v.
L. subscapular v.
L. subclavian v.
L. brachiocephalic v.
Superior vena cava

R. brachial v.
R. axillary v.
Heart

(b)

Figure 36.9   Anterior veins of the cat. (a) Diagram; (b) photograph.

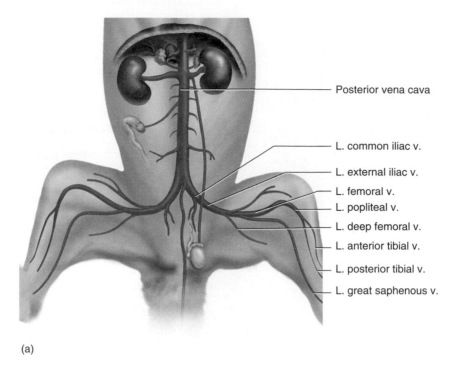

Posterior vena cava

L. common iliac v.

L. external iliac v.

L. femoral v.

L. popliteal v.

L. deep femoral v.

L. anterior tibial v.

L. posterior tibial v.

L. great saphenous v.

(a)

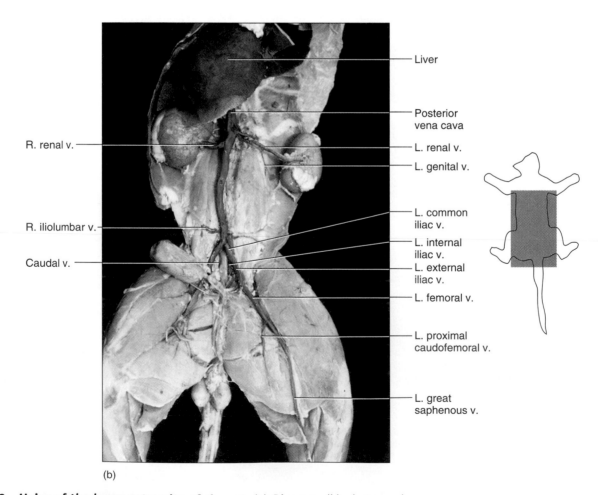

Liver

Posterior
vena cava

R. renal v.

L. renal v.

L. genital v.

R. iliolumbar v.

L. common
iliac v.

L. internal
iliac v.

Caudal v.

L. external
iliac v.

L. femoral v.

L. proximal
caudofemoral v.

L. great
saphenous v.

(b)

Figure 36.10   Veins of the lower extremity of the cat. (a) Diagram; (b) photograph.

Veins of the Pelvis

The **external iliac vein** and the **internal iliac vein** join to form the **common iliac vein.** The two common iliac veins lead to the **posterior vena cava** (inferior vena cava in humans) along with the **caudal vein,** which takes blood from the tail. Compare your cat with figures 36.10 and 36.11. Be careful as you dissect these structures in the cat. There are numerous ducts that cross the external iliac veins such as the ureter and, in males, the ductus deferens. Look for these structures and *do not cut them.* You will study them later.

Abdominal Veins

Make sure you identify the major organs in the digestive tract before you proceed with the study of the veins of this region. Pay particular attention to the stomach, liver, spleen, kidneys, adrenal glands, small intestine, and large intestine. The **posterior (inferior) vena cava** is a large vessel that travels up the right side of the body in the cat. You may have to gently move some of the digestive organs to the side to find the posterior vena cava. Examine where the common iliac veins join to form the posterior vena cava. The **caudal vein** joins the posterior vena cava at this point. It takes blood from the tail of the cat. This vein is absent in humans. The **genital veins** receive blood from the testes or ovaries. The **right genital vein** takes blood to the posterior vena cava and the **left genital vein** takes blood to the **left renal vein,** which subsequently leads to the inferior vena cava. The paired **renal veins** receive blood from each kidney and empty into the posterior vena cava. The **iliolumbar veins** take blood from the body wall and return it via the pos-terior vena cava. The body wall is also drained by the **adrenolumbar veins,** which also receive blood from the adrenal glands in addition to the body wall. Compare the veins in your cat to figure 36.11.

Hepatic Portal System

The **hepatic portal system** is a network of veins that takes blood from a number of abdominal organs and transfers it to the liver, an organ where significant metabolic processing occurs. As you examine the lower portion of the digestive tract (transverse colon and descending colon), locate the **posterior mesenteric vein** (inferior mesenteric vein in humans). The **anterior mesenteric vein** (superior mesenteric vein in humans) receives blood from the duodenum, pancreas, jejunum, and ileum and part of the large intestines as well. It also receives blood from the **posterior mesenteric vein.** The **gastrosplenic vein** joins the anterior mesenteric vein and brings blood to the **hepatic portal vein** along with other digestive veins. The gastrosplenic vein receives blood from the spleen and the stomach. The **hepatic portal vein** thus receives blood from the spleen and the digestive organs and transports that blood to the liver. After the liver receives the blood, it is transferred to the posterior vena cava by the short **hepatic vein.** This vein is usually difficult to dissect since it is at the dorsal aspect of the liver and joins the posterior vena cava at about the junction of the diaphragm. Examine the hepatic portal system of the cat while trying to maintain the integrity of the digestive organs. Compare your dissection to figure 36.12.

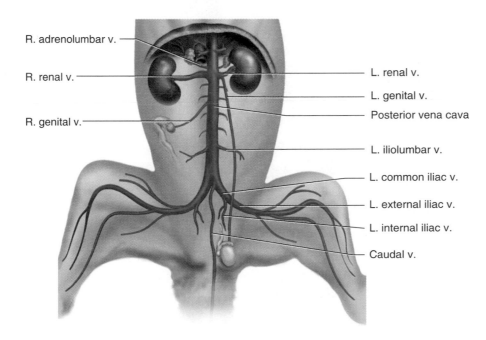

R. adrenolumbar v.
R. renal v.
R. genital v.
L. renal v.
L. genital v.
Posterior vena cava
L. iliolumbar v.
L. common iliac v.
L. external iliac v.
L. internal iliac v.
Caudal v.

Figure 36.11 Veins of the pelvis and abdomen of the cat.

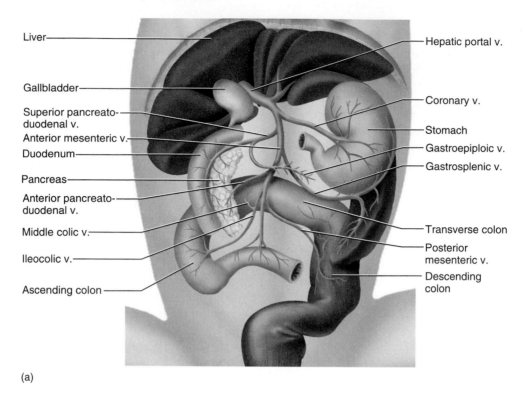

(a)

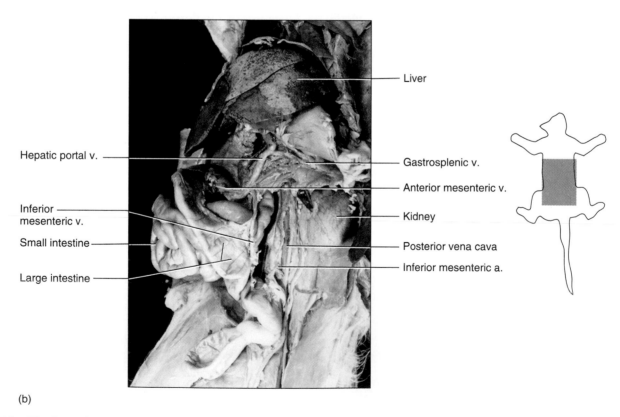

(b)

Figure 36.12 The hepatic portal system of the cat. (a) Diagram; (b) photograph.

Fetal Circulation

The pathway of fetal blood is somewhat different from that of the adult in that the lungs are nonfunctional in the fetus. Oxygen and nutrients move from the maternal side of the placenta to the fetal bloodstream, while carbon dioxide and metabolic wastes move from the fetal bloodstream to the placenta. From the **placenta** the blood flows through the **umbilical vein**, which is located in the **umbilical cord.** The blood from the umbilical vein travels through the **ductus venosus,** which serves as a shunt to the **inferior vena cava** of the fetus. The maternal blood, which is relatively high in oxygen and nutrients, mixes with the deoxygenated, nutrient-poor fetal blood and thus the fetus receives a mixture of blood. Examine figure 36.13 for an overview of fetal circulation.

The blood from the **inferior vena cava** travels to the **right atrium** of the heart. While in the right atrium the blood can travel either to the **right ventricle** (which pumps blood to the lungs) or through a hole in the right atrium called the **foramen ovale.** Since the lungs are nonfunctional in the fetus, the foramen ovale serves as a bypass route away from the lungs and to the chambers of the heart that will pump blood to the body. Blood in the right ventricle is pumped to the **pulmonary trunk,** where another shunt vessel, the **ductus arteriosus,** carries blood to the **aortic arch** bypassing the lungs. The lungs do receive some blood, but it is for the nourishment of the lung tissue as opposed to gas exchange. Locate the structures of the fetal heart in figure 36.13.

Blood from the heart exits the left ventricle and passes through the aorta to the **systemic arteries.** Blood travels down the **internal iliac arteries** to the lower extremities and some moves into the **umbilical arteries,** which carry blood to the **placenta** (see figure 36.13). At birth, the pressure changes in the newborn's heart cause the closing of a flap of tissue over the **foramen ovale,** leaving a thin spot in the interatrial septum known as the **fossa ovalis.** Lack of closure of this foramen can lead to a condition known as "blue baby."

Fetus

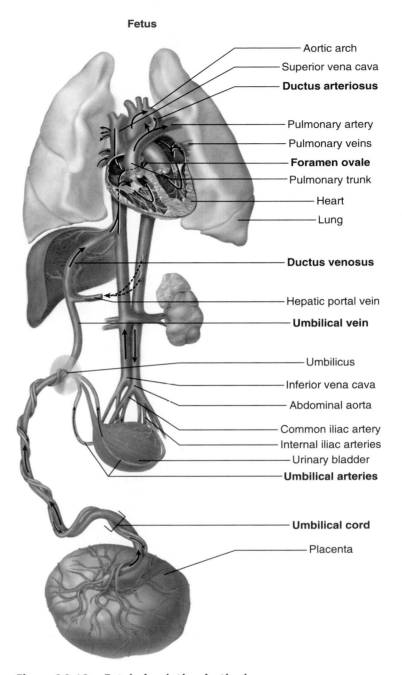

Figure 36.13 Fetal circulation in the human.

Name _____

1. Blood from the right axillary vein would next travel to what vessel?

2. What vessels take blood to the left femoral vein?

3. What area do the right and left external jugular veins drain?

4. What is the functional nature of a "portal system," and how does it differ from normal venous return flow?

5. What major vessels take blood to the hepatic portal vein?

6. Label the following illustration.

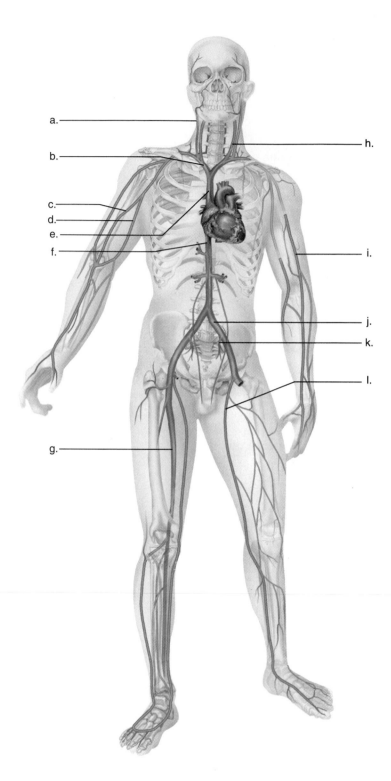

Veins of the human.

7. Which veins (superficial/deep) have names that do not correlate to arteries?

8. Where is the median cubital vein found?

9. What vessel receives blood from the ulnar vein?

10. Where do you find the cephalic vein?

11. Blood in the superior vena cava next flows to what area?

12. The internal jugular vein takes blood from what area?

13. What veins pass through the transverse vertebral foramina?

14. In what region of the body is the great saphenous vein found?

15. Where does blood flow after it leaves the femoral vein?

16. Blood in the small intestine travels to the hepatic portal vein by what vessel?

17. In the fetal heart, what is the name of the shunt between the pulmonary trunk and the aortic arch?

18. Name the opening between the atria in the fetal heart.

Functions of Vessels, Lymphatic System

Introduction

Studying the functions of blood vessels and the lymphatic system is vital to an understanding of the circulatory pattern. Fluid in the circulatory route may travel by several different pathways. Blood is pumped from the **heart** through **arteries** and distributed through the **arterioles** to the **capillaries,** where nutrients, water, and oxygen are exchanged with the cells of the body. The blood returns via **venules** to **veins** and back to the heart under relatively low pressure. The arteries are thicker than veins, which reflects the greater amount of pressure to which they are subjected. Arteries and veins function to transport blood cells and plasma. Blood cells and plasma proteins typically stay in the vessels, yet a significant amount of fluid from the **plasma** portion leaks from the capillaries and bathes the cells of the body. As this fluid flows between the cells of the body it is known as **interstitial fluid.** This fluid provides nutrients to the cells and also receives dissolved wastes from the cells along with cellular debris. Interstitial fluid is eventually picked up by **lymph capillaries,** which return the fluid, now known as lymph, to the lymphatic vessels, which subsequently flow to the venous system. In this way the fluid that has leaked from the capillaries is returned back to the cardiovascular system. This process is illustrated in figure 37.1.

In addition to the lymph serving to protect the body, the system is also instrumental in the absorption of lipids from the digestive system. Lipids in the small intestine are converted to **chylomicrons** (phospholipids and other molecules) in the cells that line the digestive tract. These chylomicrons are conducted into the lymphatic system, which then transports the material through the thoracic duct and into the subclavian vein where it enters the cardiovascular system.

As the fluid flows through the lymphatic vessels **lymph nodes** clean the lymph of cellular debris and foreign material (bacteria, viruses) that may have entered into the lymphatic system. In this exercise, we examine the nature of the lymphatic system.

Objectives

At the end of this exercise you should be able to

1. differentiate between conducting arteries and distributing arteries;
2. describe the function of valves that occur in veins;
3. identify the valves in lymph vessels;
4. locate the major structures found in a lymph node.

Materials

Microscopes
Microscope slides of lymphatics with valves
Torso models
Live grass frog (or similar-sized frog)
Dissection scopes
Paper towels
Small squeeze bottle of water

Procedure

Demonstration of Valves in Veins

One way to examine the nature of **valves** in veins is to let your arm hang at your side until the veins become engorged with blood. As you hold your arm in this position, stroke the superficial veins from distal to proximal with the index finger of your free hand, applying uniform and constant pressure (figure 37.2). Maintain pressure on the vein at its proximal end and see if the vein fills with blood.

Record your results, indicating whether or not the veins fill with blood.

Results: _____

Try the experiment again, but keep pressure on the distal part of the vein with your index finger as you push the blood towards the heart with your thumb (figure 37.3). Release the vein from the proximal area and see if blood fills the vein. Record your results.

Results: _____

Types of Arteries

Conducting (elastic) arteries are those close to the heart. Their appearance is made distinctive by the presence of significant amounts of elastic fibers in the tunica media. **Distributing (muscular) arteries** are found farther away from the heart and have more smooth muscle in the tunica media.

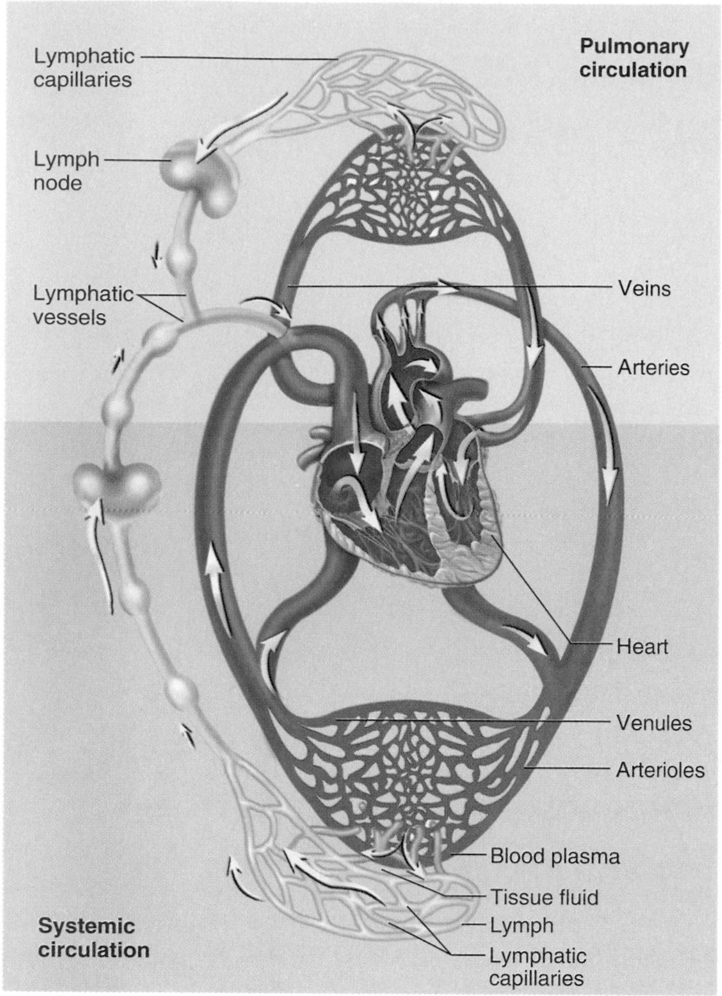
Lymphatic capillaries
Pulmonary circulation
Lymph node
Lymphatic vessels
Veins
Arteries
Heart
Venules
Arterioles
Blood plasma
Tissue fluid
Lymph
Lymphatic capillaries
Systemic circulation

Figure 37.1 Flow of interstitial fluid and lymph.

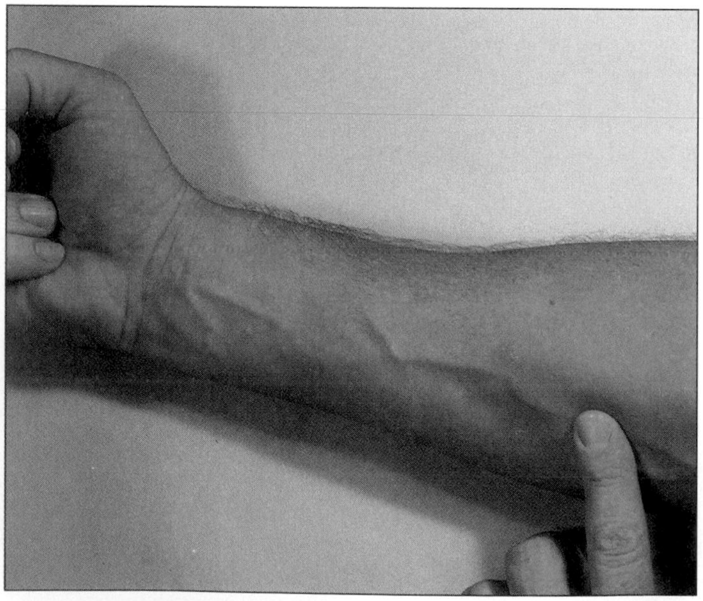

Figure 37.2 Testing for valves in veins (part 1).

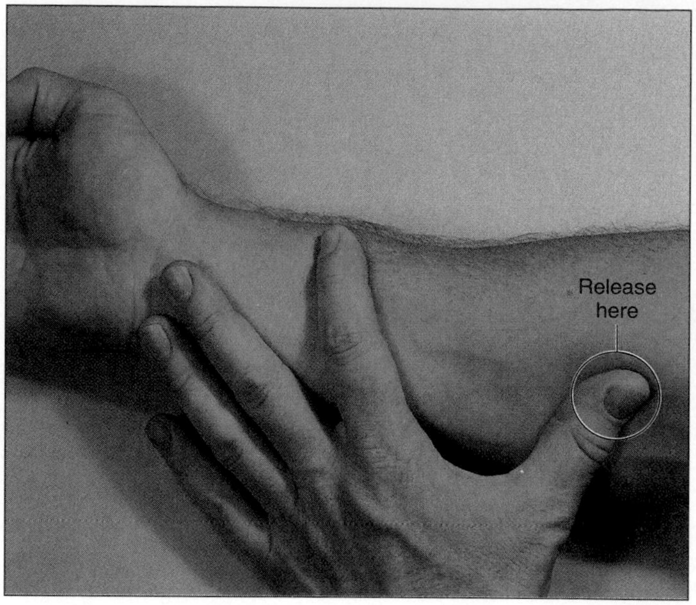
Release here

Figure 37.3 Testing valves in veins (part 2).

Capillary Action

Blood capillaries have a relatively simple structure in that they are composed of **endothelium.** The endothelium consists of a single layer of simple squamous epithelium that forms a tube slightly larger than the erythrocytes that pass through it. Some capillaries are more permeable than others. For example, the capillaries of the brain are tightly knit together and this helps form the blood/brain barrier. Other capillaries, such as those of the kidneys, have pores that allow for significant material to pass through. You can observe capillaries in living tissue by examining the blood flow through the webbed foot of a live frog. Obtain a *living* grass frog and wrap its body in a wet paper towel. Leave the head and the foot exposed. Do not let the frog dry out while you are examining it. Hold the foot of the frog under a dissecting microscope and fan out the webbing of the foot. Observe the flow of the blood through the capillaries, keeping the foot moist with water. Describe the flow of the blood through the capillaries of the foot in terms of speed and in terms of the size of the capillary relative to the diameter of an erythrocyte.

Your description:

Lymphatic System

The lymphatic system is difficult to study in preserved specimens because it collapses at death. The lymphatic system is composed of **lymph capillaries, lymphatic vessels, lymph nodes, lymph organs,** and **lymph tissue.** The lymph capillar-

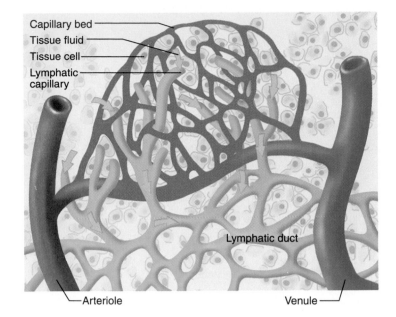

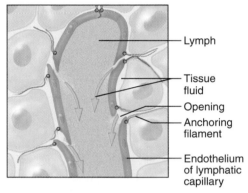

Figure 37.4   Lymph capillary.

ies have blunt ends and flaplike valves that conduct lymph away from interstitial areas. Examine figure 37.4 and describe how the valves might operate.

Valve operation: _____

Lymph Node

Look at a slide of a **lymph node.** The lymph node is enclosed by a sheath of tissue called the capsule. Locate the dark purple **lymph nodules** in the lymph node. The outer portion of the lymph node is called the **cortex** and the inner part is the **medulla.** The cortex is the outer region of the node that has numerous lymph nodules. In the medulla, medullary cords and sinuses function to cleanse the lymph of foreign particles and cellular debris. **Afferent lymphatic vessels** take lymph to the node, and **efferent lymphatic vessels** take lymph away from the node. Compare your slide to figure 37.5.

Lymphatic Vessels

Use figure 37.6 to examine the drainage pattern of lymph in the body. Note that the **lymphatic vessels** lead to regions of

the body where lymph nodes are found. Nodes are clustered in the groin (inguinal region), axilla, antecubital fossa, popliteal region, neck, thorax, and abdomen. The **thoracic duct** drains most of the body, taking lymph to the left subclavian vein, where the fluid is returned to the cardiovascular system. The **right lymphatic duct** drains the right side of the head and neck, the right thoracic region, and the right upper extremity. The right lymphatic duct returns lymph to the right subclavian vein. Identify the **cisterna chyli,** which is an enlarged portion of the thoracic duct in the abdominal region.

Examine a prepared slide of a lymphatic vessel and note the presence of the **valve.** Valves prevent the backflow of lymph away from the vascular system into which they drain. Compare your slide to figure 37.7.

Lymph Organs

Tonsils   Tonsils are lymph organs found in the oral cavity and nasopharynx. Look at illustrations in your text and models in the lab and locate the three pair of tonsils. These are the **palatine tonsils** found along the sides of the oral cavity near the oropharynx, the **lingual tonsils** at the posterior portion of the tongue, and the **pharyngeal tonsils** found in the nasopharynx. Adenoids are enlarged pharyngeal tonsils. The tonsils have crypts lined with **lymph nodules** that serve as a first line of defense against foreign matter that may be inhaled or swallowed. Compare these to figure 37.8.

Spleen   The **spleen** is an important lymph organ that filters blood, removing aging erythrocytes and foreign particles. The spleen is located on the left side of the body adjacent to the stomach. It is a highly vascular organ that functions to filter blood and produce lymphocytes. The spleen contains **red pulp,** which functions in filtering blood, and **white pulp,** which is involved in producing lymphocytes. Locate the spleen in models and charts in the lab and compare it to figure 37.9.

Thymus   The **thymus** overlays the vessels superior to the heart. It is an important site in determining immune competence. Lymphocytes travel from the bone marrow to the thymus, where they undergo maturation essential to immune responses. Once these cells mature they are known as **T cells.** Examine the charts and models in the lab and compare them to figure 37.10.

Cat Dissection

Your instructor may want you to examine the lymphatic system in the cat in this exercise if you have not already done so. If you do examine the cat in this exercise, look for small lumps of tissue in the axilla. These are the **lymph nodes** of the region. Note the large **spleen,** which is an approximately 6-inch elongated organ on the left side of the abdominal cavity, and the **thymus,** which is anterior to the heart. Compare what you find in the cat to figure 37.11.

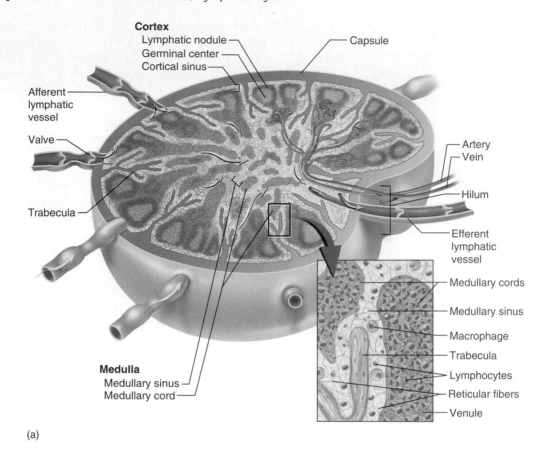

Cortex
Lymphatic nodule
Germinal center
Cortical sinus

Capsule

Afferent lymphatic vessel

Valve

Trabecula

Artery
Vein

Hilum

Efferent lymphatic vessel

Medullary cords

Medullary sinus

Macrophage

Trabecula

Lymphocytes

Reticular fibers

Venule

Medulla
Medullary sinus
Medullary cord

(a)

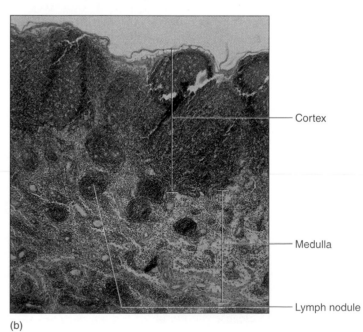

Cortex

Medulla

Lymph nodule

(b)

Figure 37.5   Section of a lymph node. (a) Diagram; (b) photomicrograph (40×). ✗

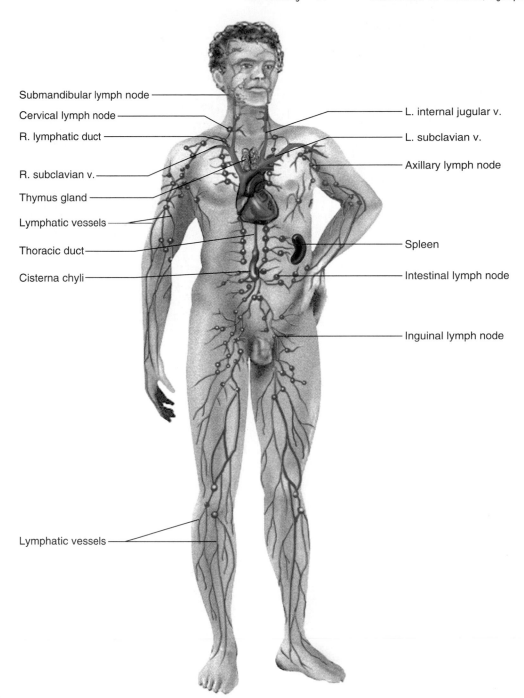

Submandibular lymph node
Cervical lymph node
R. lymphatic duct

R. subclavian v.

Thymus gland

Lymphatic vessels

Thoracic duct

Cisterna chyli

Lymphatic vessels

L. internal jugular v.

L. subclavian v.

Axillary lymph node

Spleen

Intestinal lymph node

Inguinal lymph node

Figure 37.6   Lymphatic vessels of the body.

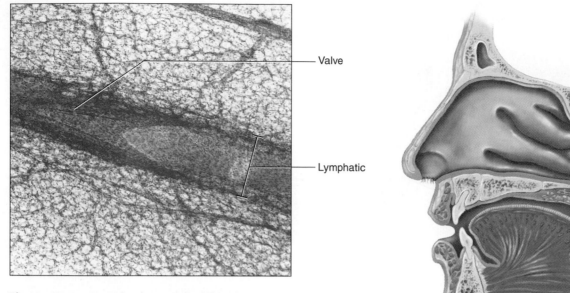

Figure 37.7   Lymphatic vessel with valve.

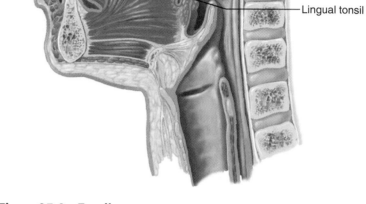

Figure 37.8   Tonsils.

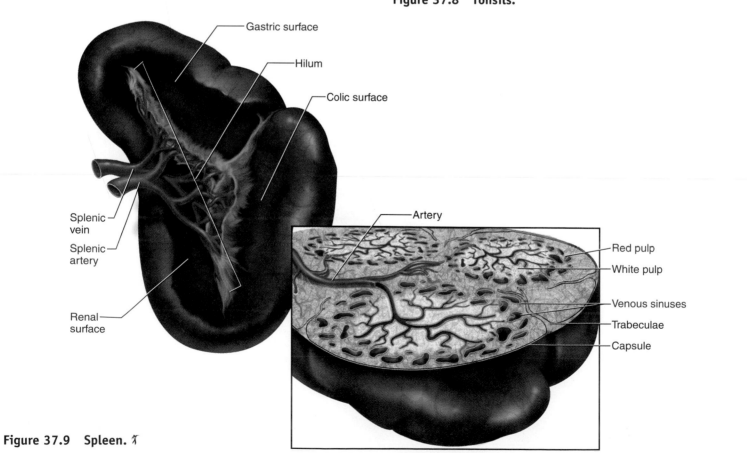

Figure 37.9   Spleen.

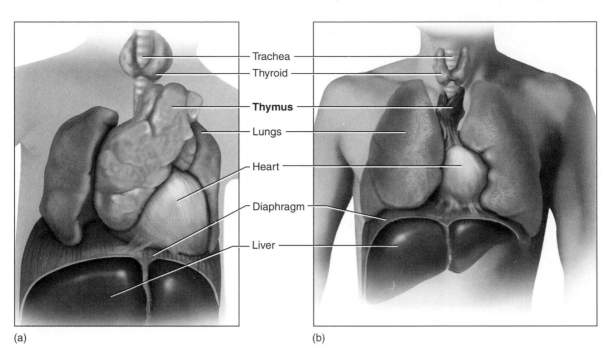

Figure 37.10   Thymus. (a) Young individual; (b) mature adult. ✗

Trachea
Thyroid
Thymus
Lungs
Heart
Diaphragm
Liver

(a)   (b)

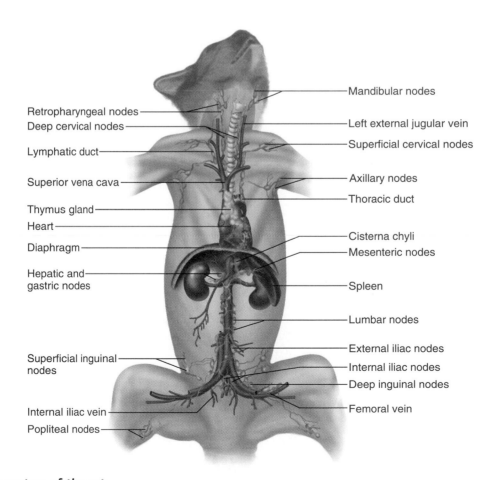

Retropharyngeal nodes
Deep cervical nodes
Lymphatic duct
Superior vena cava
Thymus gland
Heart
Diaphragm
Hepatic and gastric nodes

Superficial inguinal nodes

Internal iliac vein
Popliteal nodes

Mandibular nodes
Left external jugular vein
Superficial cervical nodes
Axillary nodes
Thoracic duct
Cisterna chyli
Mesenteric nodes
Spleen
Lumbar nodes
External iliac nodes
Internal iliac nodes
Deep inguinal nodes
Femoral vein

Figure 37.11   Lymph system of the cat.

Name _____

1. What is the name of the inner region of a lymph node?

2. Once tissue fluid enters the lymphatic vessels, what is it called?

3. What kind of vessel takes lymph away from a lymph node?

4. The adenoids are enlarged _____ tonsils.

5. Which tonsils are found on the sides of the oral cavity?

6. Blood is filtered by which lymph organ in the adult?

7. Where do T cells mature?

8. From what you know of the functions of lymph nodes, make a prediction of the difference between lymph entering a node and lymph leaving a node. What materials may be missing from the lymph leaving the node?

9. Elephantiasis is a disease that may be caused, in some cases, by a parasitic worm blocking the lymphatic vessels. Examine the following illustration and predict where the lymphatic blockage occurs.

Lymph drainage.

10. In the analysis of breast cancer, lymph nodes of the axillary region are removed and a biopsy is performed. The removal of the nodes is done to determine if cancer has spread from the breast to other regions of the body. What effect would the removal of lymph nodes have on the drainage of the pectoral region?

11. Superficial veins contain valves, yet deep veins do not. The deep veins are surrounded by muscles. People who are inactive may have problems with their veins. Can you propose a mechanism by which blood from the deep veins may be returned to the heart? (A mechanism other than standing on your head!)

12. Lymphatic vessels have a one-way flow from the extremities to the heart. Damage to the lymph system can lead to edema or an increase in tissue fluid. From the standpoint of reducing edema, how does the use of medical leeches (segmented worms which drain tissue fluid) work for a region that has suffered trauma?

Blood Vessels and Blood Pressure

Introduction

Maintenance of blood pressure is very important for the health of the heart and for proper functioning of various organs. High blood pressure increases the workload of the heart by increasing the force with which the heart must pump to provide blood to the body. High blood pressure is also a concern for proper kidney functioning. On the other hand, low blood pressure is a concern for maintaining adequate blood pressure to the brain. A decrease in pressure may lead to fainting or dizziness. In this exercise, you learn how to determine blood pressure and some factors affecting blood pressure.

Objectives

At the end of this exercise you should be able to

1. distinguish between systolic pressure and diastolic pressure;
2. properly take and record blood pressure;
3. define hypertension in terms of millimeters of mercury pressure.

Materials

Stethoscope
Alcohol wipes or isopropyl alcohol and sterile cotton swabs
Sphygmomanometer
Watch or clock with accuracy in seconds

Procedure

Measurement of Pulse Rate in Beats per Minute (BPM)

Initially record a baseline pulse rate. The pulse rate is measured in beats per minute (BPM), which is done by locating regions of the body where the pulse can be palpated. These areas include the radial artery, carotid artery, and other sites shown in figure 38.1. If you are recording your lab partner's pulse, make sure you use the tips of your fingers and not your thumb. If you use your thumb you may feel the pulse in your thumb and not the pulse of the person you wish to record. You can determine beats per minute by counting the pulse for 15 seconds and multiplying it by four.

Pulse rate of lab partner: _____ BPM

Measurement of Blood Pressure

The accuracy of a blood pressure measurement is important because extremely low blood pressure (hypotension) or extremely high blood pressure (hypertension) are health risks. Blood pressure can be indirectly measured with a **sphygmomanometer** (blood pressure cuff), which measures pressure in millimeters of mercury rising in a glass column. The greater the pressure the higher the rise of mercury.

Blood pressure is not uniform throughout the body but is influenced by gravity, therefore the pressure in the arteries of the head and neck are less than the blood pressure from the heart. The pressure in the arteries in the leg is greater than the blood pressure in the heart. Blood pressure is measured in the **brachial artery,** which is about at the same level of the heart and therefore is at the same approximate pressure. Ausculatory measurement of blood pressure involves listening to sounds as blood passes through the brachial artery. Normally, no sound is heard through a stethoscope as blood passes through the brachial artery. The blood passes smoothly through the vessel, like water through a garden hose. When pressure is applied by pinching off a garden hose the turbulence creates sound. When pressure is applied to the arm due to the constriction from the blood pressure cuff, the turbulence of the blood passing through the vessel creates sound. The significant difference between the flow of blood in the body versus water in a garden hose is that the blood pulses through the arteries as the heart contracts during ventricular systole. The sounds made by the flow of blood in a constricted artery are known as the "sounds of Korotkov."

1. To determine blood pressure, locate the pulse of the brachial artery on your lab partner. This can be done by placing two fingers on the medial side of the biceps brachii muscle near the antecubital fossa. You can place a small "X" with a felt pen on this location on the arm.
2. Place the blood pressure cuff around the arm of your lab partner at the level of the heart. Make sure the inflatable portion of the cuff is on the anterior medial side. Some cuffs have a metal bar that provides a loop through which a part of the cuff goes through. This bar should not be located on the medial side of the arm.
3. Place the bell of the stethoscope on the "X" where you located the brachial pulse and have your lab partner rest his or her forearm on the lab counter. You should *not* hear any sound at this time.

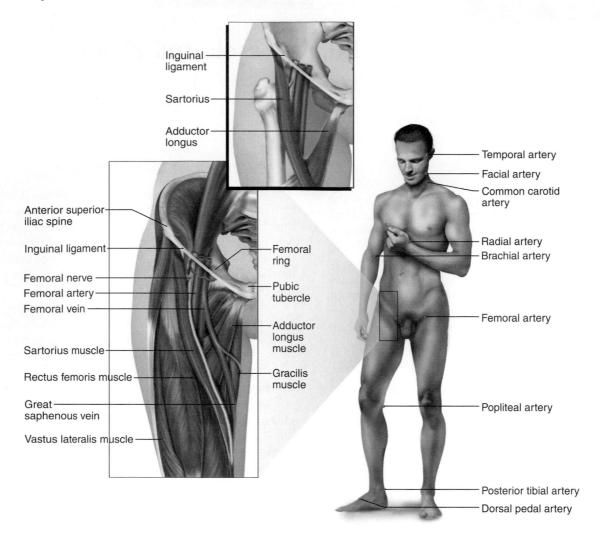

Figure 38.1 Locations where pulse can be recorded.

4. Hold on to the rubber squeeze bulb with the attached rubber tubing leading away from you. Turn the metal dial clockwise until it is closed. You can now begin to inflate the cuff.

5. Pump the cuff up to about 80 mmHg. Look at the mercury in the glass tube. If it seems to bounce up and down a little, then listen closely for the sounds in the stethoscope. (The ear pieces are inserted into the ears facing forward.) If you do not hear any sound and the mercury is still, then inflate the mercury to 100 or 120 mmHg. If you see the mercury pulsing, listen again for the sound. Make sure the bell of the stethoscope is in the right place.

6. Once you are sure you have heard the sound of the heart, remove the cuff and place it on the other arm. Excessive constriction of the arm may elevate your lab partner's blood pressure.

7. Inflate the cuff on the other arm but make sure you exceed the level where the mercury pulses up and down

(usually around 150 mmHg). Do not leave the cuff inflated for a long period of time.

8. Release the knob slowly so the mercury slowly begins to descend down the glass column of the sphygmomanometer.

9. Record the mmHg when the first sound is heard. This sound represents the **systolic pressure** of the ventricles, which is the pressure the heart generates that exceeds the pressure of the cuff. The sound may muffle a bit and then come in very strong.

10. Continue to let air out of the cuff slowly until the sound starts to muffle. Listen very carefully until the sound completely disappears. At the exact level where the sound disappears is the **diastolic pressure.**

11. Record the level of blood pressure as the systolic pressure over the diastolic pressure.

$$\frac{\text{Systolic pressure}}{\text{Diastolic pressure}} = \underline{\hspace{3cm}}$$

If you are unsure of exactly where the pressure is you can "fine-tune" your readings by increasing the cuff pressure at each approximate end to obtain an accurate reading. Do not subject your lab partner to more than two or three attempts. Ask your instructor for help if you have difficulty determining blood pressure.

Hypertension is elevated blood pressure. Normal blood pressure is approximately 120/80 mmHg. Hypertension may be due to a number of factors; however, the majority of hypertensive cases occur for reasons as yet unknown. Hypertension is typically a blood pressure in excess of 140/90 mmHg, though the greatest concern is in the diastolic reading. The diastolic pressure should be less than 90 mmHg.

Pulse Pressure

You can now determine the pulse pressure of your lab partner. The pulse pressure is determined by subtracting the diastolic pressure from the systolic pressure. It is normally around 50 mmHg.

Pulse pressure: _____

Factors Affecting Blood Pressure

Body Position

Measure the blood pressure of your lab partner while he or she lies down. Record this value.

Measurement lying down: _____

Have your lab partner stand suddenly and quickly take a new measurement. Record the results.

Blood pressure on immediately standing: _____

How can you account for these two readings?

Exercise

Have your lab partner run around the building for awhile or do strenuous physical activity for a couple of minutes. Measure the blood pressure immediately after the cessation of exercise and record your results.

Caution Do not do strenuous exercise if you are not feeling well, have a history of heart trouble, or have been advised by a physician to avoid exercise.

Blood pressure after exercise: _____

Name _____

1. The letters BPM stand for what phrase in cardiac measurement?

2. What does a sphygmomanometer measure?

3. If you were to measure blood pressure, what artery would you most commonly use?

4. If you have a blood pressure of 140/80, what does the 80 stand for?

5. The threshold of high blood pressure has what clinical threshold?

6. When the first sound is heard during measurement with a blood pressure cuff, what is measured—systolic or diastolic pressure?

7. Emotions have an effect on blood pressure. Predict the blood pressure of an individual who recently had a heated argument with a roommate about rent money.

8. Illness can also affect blood pressure. Illness tends to increase stress responses. Predict the blood pressure of an individual with a sinus headache and postnasal drip.

9. Nicotine and caffeine both elevate blood pressure. Explain how an increase in blood pressure could have a negative effect on the pumping efficiency of the heart.

10. Record your blood pressure.

LABORATORY EXERCISE 39

Structure of the Respiratory System

Introduction

The vital exchange of oxygen and carbon dioxide occurs by the respiratory system. Atmospheric oxygen moves into the lungs and diffuses into the circulatory system. It subsequently reaches the individual cells of the body while the metabolic waste product, carbon dioxide, is released from the intercellular environment and travels via the blood to the lungs, where it is released by exhalation. The respiratory system is therefore integrated with the other body systems. As you study the anatomy of the respiratory system in this exercise, be aware of the role that other systems play in respiration.

Objectives

At the end of this exercise you should be able to

1. list the organs and significant structures of the respiratory system;
2. define the role of the respiratory system in terms of the overall function of the body;
3. explain the physical reason for the tremendous surface area of the lungs;
4. identify the cartilages of the larynx;
5. distinguish between a bronchus, bronchiole, and respiratory bronchiole.

Materials

Lung models or detailed torso model, including
    midsagittal section of head
Model of larynx
Microscopes
Prepared microscope slides of lung tissue
Charts and illustrations of the respiratory system
Cats
Dissection equipment (dissection trays and instruments)

Procedure

Look at the charts and models of the respiratory system and compare them to figure 39.1. Locate the following structures:

Nose
Nasal cavity
Pharynx

Larynx
Trachea
Bronchus
Left lung
Right lung

Nose and Nasal Cartilages

Examine a midsagittal section of a model or chart of the head and look for the **nose, nasal cartilages, external nares,** and **nasal septum.** The nasal septum separates the nasal cavities and is composed of the perpendicular plate of the ethmoid bone, the vomer, and the nasal cartilage. Examine these features in figures 39.2 and 39.3.

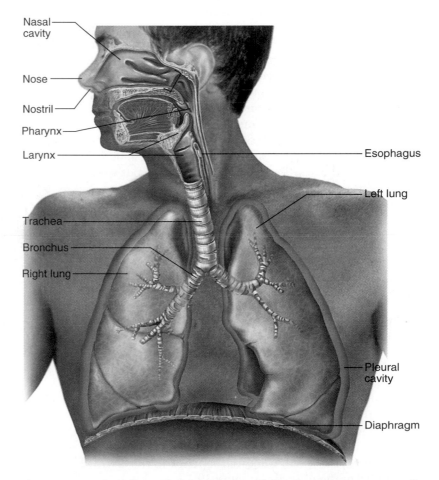

Figure 39.1   Overview of the anatomy of the respiratory system ♊.

435

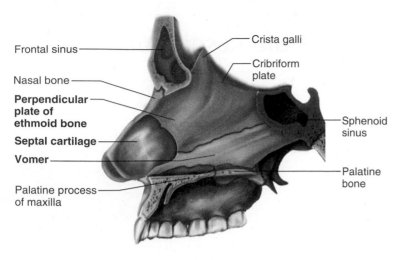

Figure 39.2 **Structures that make up the nasal septum.**

The entrance of the external nares is protected by guard hairs. The region of the nose, just posterior to the external nares, is the **nasal vestibule,** which is lined with stratified squamous epithelium. Behind the vestibule is the nasal cavity lined with a **mucous membrane** that moistens and warms air entering the respiratory system. The membrane consists of **respiratory epithelium,** composed of **pseudostratified ciliated columnar epithelium** with **goblet cells.** The lateral walls of the cavity have three protrusions that push into the nasal cavity. These are the **nasal turbinates.** The turbinates consist of the nasal conchae and the mucous membrane covering. They cause the air to swirl in the nasal cavity and come into contact with the mucous membrane. Locate the **superior, middle,** and **inferior turbinates** in the nasal cavity. The nasal cavity ends where two openings, the **internal (posterior) nares,** or **choanae** lead to the **pharynx.** These structures are seen in figure 39.3.

Pharynx

The pharynx can be divided into three regions based on location. The uppermost area is the **nasopharynx,** which is directly posterior to the nasal cavity. The nasopharynx has two openings on the lateral walls, which are the openings of the **auditory,** or **eustachian, tubes.** As the pharynx descends behind the oral cavity it becomes the **oropharynx.** The oropharynx is a common passageway for food, water, and air. The **uvula** is a small pendulous structure that partially separates the oral cavity from the oropharynx. The uvula serves to flip upwards during swallowing, helping to prevent fluids from entering the nasopharynx. The most inferior portion of the pharynx is the **laryngopharynx,** which is located superior to the larynx. Find the regions of the pharynx in figure 39.3.

Larynx

The **larynx** is commonly known as the "voice box" because it is an important organ for sound production in humans. It controls the pitch of the voice, while the shape of the oral cavity and the placement and size of the paranasal sinuses are responsible for the sonority of the voice. The larynx occurs at about the level of the fourth through sixth cervical vertebrae and consists of a number of cartilages. The most prominent cartilage in the larynx is the **thyroid cartilage,** which is a shield-shaped structure made of hyaline cartilage. Examine models and charts in the lab and compare the structures of the larynx to figure 39.4.

The thyroid cartilage is more prominent in males because it increases in size under the influence of testosterone. Inferior to the thyroid cartilage is the **cricoid cartilage.** The cricoid cartilage is also composed of hyaline cartilage, and it is relatively narrow when seen from the anterior but increases in size at its posterior surface. Superior to the cricoid cartilage in the posterior wall of the larynx are the paired **arytenoid cartilages.** These cartilages attach to the posterior end of the **true vocal folds** (vocal cords). Movement of the arytenoid cartilages pulls on the true vocal folds, causing them to stretch and thus increase the pitch of the voice. This occurs by contraction of intrinsic muscles that are attached to the arytenoid cartilages from the back, while the vocal cords are held stationary by the thyroid cartilage in the front. Above the true vocal folds are the **vestibular folds (or false vocal folds)** (figure 39.4).

At the very posterior, superior edge of the larynx are the **corniculate** and **cuneiform cartilages.** These are also made of hyaline cartilage. The last cartilage of the larynx is the **epiglottis,** which is composed of elastic cartilage. During swallowing the epiglottis protects the opening of the larynx, which is known as the **glottis.** In the swallowing reflex, muscles pull the epiglottis down over the glottis. This is not a perfect system, as anyone knows who has started swallowing a liquid and responded by laughing at a joke. Inhalation at the beginning of the laugh causes fluid to move into the larynx and trachea, irritating the respiratory lining. This causes another reflex called the **cough reflex,** which propels the liquid out of the respiratory system. Locate the structures of the larynx on models or charts in the lab and in figure 39.4.

Trachea and Bronchi

The **trachea** is commonly known as the "windpipe" because it conducts air from the larynx to the lungs. The trachea is a straight tube whose lumen is kept open by **tracheal (C-shaped) cartilages.** Examine these cartilages by running your fingers gently down the outside of your throat. Palpate the cartilage rings below the larynx. The tracheal cartilages are composed of hyaline cartilage. At the most inferior portion of the trachea is a center point known as the **carina** (keel). Locate the features of the trachea in figures 39.5 and 39.6.

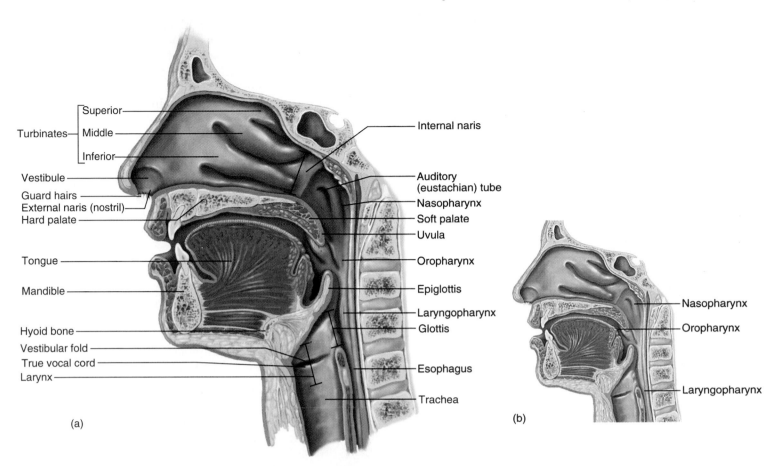

Figure 39.3   Upper respiratory tract. (a) Midsagittal section of the head with nasal septum removed; (b) three regions of the pharynx.

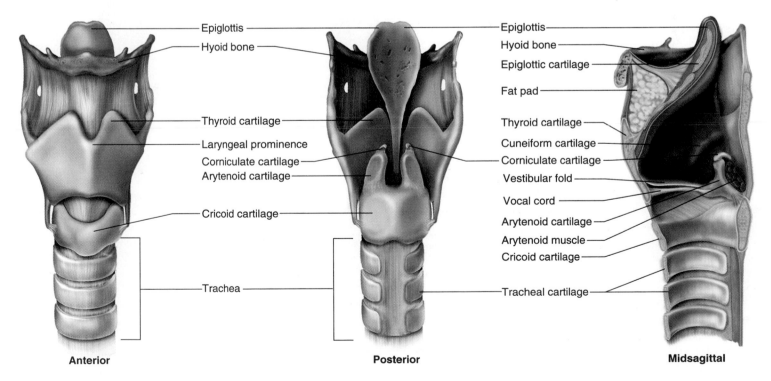

Figure 39.4   Larynx.

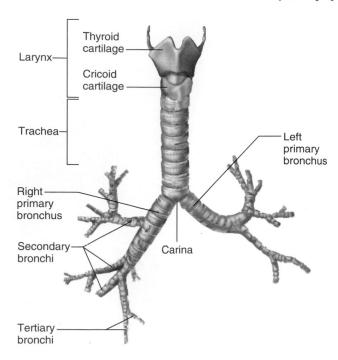

Figure 39.5   Larynx, trachea, and bronchi, anterior view.

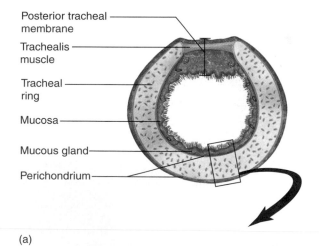

(a)

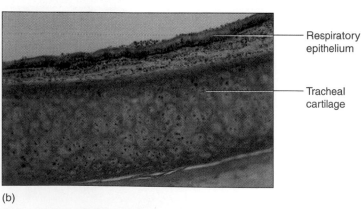

(b)

Figure 39.6   Trachea, cross section. (a) Diagram;
(b) photomicrograph (100×).

The trachea is also lined with respiratory epithelium. Obtain a prepared slide of the trachea and find the tracheal cartilage, respiratory epithelium, and the **posterior tracheal membrane** (with the trachealis muscle) (figure 39.6).

The trachea splits into two tubes, which enter the lungs. These tubes are the **primary bronchi.** Each lung receives air from a primary bronchus. These primary bronchi contain hyaline cartilage and are lined with respiratory epithelium. The primary bronchi of the lung divide into the **secondary bronchi,** and these further divide to form tertiary bronchi. The extensive branching of the bronchi produce a structure which is called the **bronchial tree** (figures 39.5 and 39.7).

Lungs

There are two **lungs** in humans; the right lung has **three lobes** and the left lung has **two lobes.** The right lung consists of **superior, middle,** and **inferior lobes** and the left lung has **superior** and **inferior lobes** as well as an indentation that is occupied by the apex of the heart. This indentation is known as the **cardiac notch.** Look at models or charts in the lab and identify the major features as shown in figure 39.8.

Figure 39.7   Cast of bronchial tree ⅄.

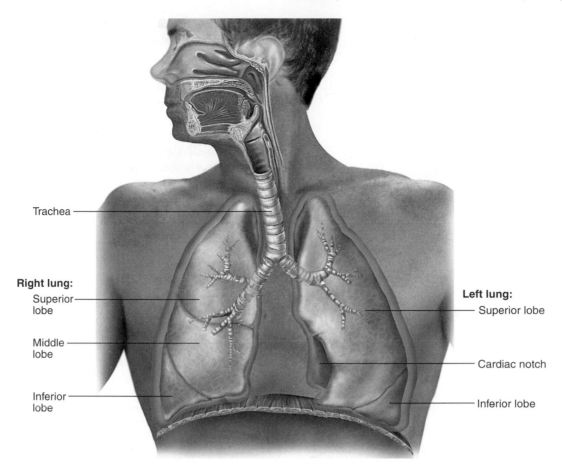

Trachea

Right lung:
Superior lobe

Middle lobe

Inferior lobe

Left lung:
Superior lobe

Cardiac notch

Inferior lobe

Figure 39.8   Thoracic region with lungs $\mathcal{X}$.

The lungs are located in the **pleural cavities** on each side of the **mediastinum.** The **parietal pleura** is the outer membrane on the chest cavity wall, and the membrane that is adhered to the surface of the lungs is the **visceral pleura.** The space between the membranes is known as the **pleural cavity.** These are shown in figure 39.9.

Histology of the Lung

The bronchi continue to divide until they become **bronchioles,** which are small respiratory tubules with smooth muscle in their walls. Obtain a prepared slide of lung and scan first under low power and then under higher powers. The bronchioles in the lung further divide into **respiratory bronchioles,** which are so named due to small structures called alveoli attached to their walls. The respiratory bronchioles lead to passageways known as **alveolar ducts,** which branch into **alveoli.** Alveoli are air sacs in the lung that exchange oxygen and carbon dioxide with the blood capillaries of the lungs. Examine your slide and locate the bronchi, bronchioles, respiratory bronchioles, alveolar ducts, and alveoli. Alveoli that are clustered around an alveolar duct are collectively known as an **alveolar sac** (figure 39.10).

The alveoli are composed of simple squamous epithelium, as are the capillaries that surround the alveoli. The division of the lung into numerous small sacs tremendously increases the surface area of the lung. This increase is vital for the rapid *diffusion* of oxygen across the respiratory membranes.

You may see other cell shapes that occur in the prepared lung sections. Some of these cells are **type II alveolar cells** (septal cells), and they function to decrease the surface tension of the lung by the secretion of surfactant.

Examine a prepared slide of smoker's lung. Note the dark material in the lung tissue and the general destruction of the alveoli. Breakdown of the alveoli leads to a disease known as **emphysema.**

Cat Dissection

Prepare your cat for dissection and remember to place all excess tissue in the appropriate waste container and not in a standard wastebasket or down the sink!

Locate the **larynx** of the cat above the **trachea.** Notice the broad hyaline cartilage wedge-shaped structure in the front.

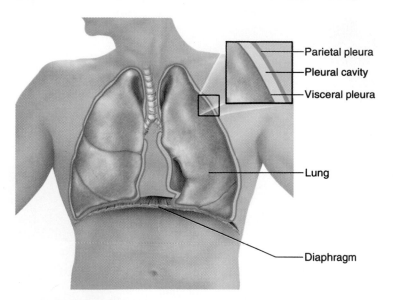

Figure 39.9 Pleural membranes and cavity.

This is the **thyroid cartilage.** Make a midsagittal incision through the thyroid cartilage and continue carefully cutting until you have cut completely through the larynx. To examine the larynx more completely you may wish to continue your mid-

sagittal cut partway down through the trachea. Open the larynx and find the **epiglottis** of the cat. Notice how the elastic cartilage is lighter in color than that of the thyroid cartilage. Find the **cricoid cartilage,** the **arytenoid cartilages,** and the **true** and **false vocal folds** (figure 39.11).

Examine the trachea as it passes from the larynx and into the thoracic cavity. Ask your instructor for permission before you cut the trachea in cross section. If you do, you should see the **tracheal cartilages** and the **posterior tracheal membrane.** The posterior portion of the trachea is located ventrally to the esophagus.

Notice how the trachea splits into the two **bronchi,** which then enter the lungs. The lobes of the lungs are different in the cat than in the human. The right lung in the cat has four lobes, while the left lung has three lobes. How does this compare to the pattern in humans?

Examine the structure of the lungs in your specimen and compare it to figure 39.12. The lungs are covered with a thin serous membrane called the visceral pleura, while the outer covering (occurring on the deep surface of the ribs and intercostal muscles) is called the parietal pleura. The space between these two membranes is the pleural cavity. Cut into one of the lungs of the cat and examine the lung tissue. Note how the lungs appear like a very fine mesh sponge. The alveoli of the lungs are microscopic.

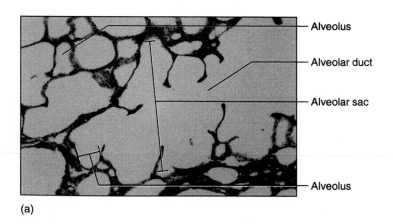

(a)

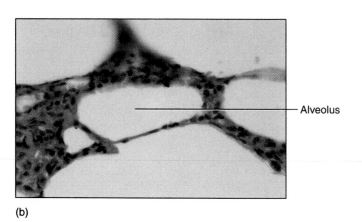

(b)

Figure 39.10 Histology of the lung. (a) Alveolar sac (100×); (b) alveolus (400×) ✗.

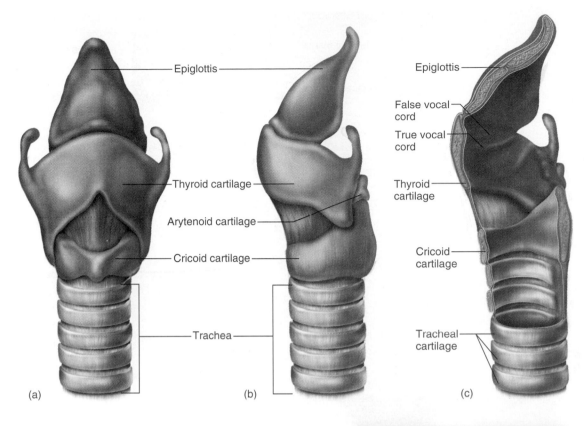

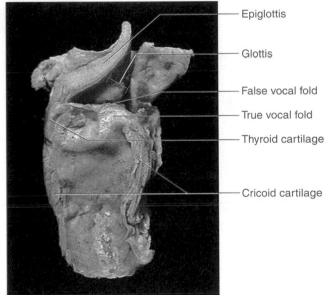

Figure 39.11   Larynx of the cat. Diagram (a) anterior view; (b) left lateral view; (c) midsagittal view; (d) photograph, midsagittal view.

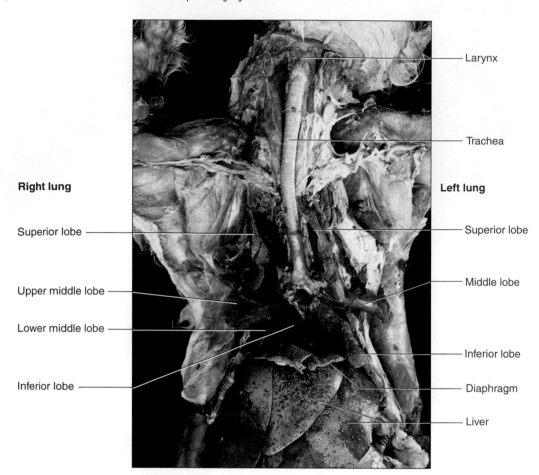

Figure 39.12   Respiratory structures of the cat.

Name _____

1. What is the common name for the external nares?

2. The nasal cavities are separated from each other by what structure?

3. Three structures make up the nasal septum. What are they?

4. What is the function of respiratory epithelium and the superficial blood vessels in the nasal cavity?

5. Name the openings that occur between the nasal cavity and the pharynx.

6. What is the name of the space found behind the oral cavity and above the laryngopharynx?

7. What is the name of the large cartilage of the anterior larynx?

8. What is the structure that protects the glottis from fluid?

9. Which lung has just two lobes in the human?

10. What membrane attaches directly to the lungs?

11. The trachea branches into two tubes that go to the lungs. What are these tubes called?

12. Where is the bronchial tree found?

13. What small structure in the lung is the site of exchange of oxygen with the blood capillaries?

14. The nasal cartilages are made of hyaline cartilage. What functional adaptation does cartilage have over bone in making up the external framework of the nose?

15. The surface area of the lungs in humans is about 70 square meters. How can this be so if the lungs are located in the small space of the thoracic cavity? What role do alveoli play in the nature of surface area?

16. Emphysema is a destruction of the alveoli of the lungs. What effect does this have on the surface area of the lungs?

17. Fill in the following illustration of the human respiratory system.

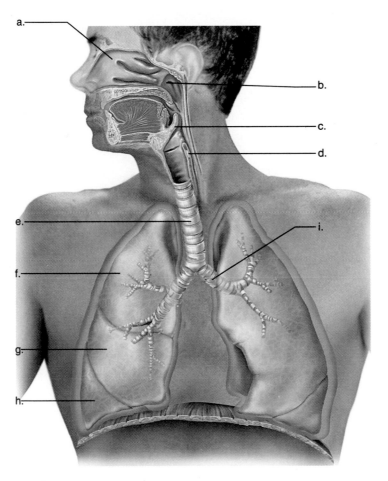

Respiratory system.

LABORATORY EXERCISE 40

Respiratory Function, Breathing, Respiration

Introduction

The function of the respiratory system is to exchange gases between the external air and the blood. Oxygen diffuses from the air into the blood of the lungs, and carbon dioxide diffuses from the blood into the air. The process can be divided into **pulmonary ventilation,** which is the mechanical process of moving air into and out of the lungs, and the **exchange of gases** across the respiratory membrane. Oxygen requirements vary with the body's need. Variations in the loading of oxygen by the lungs can be accomplished by increasing/decreasing the volume of air with each inhalation, the breathing rate, or both. Removal of carbon dioxide can also be accomplished by an increase in volume or breathing rate. This is important in maintaining an appropriate acid-base balance of the blood. In this exercise, you measure breathing rates, lung volumes and capacities, and the effects of carbon dioxide on the acid-base balance of a solution.

Objectives

At the end of this exercise you should be able to

1. measure and understand the relationship of pulmonary volumes including vital capacity, tidal volume, inspiratory reserve volume, and expiratory reserve volume;
2. describe the relationship between oxygen and carbon dioxide levels in the blood and breathing rates;
3. describe the mechanical process of breathing;
4. describe the use of the spirometer or respirometer available in your lab;
5. demonstrate the use of the stethoscope in obtaining respiratory sounds;
6. describe the nature and importance of carbon dioxide to the respiratory system.

Materials

Respiration Model
Bell jar respiration model

Pulmonary Volume Setup
Respirometer (Collins respirometer) or handheld spirometers
Disposable mouthpieces to fit respirometer or spirometers
Watch or clock with accuracy in seconds
Biohazard bag

Breathing Sounds and Breathing Rate Setup
Medium-sized brown paper sacs (2 gallon) (one per student)
Stethoscope
Alcohol wipes

Acid-Base Setup (one setup per table)
Litmus solution (2 g litmus powder in 600 ml water)
NaOH solution (1 N) in dropper bottles
Straws
100 ml Erlenmeyer flasks
Safety glasses

 Virtual Physiology Lab #6: Pulmonary Function

Procedure

Mechanics of Breathing

Air moves from regions of higher pressure to lower pressure. Normal atmospheric air pressure is measured in millimeters of mercury (mm Hg) and standard air pressure at sea level is 760 mmHg. When there is no movement of air into or out of the lungs, the pressure in the lungs is equal to that of the surrounding air as illustrated in figure 40.1a. During inspiration, the diaphragm contracts increasing the volume in the thoracic cavity. This leads to a decrease in the pressure in the lungs and air moves from the atmosphere into the lungs. This is illustrated in figure 40.1b. If the diaphragm relaxes, then the abdominal pressure forces the diaphragm upwards and increases the pressure in the thoracic cavity beyond that of the atmospheric pressure and the air moves out of the lungs as illustrated in figure 40.1c. You can use a simple bell jar model to demonstrate this in lab.

This model illustrates the movement of air into or out of the lungs, depending on air pressure differentials between the lungs and the external environment. The glass housing represents the thoracic cage, the balloons represent the lungs, the latex sheeting represents the diaphragm, and the rubber stopper, the nose. Pull *gently* on the latex diaphragm. As you do this the volume in the pleural cavity (the space between the balloons and the glass jar) increases, thus causing a decrease in the pressure of the pleural cavity.

When the pressure in the pleural cavity decreases, air in the external environment is at a higher pressure than that in the jar

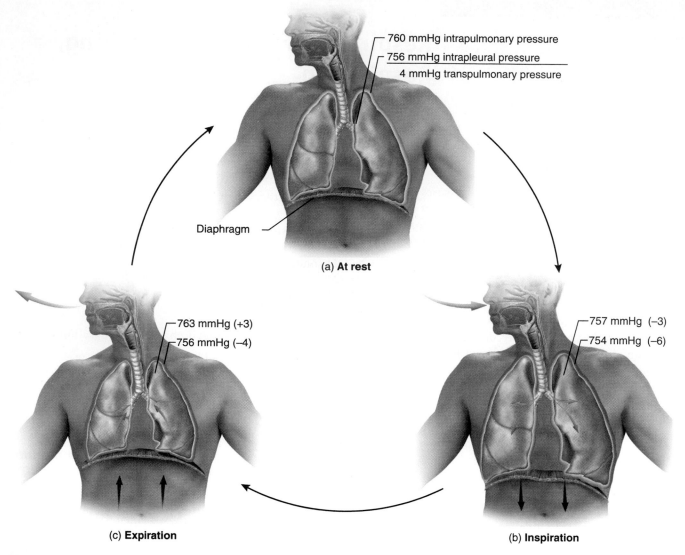

760 mmHg intrapulmonary pressure
756 mmHg intrapleural pressure
4 mmHg transpulmonary pressure

Diaphragm

(a) **At rest**

763 mmHg (+3)
756 mmHg (–4)

(c) **Expiration**

757 mmHg (–3)
754 mmHg (–6)

(b) **Inspiration**

Figure 40.1   Mechanics of breathing ⑆.

and moves into the bell jar, filling the balloon lungs. As you release the diaphragm, the volume decreases and the pressure increases in the pleural cavity bell jar, exceeding the external air pressure. Air moves out of the lungs and through the nose (rubber stopper). Although the mechanics of human breathing differ somewhat from this model, the bell jar model is valuable to demonstrate the basic principle of air moving from a region of high pressure to a region of low pressure.

Measurement of Relaxed Breathing Rate

It is important that your lab partner be distracted from thinking about breathing. Therefore your lab partner should read the remainder of the laboratory exercise while you count the number of breaths he or she takes for a total of 2 minutes.

Divide the number by two to calculate the average number of breaths per minute. Record that number for your lab partner.

Breaths per minute: _____

Breathing rate varies with oxygen demand and/or carbon dioxide levels in the blood. Before doing any exercise *estimate* the breaths per minute your lab partner might take after completing 2 minutes of strenuous exercise. Record your estimation.

Estimation: _____

Caution If you have a heart condition, family history of heart failure, or other medical condition that prevents you from doing strenuous exercise, then do not do the exercise sections of this lab.

Have your lab partner do strenuous exercise for 2 minutes (jumping jacks, running in place or outside of lab, running up and down stairs, etc.). As soon as your lab partner is finished, record the number of breaths per minute for the *first minute* after exercise and record this number.

Number of breaths per minute after 2 minutes of exercise: _____

How does this compare with your estimation?

Breath-Holding Exercises

Caution If you are prone to dizzy spells, or have a history of fainting episodes, a heart condition, or any other medical condition that may be affected by holding your breath, do not do this part of the exercise.

Carbon Dioxide Levels and Breath-Holding

1. Breathe into a medium-sized paper grocery sac (the size in between a lunch sac and a large bag) by holding the bag over your mouth and nose for one minute.
2. At the end of one minute remove the bag and inhale maximally, holding your breath.
3. Make sure you work with a partner for the experiment. Your lab partner should place his or her hands on your shoulder. The level of carbon dioxide in your blood has been elevated in the rebreathing of the air in the paper bag.
4. Record the amount of time, in seconds, you can hold your breath and record your findings.

Number of seconds of breath-holding after rebreathing: _____

Relax, take a deep breath, and wait 5 minutes before doing the next experiment.

Baseline Rate

Wait at least 5 minutes after the previous experiment before you do this experiment.

1. Sit comfortably and inhale maximally.
2. Hold your breath for as long as possible and record this time.
3. Your lab partner should monitor your condition in case you feel faint. Your lab partner should place his or her hands on your shoulders to support you in case you become dizzy. If you feel light-headed while conducting this part of the experiment, then stop holding your breath.

4. Record the maximum number of seconds you can hold your breath under normal conditions.

Breath-holding number of seconds: _____

Relax, take a deep breath, and wait 5 minutes before doing the next experiment.

Caution If you are prone to dizzy spells, black-out easily, have had seizures or have a family history of seizures, or have any medical condition that would be aggravated by hyperventilation, do not do this part of the exercise.

Hyperventilation and Breath-Holding

1. Wait 5 minutes after the last breathing experiment before continuing with this experiment.
2. Take 15 deep breaths *while seated* and then inhale maximally.
3. Hold your breath as you did in the previous experiments. If you feel dizzy or faint, discontinue the experiment.
4. Have your lab partner nearby to guard against your falling over. Record the time, in seconds, you can hold your breath.
5. Record this number.

Number of seconds of breath-holding after hyperventilation: _____

Measurement of Pulmonary Volumes and Capacities

Caution Unless directed otherwise by your instructor, *do not inhale* when using the respirometer. You should never inhale using a spirometer.

Tidal Volume with a Respirometer

The normal volume of breath exhaled is referred to as the **tidal volume (TV).** It represents the average amount of breath inhaled or exhaled when an individual is relaxed. If you are using a respirometer, you will want to insert a disposable mouthpiece (or sterile mouthpiece) into the machine. When you are finished with the mouthpiece at the end of the exercise, dispose of it in the biohazard bag. A more accurate recording is made if you close your nostrils with your fingers or use a clean nose clip.

1. If you are using a recording respirometer (figure 40.2), set the ink pen on the drum as demonstrated by your

Figure 40.2 Respirometer.

instructor, adjust the pen to rest on a line of the chart paper, and turn the respirometer on.
2. Try to exhale a "normal" breath into the respirometer.
3. Repeat this four times for a total of five tidal volumes and turn the respirometer off.
4. Calculate the average tidal volume by adding the five volumes together and dividing by five.
5. Record the data.

Trial 1: _____

Trial 2: _____

Trial 3: _____

Trial 4: _____

Trial 5: _____

Average tidal volume: _____

Tidal Volume with Handheld Spirometer

1. If you are using a handheld spirometer to make the recording, place a disposable mouthpiece on the spirometer tube.
2. Make sure you set the indicator dial to zero by twisting the knurled ring on the top of the spirometer (zero may be the same as the maximum volume, 7,000 in some spirometers) (figure 40.3).
3. As you exhale, estimate what a normal breath volume will be and exhale this amount rather forcefully. If you breathe gently into a handheld spirometer, you may not cause the vanes to spin enough to get a significant recording.
4. Exhale five total breaths and record your results.

Total of five breaths recorded by spirometer: _____

Divide the total number by five and enter the average tidal volume.

Figure 40.3 Handheld spirometer.

Average tidal volume: _____

You may find that the tidal volume is variable among members of your class. This is due to the difficulty of measuring tidal volumes with a standard lab apparatus. The average tidal volume is about 500 ml.

Expiratory Reserve Volume

You can determine your **expiratory reserve volume (ERV)** with the recording respirometer or the handheld spirometer. The expiratory reserve volume is the maximal amount of air you can exhale *after a normal exhalation*.

1. Make sure to close off your nostrils and, after a normal exhalation, forcibly expel the remainder of your breath through the mouthpiece into the respirometer or spirometer.
2. Repeat this two more times for a total of three exhalations, and calculate the average expiratory reserve by dividing the sum of the volumes by three.
3. Record your results.

Trial 1: _____

Trial 2: _____

Trial 3: _____

Average expiratory reserve volume: _____

Vital Capacity (VC)

You can now record your vital capacity (VC), which is the total volume of air that can be forcefully expelled from the lungs after a maximum inhalation. Measuring vital capacity is like participating in the national championship of exhalation.

1. Measure the vital capacity with the use of a respirometer or spirometer by first breathing in as deeply as you possibly can.
2. Close your nostrils, and then exhale through the mouthpiece completely until you cannot exhale anymore.
3. Force as much air from your lungs as you can.
4. Record your vital capacity.

Vital capacity: _____

The lung volumes are illustrated in figure 40.4.

Percent of Normal Vital Capacity

You can compare your vital capacity with other individuals of your sex, height, and age by using one of the two charts (one for females, one for males) in table 40.1. As you examine the chart for a person of your height, notice that the vital capacity decreases with age.

Calculate your percent of normal vital capacity by dividing your data by the normal data for a person of your age, sex, and height and multiplying the result by 100.

$$\frac{\text{Your vital capacity}}{\text{Normal vital capacity}} \times 100 = \text{Percent of normal vital capacity}$$

Your percent of normal vital capacity: _____

If you come out at 100% of normal, then you have an average vital capacity of individuals in your bracket. If your value is less than 100%, then you have a smaller vital capacity than normal for your bracket; if your value is larger than 100%, then you have a larger vital capacity than average.

Calculation of the Inspiratory Reserve Volume (IRV)

The vital capacity consists of the expiratory reserve volume, the tidal volume, and the inspiratory reserve volume. You can *indirectly* determine the **inspiratory reserve volume (IRV)** by subtracting the expiratory reserve volume (ERV) and the tidal volume (TV) from the vital capacity (VC). You cannot measure the inspiratory reserve volume directly with the use of a handheld spirometer (it records exhalations only). Disinfect a large

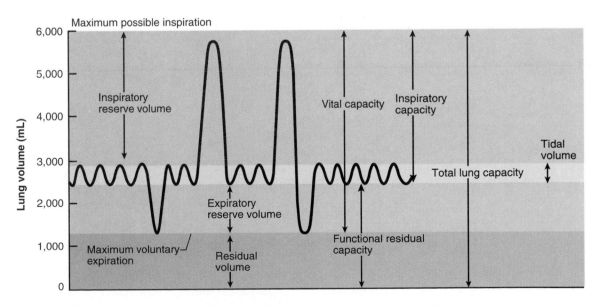

Figure 40.4   Lung volumes.

Table 40.1 Predicted Vital Capacities for Females and Males

Females

Height in Centimeters and Inches

Age	cm 152 / in 59.8	154 60.6	156 61.4	158 62.2	160 63.0	162 63.7	164 64.6	166 65.4	168 66.1	170 66.9	172 67.7	174 68.5	176 69.3	178 70.1	180 70.9	182 71.7	184 72.4	186 73.2	188 74.0
16	3,070	3,110	3,150	3,190	3,230	3,270	3,310	3,350	3,390	3,430	3,470	3,510	3,550	3,590	3,630	3,670	3,715	3,755	3,800
18	3,040	3,080	3,120	3,160	3,200	3,240	3,280	3,320	3,360	3,400	3,440	3,480	3,520	3,560	3,600	3,640	3,680	3,720	3,760
20	3,010	3,050	3,090	3,130	3,170	3,210	3,250	3,290	3,330	3,370	3,410	3,450	3,490	3,525	3,565	3,605	3,645	3,695	3,720
22	2,980	3,020	3,060	3,095	3,135	3,175	3,215	3,255	3,290	3,330	3,370	3,410	3,450	3,490	3,530	3,570	3,610	3,650	3,685
24	2,950	2,985	3,025	3,065	3,100	3,140	3,180	3,220	3,260	3,300	3,335	3,375	3,415	3,455	3,490	3,530	3,570	3,610	3,650
26	2,920	2,960	3,000	3,035	3,070	3,110	3,150	3,190	3,230	3,265	3,300	3,340	3,380	3,420	3,455	3,495	3,530	3,570	3,610
28	2,890	2,930	2,965	3,000	3,040	3,070	3,115	3,155	3,190	3,230	3,270	3,305	3,345	3,380	3,420	3,460	3,495	3,535	3,570
30	2,860	2,895	2,935	2,970	3,010	3,045	3,085	3,120	3,160	3,195	3,235	3,270	3,310	3,345	3,385	3,420	3,460	3,495	3,535
32	2,825	2,865	2,900	2,940	2,975	3,015	3,050	3,090	3,125	3,160	3,200	3,235	3,275	3,310	3,350	3,385	3,425	3,460	3,495
34	2,795	2,835	2,870	2,910	2,945	2,980	3,020	3,055	3,090	3,130	3,165	3,200	3,240	3,275	3,310	3,350	3,385	3,425	3,460
36	2,765	2,805	2,840	2,875	2,910	2,950	2,985	3,020	3,060	3,095	3,130	3,165	3,205	3,240	3,275	3,310	3,350	3,385	3,420
38	2,735	2,770	2,810	2,845	2,880	2,915	2,950	2,990	3,025	3,060	3,095	3,130	3,170	3,205	3,240	3,275	3,310	3,350	3,385
40	2,705	2,740	2,775	2,810	2,850	2,885	2,920	2,955	2,990	3,025	3,060	3,095	3,135	3,170	3,205	3,240	3,275	3,310	3,345
42	2,675	2,710	2,745	2,780	2,815	2,850	2,885	2,920	2,955	2,990	3,025	3,060	3,100	3,135	3,170	3,205	3,240	3,275	3,310
44	2,645	2,680	2,715	2,750	2,785	2,820	2,855	2,890	2,925	2,960	2,995	3,030	3,060	3,095	3,130	3,165	3,200	3,235	3,270
46	2,615	2,650	2,685	2,715	2,750	2,785	2,820	2,855	2,890	2,925	2,960	2,995	3,030	3,060	3,095	3,130	3,165	3,200	3,235
48	2,585	2,620	2,650	2,685	2,715	2,750	2,785	2,820	2,855	2,890	2,925	2,960	2,995	3,030	3,060	3,095	3,130	3,160	3,195
50	2,555	2,590	2,625	2,655	2,690	2,720	2,755	2,785	2,820	2,855	2,890	2,925	2,955	2,990	3,025	3,060	3,090	3,125	3,155
52	2,525	2,555	2,590	2,625	2,655	2,690	2,720	2,755	2,790	2,820	2,855	2,890	2,925	2,955	2,990	3,020	3,055	3,090	3,125
54	2,495	2,530	2,560	2,590	2,625	2,655	2,690	2,720	2,755	2,790	2,820	2,855	2,885	2,920	2,950	2,985	3,020	3,050	3,085
56	2,460	2,495	2,525	2,560	2,590	2,625	2,655	2,690	2,720	2,755	2,790	2,820	2,855	2,885	2,920	2,950	2,980	3,015	3,045
58	2,430	2,460	2,495	2,525	2,560	2,590	2,625	2,655	2,690	2,720	2,750	2,785	2,815	2,850	2,880	2,920	2,945	2,975	3,010
60	2,400	2,430	2,460	2,495	2,525	2,560	2,590	2,625	2,655	2,685	2,720	2,750	2,780	2,810	2,845	2,875	2,915	2,940	2,970
62	2,370	2,405	2,435	2,465	2,495	2,525	2,560	2,590	2,620	2,655	2,685	2,715	2,745	2,775	2,810	2,840	2,870	2,900	2,935
64	2,340	2,370	2,400	2,430	2,465	2,495	2,525	2,555	2,585	2,620	2,650	2,680	2,710	2,740	2,770	2,805	2,835	2,865	2,895
66	2,310	2,340	2,370	2,400	2,430	2,460	2,490	2,525	2,555	2,585	2,615	2,645	2,675	2,705	2,735	2,765	2,800	2,825	2,860
68	2,280	2,310	2,340	2,370	2,400	2,430	2,460	2,490	2,520	2,550	2,580	2,610	2,640	2,670	2,700	2,730	2,760	2,795	2,820
70	2,250	2,280	2,310	2,340	2,370	2,400	2,425	2,455	2,485	2,515	2,545	2,575	2,605	2,635	2,665	2,695	2,725	2,755	2,780
72	2,220	2,250	2,280	2,310	2,335	2,365	2,395	2,425	2,455	2,480	2,510	2,540	2,570	2,600	2,630	2,660	2,685	2,715	2,745
74	2,190	2,220	2,245	2,275	2,305	2,335	2,360	2,390	2,420	2,450	2,475	2,505	2,535	2,565	2,590	2,620	2,650	2,680	2,710

—Continued

"Predicted Vital Capacities for Females and Males (tables)" by E. A. Gaensler, M.D. and G. W. Wright, M.D., AEH, Vol. 12, pp. 146–189, February 1966. Reprinted with permission of the Helen Dwight Reid Educational Foundation. Published by Heldref Publications, 1319 18th Street NW, Washington, DC 20036–1802. Copyright © 1966.

Table 40.1. *Continued.*

Males

Height in Centimeters and Inches

Age	cm in	152 59.8	154 60.6	156 61.4	158 62.2	160 63.0	162 63.7	164 64.6	166 65.4	168 66.1	170 66.9	172 67.7	174 68.5	176 69.3	178 70.1	180 70.9	182 71.7	184 72.4	186 73.2	188 74.0
16		3,920	3,975	4,025	4,075	4,130	4,180	4,230	4,285	4,335	4,385	4,440	4,490	4,540	4,590	4,645	4,695	4,745	4,800	4,850
18		3,890	3,940	3,995	4,045	4,095	4,145	4,200	4,250	4,300	4,350	4,405	4,455	4,505	4,555	4,610	4,660	4,710	4,760	4,815
20		3,860	3,910	3,960	4,015	4,065	4,115	4,165	4,215	4,265	4,320	4,370	4,420	4,470	4,520	4,570	4,625	4,675	4,725	4,775
22		3,830	3,880	3,930	3,980	4,030	4,080	4,135	4,185	4,235	4,285	4,335	4,385	4,435	4,485	4,535	4,585	4,635	4,685	4,735
24		3,785	3,835	3,885	3,935	3,985	4,035	4,085	4,135	4,185	4,235	4,285	4,330	4,380	4,430	4,480	4,530	4,580	4,630	4,680
26		3,755	3,805	3,855	3,905	3,955	4,000	4,050	4,100	4,150	4,200	4,250	4,300	4,350	4,395	4,445	4,495	4,545	4,595	4,645
28		3,725	3,775	3,820	3,870	3,920	3,970	4,020	4,070	4,115	4,165	4,215	4,265	4,310	4,360	4,410	4,460	4,510	4,555	4,605
30		3,695	3,740	3,790	3,840	3,890	3,935	3,985	4,035	4,080	4,130	4,180	4,230	4,275	4,325	4,375	4,425	4,470	4,520	4,570
32		3,665	3,710	3,760	3,810	3,855	3,905	3,950	4,000	4,050	4,095	4,145	4,195	4,240	4,290	4,340	4,385	4,435	4,485	4,530
34		3,620	3,665	3,715	3,760	3,810	3,855	3,905	3,950	4,000	4,045	4,095	4,140	4,190	4,225	4,285	4,330	4,380	4,425	4,475
36		3,585	3,635	3,680	3,730	3,775	3,825	3,870	3,920	3,965	4,010	4,060	4,105	4,155	4,200	4,250	4,295	4,340	4,390	4,435
38		3,555	3,605	3,650	3,695	3,745	3,790	3,840	3,885	3,930	3,980	4,025	4,070	4,120	4,165	4,210	4,260	4,305	4,350	4,400
40		3,525	3,575	3,620	3,665	3,710	3,760	3,805	3,850	3,900	3,945	3,990	4,035	4,085	4,130	4,175	4,220	4,270	4,315	4,360
42		3,495	3,540	3,590	3,635	3,680	3,725	3,770	3,820	3,865	3,910	3,955	4,000	4,050	4,095	4,140	4,185	4,230	4,280	4,325
44		3,450	3,495	3,540	3,585	3,630	3,675	3,725	3,770	3,815	3,860	3,905	3,950	3,995	4,040	4,085	4,130	4,175	4,220	4,270
46		3,420	3,465	3,510	3,555	3,600	3,645	3,690	3,735	3,780	3,825	3,870	3,915	3,960	4,005	4,050	4,095	4,140	4,185	4,230
48		3,390	3,435	3,480	3,525	3,570	3,615	3,655	3,700	3,745	3,790	3,835	3,880	3,925	3,970	4,015	4,060	4,105	4,150	4,190
50		3,345	3,390	3,430	3,475	3,520	3,565	3,610	3,650	3,695	3,740	3,785	3,830	3,870	3,915	3,960	4,005	4,050	4,090	4,135
52		3,315	3,353	3,400	3,445	3,490	3,530	3,575	3,620	3,660	3,705	3,750	3,795	3,835	3,880	3,925	3,970	4,010	4,055	4,100
54		3,285	3,325	3,370	3,415	3,455	3,500	3,540	3,585	3,630	3,670	3,715	3,760	3,800	3,845	3,890	3,930	3,975	4,020	4,060
56		3,255	3,295	3,340	3,380	3,425	3,465	3,510	3,550	3,595	3,640	3,680	3,725	3,765	3,810	3,850	3,895	3,940	3,980	4,025
58		3,210	3,250	3,290	3,335	3,375	3,420	3,460	3,500	3,545	3,585	3,630	3,670	3,715	3,755	3,800	3,840	3,880	3,925	3,965
60		3,175	3,220	3,260	3,300	3,345	3,385	3,430	3,470	3,500	3,555	3,595	3,635	3,680	3,720	3,760	3,805	3,845	3,885	3,930
62		3,150	3,190	3,230	3,270	3,310	3,350	3,390	3,440	3,480	3,520	3,560	3,600	3,640	3,680	3,730	3,770	3,810	3,850	3,890
64		3,120	3,160	3,200	3,240	3,280	3,320	3,360	3,400	3,440	3,490	3,530	3,570	3,610	3,650	3,690	3,730	3,770	3,810	3,850
66		3,070	3,110	3,150	3,190	3,230	3,270	3,310	3,350	3,390	3,430	3,470	3,510	3,550	3,600	3,640	3,680	3,720	3,760	3,800
68		3,040	3,080	3,120	3,160	3,200	3,240	3,280	3,320	3,360	3,400	3,440	3,480	3,520	3,560	3,600	3,640	3,680	3,720	3,760
70		3,010	3,050	3,090	3,130	3,170	3,210	3,250	3,290	3,330	3,370	3,410	3,450	3,480	3,520	3,560	3,600	3,640	3,680	3,720
72		2,980	3,020	3,060	3,100	3,140	3,180	3,210	3,250	3,290	3,330	3,370	3,410	3,450	3,490	3,530	3,570	3,610	3,650	3,680
74		2,930	2,970	3,010	3,050	3,090	3,130	3,170	3,200	3,240	3,280	3,320	3,360	3,400	3,440	3,470	3,510	3,550	3,590	3,630

recording respirometer *with permission of your instructor* before determining the inspiratory reserve with this instrument. Calculate your inspiratory reserve.

$$IRV = VC - (ERV + TV)$$

Your IRV: _____

Respiratory Sounds

In this section, you listen to breathing sounds, which can be heard with a stethoscope.

1. Before you begin with the experiment, clean the earpieces of the stethoscope with an alcohol wipe and let them dry. The stethoscope earpieces should point towards the anterior as you insert them into your ears.
2. Locate the larynx of your lab partner and place the bell of the stethoscope just inferior to it.
3. Listen for the sound as your lab partner inhales and exhales. These are the tracheal and bronchial sounds.
4. Locate the triangle of auscultation, which is an area just below the inferior angle of the scapula. This is an ideal area for listening to sounds because the thoracic cage is not covered by muscles in this location.
5. Have your lab partner inhale and exhale deeply several times.
6. Listen for a smooth flow of air into and out of the lungs. Wheezing or rattling are indicators of congestion in the lungs.
7. Record the sounds you hear in the space below and indicate the condition of your lab partner.

Acid-Base Effects of the Respiratory Gases

Carbon dioxide combines with water to form carbonic acid, which subsequently dissociates into bicarbonate and hydrogen ions as represented in the following equation:

$$CO_2 \ + \ H_2O \longrightarrow H_2CO_3 \longrightarrow HCO_3^- \ + \ H^+$$

| Carbon dioxide | Water | Carbonic acid | Bicarbonate ion | Hydrogen ion |

You can see the effects of increasing carbon dioxide levels in the blood in the following experiment.

1. Add about 50 ml of litmus solution to a small (100 ml) Erlenmeyer flask. If the solution is red, add the sodium hydroxide solution (NaOH) drop by drop until the color just begins to turn blue.

Caution Wear safety goggles if you add NaOH to the flask.

---

2. Insert a drinking straw into the flask and gently blow air into the flask.
3. Bubble your exhaled breath into the flask and look for a color change. A blue color indicates an alkaline condition, and a red color indicates an acid condition.
4. Using the preceding formula, determine how the exhalation into the flask alters the acid-base conditions.

1. What is pulmonary ventilation?

2. What happens to the level of carbon dioxide in the blood during hyperventilation?

3. Why would carbon monoxide (it binds to hemoglobin) and carbon dioxide in cigarette smoke cause smokers to generally have a higher than average vital capacity?

4. What was your measured breathing rate in breaths per minute?

5. What is the tidal volume in liters for an average adult?

6. Define tidal volume.

7. Measurement of breathing volumes occurs by the use of what instrument?

8. If you inhale maximally, what is the name of the volume of air that you completely exhale?

9. What is the approximate percent decrease of vital capacity in the same individual from age 25 to age 75?

10. How does the decrease in vital capacity potentially influence an individual's athletic performance or aerobic condition as aging occurs?

11. Using your lecture text try to determine which gas is important in stimulating the breathing reflex in the body. How does this gas relate to your experiment of breathing into the paper bag?

12. What is the pressure difference between the external air and the pleural cavity when inhalation just begins?

13. Calculate the IRV of an individual with a vital capacity of 4,400 ml, an expiratory reserve volume of 1,300 ml, and a tidal volume of 500 ml.

14. How does carbon dioxide change the acid-base condition of a solution when present in excess?

LABORATORY EXERCISE 41

Physiology of Exercise

Introduction

An understanding of exercise physiology is important not only for athletic training but also as a part of wellness and general health. Significant changes occur during exercise that affect many organ systems of the body. The changes vary depending on the type of exercise. We can broadly classify exercise into aerobic exercise (where oxygen reaches the muscle tissue) and anaerobic exercise (where the capacity to deliver oxygen to tissues is exceeded by the demand from the tissue). In the latter case lactic acid is produced anaerobically.

Exercise has an effect on the muscular, skeletal, cardiovascular, and respiratory systems. If you consider the increased metabolic demands placed on skeletal muscles during repeated contractions, you can see how other systems are affected.

Since skeletal muscle contracts more forcefully during exercise than at rest, the muscle uses additional oxygen and nutrients. Oxygen diffuses from the blood to the muscle tissue due to differences in the concentration of oxygen between these two areas. In addition to the physical process of diffusion, the precapillary sphincters open and the arterioles dilate, providing greater blood flow to the tissue.

This additional blood flow requires an increase in the volume of blood that passes through the heart with each contraction. The lung capillaries expand, and greater diffusion occurs from the alveoli to the capillaries. The volume of air increases with each breath, and the number of breaths per minute increases.

Long-term effects of exercise include an increase in muscle and bone mass. A rigorous exercise program causes remodeling of the bone tissue in response to the muscular pull. Increases in pulmonary volumes also increase with exercise, as well as an increase in the heart size and a decrease in the pulse rate.

In this exercise, you examine the effects of exercise on the body and the comparative responses the body shows in response to aerobic exercise.

Objectives

At the end of this exercise you should be able to

1. list the major organ systems directly involved in fitness;
2. describe basic physiological differences between an individual who is "in shape" versus one who is "out of shape";
3. calculate the personal fitness index of a subject who performs the Harvard step test;
4. determine the forced expiratory volume exhaled in 1 second.

Materials

16-inch step
20-inch step
Metronome or clock with second hand
Exercise bicycle, treadmill, or similar aerobic machine
Respirometer capable of measuring volume and time
Disposable mouthpieces
Chart paper
Tape
Nose clips

 Virtual Physiology Lab #7: Respiration and Exercise

Procedure

Forced Expiratory Volume (FEV)

Indications of health can be roughly correlated with the amount of air expelled from the lungs in 1 second compared to the total vital capacity of the individual. This exercise is similar to the one in Laboratory Exercise 40 involving vital capacity, yet the measurement involved here concerns the volume of air that can be exhaled in as short a time as possible. Decreases in forced expiratory volume (FEV) may be due to asthma, emphysema, or obstructive pulmonary conditions. To perform the experiment, tape the chart paper to the drum of the respirometer and set up the apparatus as shown in Laboratory Exercise 40, figure 40.2.

You will notice that as you exhale into the respirometer the pen marks a deflection. In some respirometers an upward deflection indicates expiration; in others an upward deflection indicates inspiration. Obtain a new, disposable mouthpiece and insert it into the respirometer tube. The respirometer should be set to a speed of 1,920 mm/minute. Move the air chamber up or down until the pen registers on a line of the chart paper. Do not start the recording yet. If you are using nose clips, clean them with alcohol prior to placing them on your nose. Have your lab partner prepare to turn on the respirometer as you place the clean nose clips on your nose or pinch your nose with your fingers. Inhale maximally and signal your lab partner to turn on the recording apparatus. Exhale quickly and maximally into the mouthpiece. When the chart pen "bottoms out," turn off the respirometer.

Remove the paper from the recording drum and look for the horizontal upper line *A* (prior to respiration), the descending line *B* (during exhalation), and the lower horizontal line *C* (after respiration) (figure 41.1). Some respirometers reverse the graph (that is, as the subject exhales the line moves up on the chart paper). At a chart speed of 1,920 mm/minute, the distance between each vertical line is 1 second. You want to measure the percent of exhalation in 1 second (*D*) compared to your vital capacity or total volume of exhalation (*E*).

If you began your exhalation before or after the pen reached the vertical line on the chart paper, then measure the horizontal distance, 32 mm, from the beginning of your exhalation to the left of the chart paper (indicated by *F*). The chart paper moves at 32 mm/second at a speed of 1,920 mm/minute. Draw a vertical line (indicated by *G*) parallel to the lines on the chart paper until it intersects your exhalation line (line *B*). Examine figure 41.1 for a representation of this measurement.

Subtract the smaller volume from the larger volume to obtain the amount of air exhaled. If you are using the 9-liter Collins® respirometer you can use the numbers on the chart paper. If you are using a 13.5-liter Collins® respirometer you will need to double the value of the numbers for your vol-

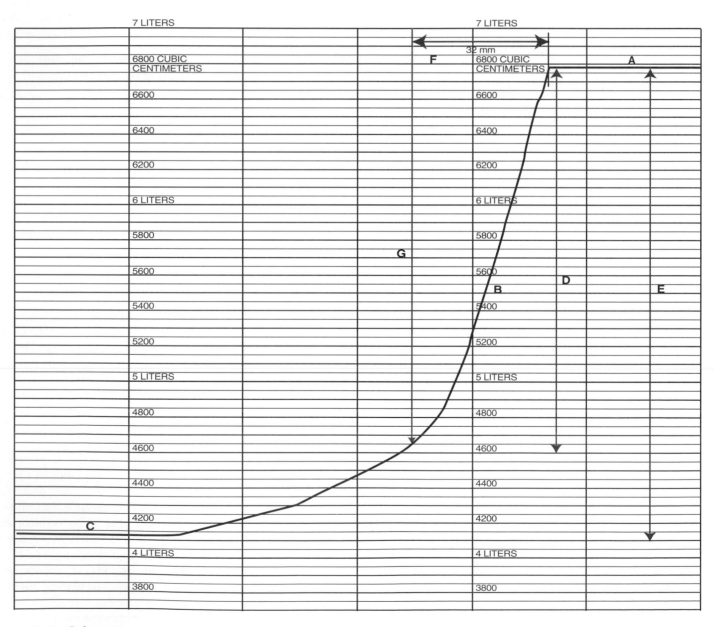

Figure 41.1   Spirogram.

ume. Record the volume of air you exhale in 1 second in the space provided.

Volume of air exhaled in 1 second: _____

Due to the changes in volume based on temperature, pressure, and water saturation you should multiply your result by 1.1 to obtain the corrected value. Enter this data below.

Corrected volume of air exhaled in 1 second
(volume × 1.1): _____

Record the vital capacity as represented by the total volume of air exhaled (subtract the value of the lesser volume from the greater volume). Record this number.

Vital capacity: _____

Multiply this number by 1.1 and record the corrected value.

Corrected vital capacity (volume × 1.1): _____

To find the FEV you need to find what percent of the vital capacity is represented by the volume of air exhaled in 1 second. Enter the values.

$$\text{Your FEV} = \frac{\text{Corrected volume of air in 1 second}}{\text{Corrected vital capacity}} \times 100$$

As an example, if the vital capacity equals 4,000 ml and FEV in 1 second equals 3,000, then the FEV is 3,000/4,000 × 100, or 75%.

If your forced expiratory volume in 1 second is greater than 75%, then this indicates you have no restrictive breathing conditions. If it is less than 75%, then this indicates potential respiratory inhibition.

Measurement of Heart Rate

The measurement of fitness in this part of the exercise is on a *voluntary basis*. It is best if the class can obtain data from a person who regularly participates in aerobic exercises (three to six times per week) and from a person who does not exercise.

Caution Do not do these exercises if you are at risk of heart disease or have a family history of heart disease or other condition, such as asthma, for which exercise is harmful. Stop if you feel exhausted or faint or have chest pain or pain radiating down the left arm.

A correlation exists between heart rate and the fitness of an individual. The resting heart rate of individuals involved in regular, active aerobic exercise is lower than the rate for sedentary individuals. In addition, the heart recovers faster in individuals who have a regular exercise program than in those who do not

exercise. In this experiment, record the measurements of at least two volunteers from the class. The greater the number of students who participate, the better the data. Read the entire exercise before beginning.

Harvard Step Test

In this exercise, students should select either the 20-inch step for males or the 16-inch step for females. With one person acting as an observer, the subject should step up on the step in 1 second and down on the floor in another second, thus completing 30 complete cycles in 1 minute. The subject should keep the body upright and keep pace with the observer's count or with a metronome set at 60 beats per minute. The subject should exercise for at least 3 minutes but no more than 5 minutes. If the subject stops due to exhaustion, then the observer should note the time of exercise. After the period of exercise, the subject should sit and rest for 1 minute. The pulse should be taken from 1 minute to 1 minute 30 seconds and recorded below.

Pulse from 1 minute to 1 minute 30 seconds: _____

The subject should rest for another 30 seconds, before the pulse is taken from 2 minutes to 2 minutes 30 seconds after exercise and recorded here.

Pulse from 2 minutes to 2 minutes 30 seconds:

_____

The subject should remain seated for another 30 seconds before the pulse is taken from 3 minutes to 3 minutes 30 seconds and recorded here.

Pulse from 3 minutes to 3 minutes 30 seconds: _____

Add the three pulse counts recorded.

Sum of three pulse counts: _____

To determine the personal fitness index (PFI) use the following formula:

$$\text{PFI} = \frac{\text{Number of seconds of exercise} \times 100}{2\ (\text{sum of three pulse counts})}$$

Therefore, if you exercised for 4 minutes and your pulse counts were 50, 48, and 45, then the PFI would be:

$$\text{PFI} = \frac{(4 \times 60) \times 100}{2\ (50 + 48 + 45)} = \frac{240 \times 100}{2\ (143)} = \frac{24,000}{286} = 84$$

Determine the fitness evaluation of the subject using the following data:

Below 55: poor physical condition
55–64: low average physical condition
65–79: high average physical condition
80–90: good physical condition
90 and above: excellent physical condition

REVIEW

Name _____

1. Record your FEV, or the one you measured in lab: _____

2. Does this value fall within normal limits?

3. What is your personal fitness index, or the one measured in lab?

4. If the heart rate after 5 minutes of exercise was 70, 68, and 66 beats in the consecutive 30-second trials, what would the personal fitness index be and what condition would that represent?

5. In figure 41.1 the experiment was conducted with a 13.5-liter respirometer. Calculate the FEV of this individual.

Anatomy of the Digestive System

Introduction

The digestive system can be divided into two major parts, the **alimentary canal** and the **accessory organs.** The alimentary canal is a long tube that runs from the mouth to the anus and comes into contact with food or the breakdown products of digestion. The alimentary canal is sometimes known as the "food tube." Some of the organs of the alimentary canal are the esophagus, stomach, intestines, rectum, and anus. The accessory organs are important in that they secrete many important substances necessary for digestion, yet these organs do not come into direct contact with food. Examples of accessory organs are the salivary glands, liver, gallbladder, and pancreas.

The functions of the digestive system are many and include ingestion of food, physical breakdown of food, chemical breakdown of food, food storage, water absorption, vitamin synthesis, food absorption, and elimination of indigestible material.

In this exercise, you examine the anatomy of the digestive system in both human and cat, correlating the structure of the digestive organs with their functions.

Objectives

At the end of this exercise you should be able to

1. list, in sequence, the major organs of the alimentary canal;
2. describe the basic function of the accessory digestive organs;
3. note the specific anatomical features of each major digestive organ;
4. describe the layers of the wall of the gastrointestinal tract;
5. describe the major functions of the stomach and small and large intestines.

Materials

Models, charts, or illustrations of the digestive system
Mirror
Cats
Dissection trays
Pins
Scalpels or razor blades
Latex gloves
Waste container
Skull, human teeth, or cast of teeth
Cadaver (if available)
Microscopes

Microscope slides:

Esophagus
Stomach
Small intestine
Large intestine
Liver

Procedure

Begin this exercise by examining a torso model or charts in the lab and compare these to figure 42.1. Locate the major digestive organs and place a check mark in the appropriate space.

_____ Oral cavity
_____ Pharynx
_____ Esophagus
_____ Stomach
_____ Small intestine
_____ Large intestine (colon)
_____ Liver
_____ Vermiform appendix
_____ Pancreas
_____ Gallbladder
_____ Salivary glands

Begin your study of the alimentary tract with the mouth and the oral cavity. Examine a midsagittal section of the head as represented in figure 42.2 and locate the major anatomical features.

At the beginning of the alimentary canal is the **mouth,** which is the opening surrounded by lips, or labia. The **labial frenulum** is a membranous structure that keeps the lip adhered to the gums, or **gingivae.** Behind the mouth is the **oral cavity,** which is a space bordered in front by the mouth, behind by the oropharynx, and on the sides by the inner wall of the cheeks. The hard and soft palates form the roof of the oral cavity, and the floor of the chin is the inferior border. The **hard palate** is composed of the palatine processes of the maxillae and palatine bones. The **soft palate** is composed of connective tissue and a mucous membrane. At the posterior portion of the oral cavity is the **uvula,** a small grapelike structure that is suspended from the posterior edge of the soft palate. The uvula helps to prevent food or liquid from moving

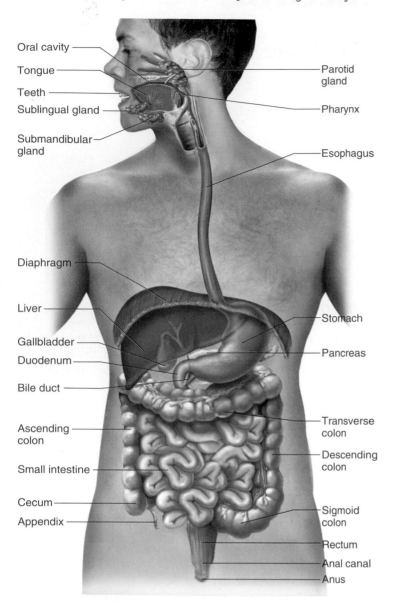

Figure 42.1 Overview of the digestive system. ✗ ▭

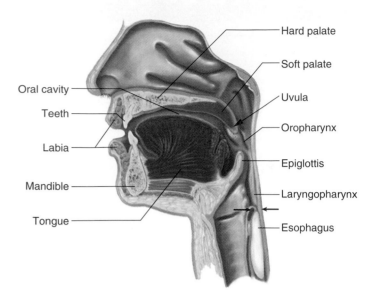

Figure 42.2 Oral cavity, midsagittal section.

the frictional surface of the tongue. **Taste buds** occur on the tongue along the sides of the papillae. The sense of taste is covered in Laboratory Exercise 25.

The oral cavity is important in digestion for the physical breakdown of food. This process is driven by powerful muscles called the **muscles of mastication.** The **masseter** and the **temporalis muscles,** are involved in the closing of the jaws, and the **pterygoid muscles** are important in the sideways grinding action of the molar and premolar teeth.

Structure of Teeth

Examine models of teeth, dental casts, or real teeth on display in the lab and compare these to figure 42.3. A tooth consists of a **crown, neck,** and **root.** The crown is the exposed part of the tooth; the neck is a constricted portion of the tooth that normally occurs at the surface of the gingivae; and the root is embedded in the jaw. Examine a model or illustration of a longitudinal section of a tooth and find the outer **enamel,** an extremely hard material. Inside this layer is the **dentin,** which is made of bonelike material. The innermost portion of the tooth consists of the **pulp cavity,** which leads to the **root canal,** a passageway for nerves and blood vessels into the tooth. The nerves and blood vessels enter the tooth through the **apical foramen** at the tip of the root of the tooth. The teeth occur in depressions in the mandible or maxilla called **alveolar sockets** and are anchored into the bone by periodontal membrane.

There are four different types of teeth in the adult mouth:

- **Incisors** are flat, bladelike front teeth that serve to nip food. There are eight incisors in the adult mouth.
- **Cuspids,** or **canines,** are the pointed teeth just lateral to the incisors that function in shearing food. There are four

into the nasal cavity during swallowing. The oral cavity is lined with **nonkeratinized stratified squamous epithelium,** which serves to protect the underlying tissue from abrasion. Using a mirror, examine your **tongue,** which is made of skeletal muscle. One of the major muscles of the tongue is the **genioglossus.** The tongue is important in speech, taste, the movement of food towards the teeth for chewing, and swallowing. The tongue acts as a piston to propel food to the **oropharynx,** the space behind the oral cavity. The tongue is held down to the floor of the mouth by a thin mucous membrane called the **lingual frenulum.** There are three types of **papillae,** or raised areas, on the tongue. These are **fungiform, filiform,** and **vallate** (circumvallate) papillae. Papillae increase

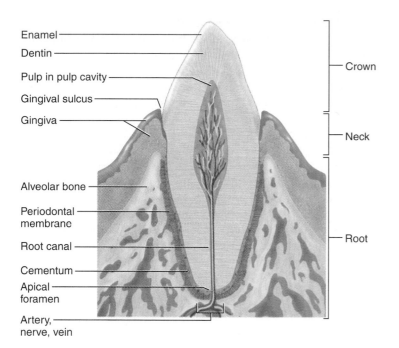

Figure 42.3   Tooth, longitudinal section.

Labels (top to bottom, left):
Enamel
Dentin
Pulp in pulp cavity
Gingival sulcus
Gingiva
Alveolar bone
Periodontal membrane
Root canal
Cementum
Apical foramen
Artery, nerve, vein

Labels (right):
Crown
Neck
Root

cuspids in the adult, and they can be identified as the teeth with just one cusp or point.

• **Premolars,** or **bicuspids,** are lateral to the cuspids and serve to grind food. There are typically eight premolars in adults, and they can be identified by the teeth with two cusps.

• **Molars** are found closest to the oropharynx. These serve to grind food, and there are 12 molar teeth in the adult mouth (including the wisdom teeth). Molars typically have three to five cusps.

Examine a human skull or dental cast and identify the characteristics of the four different types of teeth (figure 42.4).

Humans have two sets of teeth: the **deciduous,** or "milk," **teeth** appear first, and these are replaced by the **permanent teeth.** There are 20 deciduous teeth. Children have no pre-molars and eight molar teeth. Compare figure 42.4, the adult pattern, to figure 42.5, the deciduous teeth.

Frequently the pattern of tooth structure is represented by a **dental formula,** which describes the teeth by quadrants. The dental formula for the deciduous teeth is illustrated here. I = incisor, C = cuspid, P = premolar, and M = molar.

Deciduous teeth (20 total)

dental formula	I	C	P	M	One side (quadrant)
	2	1	0	2	Top
	2	1	0	2	Bottom

The upper numbers refer to the number of teeth in the maxilla on one side of the head. The lower numbers refer to the number of teeth in the mandible on one side of the head. The adult dental formula is as follows:

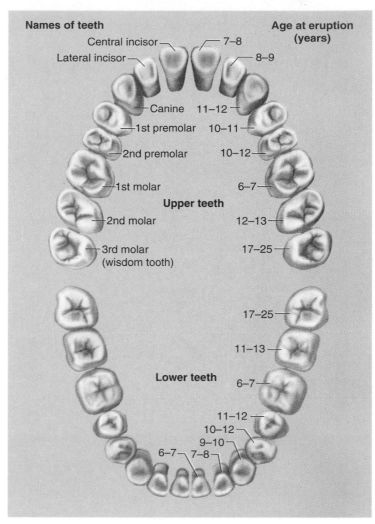

Figure 42.4   Upper and lower teeth of an adult.

Names of teeth (labels):
Central incisor — 7–8
Lateral incisor — 8–9
Canine   11–12
1st premolar   10–11
2nd premolar   10–12
1st molar   6–7
Upper teeth
2nd molar   12–13
3rd molar (wisdom tooth)   17–25

Lower teeth labels (ages): 17–25, 11–13, 6–7, 11–12, 10–12, 9–10, 6–7, 7–8

Age at eruption (years)

Adult (permanent) teeth (32 total)

I	C	P	M
2	1	2	3
2	1	2	3

Oropharynx

The **oropharynx** is the space behind the oral cavity that serves as a common passageway for air, food, and water. Above the oropharynx is the **nasopharynx,** which leads to the nasal cavity, and below the oropharynx is the **laryngopharynx,** which leads to the larynx and the esophagus. The oropharynx is composed of nonkeratinized stratified squamous epithelium. Muscles around the wall of the oropharynx are the **pharyngeal constrictor muscles** and are involved in swallowing. Food is moved by the tongue to the region of the pharynx where it is propelled into the esophagus. Locate the oropharynx and the esophagus in figure 42.2.

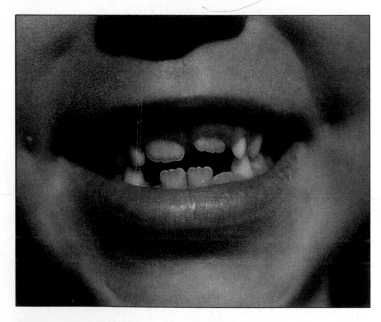

Names of teeth

Age at eruption (months)

Central incisor — 7–9
Lateral incisor — 9–11
Canine — 18–20
1st molar — 14–16
Upper teeth
2nd molar — 24–26

20–22
Lower teeth 12–14
16–18
7–9
6–8

Figure 42.5   Upper and lower teeth of a child.

Esophagus

The esophagus is also known commonly as the "food tube." It conducts food from the oropharynx, through the diaphragm, and into the stomach. Normally the esophagus is a closed tube that begins at about the level of the sixth cervical vertebra. As a lump of food, or **bolus,** enters the esophagus, skeletal muscle begins to move it towards the stomach. The middle portion of the esophagus is composed of both **skeletal** and **smooth muscle,** while the lower portion of the esophagus is made of smooth muscle. In the lower region of the esophagus the smooth muscle contracts, moving the bolus by a process known as **peristalsis.** The esophagus has an inner epithelial lining of **stratified squamous epithelium** and an outer connective tissue layer called the **adventitia.** Identify the four layers of the esophagus—mucosa, submucosa, muscularis, and adventitia—under the microscope. The portion of the esophagus that comes into contact with food is called the **lumen,** which continues through the gastrointestinal tract. The lower portion of the esophagus has an **esophageal sphincter,** which prevents the backflow of stomach acids. **Heartburn** occurs if the stomach contents pass through the esophageal sphincter and irritate the esophageal lining.

Abdominal Portions of the Alimentary Canal

The inner structure of the body has been referred to as a "tube within a tube." The body wall forms the outer tube and the gastrointestinal tract, including the stomach, small intestine, and large intestine, forms the inner tube. Specialized serous membranes cover the various organs and line the inner wall of the **coelom** (the body cavity). The membrane lining the outer surface of the gastrointestinal tract is called the **visceral peritoneum** (serosa) and continues as a double-folded membrane called the **mesentery,** which attaches the tract to the back of the body wall. The mesentery is continuous with the membrane on the inner side of the body wall where it is called the **parietal peritoneum.** Locate these three membranes, in figure 42.6.

The gastrointestinal tract has a particular series of layers that can be examined microscopically. In general, the stomach, small intestine, and large intestine have the same layers from the lumen to the coelom. The innermost layer is the **mucosa,** which consists of a mucous membrane closest to the lumen, a connective tissue layer called the **lamina propria,** and an outer muscular layer called the **muscularis mucosa.**

The next layer is called the **submucosa,** which is mostly made of connective tissue and contains numerous blood vessels. The next layer is the **muscularis** (or **muscularis externa**), which is typically made of two or three layers of smooth muscle. The muscularis functions to propel material through the gastrointestinal tract and to mix ingested material with digestive juices. The outermost layer is called the **serosa,** or **visceral peritoneum,** and this layer is closest to the coelom. Examine a microscope slide of the gastrointestinal tract (small intestine) and locate these three membranes in figure 42.7.

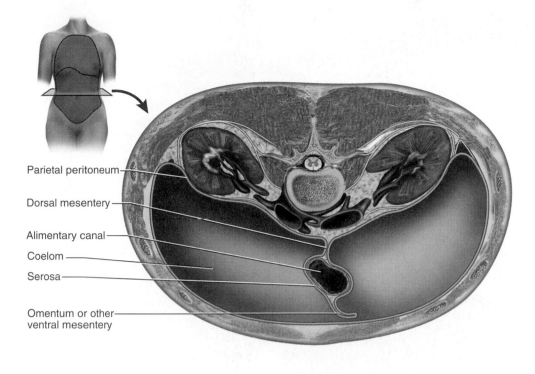

Parietal peritoneum

Dorsal mesentery

Alimentary canal

Coelom

Serosa

Omentum or other
ventral mesentery

Anterior

Figure 42.6   Membranes of the gastrointestinal tract.

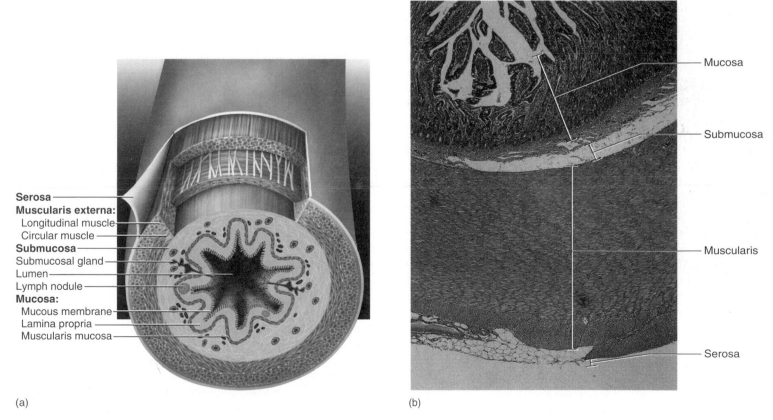

Serosa
Muscularis externa:
  Longitudinal muscle
  Circular muscle
Submucosa
Submucosal gland
Lumen
Lymph nodule
Mucosa:
  Mucous membrane
  Lamina propria
  Muscularis mucosa

(a)

Mucosa

Submucosa

Muscularis

Serosa

(b)

Figure 42.7   Gastrointestinal tract, cross section. (a) Diagram; (b) photomicrograph (100×).

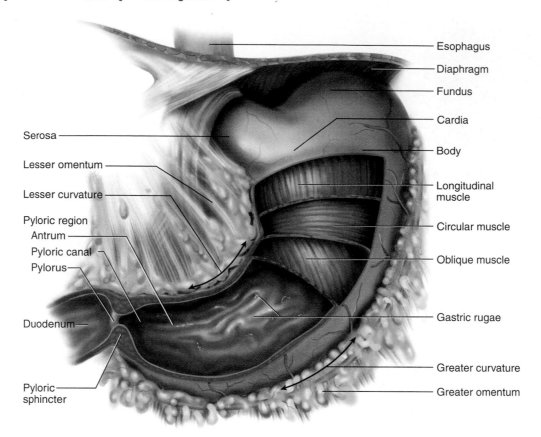

Figure 42.8 Gross anatomy of the stomach.

Stomach

The **stomach** is located on the left side of the body and receives its contents from the esophagus. The food that enters the stomach is stored and mixed with enzymes and hydrochloric acid to form a soupy material called **chyme**. The stomach can have a pH as low as 1 or 2. Chyme remains in the stomach as the acids denature proteins and enzymes reduce proteins to shorter fragments. The acid of the stomach also has an antibacterial action as most microbes do not grow well in conditions of low pH.

Examine a model of the stomach or charts in the lab and compare these to figure 42.8. Locate the upper portion of the stomach called the **cardia**, or **cardiac region**. A part of the cardia extends superiorly as a domed section called the **fundus**. The main part of the stomach is called the **body** and the terminal portion of the stomach, closest to the small intestine, is called the **pyloric region**. The pyloric region has an expanded area called the **antrum** and a narrowed region called the **pyloric canal**. The opening to the duodenum is by the pylorus and is closed by the **pyloric sphincter**. The left side of the stomach is arched and forms the **greater curvature**, while the right side of the stomach is a smaller arch forming the **lesser curvature**. The inner surface of the stomach has a series of folds called **rugae**, which allow for significant expansion of the stomach. The contents are held in the stomach by two sphincters. The upper, or **esophageal, sphincter** prevents stomach

contents from moving into the esophagus. The lower sphincter, the pyloric sphincter, prevents the premature release of stomach contents into the small intestine. Locate the pyloric sphincter at the terminal portion of the stomach.

Stomach Histology

Examine a prepared slide of stomach. Identify the four primary layers, **mucosa, submucosa, muscularis,** and **serosa.** Notice how the mucosa in the prepared slide has a series of indentations. These depressions are **gastric glands,** which occur in the inner lining of the stomach. The mucous membrane is composed predominantly of **simple columnar epithelium** along with several specialized cells. **Goblet cells** occur in the membrane and secrete **mucus,** which protects the stomach lining from erosion by stomach acid and **proteolytic** (protein-digesting) **enzymes.** Other specialized cells that you might find in the mucosa are **chief cells,** which secrete **pepsinogen** (the inactive state of a proteolytic enzyme). Chief cells contain blue-staining granules in some prepared slides. Other cells are **parietal cells,** which secrete HCl. They typically contain orange-staining granules. When pepsinogen comes into contact with HCl it is activated as pepsin.

Deeper to the mucous membrane locate the **lamina propria,** which is usually lighter in color. The **muscularis mucosa** is even farther away from the lumen and serves to move

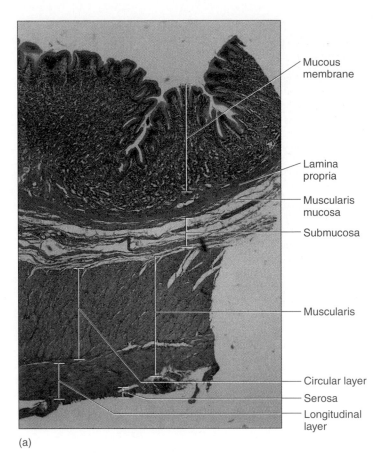

(a)

(b)

Figure 42.9 Histology of the stomach. (a) Overview (40×); (b) mucosa (100×).

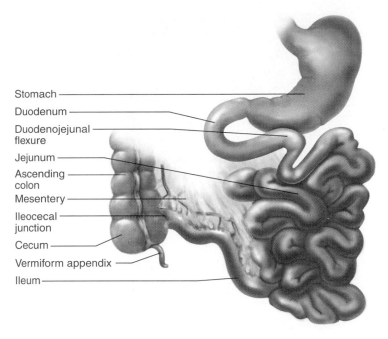

Figure 42.10 Gross anatomy of the small intestine.

Small Intestine

The **small intestine** is approximately 5 m (17 ft) long. It is called the small intestine because it is small in diameter. The small intestine is typically 3 to 4 cm (1.5 in.) in diameter when empty. The primary function of the small intestine is nutrient absorption.

Locate the three major regions of the small intestine on a model or chart and compare them to figure 42.10. The first part of the small intestine is the **duodenum,** which is a C-shaped structure attached to the pyloric region of the stomach. The duodenum begins in the body cavity and then moves behind the parietal peritoneum and returns to the body cavity. The duodenum is approximately 25 cm (10 in.) long. It receives fluid from both the **pancreas** and the **gallbladder.** The gallbladder releases **bile,** which emulsifies lipids, into the duodenum. The lipids break into smaller droplets, which increase the surface area for digestion. The bile may be transported to the duodenum by the common bile duct. The pancreas secretes many digestive enzymes and buffers, which neutralize the stomach acids. These are secreted into the duodenum by the **pancreatic duct.** In some cases the pancreatic duct joins with the common bile duct to form the **hepatopancreatic ampulla** (ampulla of Vater).

The second portion of the small intestine is the **jejunum,** which is approximately 2 m (6.5 ft) in length. The junction between the jejunum and the duodenum is known as the **duodenojejunal flexure.** The terminal portion of the small intestine is the **ileum,** which is approximately 3 m (10 ft) in length. A closure between the small intestine and large intestine is called the **ileocecal valve.** This valve keeps material in the large intestine from reentering the small intestine. Locate the small intestine and associated structures in figure 42.10.

the mucous membrane. The **submucosa** is the next layer and is typically lighter in color in prepared slides.

The next layer of the stomach is the **muscularis.** In some parts of the stomach there are three layers of the muscularis—an **inner oblique layer,** a **middle circular layer,** and an **outer longitudinal layer.** The muscularis serves to move **chyme** from the stomach through the pyloric sphincter and into the small intestine.

The outermost layer is the **serosa,** and it is composed of a thin layer of connective tissue and **simple squamous epithelium.** Locate these structures and compare them to figure 42.9.

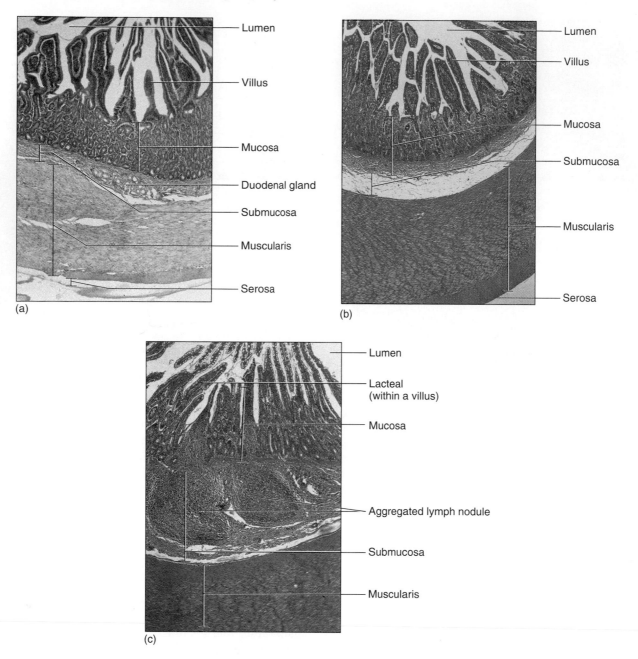

Figure 42.11   **Three sections of small intestine (40×).** (a) Duodenum; (b) jejunum; (c) ileum.

Histology of the Small Intestine

The small intestine, due to the presence of **villi,** can be distinguished from both the stomach and the large intestine. Villi are fingerlike projections that increase the surface area of the mucosa. Each villus contains **blood vessels,** which transport sugars and amino acids from the intestine to the liver. In addition to this the villi contain **lacteals,** which transport fatty acids via lymphatics to the venous system. The villi give the lining of the small intestine a velvety appearance to the naked eye. As with the stomach, the inner lining of the small intestine consists of **simple columnar epithelium** with **goblet cells.** You may distinguish the three sections of the small intestine by noting that the duodenum has **duodenal glands** in the portion of the mucosa away from the lumen. The jejunum and the ileum lack these glands. The ileum is distinguished by the presence of **aggregated lymph nodules,** or **Peyer's patches,** present in the submucosa. These lymph nodules produce lymphocytes, which protect the body from the bacterial flora in the lumen of the small intestine. Compare the prepared slides of the small intestine to figure 42.11.

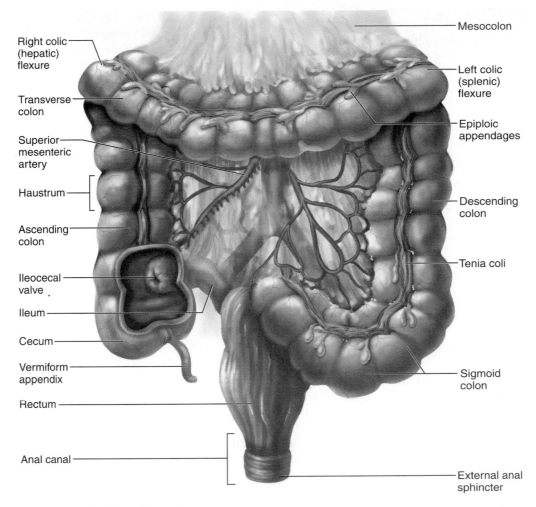

Right colic (hepatic) flexure

Transverse colon

Superior mesenteric artery

Haustrum

Ascending colon

Ileocecal valve

Ileum

Cecum

Vermiform appendix

Rectum

Anal canal

Mesocolon

Left colic (splenic) flexure

Epiploic appendages

Descending colon

Tenia coli

Sigmoid colon

External anal sphincter

Figure 42.12   Gross anatomy of the large intestine.

Large Intestine

The **large intestine** is so named because it is large in diameter. The large intestine is approximately 7 cm (3 in.) in diameter and 1.4 m (4.5 ft) in length. The function of the large intestine is absorption of water, and formation of feces. The mucosa of the large intestine is made of **simple columnar epithelium** with a large number of **goblet cells.** There are no villi present, nor aggregated lymph nodules, yet the wall of the large intestine has solitary lymph nodules.

Look at a model or chart of the large intestine and note the major regions. Compare these to figure 42.12.

- **Cecum:** first part of the large intestine. The cecum is a pouchlike area that articulates with the small intestine at the level of the ileocecal valve.
- **Ascending colon:** found on the right side of the body. It becomes the transverse colon at the hepatic (right colic) flexure.
- **Transverse colon:** traverses the body from right to left. It leads to the descending colon at the splenic (left colic) flexure.

- **Descending colon:** passes inferiorly on the left side of the body and joins with the sigmoid colon.
- **Sigmoid colon:** an S-shaped segment of the large intestine in the left inguinal region.
- **Rectum:** a straight section of colon found in the pelvic cavity. The rectum has superficial veins in its wall called **hemorrhoidal veins.** These may enlarge and cause the uncomfortable condition known as **hemorrhoids.**

The large intestine has some unique structures, including small areas that form pouches or puckers in the intestinal tract. The longitudinal layer of the muscularis of the large intestine is not found as a continuous sheet but occurs along the length of the large intestine as three bands called **tenia coli.** These muscles contract and form pouches or puckers in the intestinal tract called **haustra** (singular, *haustrum*). Another unique feature of the outer wall of the large intestine are fat lobules called **epiploic appendages.** Locate these structures in figure 42.12.

Fecal material passes through the large intestine by peristalsis and is stored in the rectum and sigmoid colon. **Defecation** occurs as **mass peristalsis** causes a bowel movement.

Histology of the Large Intestine

Examine a prepared slide of the large intestine and compare it to figure 42.13. The large intestine is distinguished from the small intestine by the absence of villi, and from the stomach by the presence of large numbers of goblet cells. Examine your slide for these characteristics.

Anal Canal

The **anal canal** is not part of the large intestine but is a short tube that leads to an external opening termed the anus. Locate the anal canal in figure 42.12.

Accessory Structures

The accessory structures are frequently associated with segments of the alimentary canal. The first of these structures, the **salivary glands,** are located in the head and secrete **saliva** into the oral cavity. Saliva is a watery secretion that contains **mucus** (a protein lubricant) and **salivary amylase,** a starch-digesting enzyme. The average adult secretes about 1.5 liters of saliva per day.

There are three pairs of salivary glands. The **parotid glands** are located just anterior to the ears. Each gland secretes saliva through a **parotid duct,** a tube that traverses the buccal (cheek) region and enters the oral cavity just posterior to the upper second molar. The second pair, the **submandibular glands,** are located just medial to the mandible on each side of the face. The submandibular glands secrete saliva into the oral cavity by a single duct on each side of the oral cavity inferior to the tongue. The third pair, the **sublingual glands,** are located inferior to the tongue, which open into the oral cavity by several ducts. Locate these salivary glands on a model of the head and in figure 42.14.

Vermiform Appendix

The **vermiform appendix** is about the size of your little finger and is located near the junction of the small and large intestines (at the region of the ileocecal valve). Locate the appendix on a torso model or chart in the lab and compare it to figure 42.15. The appendix is absent in the cat.

Omenta

The **lesser omentum** is an extension of the peritoneum that forms a double fold of tissue between the stomach and the liver. The **greater omentum** is a section of peritoneum that originates between the stomach and transverse colon and drapes over the intestines as a fatty apron. Locate the lesser and greater omenta in figure 42.16.

Liver

The **liver** is a complex organ with numerous functions, some of which are digestive and many of which are not. The liver

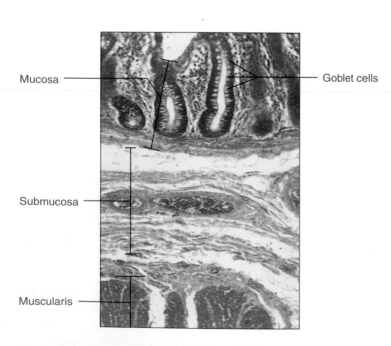

Figure 42.13   Histology of the large intestine (100×).

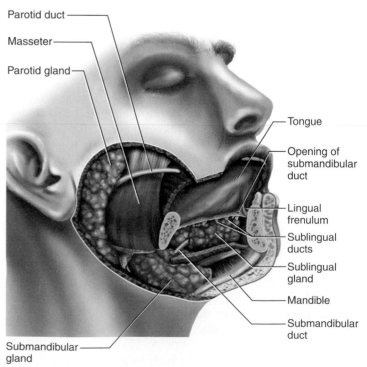

Figure 42.14   Salivary glands.

processes digestive material from the vessels returning blood from the intestines and has a role in either moving nutrients into the bloodstream or storing them in the liver tissue. The liver also produces blood plasma proteins, detoxifies harmful material that has been produced by, or introduced into, the body and produces bile.

Examine a model or chart of the liver and note its relatively large size. The liver is located on the right side of the body and is divided into four lobes, the **right, left, quadrate,** and **caudate lobes.** The right lobe of the liver is the largest lobe; the left is also fairly large. The quadrate lobe is located in the middle portion of the liver ventral to the caudate lobe. Traversing through the liver is the **falciform ligament,** which attaches the liver to the inferior side of the diaphragm. Locate the **gallbladder,** which is on the inferior aspect of the liver, and compare the gallbladder and liver structures to figure 42.17.

Liver Histology

Examine a prepared slide of liver tissue. Note the hexagonal structures in the specimen. These are known as **liver lobules.** Each lobule has a blood vessel in the middle called the **central vein.** Vessels that carry blood to the central vein are the liver **sinusoids,** and these are lined with a double row of cells called **hepatocytes.** Hepatocytes carry out the various functions of the liver as just described. The tissue of the liver is extremely vascular and functions like a great sponge. Fresh, oxygenated blood from the hepatic artery and deoxygenated blood from the hepatic portal vein mix in the liver. **Kupffer cells** occur

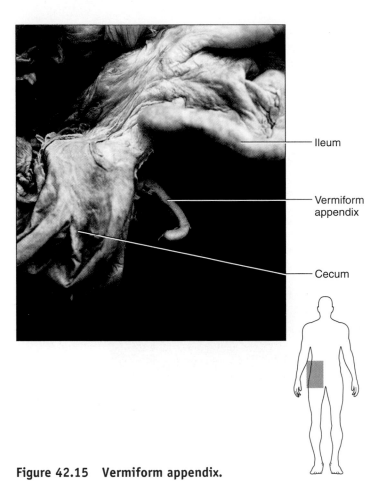

Ileum

Vermiform appendix

Cecum

Figure 42.15   Vermiform appendix.

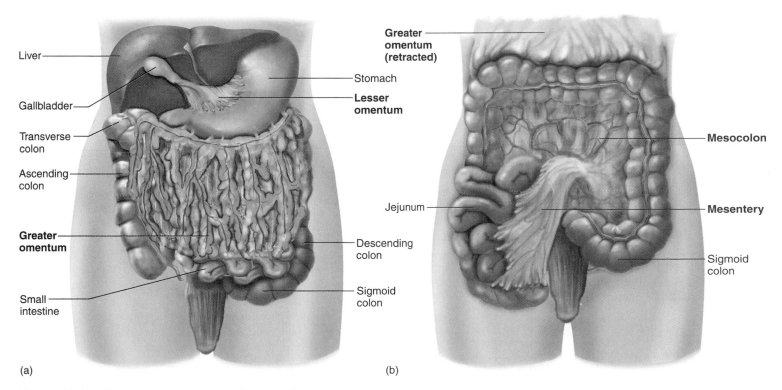

Liver

Gallbladder

Transverse colon

Ascending colon

Greater omentum

Small intestine

Stomach

Lesser omentum

Descending colon

Sigmoid colon

(a)

Greater omentum (retracted)

Jejunum

Mesocolon

Mesentery

Sigmoid colon

(b)

Figure 42.16   Omenta, mesentery, and mesocolon.

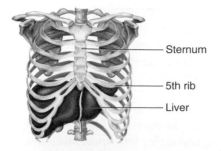

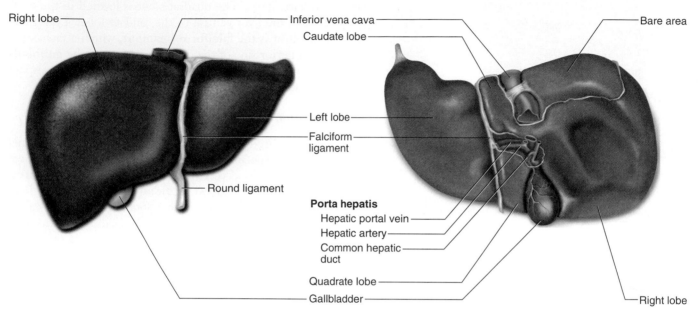

Figure 42.17 Liver.

throughout the liver tissue and function as phagocytic cells in the liver. Locate the lobules, central vein, sinusoids, and hepatocytes in a prepared slide using figure 42.18 to aid you.

Pancreas

The **pancreas** is located inferior to the stomach and is located on the left side of the body. It has both endocrine and exocrine functions. The hormonal function of the pancreas is covered in Laboratory Exercise 28. The exocrine function of the pancreas is studied in Laboratory Exercise 43. The pancreas consists of a **tail,** found near the spleen, an elongated **body,** and a rounded **head** found near the duodenum. Enzymes and buffers pass from the tissue of the pancreas into the **pancreatic duct** and then into the duodenum. Locate the pancreatic structures as represented in figure 42.19.

Gallbladder

The **gallbladder** is located just inferior to the liver. The liver is the site of bile production, and the bile flows from the liver

through **left** and **right hepatic ducts** to enter the **common hepatic duct.** Once in the common hepatic duct the bile flows into the **cystic duct** and then is stored in the gallbladder. As the stomach begins to empty its contents into the duodenum the gallbladder constricts, and bile flows from the gallbladder back into the cystic duct and into the **bile duct,** which empties into the duodenum. Locate these structures on a model and compare them to figure 42.19.

Cat Dissection

Prepare for the cat dissection by obtaining a dissection tray, scalpel, scissors or bone cutter, string, forceps, and plastic bag with label. Remember to place all excess tissue in the appropriate waste container and not in a standard wastebasket or down the sink!

Begin your study of the digestive system of the cat by locating the large **salivary glands** around the face. You must first remove the skin from in front of the ear. If you have dissected the face for the musculature you may have already removed the **parotid gland,** which is a spongy, cream-colored

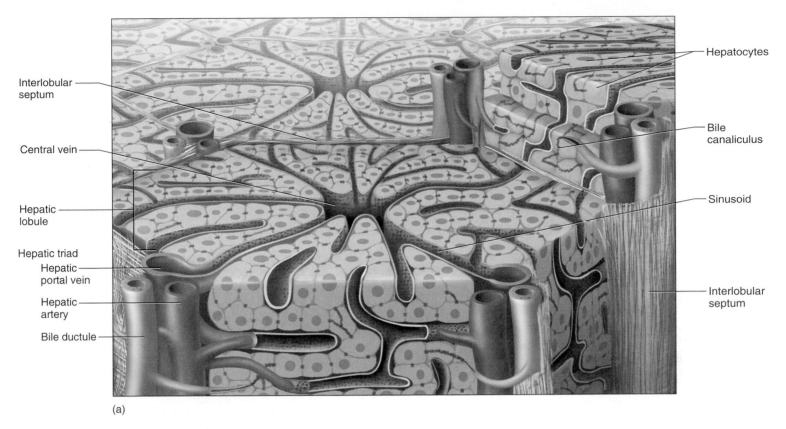

Interlobular septum

Central vein

Hepatic lobule

Hepatic triad

Hepatic portal vein

Hepatic artery

Bile ductule

Hepatocytes

Bile canaliculus

Sinusoid

Interlobular septum

(a)

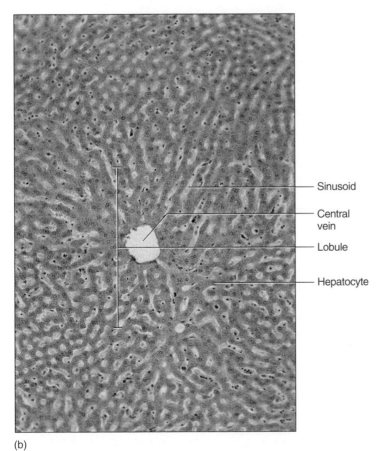

Sinusoid

Central vein

Lobule

Hepatocyte

(b)

Figure 42.18   Histology of the liver. (a) Diagram; (b) photomicrograph (100×).

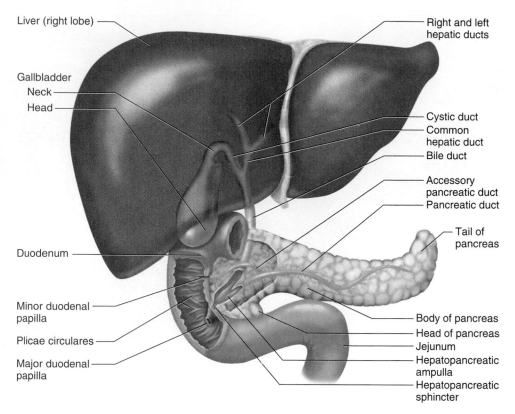

Figure 42.19   **Liver, gallbladder, and pancreas.**

pad anterior to the ear. You can also locate the **submandibular gland** as a pad of tissue slightly anterior to the angle of the mandible and lateral to the digastric muscle. The **sublingual gland** is just anterior to the submandibular gland and is an elongated gland that parallels the mandible. Use figure 42.20 to help you locate these glands.

Dissection of the head of the cat should be done only if your instructor gives you permission to do so. To dissect the head you should first use a scalpel and make a cut in the midsagittal plane from the forehead of the cat to the region of the occipital bone. Cut through any overlying muscle in the cat. You should then use a small saw and gently cut through the cranial region of the skull being careful to stay in the midsagittal plane. Once you make an initial cut through the dorsal side of the head, you should cut the mental symphysis of the cat thus separating the mandible.

You can then use a long knife (or scalpel) to cut through the softer regions of the skull, brain and tongue. Use a scalpel to cut through the floor of the oral cavity to the hyoid bone. It is best to use a scissors or small bone cutter to cut through the hyoid bone. This should allow you to open the head and examine the structures seen in a midsagittal section as seen in figure 42.21. Locate the hard palate, tongue, oral cavity, and oropharynx. Note how the larynx in the cat is closer to the tongue than in humans.

Examine the **tongue** of the cat and note the numerous papillae on the dorsal surface. These are **filiform papillae** and have both a digestive and a grooming function. You can lift up the tongue and examine the **lingual frenulum** on the ventral surface.

Look for the muscular tube of the esophagus by carefully lifting the trachea ventrally away from the neck. You should be able to insert your blunt dissection probe into the oral cavity and gently wiggle it into the esophagus. The tongue and esophagus are represented in figure 42.21.

Abdominal Organs

To get a better view of the abdominal organs it is best to cut the **diaphragm** away from the body wall. Carefully cut the lateral edges of the diaphragm from the ribs. You should see a fatty drape of material covering the intestines. This is the **greater omentum.** Just caudal to the diaphragm note the dark brown multilobed **liver.** In the middle of the liver is the green **gallbladder.** To the left of the liver (in reference to the cat) is the J-shaped **stomach.** If you lift the liver you should be able to see the **lesser omentum,** which is a fold of tissue that connects the stomach to the liver. Locate these structures in figure 42.22.

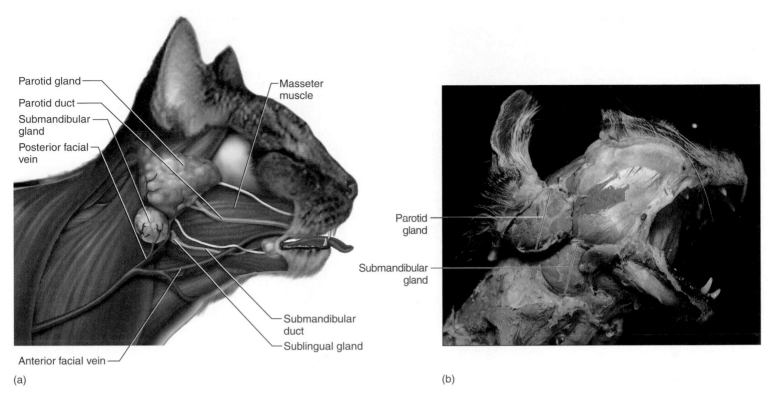

Parotid gland
Parotid duct
Submandibular gland
Posterior facial vein
Masseter muscle
Submandibular duct
Sublingual gland
Anterior facial vein
(a)

Parotid gland
Submandibular gland
(b)

Figure 42.20   Salivary glands of the cat.

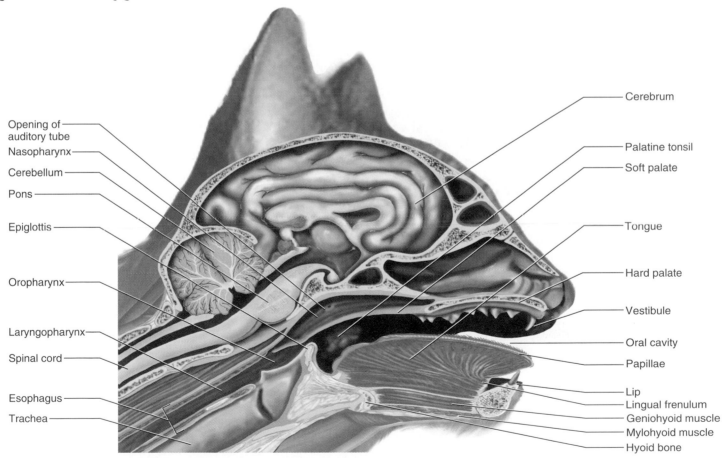

Opening of auditory tube
Nasopharynx
Cerebellum
Pons
Epiglottis
Oropharynx
Laryngopharynx
Spinal cord
Esophagus
Trachea

Cerebrum
Palatine tonsil
Soft palate
Tongue
Hard palate
Vestibule
Oral cavity
Papillae
Lip
Lingual frenulum
Geniohyoid muscle
Mylohyoid muscle
Hyoid bone

Figure 42.21   Midsagittal section of the head and neck of a cat.

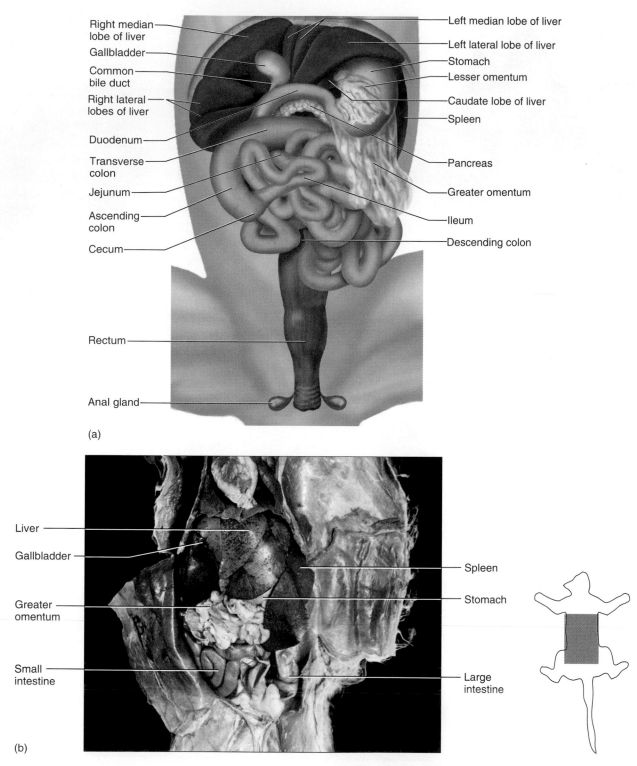

Right median lobe of liver

Gallbladder

Common bile duct

Right lateral lobes of liver

Duodenum

Transverse colon

Jejunum

Ascending colon

Cecum

Rectum

Anal gland

Left median lobe of liver

Left lateral lobe of liver

Stomach

Lesser omentum

Caudate lobe of liver

Spleen

Pancreas

Greater omentum

Ileum

Descending colon

(a)

Liver

Gallbladder

Greater omentum

Small intestine

Spleen

Stomach

Large intestine

(b)

Figure 42.22 Abdominal organs.

If you make an incision into the stomach you should be able to see the folds known as **rugae.** Place a blunt probe inside the stomach and move it anteriorly. Notice how the **esophageal sphincter** makes it difficult to pierce the diaphragm. If you do move the probe into the esophagus you should be able to see it as you look anterior to the diaphragm.

The stomach has an anterior **cardia,** a domed **fundus, greater** and **lesser curvatures,** and a **pyloric region.** Cut into the pyloric region of the stomach and through the duodenum. You should be able to locate a tight sphincter muscle between the stomach and the duodenum. This is the **pyloric sphincter.** As you move into the duodenum you may have to scrape some of the chyme away from the wall of the small intestine to be able to see the fuzzy texture of the intestinal wall. This texture is due to the presence of **villi.** As in the human, the small intestine is composed of three regions, the **duodenum,** the **jejunum,** and a terminal **ileum.** Compare the stomach and small intestine to figure 42.22.

If you elevate the greater omentum you should be able to see the **pancreas,** along the caudal side of the stomach. The pancreas appears granular and brown. The **tail** of the pancreas is near the spleen, which is a brown, elongated organ on the left side of the body.

The **small intestine** is an elongated coiled tube about the diameter of a wooden pencil. Note the **mesentery,** which holds the small intestine to the posterior body wall near the vertebrae. The small intestine is extensive in the cat and rapidly expands into the large intestine. At the junction of the small and large intestines note that there is no appendix. The **large intestine** in the cat is a much shorter tube with a diameter slightly larger than your thumb. The first part of the large intestine is a pouch called the **cecum.** The remainder of the large intestine can be further divided into the **ascending, transverse, descending colon,** and **rectum.** Compare these to figure 42.22. Examine the **parietal peritoneum** along the inner surface of the body wall and the visceral peritoneum that envelops the intestines.

Clean Up When you are done with your dissection carefully place your cat back in the plastic bag.

Name _____

1. The pancreas belongs to what part of the digestive system (alimentary canal/accessory organ)?

2. The descending colon belongs to what part of the digestive system (alimentary canal/accessory organ)?

3. Name the layer of material that is found on the outer surface of the stomach and small intestine.

4. In the stomach, what is the semidigested food called?

5. The portion of the stomach closest to the small intestine is what region?

6. What cell type is found as the inner lining of the mucosa (near the lumen)?

7. How many meters in length is the large intestine?

8. What is the middle part of the small intestine called?

9. Where do you find lacteals in the digestive tract?

10. What membrane holds the tongue to the floor of the oral cavity?

11. What part of the tooth is found above the neck?

12. What is the layer of a tooth superficial to the dentin?

13. What are the adult teeth that are directly posterior to the canine teeth called?

14. What muscle cell type lines the small intestine?

15. The segments or pouches of the large intestine have what particular name?

16. What are the names of the salivary glands found anterior to the ear?

17. Where is the lesser omentum found?

18. Where does the cystic duct take bile for storage?

19. Stomach acidity is in the range of pH 1 to 2. What possible benefits could this have in terms of growth of ingested bacteria?

20. Trace the flow of bile from the liver to the duodenum listing all of the structures that come into contact with the bile on its journey.

21. How does the large intestine differ from the small intestine in terms of length?

22. How does the large intestine differ from the small intestine in terms of diameter?

23. Name two functions of the pancreas.

24. Label the following illustration.

ascending colon	vermiform appendix	sigmoid colon
rectum	tongue	small intestine
oral cavity	duodenum	esophagus
stomach	parotid gland	liver

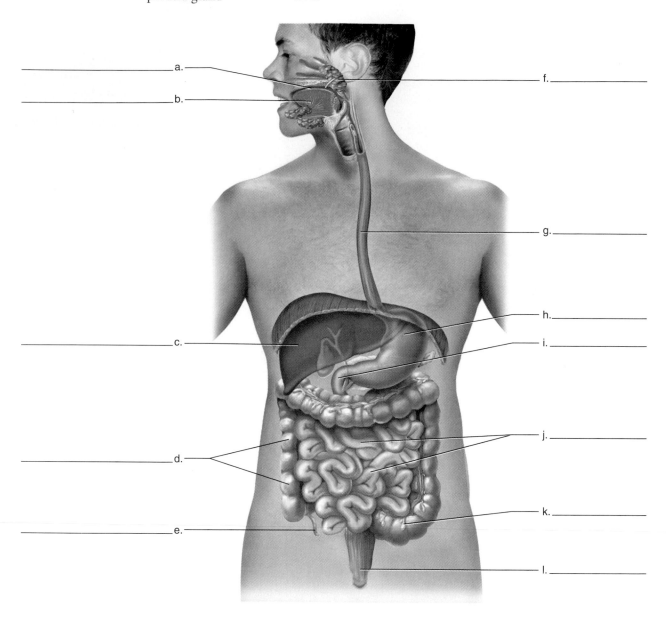

Digestive organs

LABORATORY EXERCISE 43

Digestive Physiology

Introduction

An understanding of the digestive process is fundamental to the study of human physiology. From digestion we obtain almost all of the raw materials the body uses for growth, development, life-sustaining energy, and other metabolic functions. Digestive physiology involves the physical and chemical breakdown of material, absorption of nutrients, coordination between sections of the digestive tract and accessory organs (liver secretions and processing of absorbed foods), and elimination of fecal material. In this exercise, we look at the physical (mechanical) and chemical digestion of food.

To develop some understanding of the digestive process we examine the functions of digestive enzymes. This function is to break down large **macromolecules** in the stomach and small intestine into smaller biomolecules in a process known as **catabolism.** This enzymatic process is shown in figure 43.1.

In this experiment, we analyze several enzymes important in digestion and their effectiveness in digesting four materials commonly found in plant or animal tissue: starch, lipid, cellulose, and protein. The goal of these experiments is to determine if various enzymes in the human digestive tract can break down the four substances into smaller products. If the enzymes are effective then starch should be broken down into sugars, lipids broken down into glycerol and fatty acids, cellulose reduced to sugars, and proteins split into amino acids.

Furthermore, we want to examine the mechanical breakdown of food that precedes most of the chemical processes. By taking solid food and increasing its surface area, we will determine if the rate of digestion increases.

Objectives

At the end of this exercise you should be able to

1. discuss the importance of catabolism in the digestive process;
2. describe how an enzyme is important in chemical digestion;
3. list possible human digestive enzymes as determined by experiment;
4. demonstrate the effectiveness of enzymes in food digestion;
5. describe the mechanisms involved in the mechanical digestion of food;
6. state one of the limitations of human digestive enzymes.

Materials

Microscopes

General Supplies
80 test tubes (15 ml each)
12 test tube holders
6 test tube racks
12 test tube brushes and soap
Warm water bath, set at 35° to 40°C with test tube racks and thermometer
Hot water bath or hot plates and 400 ml beakers for hot bath (100°C) and thermometer

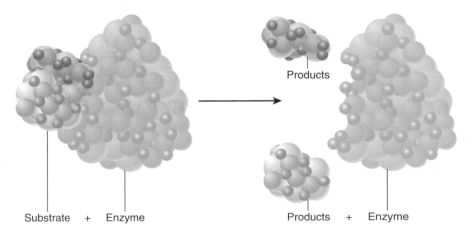

Substrate + Enzyme   Products

                            Products + Enzyme

Figure 43.1 Catabolic reactions with enzymes.

6 permanent markers
18 pipettes of 5 ml each
6 pipette pumps
7 100 ml beakers to pour from stock bottle for distribution
1 250 ml bottle of tap water
6 small spatulas
100 ml 1% pancreatin solution, fresh
300 ml 1% alpha amylase solution, fresh
100 ml 0.1% alpha amylase solution, fresh
300 ml Benedict's reagent (6 50 ml bottles)
Parafilm squares (10 per table)

Starch Digestion
6 dropper bottles of iodine solution
250 ml of 0.5% potato starch solution
6 boxes of microscope slides (1 per table)
6 boxes of coverslips (1 per table)
6 eyedroppers

Sugar Test
100 ml of 1% maltose solution

Cellulose Digestion
Cellulose—2 cotton balls cut into small pieces

Lipid Digestion
200 ml litmus cream
6 dropper bottles of "acid solution" (lemon juice or vinegar)
6 dropper bottles of 1% sodium hydroxide solution (NaOH)
12 pair of goggles (during use of NaOH)

Protein Digestion
25 ml of 1% BAPNA solution; BAPNA (benzoyl-DL-
    arginine-p-nitroanilide) is available from Sigma B4875 or
    Aldrich 85,711-4

Surface Area and Digestion
2 large potatoes per class
2 breadboards for potato experiment
2 #4 cork borers
2 small rods (applicator sticks) to push through the borers
12 rulers (15 cm)
6 razor blades
2 mortars and pestles

Procedure

Read all of the experiment before beginning! Your instructor may want you to do these experiments in student pairs or as members of a larger group. If you do these experiments as part of a larger group, make sure that you *witness* and *record* the results of your group. **Do not discard any material until everyone in your group has seen the result** or until your instructor directs you to do so. While you are waiting for one set of materials to incubate you can begin the next set of experiments.

Label all test tubes with your *group name* and the *test tube number*. This is very important as removal of the wrong tube will give you erroneous results, and the removal of a tube be-

longing to another group will not only produce the wrong result for your group but for the other group as well.

Tests in this exercise are divided into **reagent tests** and **experimental tests.** The reagent tests allow you to see the results of an experiment when using known samples. The experimental tests determine if a digestive reaction has taken place.

Starch Digestion

Starch digestion occurs by the breakdown of a large molecule of starch into smaller molecules of sugar by enzymes called **amylases.** Amylases catabolize starches as represented by figure 43.2. One way to analyze the effectiveness of amylases is to see if the enzyme removes starch from solution. The decrease or absence of starch in solution indicates that amylases are present.

Experiment 1: Reagent Test—Iodine

We use a *reagent test* to determine the presence of starch. This particular test uses **iodine.** If starch is present, iodine produces a *blue-black color* which is a **positive result.** If starch is absent, then the solution remains *yellow* and this is a **negative result.**

1. Label a test tube with your group name and the number 1.
2. Pour a small amount of starch solution *from* the stock bottle *into* a small beaker. *Never return excess solution to the stock bottle!*
3. Take a pipette pump and pipette and place 1 to 2 ml of starch solution from the beaker in the test tube labeled #1.

Caution   Never pipette material by mouth—use a pipette pump or bulb.

---

4. Add 4 to 5 drops of iodine solution to the starch solution in test tube #1. The solution should turn blue or black if starch is present. Save this test result for future comparison and record the outcome of your test.

   Color of reaction (blue/black or yellow): _____

   Test result (positive/negative for starch): _____

Experiment 2: Determination of Starch Digestion by Amylase

1. Using a pipette, place 5 ml of starch solution in a small test tube that you have labeled with your group name and the number 2.

Figure 43.2   Starch digestion.

2. Pour a small amount of 1% amylase from the stock bottle into a small beaker. With a new pipette, withdraw 2 ml of amylase solution from the beaker and add it to the starch solution. Stir the solutions in the test tube by flicking the bottom of the tube gently with your finger.

3. Incubate the mixture in test tube #2 in a **warm** water bath (35°–40°C) for 50 minutes.

4. After 50 minutes take 1 to 2 ml of the solution from test tube #2 and place it in a new test tube.

5. Test the solution with 4 to 5 drops of iodine solution and record your results.

6. Compare the results of your experiment with test tube #1.

Color of the solution with iodine: _____

Test results (positive/negative): _____

7. From test tube #2, withdraw a small sample with an eye dropper and place a drop on a microscope slide.

8. Add a drop of iodine solution to the sample and add a coverslip. Examine the slide under low power and look for starch grains. Compare this slide to one that you made from a drop of starch from the *undigested* starch solution (from test tube 1). Is there a difference? Record your observations in the following space.

Observations:

Sugar Production

The preceding experiment tested for the *decrease* in starch as a way to determine if a catabolic reaction had taken place. Another way to look at the effectiveness of amylase is to determine the *presence* of sugar after the exposure of starch to the enzyme. In this experiment, instead of looking for the absence of the **substrate** (starch) to determine enzyme effectiveness, you can test for the formation of the **product** (sugar). If amylase is effective, then the presence of sugar in solution, after exposure to the enzyme, indicates the conversion of starch to sugar.

Experiment 3: Reagent Test—Benedict's Test

The first test in this section involves the detection of sugar in a solution. This is done with the Benedict's test.

Caution   Benedict's reagent is poisonous. Use caution when mixing the solutions in the test tube.

---

1. Label a test tube with the number 3 and your group name and then pipette 1 to 2 ml of 1% maltose solution into the test tube.

2. Add 1 ml of Benedict's reagent to the maltose solution in the test tube. Hold the test tube by the top and gently flick the bottom of the tube with the pad of your index finger to mix the contents.

3. Place this mixture in a very **hot** (100°C) water bath for about 10 minutes. If the mixture turns green, yellow, orange, or brick red, then sugar is present.

The sequence of colors indicates the presence of sugar in increasing amounts (little sugar indicated by green; significant amounts by brick red). If there is no color change (for example, if the solution stays blue) after 10 minutes, then remove the tube from the bath and note the absence of sugar (a negative result). Record your results.

Color of test solution: _____

Test result (positive/negative for sugar): _____

Experiment 4: Determination of Sugar (Maltose) Production by Amylase

1. Place 3 ml of the starch solution from the stock bottle in a test tube labeled with your group name and the number 4.

2. Add 3 ml of 1% amylase solution to the starch solution in the tube. Stir the mixture by lightly flicking the bottom of the test tube.

3. Incubate the mixture in the **warm** water bath for 30 minutes.

4. After 30 minutes remove the solution in test tube #4 from the bath and add 4 ml of Benedict's reagent.

5. Place the tube in a **hot** water bath (100°C) for 10 minutes. Record your results.

Color of solution with Benedict's reagent:

Presence of sugar: _____

Blue = 0

Green = trace

Yellow = moderate

Orange, red = large amounts

Effectiveness of digestion: _____

Cellulose Digestion

Cellulose is composed of repeating **glucose** units. Starch is also composed of repeating glucose units. If amylase converts cellulose to sugar, then the presence of sugar in solution indicates that the enzyme is effective. This is the same process as experiment 4 except that the substrate is changed from starch to cellulose. To test for the effectiveness of amylase in digesting cellulose you will incubate a cellulose mixture with amylase in a warm water bath. A positive reaction indicates the presence of sugar after incubation.

Experiment 5: Determination of Sugar Production from Cellulose

1. Place a small pinch of cellulose fibers (the size of a green pea) in a test tube and add 2 ml of water and 3 ml of 1%

amylase solution. Make sure the majority of cellulose fibers are not stuck to the sides of the test tube. You can use a small spatula to push the fibers into the water, if necessary. The test tube should be labeled with your group name and the number 5.

2. Incubate for 30 minutes in a **warm** water bath.
3. After 30 minutes remove the solution from the bath and add 4 ml of Benedict's reagent.
4. Place the tube in a **hot** water bath (100°C) for 10 minutes. Record the presence or absence of sugar.

Color of Benedict's test: _____

Presence of sugar: _____

Effectiveness of digestion: _____

Lipid Digestion

In the digestive process some **lipids (triglycerides)** are broken down into **glycerol** and **fatty acids.** The fatty acids make the solution more **acidic,** thus lowering the **pH.** The decrease in pH can be used as a measure of digestion. The greater the acidity of the solution the greater the digestion.

Experiment 6: Reagent Test—Litmus Test

1. To determine the effectiveness of the litmus reagent pour 3 ml of **litmus cream** into a test tube labeled with your group name and the number 6. Litmus cream consists of dairy cream (containing fat) and litmus powder (a pH indicator).
2. Add the acid solution (lemon juice or vinegar) drop by drop while flicking the test tube until the color changes from *blue* to *pink*. Do not add too much acid but just enough to change the color. A pink color indicates an acidic condition.
3. Now add 1% sodium hydroxide solution (NaOH) a drop at a time until you reverse the color. This indicates that the solution is alkaline.

Caution Sodium hydroxide is a caustic solution that can dissolve your skin, so be careful! Wear protective goggles as you add the sodium hydroxide solution.

Experiment 7: Determination of Lipid Digestion by Pancreatin

1. Pipette 3 ml of litmus cream into a small test tube labeled with your group name and the number 7.
2. Add 1 ml of pancreatin solution (pancreatin contains many digestive enzymes) to the litmus cream and stir well with a spatula.
3. Incubate for 30 minutes in the **warm** (35° to 40°C) water bath.

4. After 30 minutes look for a color change. If the cream turns pink, then digestion has occurred. If the color remains blue, then no digestion occurred. Record your results.

Color of solution after incubation (pink/blue): _____

Condition of the litmus cream (acidic/alkaline): _____

Extent of digestion: _____

Change in smell from normal cream: _____

Protein Digestion

Proteins are made of **amino acids.** These amino acids are linked by **peptide bonds** to form **polypeptide chains.** If the chains are long enough they form **proteins.** Proteins are split into amino acids by a number of digestive enzymes called proteases. The effectiveness of proteases can be studied with the use of a chromogenic (color-producing) substance known as BAPNA (Benzoyl-DL-arginine-p-nitroanilide). If proteases are present they release a yellow aniline dye from the larger BAPNA molecule.

Experiment 8: Determination of Protein Digestion by Pancreatin

1. Label a test tube with your group name and the number 8.
2. Pipette 1 ml of BAPNA solution into the tube. To this add 1 ml of pancreatin solution and stir by flicking the bottom of the test tube.
3. Place the tube in the **warm** water bath for 15 minutes.
4. After 15 minutes, remove the tube from the bath and examine the test tube to see if it has turned yellow. Yellow indicates protein digesting abilities of the enzyme. Record your results.

Color of the tube (yellow/clear): _____

Digestion capability (yes/no): _____

Experiment 9: Surface Area and Digestion

As food is broken down physically, the surface area increases relative to the volume of the food. This experiment examines the effectiveness of an increase in surface area for digestion.

1. Label two test tubes, each having your group name and the number 9. Write "mashed" on one tube and "entire" on the other.
2. Carefully plunge a #4 cork borer into the center of a raw potato. *Make sure your hand is not on the receiving end of the cork borer! Use a breadboard.*
3. Remove the cork borer from the potato and, using a small rod, push the potato cylinder from the borer.
4. Using a ruler and a razor blade, cut the potato cylinder into two pieces, each 1 cm in length. *It is important that the two pieces are the same length!*

5. Place one piece of potato in the test tube labeled "entire."
6. Chop the other piece into several little pieces and mash these with a mortar and pestle until they are well pulverized.
7. Carefully place all of the mashed potato in the other test tube with the "mashed" label.
8. Pipette exactly 4 ml of **0.1%** amylase solution* into each test tube and incubate them for 20 minutes in the **warm** water bath.

---

*You can use a solution of 0.1% alpha amylase or dilute the 1% stock solution by adding 1 ml of 1% amylase solution to 9 mls of water.

9. After incubation, place exactly 2 ml of Benedict's reagent into each tube and place these two tubes in the hot water bath for 10 minutes.

The amount of digestion can be estimated by the color of the test. As described in experiment 4, blue indicates no sugar, with green, yellow, orange, and brick red indicating increasing amounts of sugar. In which tube did the greatest amount of digestion take place? In which tube did the least amount of digestion take place? Record your results.

Color of solution with intact potato piece: ⎯⎯⎯⎯⎯⎯

Color of solution with mashed potato piece: ⎯⎯⎯⎯⎯⎯

Relative amount of digestion of intact piece: ⎯⎯⎯⎯⎯⎯

Relative amount of digestion of mashed piece: ⎯⎯⎯⎯⎯⎯

Name _____

1. Why is pancreatin used in a digestion experiment?

2. Substrates are converted into what substances by enzymes?

3. Name the group of enzymes that digest starch.

4. Starch is reduced by enzymes into smaller units known as _____.

5. What is catabolism?

6. What are the bonds that hold amino acids together?

7. Explain why a negative iodine test for starch would indicate a positive result for the enzymatic degradation of starch by amylase.

8. What would a positive iodine test indicate in the preceding reaction?

9. Record all your negative results. Determine if these negative results indicate digestion or no digestion.

10. Explain why cellulose could or could not be digested by amylase.

11. What effect does chewing your food have on digestion? What experiment did you perform to evaluate the effectiveness of chewing for digestion?

12. Lipase is an enzyme that converts some lipids to glycerol and fatty acids. Which substance used in lab probably contained lipase?

13. What does a negative Benedict's test (a blue color) indicate?

14. What does a pink color in the lipid digestion experiment indicate?

15. Pancreatin (which has amylase) is now frequently prepared with lactose or other sugars as extenders. Determine which experiments would give erroneous results if pancreatin was used instead of amylase.

Urinary System

Introduction

The urinary system filters dissolved material from the blood, regulates electrolytes and fluid volume, concentrates and stores waste products, and reabsorbs metabolically important substances back into the circulatory system. The system consists of two kidneys, two ureters, a single urinary bladder, and a single urethra. Filtration in the kidney involves both metabolic waste products (urea) and material beneficial to the body. The urinary system also reabsorbs sugars, ions, and other materials and returns them to the cardiovascular system. In this exercise, you examine the gross and microscopic anatomy of the urinary system, study the major organs as represented in humans, and dissect a mammal kidney if available.

Objectives

At the end of this exercise you should be able to

1. list the major organs of the urinary system;
2. describe the blood flow through the kidney;
3. describe the flow of filtrate through the kidney;
4. name the major parts of the nephron;
5. trace the flow of urine from the kidney to the exterior of the body;
6. distinguish between the parts of the nephron in histological sections.

Materials

Models and charts of the urinary system
Models and illustrations of the kidney and nephron system
Microscopes
Microscope slides of kidney and bladder
Samples of renal calculi (if available)
Preserved specimens of sheep or other mammal kidney
Dissection trays and materials

Procedure

You can begin the study of the urinary system by locating its principal organs. Look at figure 44.1 and compare it to the material available in the lab. Find the **kidneys**, the **ureters**, the **urinary bladder**, and the **urethra**.

Kidneys

The kidneys are **retroperitoneal** (found posterior to the parietal peritoneum) and are imbedded in **renal adipose (fat) pads.** The adipose pads cushion the kidneys, which are found mostly below the protection of the rib cage. The kidneys are located adjacent to the vertebral column about at the level of T12 to L3. The right kidney is slightly more inferior than the left.

Examine a model of the kidney and compare it to figure 44.2. Locate the outer **renal capsule,** a tough connective tissue layer, the outer **cortex,** and the inner **medulla** of the kidney. The kidney has a depression on the medial side where the renal artery enters the kidney and the renal vein and the ureter exit the kidney. This depression is called the **hilum.**

If you examine a coronal section of the kidney, you can find the **renal (medullary) pyramids,** which are separated by the **renal columns.** Each renal pyramid ends in a blunt point called the **renal papilla** (figure 44.2). Urine drips from many papillae toward the middle of the kidney.

The urine drips into the **minor calyces** (singular, calyx), which enclose the renal papillae. Minor calyces are somewhat

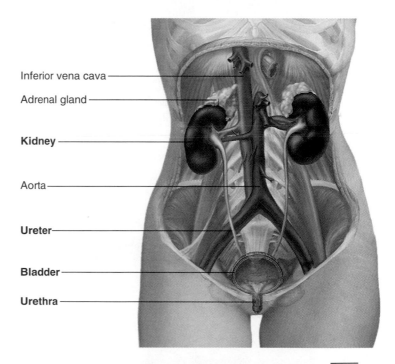

Inferior vena cava
Adrenal gland
Kidney
Aorta
Ureter
Bladder
Urethra

Figure 44.1   Major organs of the urinary system. ⚷ 📼

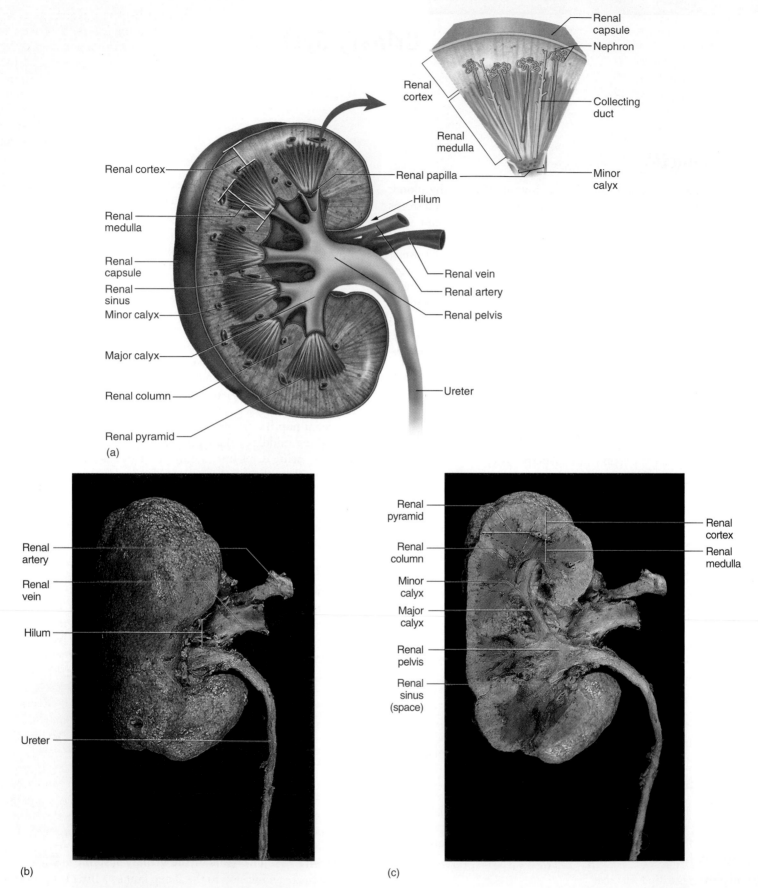

Figure 44.2 Gross anatomy of the kidney, entire and coronal section. (a) Diagram of coronal section of the kidney and details of renal pyramid; photograph (b) entire kidney; (c) coronal section.

like funnels that collect fluid. Minor calyces lead to the **major calyces** and these, in turn, conduct urine into the large **renal pelvis.** The renal pelvis is found in the **renal sinus.** The renal pelvis is like a glove in a coat pocket. The pocket is the renal sinus, and the membranous glove that occupies the space is the renal pelvis. Locate these structures in figure 44.2. The renal pelvis is connected to the ureter at the medial side of the kidney.

Blood Flow through the Kidney

The blood flows through the kidneys by way of the **renal artery,** the **interlobar arteries,** the **arcuate arteries,** and the **interlobular arteries.** These blood vessels can be seen in figure 44.3. From an interlobular artery, blood flows into an **afferent arteriole,** through the **glomerulus,** and to the **efferent arteriole.** The blood then enters the **peritubular capillaries** and returns via the **interlobular veins,** the **arcuate veins,** and the **interlobar veins** to the **renal vein.** The separation between the cortex and the medulla occurs at the level of the arcuate veins.

There are vessels that branch off of the efferent arterioles and form a network of blood vessels known as the **vasa recta.** The vasa recta vessels are found primarily in the medullary area, and though they represent only a small fraction of the capillary bed of the kidney, they are important in producing a concentrated urine. The vasa recta are found in association with **juxtamedullary nephrons.**

The blood flow in the kidney forms a portal system where blood flows from one capillary bed (the glomerulus) to another capillary bed (the peritubular capillaries) prior to returning to the heart. The capillaries of the glomerulus are the site of filtration, whereas the peritubular capillaries are the primary site of reabsorption.

Ultrastructure of the Kidney

Before you examine the sections of kidney under the microscope, first become familiar with the structure of the **nephron.** Examine models and charts in the laboratory and compare them to figure 44.3. Locate the **glomerular capsule (Bowman's capsule),** the **proximal convoluted tubule,** the **nephron loop (loop of Henle),** and the **distal convoluted tubule.** These four structures make up the nephron. The glomerulus along with the glomerular capsule are known as the **renal corpuscle.** Blood travels to the nephron via the afferent arteriole. When the blood reaches the glomerulus, a cluster of capillaries in the renal cortex, the plasma is filtered by blood pressure forcing fluid across the capillary membranes. Blood pressure is the driving force for renal filtration and this pressure is opposed by hydrostatic pressure in the Bowman's capsule space outside of the glomerular capillaries and the osmotic pressure that occurs due to the colloidal proteins left in the blood plasma. This filtered fluid, which is present in the nephron, is called **filtrate.**

As the filtrate flows through the nephron, water, glucose, and many electrolytes are returned to the blood. The urea is concentrated as it passes through the entire nephron and the **collecting duct,** a tube that receives the end product of the nephrons. The efficiency of the kidney varies with the material that flows through it. In normal conditions, glucose is completely reabsorbed back into the body. About 99% of the water in the filtrate is reabsorbed. Only 50% of the urea is excreted into the urine. The remainder of the urea returns to the blood and is filtered again.

There are two types of nephrons in the kidney. The majority of the nephrons are found in the cortex of the kidney and are thus called **cortical nephrons.** Those that are found in the medulla are called **juxtamedullary nephrons** and are far fewer in number, some extending to near the tip of the renal pyramids.

Microscopic Examination of the Kidney

Examine a kidney slide under low power. You should see the cortex of the kidney, which has a number of round structures scattered throughout. These are the **glomeruli** and they are composed of capillary tufts. The medullary region of the kidney slide should have fairly open parallel spaces. These spaces are the collecting ducts and occur in the medulla. Compare the slide with figure 44.4a.

Examine the slide under higher magnification and locate the glomerulus and the glomerular capsule. The capsule is composed of simple squamous epithelium and specialized cells called podocytes. Examine the outer edge of the capsule around the glomerulus. If you move the slide around in the cortex, you should find the proximal convoluted tubules with the **brush border** or microvilli on the inner edge of the tubule. The inner surface of the tubule appears fuzzy. The microvilli increase the surface area of the proximal convoluted tubule. What functional value could be ascribed to having such a surface area?

The distal convoluted tubules do not have brush borders, therefore the inner surface of a tubule does not appear fuzzy. The cells of the distal convoluted tubules generally have darker nuclei and cytoplasm that is relatively clear when compared to the cells of the proximal convoluted tubules (figure 44.4b). Examine the medulla of the kidney under high magnification and locate the thin-walled nephron loop and the larger diameter collecting ducts as seen in figure 44.4c.

Dissection of the Sheep Kidney

Place a sheep kidney and dissection equipment on your table. Examine the outer **capsule** of the kidney. You may see some tubes coming from a dent in the kidney. The dent is the **hilum,** and the tubes are the **renal artery, renal vein,** and **ureter.** The renal artery is smaller in diameter and has a thicker wall than the renal vein. The ureter has an expanded portion near the hilum. Make an incision in the sheep kidney a little off center in

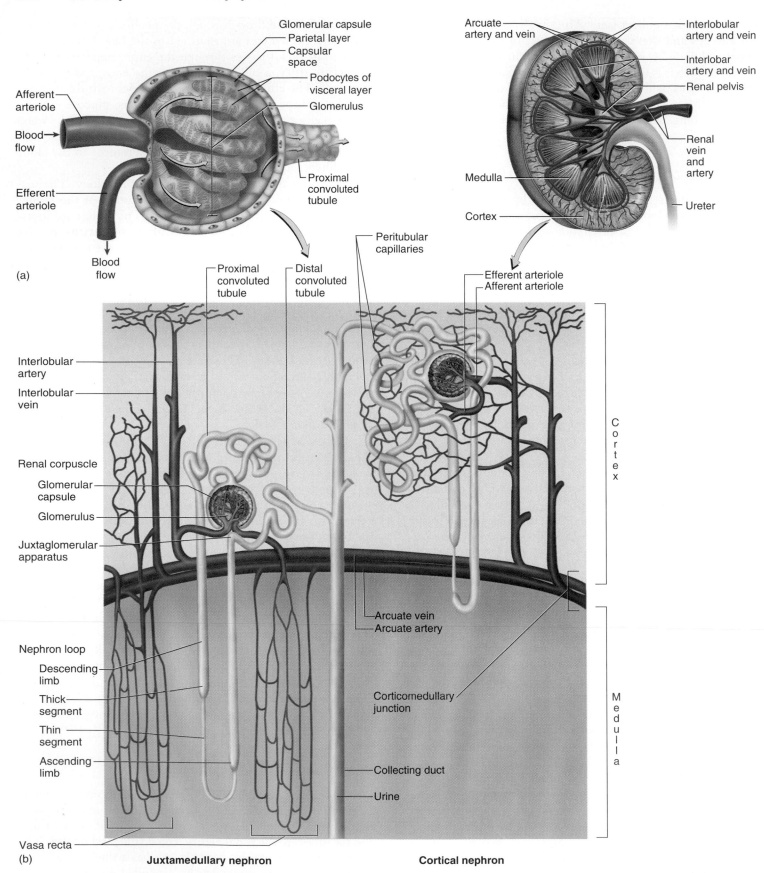

(a)

(b)

Juxtamedullary nephron **Cortical nephron**

Figure 44.3 Blood vessels and ultrastructure of the kidney. (a) Major vessels; (b) nephron and associated vessels.

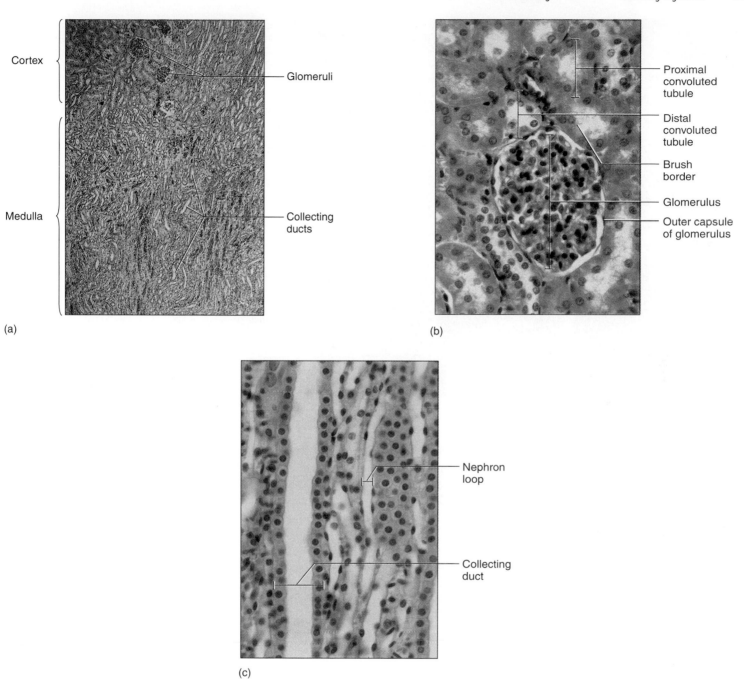

Cortex

Medulla

(a)

Glomeruli

Collecting ducts

(b)

Proximal convoluted tubule

Distal convoluted tubule

Brush border

Glomerulus

Outer capsule of glomerulus

(c)

Nephron loop

Collecting duct

Figure 44.4   Photomicrographs of a kidney. (a) Overview (40×); (b) cortex (400×); (c) medulla (400×). ✗

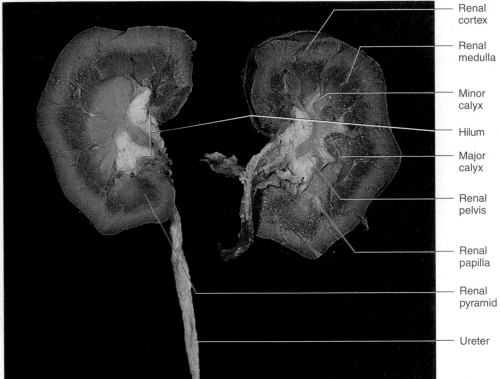

Renal
cortex

Renal
medulla

Minor
calyx

Hilum

Major
calyx

Renal
pelvis

Renal
papilla

Renal
pyramid

Ureter

Figure 44.5 Dissection of a sheep kidney.

the coronal plane (figure 44.5). This section allows you to better see the interior structures of the kidney. Locate the **renal cortex,** which is the outer layer of the kidney. You should also locate the **renal medulla,** which has triangular regions known as the **renal pyramids.** At the tip of each pyramid is a **papilla.** Urine from the papillae drips into the **minor calyx.** Many minor calyces lead to a **major calyx.** The major calyces take fluid to the **renal pelvis.**

Lift the renal pelvis somewhat to pull it away from the **renal sinus.** The sinus is the space in the kidney. You should also examine the exit of the renal pelvis as it becomes the **ureter.** When you are finished with the dissection, place the material in the proper waste container provided by your instructor.

Ureters

The **ureters** are long, thin tubes that conduct urine from the kidneys to the urinary bladder. The ureters have **transitional epithelium** as an inner lining and smooth muscle in their outer wall. Urine is expressed by peristalsis from the kidney to the urinary bladder. Examine the models in the lab and compare them to figure 44.1.

Urinary Bladder

The **urinary bladder** is located anterior to the parietal peritoneum and is thus described as being **anteperitoneal.** Locate the urinary bladder in a torso model in lab and note that it is found just posterior to the symphysis pubis. On the posterior

wall of the urinary bladder is a triangular region known as the **trigone.** The trigone is defined by the superior entrances of the ureters and the inferior exit of the urethra. Transitional epithelium lines the inner surface of the bladder, while layers of smooth muscle known as **detrusor muscles** are found in the wall of the bladder. Compare models in lab to figure 44.6.

Histology of the Bladder

Transitional epithelium has a special role in the urinary bladder. This particular epithelium can withstand a significant amount of stretching (distention) when the bladder fills with urine. Examine a slide of transitional epithelium under the microscope and compare it to figure 44.7. Transitional epithelium is also presented and illustrated in Laboratory Exercise 6. Look at the inner surface of the section for the epithelial layer. Note that the cells are shaped somewhat like teardrops. Transitional epithelium can be distinguished from stratified squamous epithelium in that the cells of transitional epithelium from an empty bladder do not flatten at the surface of the tissue. You should also examine the smooth muscle layers of the urinary bladder that make up the detrusor muscle.

Urethra

The terminal organ of the urinary system is the **urethra.** The urethra is approximately 3 to 4 cm long in females. It passes

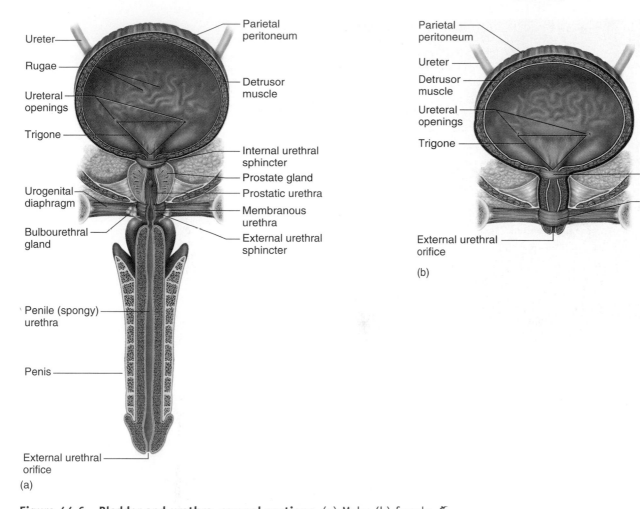

Ureter

Rugae

Ureteral openings

Trigone

Urogenital diaphragm

Bulbourethral gland

Penile (spongy) urethra

Penis

External urethral orifice

(a)

Parietal peritoneum

Detrusor muscle

Internal urethral sphincter

Prostate gland

Prostatic urethra

Membranous urethra

External urethral sphincter

Parietal peritoneum

Ureter

Detrusor muscle

Ureteral openings

Trigone

Internal urethral sphincter

External urethral sphincter

External urethral orifice

(b)

Figure 44.6   Bladder and urethra, coronal sections. (a) Male; (b) female. ⟍

Transitional epithelium

Lamina propria

Smooth muscle

Figure 44.7   Histology of the bladder (100×). ⟍

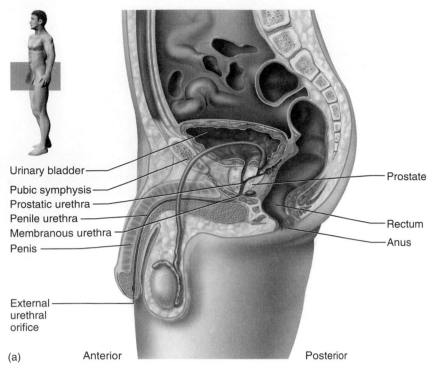

Urinary bladder

Pubic symphysis

Prostatic urethra

Penile urethra

Membranous urethra

Penis

Prostate

Rectum

Anus

External
urethral
orifice

(a)   Anterior   Posterior

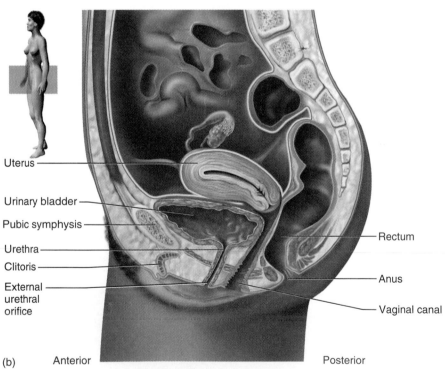

Uterus

Urinary bladder

Pubic symphysis

Urethra

Clitoris

External
urethral
orifice

Rectum

Anus

Vaginal canal

(b)   Anterior   Posterior

Figure 44.8   Midsagittal section of the male and female pelves. (a) Male pelvis; (b) female pelvis.

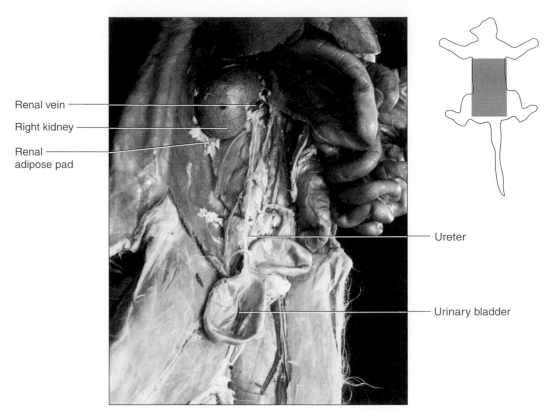

Renal vein

Right kidney

Renal adipose pad

Ureter

Urinary bladder

Figure 44.9   Urinary system in the cat.

from the urinary bladder to the **external urethral orifice** located anterior to the vagina and posterior to the clitoris (figures 44.6 and 44.8*b*). From figure 44.8 you can also see the position of the urinary bladder in relationship to the symphysis pubis. The urethra is about 20 cm long in males. It begins at the urinary bladder and passes through the prostate gland as the **prostatic urethra.** It continues and passes through the body wall as the **membranous urethra,** and then exits through the penis to the external urethra orifice at the tip of the glans penis. This terminal portion of the urethra is known as the **penile** or **spongy urethra.** Urinary bladder infections are more common in females than in males. How can you explain this in terms of microorganism movement into the bladder through the urethra?

Cat Dissection

You should open the abdominal region of the cat if you have not done so already. The kidneys are on the dorsal body wall of the cat and are located dorsal to the parietal peritoneum. Examine the kidneys and the structures that lead to and from the hilum of the kidney. You should locate the renal veins that take blood from the kidney. The veins are larger in diameter than the renal arteries and they are attached to the posterior vena cava of the cat which runs along the ventral, right side of the vertebral column. Find the renal arteries that take blood from the aorta to the kidneys. The aorta lies to the left of the posterior vena cava. You should be able to locate the ureters as they run posteriorly from the kidney to the urinary bladder. Do not dissect the urethra at this time. You can locate the urethra during the dissection of the reproductive structures in Exercises 46 and 47. You should compare your dissection to figure 44.9.

REVIEW

Name _____

1. What is the outer region of the kidney called?

2. Describe the kidneys with regard to their position to the parietal peritoneum.

3. What takes urine directly to the urinary bladder?

4. What takes urine from the bladder to the exterior of the body?

5. What blood vessel takes blood from the kidney?

6. On the posterior bladder there is a triangular region. What is it called?

7. What is the name of the cluster of capillaries in the kidney where filtration occurs?

8. Distal convoluted tubules flow directly into what structures?

9. What is a renal papilla?

10. Name the connective tissue that covers a kidney.

11. Urine flows from the tip of a renal pyramid into what structure?

12. Blood in the glomerulus next flows to what arteriole?

13. Which shows greater anatomical difference between the sexes: ureters, urinary bladder or urethra?

14. Blood in an arcuate vein would next flow into what structure?

15. What histological feature distinguishes a proximal convoluted tubule from a distal convoluted tubule?

16. Name the four parts of the nephron.

17. What type of cell lines the bladder?

18. Fill out the following illustration.

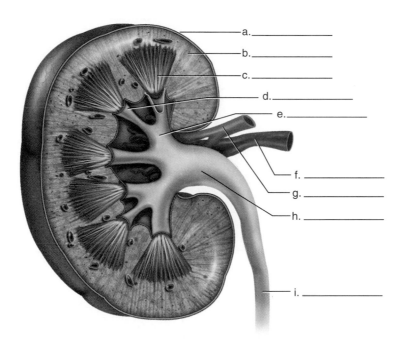

a._____

b._____

c._____

d._____

e._____

f._____

g._____

h._____

i._____

Structure of the kidney.

Urinalysis

Introduction

Urinalysis is important in clinical assessment of an individual's physical condition. Traces of blood in the urine can indicate the presence of kidney stones or renal damage; elevated levels of glucose can indicate diabetes mellitus; high bilirubin levels can be indicative of liver problems; and high bacterial counts can indicate bladder or kidney infections. The analysis of urine provides a very broad diagnostic indicator of general health.

Normal values of urine vary based on water intake and diet. A person who drinks 3 quarts of water per day will have a very different urine composition than one who drinks 3 cups of water per day. Not only will the volume of water in the urine vary, but the concentration of other urinary solutes will vary as well. Some cations such as potassium and sodium are also diet-dependent.

Objectives

At the end of this exercise you should be able to

1. determine the specific gravity of a urine sample and compare it to normal values;
2. list the sediments commonly found in urine;
3. discuss the importance of urinalysis as a general diagnostic tool;
4. distinguish between casts, crystals, and microbes in a urine sample;
5. prepare a stained sediment slide and identify major components of the sediment.

Materials

Sterile urine collection containers
Permanent marker or wax pencil
Microscope slides
Coverslips
Urine sediment stain (Sedistain, Volusol, etc.)
Chemstrip or Multistix 10SG urine test strips
Tapered centrifuge tubes
Test tube racks
Pasteur pipettes and bulbs
Centrifuge
Protective gloves (latex or plastic)
Protective eyewear

Biohazard bag
Microscopes
Urine specimen (yours or synthetic/sterilized urine)
10% bleach solution

 Caution *Urine is potentially contaminated with pathogens.* **Wear latex barrier gloves and protective eyewear during the entire exercise.** Place all disposable material that come into contact with urine in the biohazard bag. Work only with your own urine and avoid any contact between urine and an open wound or cut. If you spill your sample, notify your instructor and wipe up the spill and swab the countertop with a 10% bleach solution. When you are finished with the exercise place reusable glassware in a 10% bleach solution. Read all procedures before beginning this exercise.

Procedure

You can use either your own urine for urinalysis or simulated urine. If you use simulated urine you can test for color, pH, specific gravity, glucose, and protein. Some tests are provided in kit form from biological supply houses. If you use your own urine follow the directions given next.

Using a marker or wax pencil, write your name on a sterile collection container. Proceed to the restroom and obtain a urine sample for testing. If you collect your urine at home, do so in the morning and store the sample in the refrigerator. The sample should be a **midstream** collection because the urethra is initially full of microbes and the first volume of urine will contain abnormally elevated levels of microorganisms. Bring the specimen back to the lab and note the color of the sample.

Urine color: _____

The color of urine should normally be a very pale yellow. This is due to the pigment **urochrome,** which is a metabolic product of hemoglobin breakdown. Higher levels of B vitamins may artificially color the urine brighter yellow, and a low fluid intake may also cause the urine to be a deeper color yellow. The minimum adult intake of water should be 8 cups of water per day (½ gallon); twice that much is recommended by some urologists, particularly if you are prone to some types of kidney stones. Urine should normally be clear, but the sample may be cloudy due to bacterial infection or other contaminants. Record the transparency of your urine sample.

Transparency: _____

Smell the urine sample and record the odor.

Odor: _____

Urine should normally have a faint but characteristic odor. Consumption of certain foods such as asparagus may produce sulfur compounds in urine, which produce stronger odors.

Urine Characteristics

Gently swirl the urine before testing and pour a 10 ml sample into a clean test tube. Follow your instructor's directions for determining standard values in urine. You can test for specific compounds in urine individually, or you can use a Chemstrip or Multistix to run a battery of tests in a few minutes. The Multistix procedure is outlined next. You should have the urine sample, the test strip, a pencil or pen, and a container (so you can read the values) handy before you proceed with the test. The test covers 10 specific exams in 2 minutes, so you have to be organized to record the results. Consider any result read after 2 minutes invalid. Multistix test strips are composed of sections of paper with test reagents embedded into the fibers. They react with urine components if present.

Multistix/Chemstrip Procedure

If you are using a Multistix or Chemstrip be sure to pay attention to the time limits provided on the container for reading the results. Wear gloves to dip the strip into the urine sample and leave the test strip in the urine for no longer than 1 second. Follow directions precisely on the test container.

Glucose: _____

Although glucose is present in blood plasma there is normally no glucose in urine. **Glycosuria** is the condition of having glucose in the urine. Trace amounts can be found after meals high in carbohydrates especially sugar and significant levels can be found in individuals with diabetes mellitus.

Bilirubin: _____

The bilirubin test should be negative, since bilirubin is not normally present in urine. Bilirubin is a metabolic waste product from the destruction of erythrocytes. In the condition of **bilirubinuria** there is the presence of the bile pigments and bilirubin in urine. This can be due to erythrocyte destruction (hemolytic anemia), blockage of the bile duct, or liver damage such as hepatitis or cirrhosis.

Ketones: _____

Normally there should be trace amounts or no ketones (acetone) in the urine. Ketone bodies in urine (ketonuria) rep-

resent the general body condition known as **ketosis.** This is due to mobilization and use of fat stores in people during periods of starvation, in individuals with diabetes mellitus, or in people with an abnormally high-fat diet.

Specific gravity: _____

The specific gravity of pure water is 1.000. The specific gravity of urine varies from a dilute 1.001 to 1.035 in concentrated urine. When urine has a high specific gravity there are more dissolved solutes in the urine. A high specific gravity may be due to lower water intake or such diseases as diabetes mellitus.

Hematuria: _____

There should normally be no blood present in urine. The presence of erythrocytes or hemoglobin in urine may be due to some forms of anemia, erythrocyte destruction after incompatible blood transfusions, renal disease or infection, kidney stones, or urinary contamination during the menses (bleeding period) of a female's menstrual cycle. If the indicator strip shows a spotted pattern, then the erythrocytes are intact (nonhemolyzed). If the red blood cells have ruptured, then the presence of **hematuria** shows up as uniform color changes on the indicator strip.

pH: _____

The pH of urine can vary from around 4.5 to 8.2. This represents over 1,000-fold change in hydrogen ion concentration. The dramatic fluctuation in pH is seen as the kidneys regulate blood pH by removal of ions from the blood. Urine is normally slightly acidic. In a diet high in protein, the urine is more acidic, while a diet high in vegetable material yields a urine that is more alkaline.

Protein: _____

Normally there should be no protein found in urine. The most common blood protein is albumin, and its presence in urine is known as **albuminuria.** This may be due to diabetes mellitus, renal damage (kidney disease), extreme physical activity, or hypertension.

Urobilinogen: _____

Normal ranges of urobilinogen are 0.2 to 1 mg/ml. Increases in the secretion of urobilinogen indicate significant **hemolysis** of erythrocytes to the point that the liver cannot process the bilirubin. The bilirubin increases in the plasma, and the formation of urobilinogen in the intestines increases as well. The urobilinogen diffuses into the blood, where it is filtered by the kidneys.

Nitrites: _____

The normal blood value of nitrates should be negative. A positive value for nitrates indicates the presence of large numbers of bacteria in the urinary tract.

Leukocytes: _____

Table 45.1

Characteristic	Normal Value or Range	Your Results
Appearance	Almost clear to deep amber	
Odor	Faint odor	
Glucose	0–trace	
Bilirubin	None	
Ketones	None–trace (5 mg/ml)	
Specific gravity	1.001–1.035	
Free hemoglobin	None	
pH	4.5–8.2	
Albumin	None–trace	
Urobilinogen	Trace (0.2–1 mg/ml)	
Nitrates	None	
Leukocytes	None–trace	

The normal value for leukocytes is negative or in trace amounts. Elevated levels of leukocytes in urine is known as **pyuria.** This indicates a potential urinary tract infection, typically bladder or kidney infection.

Note that some drugs and vitamins can give false positive results in urinalysis tests. Any test that indicates renal disease should be further validated by professionals prior to any course of action.

After you have finished your results, complete table 45.1.

Sediment Study

1. Obtain a centrifuge tube from the supply area and write your name on the tube for identification.
2. Pour 10 ml of urine into the tube and place it in the centrifuge opposite another tube containing an equal volume. If there are an uneven number of tubes, make sure to fill a tube with water to the same level as the unpaired tube and place it opposite that tube in the centrifuge. This balances the centrifuge while it is operating.
3. Spin the tubes for about 4 minutes at a slow speed (1,500 rpm) and let the rotor of the centrifuge slow down over time. If you brake the spinning you may resuspend the sediments.
4. Carefully remove your tube and place it in a test tube rack on your desk.
5. Examine the tube. You should have an upper fluid layer and a small amount of urine sediment at the bottom of the tube.
6. Place one drop of urine sediment stain (Sedistain or other urine stain) on a clean microscope slide.

7. Withdraw a small sample from the bottom of the centrifuge tube with a pasteur pipette and place a drop of the sediment on the slide. **Do not put the urine on the slide before you put on the stain because you may contaminate the stain if the stain bottle comes in contact with the sediment.**
8. Place a coverslip on the sample and examine the slide under the microscope. Examine the slide under low power first and look at the free edge of the coverslip. Red-orange, elongated crystals may appear here later as the stain evaporates. Do not confuse the stain crystals with sediment crystals.
9. Now examine the slide under high power and compare your sample with the common urine sediments in figure 45.1.
10. Record the material you find in urine sediment in the review section at the end of this laboratory exercise. Some of the common items found in urine are as follows:

Cells

Epithelial cells
> Cells of urethra (squamous cells)—normal constituent
> Cells of bladder, ureter (transitional cells)—normal constituent

Bacterial cells in urine indicate a condition known as bacteriuria

Erythrocytes (RBCs)—menstrual blood in females or blood representative of the passage of kidney stones

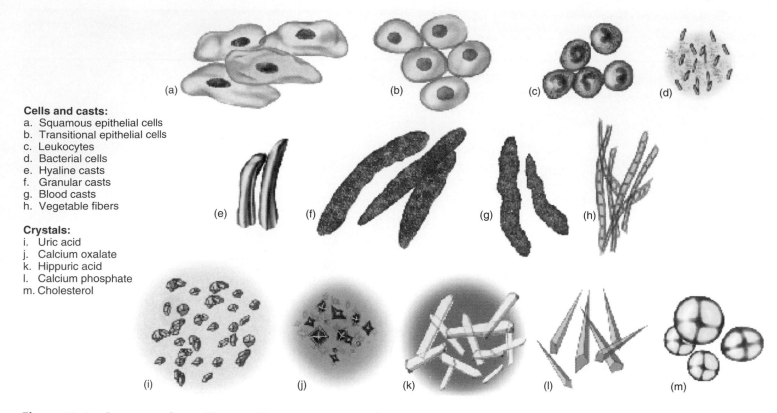

Cells and casts:
a. Squamous epithelial cells
b. Transitional epithelial cells
c. Leukocytes
d. Bacterial cells
e. Hyaline casts
f. Granular casts
g. Blood casts
h. Vegetable fibers

Crystals:
i. Uric acid
j. Calcium oxalate
k. Hippuric acid
l. Calcium phosphate
m. Cholesterol

Figure 45.1   Common urine sediments. Figures are not to scale. For example, casts in (f) and (g) are much larger than cells (a), (b), and (c).

Leukocytes (WBCs)—indicative of urinary tract (bladder/kidney) infections

Crystals (usually indicative of reduced water intake)

Calcium oxalate—small green crystals

Calcium carbonate

Calcium phosphate

Uric acid

Ammonium ureates

Other Crystals

Cholesterol

Tyrosine—amino acid

Cystine—amino acid

Casts

Conglomerations of cells, usually from the kidneys

Leukocytes (indicate pyelonephritis, inflammation of the kidney)

Erythrocytes (indicate glomerulonephritis, inflammation of glomeruli)

Hyaline

Artifacts—clothing fibers, skin oil, bath powder

Renal Calculi

Kidney stones or **renal calculi,** are due to either mineral buildup in the kidneys or to amino acid deposition such as tyrosine or cystine crystals. The calculi are masses of crystals that fuse into a larger form and can block the ureter, causing a retention of fluid in the kidney. Calculi can be exceptionally painful. Examine renal calculi in the lab, if available.

Clean Up Place urine-contaminated material to be disposed of in the biohazard container and all the urine-contaminated material to be reused in the 10% bleach container. Swab the countertops with 10% bleach, which destroys infectious agents.

REVIEW

Name _____

1. What metabolic by-product from hemoglobin colors the urine yellow?

2. How can adequate water intake be judged by the color of urine?

3. What is hematuria?

4. What is the normal value for sugar in urine? What mechanism keeps glucose out of the urine?

5. Which urine sediment material listed below is probably due to reduced water intake?
   a. vegetable fibers   b. crystals   c. epithelial cells   d. bacterial cells

6. There are about 200 grams of protein in the blood plasma and normally none in urine. What mechanism keeps protein out of the urine, and what structure would be potentially damaged (review Laboratory Exercise 44 if necessary) if protein was found in significant amounts in urine?

7. The kidneys are very efficient at balancing the blood pH. It is critical that blood acidity remain constant. If excess hydrogen ions are present in the blood and increase blood acidity, the kidneys secrete the hydrogen ions. What effect would this have on the pH of urine?

8. If you process 180 liters of water through the kidney each day, yet produce only 1.8 liters of urine, approximately how efficient are your kidneys at reabsorbing the water passing through them?

9. Assume that a person did not collect a midstream sample of urine but collected a sample from the beginning of urination. What additional materials might be in greater numbers in this particular sample?

10. List the materials you found in the urine sample.

11. What was the specific gravity of your urine sample?

12. How much water do you normally drink each day?

Male Reproductive System

Introduction

The male reproductive system produces male gametes (spermatozoa), transports the gametes to the female reproductive tract, and secretes the male reproductive hormone, testosterone. The gonad- or gamete-producing structure of the male reproductive system is the testis (plural, *testes*). Since gametes are secreted through ducts or tubules, the testes have an exocrine function. The testes are considered mixed glands because in addition to this exocrine function, they also have an endocrine function—they produce testosterone. In this exercise, we examine the gross anatomy of the male reproductive system, the histology of the system, and the male reproductive system of the cat.

Objectives

At the end of this exercise you should be able to

1. identify the gamete-producing organ of the reproductive system;
2. list the pathway that spermatozoa follow from production to expulsion;
3. list the four components of semen;
4. describe the endocrine function of the testes;
5. name the three cylinders of erectile tissue in the penis.

Materials

Charts, models, and illustrations of the male reproductive system
Microscopes
Prepared slides of a cross section of testis
Cat
Dissection trays and equipment

Procedure

Gross Anatomy of the Male

Reproductive System

Examine the models and charts of the male reproductive system available in the lab and locate the following structures in figure 46.1:

Testis
Epididymis

Scrotal sac
Ductus (vas) deferens
Seminal vesicle
Prostate gland
Bulbourethral gland
Penis

Testes

The **testes** are paired organs wrapped in a tough connective tissue sheath called the **tunica albuginea** (figure 46.2). They are surrounded by the **scrotal sac,** which keeps the testes on the exterior of the body cavity, where the temperature is somewhat cooler. The testes are the site of **spermatozoa** (sperm) production, and this process must occur at about 35°C. The scrotal sac is lined with a layer of muscle called the **dartos muscle.** It is composed of smooth muscle fibers that contract when the testes are cold, thus bringing them closer to the body. When the environment around the testes is warm, the dartos muscles relax and the testes descend from the body, thus becoming cooler.

Sperm

The structure of an individual sperm cell consists of a **head, midpiece** and **tail.** The head contains the genetic information (DNA) as well as a cap known as the **acrosome.** The acrosome contains digestive enzymes that digest the exterior covering of the female gamete. The midpiece of the sperm contains mitochondria that provide ATP to the sperm cell tail. The tail of the sperm is a flagellum that propels the sperm forward. Examine a prepared slide of sperm and compare it to figure 46.3.

Epididymis

Spermatozoa (or sperm cells) from each testis travel from tubules in the testis to the **rete testis** and into the **epididymis,** where they are stored and mature. Each epididymis has a blunt, rounded **head,** an elongated **body,** and a tapering **tail** that leads to the **ductus deferens.** Sperm maturation, or **capacitation,** occurs in the epididymis. If spermatozoa are removed from the testis proper, they are not capable of fertilizing the female oocyte (egg). Spermatozoa move slowly through coiled tubules of the epididymis. Examine a model or chart of the longitudinal section of a testis and epididymis and locate the structures by comparing them to figure 46.4.

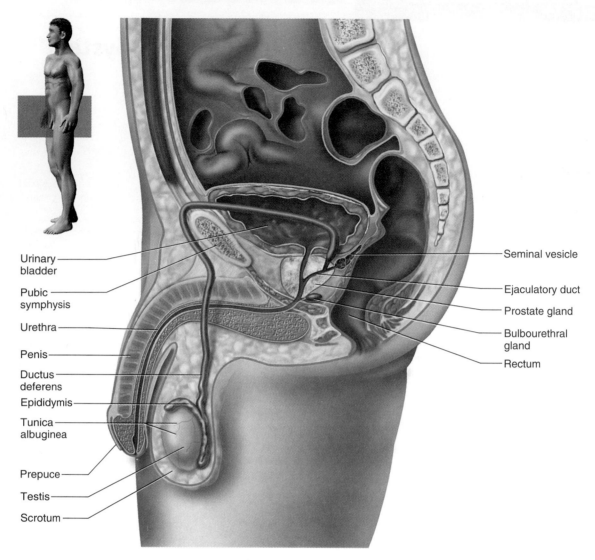

Urinary bladder

Pubic symphysis

Urethra

Penis

Ductus deferens

Epididymis

Tunica albuginea

Prepuce

Testis

Scrotum

Seminal vesicle

Ejaculatory duct

Prostate gland

Bulbourethral gland

Rectum

Figure 46.1   Male reproductive system, midsagittal view ♂.

Spermatic Cord

Spermatozoa travel from the epididymis into the ductus deferens. The ductus deferens is enclosed in the **spermatic cord,** which is a complex cable consisting of the ductus deferens, the **testicular artery** and **vein,** the **nerves,** and the **cremaster muscle.** The cremaster muscle is a cluster of skeletal muscle fibers. The spermatic cord is longer on the left side than on the right, therefore the left testis is lower than the right. Locate the structures of the spermatic cord in figures 46.2 and 46.4.

As the spermatic cord reaches the **inguinal canal** the ductus deferens travels around the posterior of the urinary bladder. You can trace the course of the ductus deferens until it reaches the inferior portion of the bladder. The ductus deferens enlarges somewhat here to form the **ampulla.**

The ductus deferens joins with the **seminal vesicle,** which is a gland that adds fluid to the spermatozoa. The union of the ductus deferens and the seminal vesicle produces the **ejaculatory duct.** Locate the seminal vesicle and the ejaculatory duct in figure 46.5. The seminal fluid adds about 60% to the final volume of semen.

From this location the ejaculatory duct leads to the inferior portion of the bladder and joins with the urethra, which passes through the **prostate gland.** The prostate gland is located just inferior to the urinary bladder, and the urethra that passes through the gland is known as the **prostatic urethra.** The prostate gland adds a buffering fluid to the secretions of the testes and seminal vesicles. The prostate fluid makes up slightly less than 40% of the final semen volume.

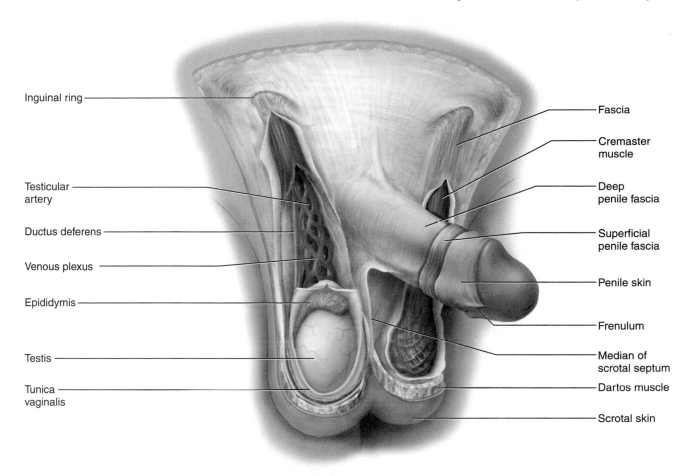

Inguinal ring

Testicular artery

Ductus deferens

Venous plexus

Epididymis

Testis

Tunica vaginalis

Fascia

Cremaster muscle

Deep penile fascia

Superficial penile fascia

Penile skin

Frenulum

Median of scrotal septum

Dartos muscle

Scrotal skin

Figure 46.2   Scrotum, testis, spermatic cord, and penis, anterior view. ✗

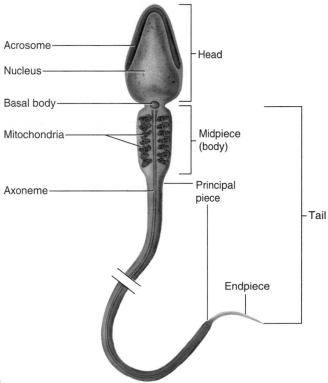

Acrosome

Nucleus

Basal body

Mitochondria

Axoneme

Head

Midpiece (body)

Principal piece

Tail

Endpiece

Figure 46.3   Isolated sperm cell.

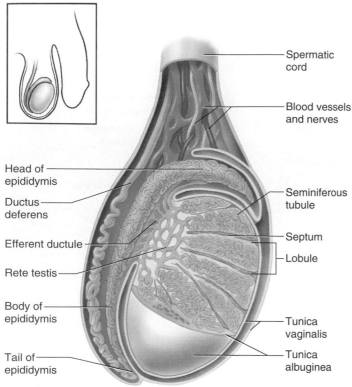

Figure 46.4   Testis and epididymis, lateral view.

As the prostatic urethra exits the prostate gland it becomes the **membranous urethra** and passes through the body wall. Here the paired **bulbourethral (Cowper's) glands** are found, which add a lubricant to the seminal fluid. **Seminal fluid** consists of secretions from the seminal vesicles, prostate gland, and bulbourethral glands. **Semen** consists of seminal fluid plus the addition of spermatozoa from the testes. Spermatozoa make up much less than 1% of the total volume of semen. The urethra passes out of the body cavity and becomes the **penile (spongy) urethra** of the penis. Locate the portions of the urethra and the accessory glands in figure 46.5.

Penis

The **penis** consists of an elongated **shaft** and a distally expanded **glans penis**. The glans is covered with the **prepuce,** or **foreskin,** which is removed in some males by a procedure called a **circumcision.** At the inferior portion of the glans is a region richly supplied with nerve endings called the **frenulum.** The glans penis is an expanded region that serves to stimulate the genitalia of the female.

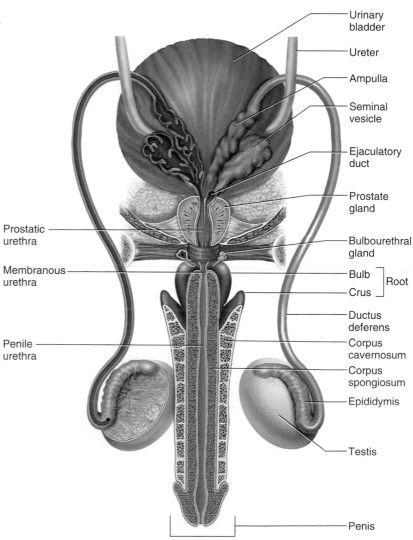

Figure 46.5   Urinary bladder with seminal vesicles, posterior view.

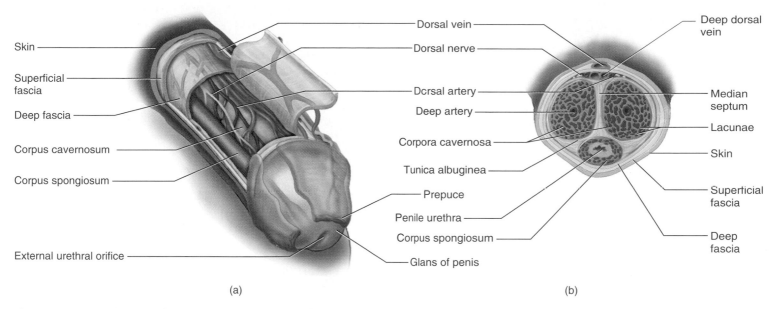

Figure 46.6 labels (a):
- Skin
- Superficial fascia
- Deep fascia
- Corpus cavernosum
- Corpus spongiosum
- External urethral orifice
- Dorsal vein
- Dorsal nerve
- Dorsal artery
- Deep artery
- Corpora cavernosa
- Tunica albuginea
- Prepuce
- Penile urethra
- Corpus spongiosum
- Glans of penis

Figure 46.6 labels (b):
- Deep dorsal vein
- Median septum
- Lacunae
- Skin
- Superficial fascia
- Deep fascia

(a)   (b)

Figure 46.6   Penis. (a) 3/4 view; (b) cross section. ↗

The penis contains three cylinders of **erectile tissue.** The **corpus spongiosum** is the cylinder of erectile tissue that contains the **penile urethra.** The two **corpora cavernosa** are located anterior to the corpus spongiosum. Examine a model or chart of a cross section of penis and compare it to figure 46.6. The proximal parts of the cylinders of erectile tissue are anchored in the body. Locate the **crus,** which is an expansion of the corpora cavernosa. The **bulb** of the penis is an extension of the corpus spongiosum. The crus and the bulb form the **root** of the penis as illustrated in figure 46.5. The corpus spongiosum expands distally to form the glans penis. Note the **dorsal arteries** and **deep arteries** of the penis. These bring blood to the penis. Locate the **dorsal vein** and the **deep dorsal vein** of the penis. When muscles contract at the base of the penis, the venous flow back to the heart is reduced and the erectile tissues engorge with blood. The penis then becomes erect. The erection subsides as the muscles at the base of the penis relax and the veins allow the blood to flow out of the penis. Examine a model of the penis and find the features listed in figure 46.6.

Histology of the Testis

The testis, as the gamete-producing organ of the male reproductive system, should be viewed in a prepared slide. Numerous tubules are seen in cross section. These are the **seminiferous tubules.** The gametes, or spermatozoa, are produced in seminiferous tubules in the testis (figure 46.7).

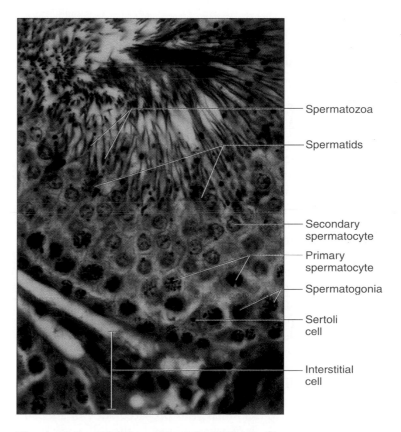

Figure 46.7 labels:
- Spermatozoa
- Spermatids
- Secondary spermatocyte
- Primary spermatocyte
- Spermatogonia
- Sertoli cell
- Interstitial cell

Figure 46.7   Histology of testis (400×). ↗

Cross section of
seminiferous tubules

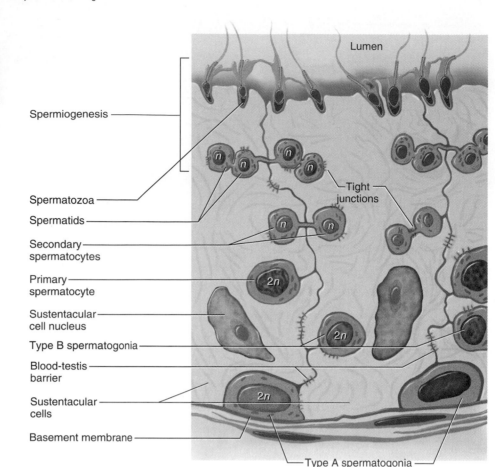

Figure 46.8   Spermatogenesis in the seminiferous tubule.

Find the clusters of cells that frequently appear as triangles in between the tubules. These are called **interstitial cells.** They produce the male sex hormone, **testosterone.** Examine the seminiferous tubules under high magnification. You should be able to see the outer row of cells called the **spermatogonia.** These cells reproduce by mitosis to produce **primary spermatocytes.** The primary spermatocytes undergo meiosis, or reduction division, to eventually produce the sex cells (spermatozoa). The primary spermatocytes divide to form **secondary spermatocytes,** which are found closer to the lumen. The secondary spermatocytes become **spermatids.** Spermatids lose their remaining cytoplasm and mature into **spermatozoa.** Compare this process to figures 46.7 and 46.8 and locate the spermatogonia, primary and secondary spermatocytes, spermatids, and spermatozoa. **Sustentacular (Sertoli) cells** assist in the movement of the primary spermatocytes.

After you have examined your prepared slide of a section of testis, you should draw what you see in the following space. You should include the interstitial cells, seminiferous tubules, spermatogonia, spermatocytes, spermatids, and spermatozoa.

You may need to look at several sections to see all of these items. Your drawing may be a composite of many tubules.

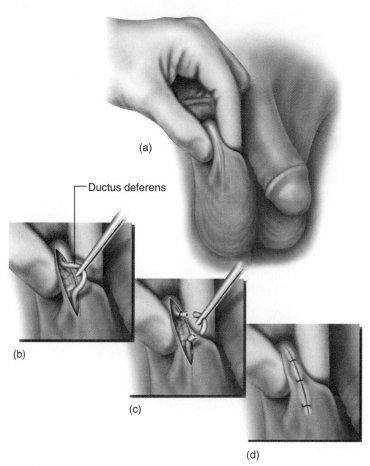

Figure 46.9   **Vasectomy procedure.** ✗

Male Contraception

The main goal in male contraception is the prevention of spermatozoa from reaching the female gametes. Abstention, or refraining from sexual intercourse, is the most effective method. The next most effective method is male sterilization, which is most commonly performed by a procedure called a **vasectomy** (the cutting and tying of the two ductus deferens). In a vasectomy an incision is made in the side of the scrotal sac and both ductus deferens are cut. The free ends of the cut ductus deferens are tied, preventing sperm from traveling from the testes to the spermatic cords (figure 46.9). Use of a barrier, such as a condom, is relatively effective if used properly, since it prevents ejaculated sperm from entering the female reproductive tract. Coitus interruptus, or preejaculatory withdrawal, is not a very effective method since sperm may be present in seminal fluid prior to ejaculation and pregnancy can result.

Cat Dissection

Prepare for the cat dissection by obtaining a dissection tray, scalpel, scissors or bone cutter, string, forceps, and plastic bag with label. Remember to place all excess tissue in the appro-priate waste container and not in a standard wastebasket or down the sink!

Check to see whether your cat is male or female. The penis may be retracted in your specimen, so look for the opening of the penile urethra and the scrotal sac. Team up with a lab partner or group that has a cat of a different sex than your specimen so you can learn both male and female reproductive systems. Once you are sure you have a male cat, locate the **scrotal sac** and paired **testes.** Make an incision on the lateral side of the scrotal sac and locate the testis. If your cat was neutered, then you will not be able to locate the testis. If the testes are present, cut through the connective tissue of the scrotal sac (the **tunica vaginalis**) and observe both the testis and the **epididymis.** The testis is covered by a tough connective tissue membrane called the **tunica albuginea.** Use figures 46.10 and 46.11 as a guide. Spermatozoa move from the testis and into the epididymis, where the sperm cells mature.

Once you have located the epididymis, proceed in an anterior direction and trace the thin **ductus deferens** from the epididymis into the spermatic cord. The spermatic cord traverses the body wall on the exterior and enters the body of the cat at the inguinal canal (a tube that pierces the inguinal

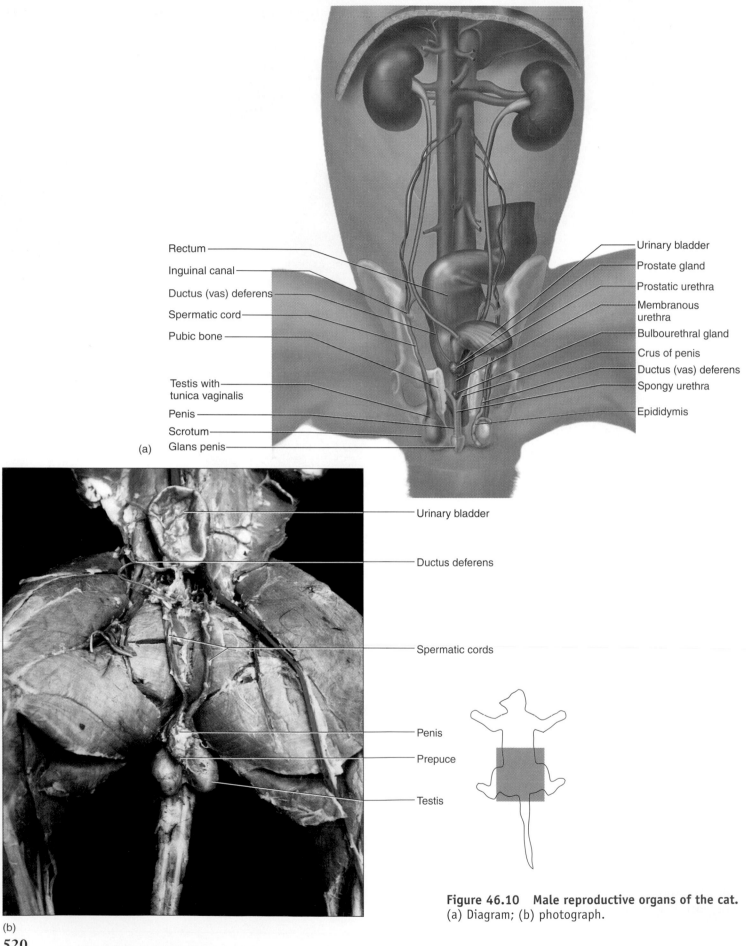

(a)

Rectum

Inguinal canal

Ductus (vas) deferens

Spermatic cord

Pubic bone

Testis with tunica vaginalis

Penis

Scrotum

Glans penis

Urinary bladder

Prostate gland

Prostatic urethra

Membranous urethra

Bulbourethral gland

Crus of penis

Ductus (vas) deferens

Spongy urethra

Epididymis

Urinary bladder

Ductus deferens

Spermatic cords

Penis

Prepuce

Testis

(b)

Figure 46.10   Male reproductive organs of the cat.
(a) Diagram; (b) photograph.

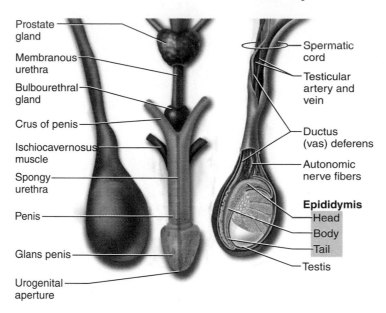

Prostate gland
Membranous urethra
Bulbourethral gland
Crus of penis
Ischiocavernosus muscle
Spongy urethra
Penis
Glans penis
Urogenital aperture

Spermatic cord
Testicular artery and vein
Ductus (vas) deferens
Autonomic nerve fibers
Epididymis
Head
Body
Tail
Testis

Figure 46.11   Testis, epididymis, spermatic cord, and penis of the cat.

ligament). You may want to gently insert a blunt probe into the inguinal canal so you can locate the ductus deferens as it passes into the coelom. Locate the ductus deferens as it enters the body cavity and notice how it arches around the ureter on the dorsal side of the urinary bladder. You may have cut the ductus deferens in an earlier exercise, so if you cannot find it on one side look for it on the other side.

You may want to look for the accessory organs of the male reproductive system, but this takes some significant dissection. Check with your instructor before cutting through the pelvis of your cat. If your instructor directs you to do so, then begin by cutting through the musculature of the cat at the level of the symphysis pubis. Make a midsagittal incision

through the groin muscles, and carefully cut through the cartilage of the symphysis pubis. You should be able to now open the pelvic cavity and locate the single **prostate gland** and the paired **bulbourethral glands** (figure 46.11). Much of the anatomy of the cat is similar to the human except that there are no seminal vesicles in the cat. Trace the ductus deferens from the posterior surface of the bladder through the prostate gland and the penis. You can make either a longitudinal section through the penis to trace the urethra in the erectile tissue or you can make a cross section of the penis to see the three cylinders of erectile tissue—the corpus spongiosum and the two corpora cavernosa.

1. The testes are considered mixed glands because they have both an endocrine and exocrine function. Describe the endocrine and exocrine products that come from the testes.

2. Male sterility can result from excessively high temperatures around the testes. What mechanism occurs in the scrotum to counteract the effects of high temperature?

3. List all of the structures involved in producing semen.

4. How does spermatozoa differ from seminal fluid?

5. A vasectomy is the cutting and tying of the two ductus deferens at the level of the spermatic cords. Review the percent of spermatozoa that composes semen and determine what effect a vasectomy has on semen volume.

6. What male reproductive gland is missing in the cat but present in the human?

7. What is the organ in the male reproductive system that functions as the gonad?

8. Where does sperm move after being in the epididymis?

9. Proper sperm production must occur at what temperature?

10. Name the lining of the scrotal sac consisting of a smooth muscle layer.

11. Where is the cremaster muscle found?

12. Name the three organs that produce seminal fluid.

13. Name of the section of urethra that passes through the prostate gland.

14. Where is the glans penis located?

15. What is the cylinder of erectile tissue below the corpora cavernosa?

16. Where is spermatozoa produced in the testis?

17. What cells initiate spermatozoa production?

18. Which one of the seminal fluid glands is not a paired gland?

Female Reproductive System

Introduction

The female reproductive system is functionally more complex than the male reproductive system. In the male, the reproductive system produces gametes and delivers them to the female reproductive system. The female reproductive system not only produces gametes and receives the gametes from the male but also, under hormonal influence of HCG from the early cell mass and later from the placenta, provides space and maternal nutrients for the developing conceptus. Finally, the female reproductive system delivers the child into the outer environment. The ovaries are the gamete-producing organs of the female reproductive system. The ovaries have both an endocrine and an exocrine function. The exocrine function is the production of oocytes, and the endocrine function is the production of the female sex hormones, estrogen (estradiol) and progesterone. In this exercise, you learn about the structure and function of the female reproductive system.

Objectives

At the end of this exercise you should be able to

1. identify the gamete-producing organ of the female reproductive system;
2. trace the pathway of a gamete from the ovary to the usual site of implantation;
3. list the structures of the vulva;
4. describe the function of each structure in the female reproductive system;
5. name the layers of the uterus from superficial to deep;
6. compare and contrast the anatomy of the cat reproductive system with that of the human.

Materials

Charts, models, and illustrations of the female reproductive system
Microscopes
Prepared slides of ovary and uterus
Dissection trays and equipment
Cat

Procedure

Gross Anatomy of the Female Reproductive System

Examine a model or chart of the female reproductive system and locate the following major reproductive organs there and in figure 47.1.

Ovary
Uterine (fallopian) tube
Uterus
Vaginal canal
Clitoris
Labia minora (singular, labium minus)
Labia majora (singular, labium majus)

Ovaries and Uterine Tubes

Each **ovary** is approximately 3 to 4 cm long and oblong in form. The ovaries produce **oocytes,** which are shed from the outer surface of the ovary during **ovulation.** From here the oocytes (which may mature into ova, if fertilized) move into the **uterine (fallopian) tube.** The ovaries are not directly attached to the uterine tube, and the oocytes must move from the surface of the ovary into the uterine tube. The uterine tube has a small fringe on the distal region known as the **fimbriae.** These are small fingerlike projections attached to an expanded region known as the **infundibulum.** The uterine tube also has an enlarged region known as the **ampulla** and a narrower portion towards the uterus. Examine a model or chart of the female reproductive system and compare it to figure 47.2.

Uterus

The **uterus** is a pear-shaped organ with a domed **fundus,** a **body,** and a circular, inferior end called the **cervix.** The uterine tubes enter the uterus at about the junction of the fundus with the uterine body. The uterine wall is composed of three layers. The outer surface of the uterus is called the **perimetrium.** The majority of the uterine wall consists of the **myometrium,** a thick layer of smooth muscle, and the

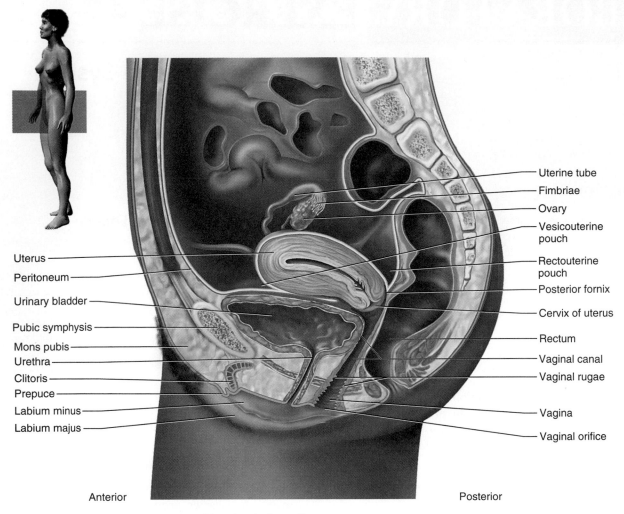

Uterine tube
Fimbriae
Ovary
Vesicouterine pouch
Rectouterine pouch
Posterior fornix
Cervix of uterus
Rectum
Vaginal canal
Vaginal rugae
Vagina
Vaginal orifice

Uterus
Peritoneum
Urinary bladder
Pubic symphysis
Mons pubis
Urethra
Clitoris
Prepuce
Labium minus
Labium majus

Anterior

Posterior

Figure 47.1   Female reproductive system, midsagittal view. ✗

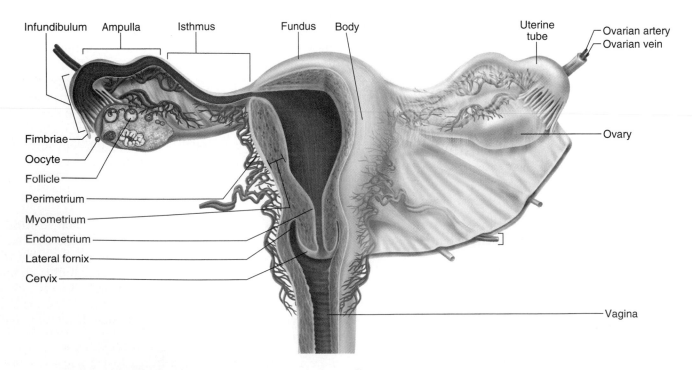

Infundibulum   Ampulla   Isthmus   Fundus   Body   Uterine tube   Ovarian artery
Ovarian vein

Ovary

Fimbriae
Oocyte
Follicle
Perimetrium
Myometrium
Endometrium
Lateral fornix
Cervix

Vagina

Figure 47.2   Female reproductive system, posterior view. ✗

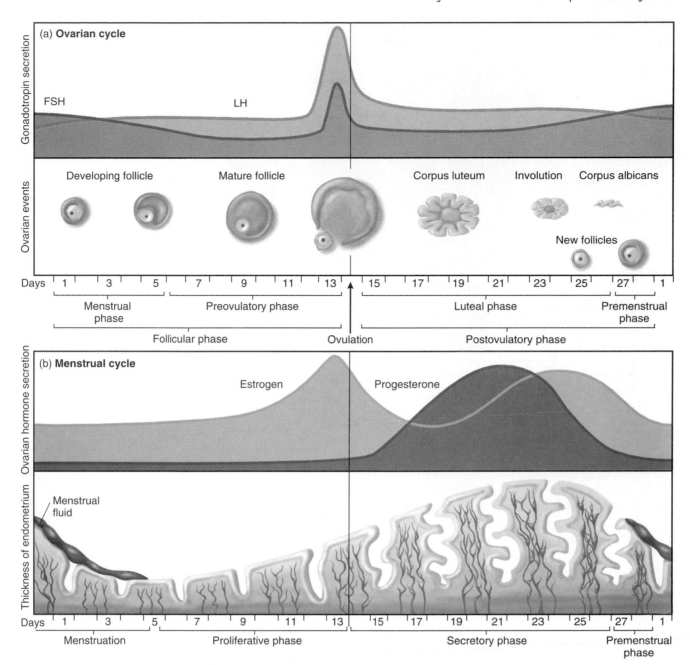

Figure 47.3   Ovarian/menstrual cycle.

innermost layer of the uterus is the **endometrium.** Examine the models or charts in the lab and locate the structures in figure 47.2.

Ovarian and Menstrual Cycles

The endometrium undergoes dynamic changes during the **menstrual cycle.** The effects of **luteinizing hormone (LH)** and **follicle-stimulating hormone (FSH),** from the anterior pituitary, and subsequently **estrogen** and **progesterone** from the ovary, have a significant effect on the endometrium. FSH stimulates the ovarian follicles to secrete estrogens and some progesterone. Elevated levels of these hormones promote the thickening of the endometrium. After ovulation, progesterone levels increase and the endometrial cells remain intact. Toward the end of a woman's menstrual cycle, estrogen and progesterone levels drop and this causes the functional layer of the endometrium to slough off. The loss of the endometrial layer is the beginning of a woman's period, or **menstruation.** These events are briefly outlined in figure 47.3. Examine the figure and note how the endothelial layer begins to decrease when estrogen and progesterone levels fall (about day 25 in figure 47.3).

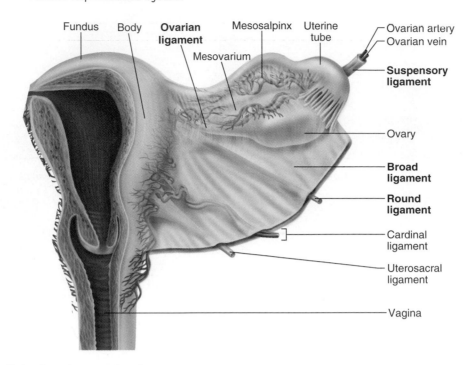

Fundus   Body   **Ovarian ligament**   Mesosalpinx   Uterine tube   Ovarian artery   Ovarian vein

Mesovarium   **Suspensory ligament**

Ovary

Broad ligament

Round ligament

Cardinal ligament

Uterosacral ligament

Vagina

Figure 47.4   Ligaments of the female reproductive system.

Ligaments

The uterus and ovaries are suspended in the pelvic cavity by a number of connective tissue sheaths called ligaments. The **broad ligament** anchors the uterus to the anterior body wall. The **round ligament** attaches the uterus to the anterior body wall at about the region of the inguinal canal. The **ovarian ligament** directly attaches the ovary to the uterus, and the **suspensory ligament** attaches the ovaries to the lumbar region. Locate these structures in models or charts in the lab and in figure 47.4.

Vagina

The vagina consists of the **vaginal canal** and the **vaginal orifice.** The uterus joins with the vaginal canal at the cervix. The vaginal canal is a tough, muscular tube with a recessed region around the cervix known as the **fornix.** The vaginal canal is poorly supplied with nerves, and the posterior wall of the vagina has cross ridges called **rugae.** The vaginal canal and orifice can expand greatly during the delivery of a child. Locate the vaginal canal, fornix, rugae, and vaginal orifice in figures 47.1 and 47.2.

External Genitalia

The floor of the pelvis as seen from the outside is referred to as the **perineum.** It can be divided into a posterior **anal** (or **rectal) triangle** and an anterior **urogenital triangle** consisting of the reproductive and urinary structures (figure 47.5).

The external female reproductive structures are called the outer genitalia and are collectively referred to as the **vulva.**

The **mons pubis** is the anterior-most structure of the vulva and is an adipose pad that overlies the symphysis pubis. Posterior to the mons pubis is the **clitoris,** which is a cylinder of erectile tissue embedded in the body wall that terminates anterior to the urethral orifice as the **glans clitoris.** The clitoris has the same embryonic origin as the penis in males, and like the penis it is richly supplied with nerve endings. The body of the clitoris is a curved structure illustrated in figure 47.1, and the glans clitoris is the terminal portion. The glans clitoris is enclosed by the **prepuce,** which is an extension of the **labia minora.** Posterior to the clitoris is the external **urethral orifice** and posterior to this is the **vaginal orifice.** The vaginal orifice is partially enclosed by a mucous membrane structure known as the **hymen.** The hymen is variable anatomically and has historically (and sometimes incorrectly) been used as an indicator of virginity. Lateral to the vaginal orifice are the labia minora (singular, labium minus). The space between the labia minora is known as the **vestibule,** and located laterally and posteriorly to the vestibule are the **greater vestibular glands** (or **Bartholin's glands**). Lateral to the labia minora are the paired **labia majora.** Locate these structures on models or charts in the lab and in figure 47.5.

Histology of the Ovary

Examine a prepared slide of the ovary under the microscope on low power. Locate the background substance of the ovary, which is known as the **stroma.** Look for circular structures in the ovary. These are the **ovarian follicles.** Locate the **pri-**

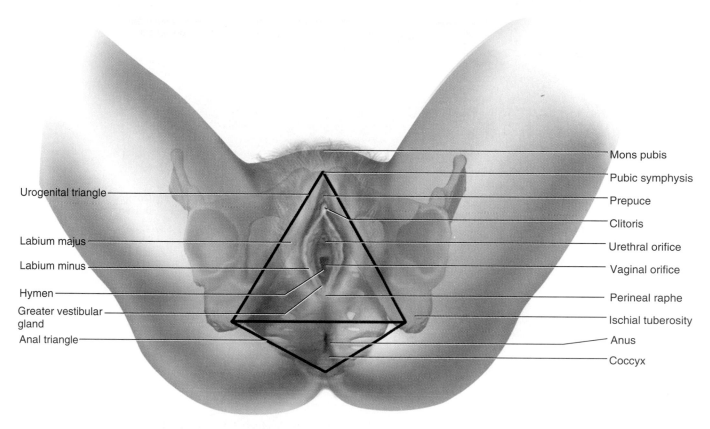

Figure 47.5   **Female perineum and external genitalia.**

mordial **follicles** in your slide and compare these to the follicles in figure 47.6.

You should also locate the **primary** and **secondary follicles.** Some of the follicles may contain **oocytes.** Primary follicles contain **primary oocytes,** secondary follicles contain **secondary oocytes.** The largest follicles in the ovary are the **mature ovarian follicles (Graafian follicles),** and you may be able to see one if it is present in your slide. One of these secondary oocytes is shed from the ovary during **ovulation.** Examine your slide and compare it to figures 47.6 and 47.7. Draw what you see in your slide in the following space.

After ovulation the remains of a mature ovarian follicle become a **corpus luteum,** which primarily secretes **progesterone.** If pregnancy does not occur the corpus luteum de-

creases in size and becomes the **corpus albicans.** Examine a prepared slide of the ovary with a corpus luteum or corpus albicans and compare it to figures 47.6 and 47.8.

Histology of the Uterus

Examine a prepared slide of the **uterus** and locate its three layers. Locate the outer **perimetrium,** the smooth muscle of the **myometrium,** and the inner **endometrium.** Compare these to figures 47.2 and 47.9.

Now examine the two layers of the endometrium of the uterus under higher magnification. The **functional layer** is the one which is shed during menstruation. It is composed of **spiral arterioles** and **uterine glands,** which appear as wavy lines towards the edge of the tissue. The **basal layer** is deeper and contains **straight arterioles.** Deep to the endometrium is the myometrium, which can be distinguished by the smooth muscle found in it. As you study the layers of the endometrium, refer to the cyclic changes that occur with the endometrium as illustrated in the bottom part of figure 47.3.

Anatomy of the Breast

The structure of the breast derives from the integumentary system, yet the role of the female breast in reproduction is

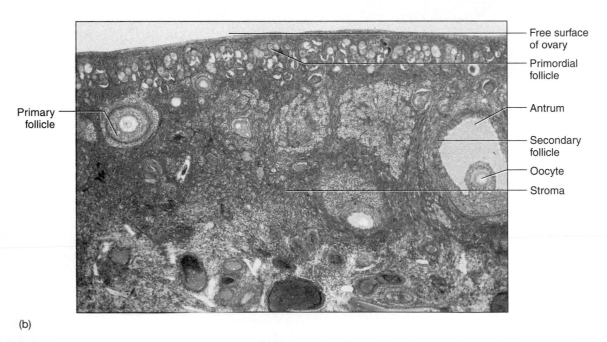

Figure 47.6 Histology of the ovary. (a) Diagram. Arrows indicate a timeline of development from primordial follicles to the corpus albicans; (b) photomicrograph (40×).

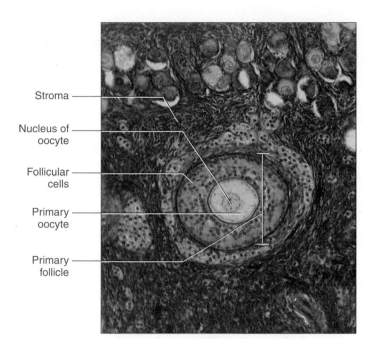

Figure 47.7 **Primary oocyte in follicle (100×).**

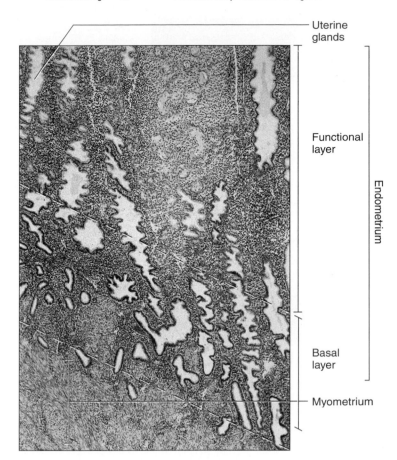

Figure 47.9 **Histology of the uterus (40×).**

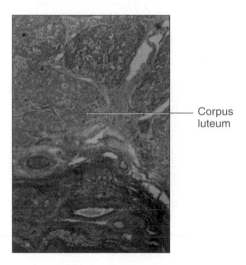

Figure 47.8 **Post-ovulatory ovary (40×).**

important as a source of nourishment for the offspring. The major structures of the external breast are the pigmented **areola,** the protruding **nipple,** the **body** of the breast, and the **axillary tail (tail of Spence).** The axillary tail is of clinical importance in that breast tumors frequently occur there. Examine the surface features of the breast in figure 47.10 and locate the structures listed.

Compare models or charts in the lab and locate the internal structures of the breast. Note that the breast is anchored to the pectoralis major muscle with **suspensory ligaments.** Much of the breast is composed of **adipose tissue,** and embedded in the adipose tissue are the **mammary glands.** The mammary glands are responsible for the production of milk in lactating females. The glands are clustered in **lobes,** and there are about 15 to 20 lobes in each breast. The mammary glands increase in size in nursing women and lead to **lactiferous ducts,** which subsequently lead to **lactiferous sinuses (ampullae)** that exit via the nipple. Humans have several ampullae leading to each nipple. The mammary glands in females begin to undergo changes prior to puberty and become functional glands after delivery of a child. Note the features illustrated in figure 47.11.

Female Contraception

As with male contraception the prevention of male and female gametes from uniting is the goal in female contraception. Abstention, or refraining from intercourse, is the most effective method of birth control. Other methods involve oral

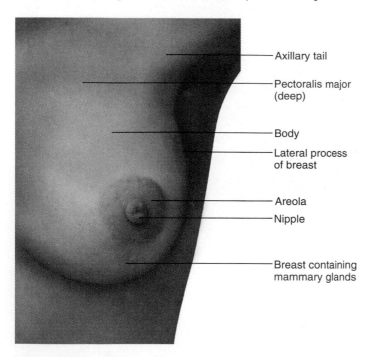

Figure 47.10 Surface features of the female breast.

Axillary tail

Pectoralis major (deep)

Body

Lateral process of breast

Areola

Nipple

Breast containing mammary glands

contraceptives or hormonal implants, tubal ligation (cutting and tying of the uterine tubes), use of barrier methods such as the condom or diaphragm, and use of spermicidal foams. The use of oral contraceptives, which are synthetic estrogens and progesterones, decrease the levels of LH and FSH in the pituitary (by negative feedback mechanisms) thus preventing ovulation from occurring. Progesterone implants elevate hormone levels which also prevent ovulation. Examine the levels of progesterone in figure 47.3. When these levels are elevated note how the levels of FSH and LH are low.

Cat Dissection

Prepare for the cat dissection by obtaining a dissection tray, scalpel, scissors or bone cutter, string, forceps, and plastic bag with label. Remember to place all excess tissue in the appropriate waste container and not in a standard wastebasket or down the sink!

You probably have already removed the multiple mammary glands of the female cat during the removal of the skin. If this is the case, then examine the external genitalia for the **urogenital orifice.** Make an incision in the groin and cut through the muscles that cover the symphysis pubis. Continue to cut in the midsagittal plane through the cartilage of the symphysis pubis and then cranially through the abdominal muscles as described in the dissection of the male cat. This should expose the reproductive organs (figure 47.12).

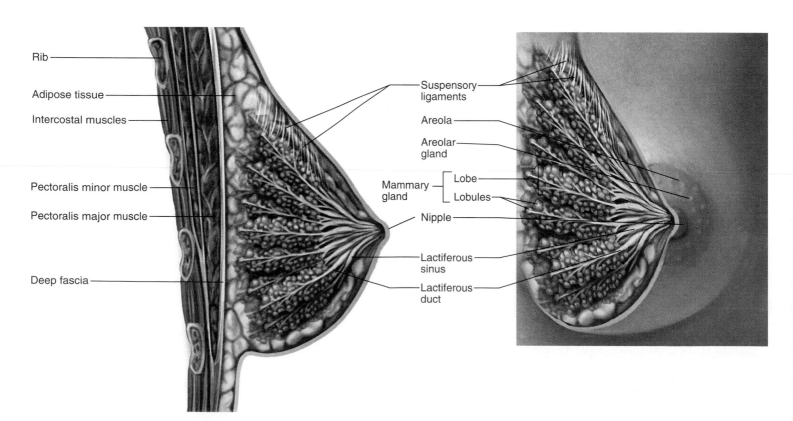

Rib

Adipose tissue

Intercostal muscles

Pectoralis minor muscle

Pectoralis major muscle

Deep fascia

Suspensory ligaments

Areola

Areolar gland

Mammary gland

Lobe

Lobules

Nipple

Lactiferous sinus

Lactiferous duct

Figure 47.11 Interior of the female breast.

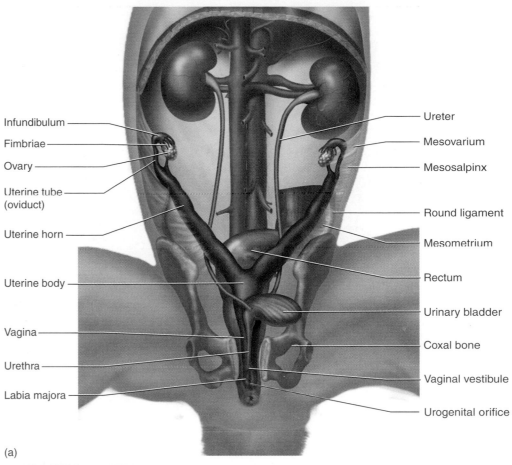

Infundibulum
Fimbriae
Ovary
Uterine tube (oviduct)
Uterine horn
Uterine body
Vagina
Urethra
Labia majora

Ureter
Mesovarium
Mesosalpinx
Round ligament
Mesometrium
Rectum
Urinary bladder
Coxal bone
Vaginal vestibule
Urogenital orifice

(a)

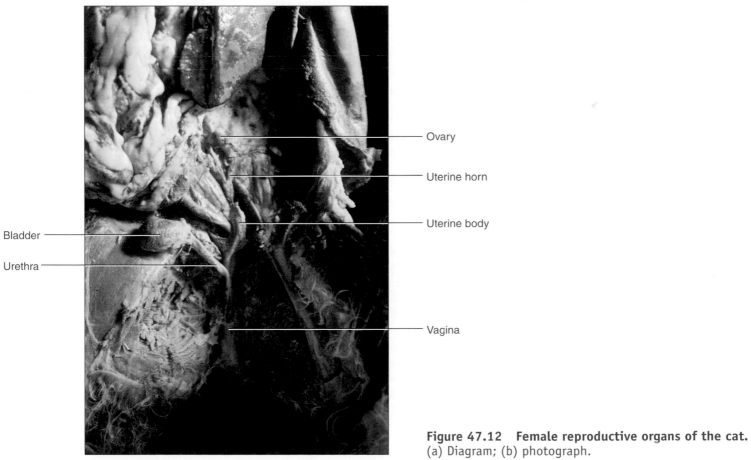

Ovary
Uterine horn
Uterine body
Bladder
Urethra
Vagina

(b)

Figure 47.12   Female reproductive organs of the cat.
(a) Diagram; (b) photograph.

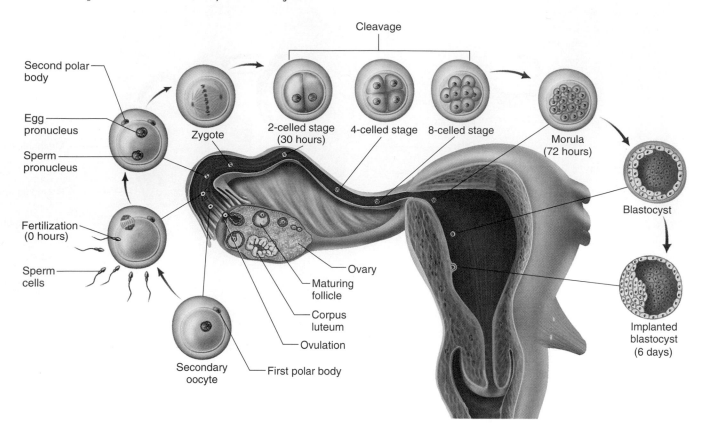

Figure 47.13 Early stages of human development. ▭

Unlike the human female the cat has a **horned (bipartite) uterus.** The uterus of a cat has an appearance of a Y with the upper two branches being the **uterine horns** and the stem of the Y the **body** of the uterus. Humans normally have single births from a pregnancy, while the expanded uterus in cats facilitates multiple births. If your cat is pregnant the uterus will be greatly enlarged, and you may find many fetuses inside.

The **ovaries** in cats are caudal to the kidneys and are relatively small organs. Examine the paired ovaries and the short **uterine tubes (oviducts)** in the cat. If you have difficulty locating them, trace the uterus towards the uterine horns and locate the ovaries. The uterine tubes in the human female are proportionally longer than those in the cat. The opening of the uterine tube near the ovary is called the **ostium,** and it receives the **oocytes** during **ovulation.**

At the termination of the uterus is the **cervix,** which leads to the **vaginal canal.** The vagina in cats is different than in humans in that the **urethral opening** is internally enclosed in the vaginal canal. This region where the vagina and the urethral opening occurs is called the **vaginal vestibule.** Thus the opening to the external environment is a common urinary and reproductive outlet called the **urogenital orifice.** Locate these structures in figure 47.12.

Stages of Development

Early Development

The union of the sperm and egg in a process known as **fertilization** initiates a remarkable phenomenon of growth and differentiation from the single-celled **zygote** to the adult human. Fertilization usually occurs in the uterine tube, and the zygote divides into 2 cells, then 4, 8, 16, etc., until a solid cluster of cells called a **morula** is formed. The morula continues to divide until it becomes a hollow ball of cells known as the **blastocyst.** The covering of cells on the outside of the blastocyst is called the **trophoblast,** and the cluster of cells on the inside is known as the **inner cell mass.** Review these stages in figure 47.13.

Embryonic Tissues

The early stage of development continues to progress and the formation of three embryonic tissues occurs. These are the **ectoderm,** the **mesoderm,** and the **endoderm.** The ectoderm gives rise to the outer layer of skin and the nervous tissue, the mesoderm gives rise to bones and muscles, and the endoderm gives rise to many internal organs such as digestive and respiratory organs.

Name _____

1. What is the gonad in the female reproductive system?

2. What is the inner layer of the uterus called?

3. What happens to estrogen and progesterone levels just prior to menstruation?

4. The uterus is connected to the ovaries by what structure?

5. What three hormones are at elevated levels just prior to ovulation?

6. Where is the fornix in the female reproductive system?

7. Which is more anterior, the urethral opening or the clitoris?

8. What is the background substance of the ovary called?

9. What is the name for the expulsion of the secondary oocyte from the ovary?

10. What is the deepest layer of the endometrium called?

11. What is the name of the part of the breast that is near the shoulder?

12. What are the milk-producing glands of the breast called?

13. A zygote is formed from the fusion of what two substances?

14. Embryonic tissue consists of three layers. What are these called?

15. What is a morula?

16. Trace the pathway of milk from the mammary glands to expulsion.

17. Ectopic pregnancies are those that occur outside of the endometrial layer of the uterus. Provide an explanation for how pregnancies may occur in the uterine tube (thus a tubal pregnancy) or in the abdominopelvic cavity.

18. How does the uterus of the human female differ from that of the cat?

19. How does the structure of the uterus in the cat correlate to multiple births from each pregnancy?

Measurement Conversions

The **metric system** is universally used in science to measure certain **values** or **quantities.** These values are **length, volume, mass** (weight), **time,** and **temperature.** The metric system is based in units of 10, and conversion to higher or lower values is relatively easy when compared to using the U.S. customary system. In the United States, length is typically measured in inches, feet, yards, or miles. The **base unit** for length in the metric system is the meter. As shown in table A.1, 1 meter is equivalent to 1.09 yards or 39.4 inches. When the measurement of great lengths makes the use of meters impractical, then the kilometer (1,000 meters) is used. When the measurement of extremely small lengths is needed, then the centimeter (1/100 of a meter), millimeter, micrometer, nanometer, or angstrom are used. The common multiples or fractions for the values are outlined in table A.1.

Table A.1 Conversions of Standard and Metric Systems

Value	Base Unit	×1,000	1/100	1/1,000	1/1,000,000
Length	Meter (m) 1 m = 39.4 in 1 m = 1.09 yd	Kilometer (km) 1 km = ⅝ mi	Centimeter (cm) 2.54 cm = 1 in	Millimeter	Micrometer 1 nanometer = 10^{-9} m 1 angstrom = 10^{-10} m
Mass	Gram (g) 454 g = 1 lb 1 g = .036 oz	Kilogram (kg) 1 kg = 2.2 lb	Centigram	Milligram	Microgram
Volume	Liter (L) 1 liter = 1.06 qt 1 gal = 3.78 liters	Kiloliter	Centiliter	Milliliter (ml) 1 ml = 1 cc	Microliter
Time	Second (sec)	Kilosecond	Centisecond	Millisecond	Microsecond
Temperature	Degrees 0°C = freezing 100°C = boiling °F = ⅘ °C + 32	Celsius			

Notes

Periodic Table of the Elements

Nineteenth-century chemists discovered that when they arranged the known elements by atomic weight, certain properties reappeared periodically. In 1869, Russian chemist Dmitri Mendeleev published the first modern periodic table of the elements, leaving gaps for those that had not yet been discovered. He accurately predicted properties of the missing elements, which helped other chemists discover and isolate them.

Each row in the table is a *period* and each column is a *group (family)*. Each period has one electron shell more than the period above it, and as we progress from left to right within a period, each element has one more proton and electron than the one before. The dark steplike line from boron (5) to astatine (85) separates the metals to the left of it (except hydrogen) from the nonmetals to the right. Each period begins with a soft, light, highly reactive *alkali metal*, with one valence electron, in family IA. Progressing from left to right, the metallic properties of the elements become less and less pronounced. Elements in family VIIA are highly reactive gases called *halogens*, with seven valence electrons. Elements in family VIIIA, called *noble (inert) gases*, have a full valence shell of eight electrons, making them chemically unreactive. Elements 1–96 occur naturally, whereas 97–109 are short-lived elements known only in the laboratory. The names of elements 104–109 are still in dispute. The names shown in this table are the ones recommended by their discoverers, but the International Union of Pure and Applied Chemistry has recommended different names for these elements; the disagreement is under review.

The 24 elements with normal roles in human physiology are color-coded according to their relative abundance in the body (see Laboratory Exercise 2). Others, however, may be present as contaminants with very destructive effects (such as arsenic, lead, and radiation poisoning).

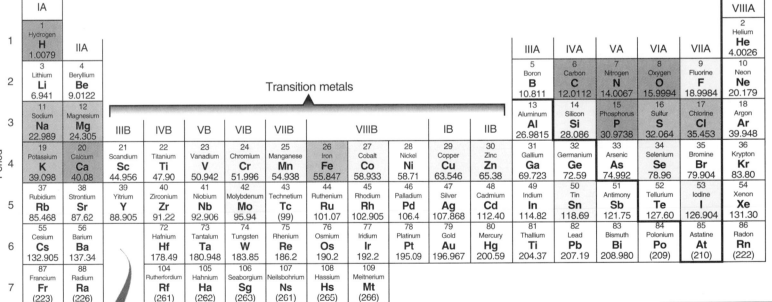

57–71, Lanthanides

*57 Lanthanum La 138.91	58 Cerium Ce 140.12	59 Praseodymium Pr 140.907	60 Neodymium Nd 144.24	61 Promethium Pm 144.913	62 Samarium Sm 150.35	63 Europium Eu 151.96	64 Gadolinium Gd 157.25	65 Terbium Tb 158.925	66 Dysprosium Dy 162.50	67 Holmium Ho 164.930	68 Erbium Er 167.26	69 Thulium Tm 168.934	70 Ytterbium Yb 173.04	71 Lutetium Lu 174.97

89–103, Actinides

**89 Actinium Ac (227)	90 Thorium Th 232.038	91 Protactinium Pa (231)	92 Uranium U 238.03	93 Neptunium Np (237)	94 Plutonium Pu 244.064	95 Americium Am (243)	96 Curium Cm (247)	97 Berkelium Bk (247)	98 Californium Cf 242.058	99 Einsteinium Es (254)	100 Fermium Fm 257.095	101 Mendelevium Md 258.10	102 Nobelium No 259.10	103 Lawrencium Lr 260.105

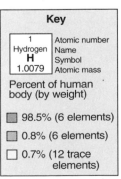

Key

1	Atomic number
Hydrogen	Name
H	Symbol
1.0079	Atomic mass

Percent of human body (by weight)

- ▓ 98.5% (6 elements)
- ▒ 0.8% (6 elements)
- ☐ 0.7% (12 trace elements)

Notes

Preparation of Solutions

The following solutions are designed for a lab of 24 students. It is best to estimate how many milliliters of stock solutions are required for all of the students in a lab and then double that amount. The preparations are listed alphabetically and the solutions are the approximate volume for the needs of a class of 24. The number of the lab exercise follows the solution description for cross referencing.

Acetylcholine Chloride Solution (0.1%)

Add 0.1 g of acetylcholine chloride to 100 ml frog Ringer's solution. Pour into a small, labeled dropper bottle. (Laboratory Exercise 33)

Acid Solution

Pour 200 ml lemon juice or vinegar into dropper bottles labeled "Acid Solution." (Laboratory Exercise 43)

Agar Plates

Add 15 g agar to enough water to make 1 liter of solution. Boil and stir the agar until it all dissolves. Pour into petri dishes for three dishes per table. (Laboratory Exercise 5)

Alpha Amylase Solution (1%)

Add 3 g of alpha amylase to a graduated cylinder and add water to the 300 ml mark. Pour into a small labeled bottle or beaker. (Laboratory Exercise 43)

Alpha Amylase Solution (0.1%)

Add 0.1 g of alpha amylase to a graduated cylinder and fill to 100 ml with water or make a 0.1 serial dilution by taking 10 ml of the above solution and adding 90 ml of water. Pour into a small labeled bottle or beaker. (Laboratory Exercise 43)

BAPNA Solution (1%)

N-alpha-Benzoyl-DL-arginine-4-nitroanilide hydrochloride

Add 0.5 g of BAPNA powder in 50 ml of water. Pour into 2 bottles of 25 ml each. BAPNA is an expensive material but you do not need much of it for the lab. Available from Sigma B4875 or Aldrich 85,711-4. (Laboratory Exercise 43)

Benedict's Reagent

A copper sulfate solution that turns color if reducing sugars are present and remains blue in the absence of reducing sugars. Add 35 g sodium citrate and 20 g sodium carbonate (Na_2CO_3) to 160 ml water. Filter through paper into a glass beaker. Dissolve 3.5 g copper sulfate ($CuSO_4$) in 40 ml water. Pour the copper sulfate solution into the 160 ml, stirring constantly. (Laboratory Exercise 43)

Bleach Solution (10%)

Mix 100 ml household bleach (sodium hypochlorite) with 900 ml tap water. (Laboratory Exercises 29, 30, and 45)

Caffeine Solution, Saturated

Add small amounts of caffeine to 50 ml water until no more will dissolve. Decant the solution into small dropper bottles. (Laboratory Exercise 33)

Calcium Chloride Solution (2%)

Weigh 5 g calcium chloride and place in a graduated cylinder. Add frog Ringer's solution to make 250 ml. Pour in dropper bottles. (Laboratory Exercise 33)

Cat Wetting Solution

Numerous formulations are available for keeping preserved specimens moist. Some commercial preparations are available that reduce the exposure of students to formalin or phenol. You may not need any wetting solution at all if the cats are kept in a plastic bag that is securely closed. You can make a wetting solution by putting 75 ml formalin, 100 ml glycerol, and 825 ml distilled water in a 1-liter squeeze bottle. Another mixture consists of equal parts Lysol and water. (Laboratory Exercises 13–18)

Cellulose

Cut several (3–4) g pure cotton (cotton wool, cotton balls) into fine pieces (0.5 cm or less). Label "Cellulose." (Laboratory Exercise 43)

Epinephrine Solution (0.1%)

Add 0.1 g of adrenalin chloride in 100 ml frog Ringer's solution. Label and pour the solution into small dropper bottles. (Laboratory Exercise 33)

Essential Oils Preparation

Fill several small screw-top vials with peppermint, almond, wintergreen, and camphor oils (available from local drug stores). Label "Peppermint," "Almond," "Wintergreen," and "Camphor," respectively, and keep vials in separate wide-mouthed jars to prevent cross-contamination of scent. (Laboratory Exercise 25)

Fill four small vials halfway to the top with cotton and color them red with food coloring. Label the vials "Wild Cherry" and add benzaldehyde solution until the cotton is moist. (Laboratory Exercise 25)

Fat Testing Solution (1% Sudan III Solution)

Dissolve 1 g Sudan III in 100 ml absolute alcohol. (Laboratory Exercise 5)

Filtration Solution (1% Starch, Charcoal, and Copper Sulfate Solution)

Place 5 g starch, 5 g powdered charcoal, and 5 g copper sulfate ($CUSO_4$) in a 1-liter beaker. Add enough water to make 500 ml. Stir well and pour into a 500 ml bottle. Label "Filtration Solution." (Laboratory Exercise 5)

Frog Ringer's Solution

Weigh and place the following materials in a 1-liter graduated cylinder:

> 6.5 g NaCl (sodium chloride)
>
> 0.2 g $NaHCO_3$ (sodium bicarbonate)
>
> 0.1 g $CaCl_2$ (calcium chloride)
>
> 0.1 g KCl (potassium chloride)

To these add enough water to make 1,000 ml. This solution should be prepared fresh and used within a few weeks. (Laboratory Exercises 23 and 33) (In Laboratory Exercise 33, three solutions are needed—one at room temperature, one at 37°C, and one in an ice bath.)

Hydrochloric Acid Solution (0.1%)

Add 1 ml concentrated HCl to 1 liter of water. (Laboratory Exercise 23)

Iodine Solution

Prepare by adding 10 g I_2 (iodine) and 20 g KI (potassium iodide) to 1 liter of distilled water. Store in small dark dropper bottles. Label "Iodine Solution." (Laboratory Exercises 5 and 43)

Litmus Cream

Use approximately 250 ml heavy cream. To this add powdered litmus until the cream is a light blue. Pour into two separate bottles and label "Litmus Cream." (Laboratory Exercise 43)

Litmus Solution

Weigh 2 g litmus powder and dissolve in 600 ml water. Titrate HCl into the solution until it begins to turn red; then add NaOH solution by drops until it just turns back to blue. Pour into two bottles. (Laboratory Exercise 40)

Maltose Solution (1%)

Add 1 g maltose in enough water to make 100 ml. Stir until dissolved and pour into two clean bottles. Label "1% Maltose Solution." (Laboratory Exercise 43)

Methylene Blue (1%)

Add 5 g methylene blue powder to 500 ml distilled water. Pour into dropper bottles. (Laboratory Exercise 3)

Methylene Blue Solution (0.01 M)

Add 3.2 g methylene blue (MW 320) to distilled water to make 1 liter of solution. Place in dropper bottles. (Laboratory Exercise 5)

Molasses or Concentrated Sucrose Solution (20%)

Use undiluted molasses or a 20% sugar solution. To make the sugar solution add 100 g table sugar (sucrose) to water to make 500 ml of solution. Make sure the sucrose is completely dissolved. (Laboratory Exercise 5)

Nitric Acid (1 N) (for Decalcifying Bones)

Add 64 ml concentrated nitric acid (70%) slowly to water to make 1 liter of solution. (Laboratory Exercise 8)

Pancreatin Solution (1%)

Place 1 g pancreatin powder in a graduated cylinder and add water to make 100 ml. Stir well. Adjust the pH with 0.05 M sodium bicarbonate until neutral (pH 7). Pour into different stock bottles and label "1% Pancreatin Solution." Preparation note: Use fresh pancreatin. Pancreatin may be stored frozen (not in a frost-free freezer that regularly cycles between freezing and defrosting). Also note that commercially prepared pancreatin has an optimum pH. If the pH is too low, the reaction will be slowed or stopped. (Laboratory Exercise 43)

Perfume, Dilute Solution

Add 10 ml inexpensive perfume to 50 ml isopropyl alcohol. (Laboratory Exercise 25)

Phosphate Buffer Solution

Add 3.3 g potassium phosphate (monobasic) and 1.3 g sodium phosphate (dibasic) to 500 ml water. Place in squeeze bottles. (Laboratory Exercise 29)

Potassium Dichromate Solution (0.01 M)

Add 2.94 g potassium dichromate (MW 294) crystals to water to make 1 liter of solution. Label and pour into dropper bottles. (Laboratory Exercise 5)

Potassium Permanganate Solution (0.01 M)

Add 1.58 g potassium permanganate (MW 158) crystals to water to make 1 liter of solution. Label and pour into dark brown dropper bottles. (Laboratory Exercise 5)

Procaine Hydrochloride

Place 1 g procaine hydrochloride solution in 1 ml water. Add to this 30 ml pure ethanol. Place in small screw-capped bottle. (Laboratory Exercise 23)

Quinine Solution (0.5% Quinine Sulfate Solution)

Measure 2.5 g pharmaceutical grade quinine sulfate (food grade) and place it in a clean container. Add 500 ml water to the quinine sulfate and stir well. Pour into 12-ounce paper cups for the lab. (Laboratory Exercise 25)

Saline Solution (0.9%) (Physiological Saline)

Put 9.0 g NaCl in 1,000 ml water and pour into small dropper bottles. (Laboratory Exercise 5)

Saline Solution (5%)

Add 25 g NaCl crystals to water to make 500 ml of solution. Label the solution and place in small dropper bottles. (Laboratory Exercise 5)

Saltwater Solution (3%)

Add 15 g NaCl crystals (food grade table salt) to water to make 500 ml of solution. Pour into 12-ounce paper cups for the lab. (Laboratory Exercise 25)

Sodium Chloride Solution (5%)

Add 2.5 g NaCl crystals to water to make 50 ml of solution. Label the solution and place in small beaker. (Laboratory Exercise 23)

Sodium Hydroxide Solution (1%)

Add water to 2 g NaOH to make 200 ml of solution. Pour into 6 small dropper bottles. (Laboratory Exercise 43)

Caution   Label as "1% NaOH Solution."

Sodium Hydroxide Solution (1 Normal)

Add 40 g NaOH crystals or powder to water to make 1 liter of solution. Pour into small dropper bottles with a caution label. (Laboratory Exercise 40)

Caution   The reaction is exothermic and generates heat. Wear protective gloves and eyewear. If you spill this on your skin make sure you flush your skin immediately with cold water.

Starch Solution (0.5%)

A potato starch solution is made by first boiling 500 ml water. Remove the water from the hot plate and add 2.5 g potato starch powder. Stir and cool the mixture. Do not boil the starch and water mixture because this will lead to some hydrolysis of starch to sugar. Test for the presence of sugar by using Benedict's reagent. There should be no sugar present. Place into 250 ml bottles and label "0.5% Starch Solution." (Laboratory Exercise 43)

Starch Solution (1%)

Boil 1 liter of distilled water. Remove the water from the heat and add 10 g cornstarch (or 10 g potato starch). Filter the mixture through cheesecloth into bottles. (Laboratory Exercise 5)

Sugar Solution (3%)

Using a clean container, for food use, dissolve 15 g table sugar in enough water to make 500 ml of solution. Pour into a 12-ounce paper cup for the lab. (Laboratory Exercise 25)

Sugar Solutions

Four table sugar solutions of 2 liters each. (Laboratory Exercise 5)

0%—2 liters of water

5%—dissolve 100 g sugar in enough water to make 2 liters of solution

15%—dissolve 300 g sugar in enough water to make 2 liters of solution

30%—dissolve 600 g sugar in enough water to make 2 liters of solution

Vegetable Oil

Add vegetable oil from the grocery store to 6 small dropper bottles. (Laboratory Exercise 5)

Vinegar Solution

Use household vinegar or make a 5% food-grade acetic acid solution by adding 5 ml concentrated acetic acid to about 50 ml water and then adding additional water to make 100 ml. (Laboratory Exercise 25)

Wright's Stain

Wright's stain is available as a commercially prepared solution from a number of biological supply houses. (Laboratory Exercise 29)

Notes

APPENDIX D

Lab Reports

Part of working in science involves writing lab reports. You should write lab reports in a certain style and follow basic guidelines that are generally accepted in the field. The general format for the lab reports falls into four categories: **introduction, materials and methods, results,** and **conclusion** sections. Each of these parts is important in the write-up, and each of these four sections must be included in the lab report. Your report should have a title, your name, your instructor's name, the course name and semester, and your lab section.

The purpose of a scientific report is to explain an investigation. Your instructor is your primary audience for your report in this class so you should communicate that you understand the basic information. Part of a grade in a lab report is dependent on doing the experiment correctly and demonstrating that you are aware of the outcome of the experiment and its relationship to theory or applications you have learned in lecture.

If your experiment did not come out as you thought it might, you still can do well in your lab report. Results from your experiment may have come out satisfactorily and be perfectly fine even if you think that the data should have been different from what you obtained.

Introduction

The introduction section consists of a description of the problem and the subject of your study. In professional journals the introduction often includes a history of past experimentation or current knowledge in the field, but you will probably not have this in your lab report. You should pose the question or hypothesis that your experiment is trying to resolve in the introduction. You may be conducting an experiment to determine whether an enzyme functions on a substrate or whether a particular effect occurs when you perform an action, etc.

Materials and Methods

The materials and methods section involves a clear description of what equipment, animals, chemicals, etc., were used and the experimental procedure followed. You must write this section in very clear and precise terms. From this section you should expect that a person could repeat your experiment and produce the same results.

In one way this is the 'recipe' for your experiment. As in baking a cake it is not good enough simply to list the materials. You must include quantity, what sequence the materials were added, how long things were stirred, baked, etc. You must provide specific details to your procedure. Make sure that you state how much material was added to a sample. For example, you should write: "We added 5 ml." This is a known quantity that people can duplicate; whereas, "We added a little" is vague. You must make sure that you include all steps in your procedure. If you leave something out in your description then a person following your directions might get very different results.

Results

The results section is where the outcome from your experiment is listed in a clear and defined way. Your data must be clearly presented. This may consist of the tabulation of data that you acquired from your experiment in the form of a line graph, bar graph, or table. Remember that any graph should have a complete description. If you were studying the effect of exercise on heart rate, then exercise is the independent variable and is listed on the horizontal axis, while heart rate is the dependent variable and it is listed on the vertical axis as illustrated in the following graph.

The data for your lab result section may be in raw form (direct measurements from your experiment) or you may want to organize the data by calculating the mean (average) data, and range (high and low), etc., as determined from your experiment. There should be no interpretation of the results at this point. You should not try to explain your results but simply list what the outcome was. Save the interpretation of the data for the conclusion.

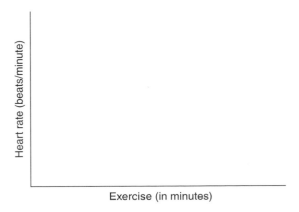

Conclusion

In the conclusion you may analyze and interpret your data and determine whether the results of your experiment proved your initial question (make sure you remember to answer your original question). This is the most important part of your lab report in terms of finding out if you understood the lab. Did you prove your hypothesis, did you disprove your hypothesis, or are the results inconclusive? Where should you go from here?

Sometimes experiments go awry and the results do not support the original question. This can be due to many things such as experimental error, which is a big factor in lab experiments especially at the undergraduate level. Someone may have not followed the experimental procedure correctly and added too much, too little or, in some cases, none of the materials that should have been part of the procedure. Sometimes it is a matter of timing; the process did not go on as long as it should or went on too long. In some cases the data are taken down incorrectly or read wrong. A '3' in the data might be mistaken for an '8,' or the data are entered in the wrong location. There are hundreds of possibilities as to why error plays a part in experiments and your instructor is probably very familiar with many of them.

Another factor is due to faulty equipment, supplies, or set-up. Other problems may be due to the variation in living organisms. Humans and other animals have their own idiosyncratic physiologic responses. Not every person and not all experimental animals respond in the same way. For example, almost all product information sheets that come with pharmaceutical drugs list adverse reactions seen in some people. If everyone responded the same way there would be no need for warnings on drug products.

If you obtain variances from the results you expect to get, then you should write this in your report. Sometimes experimental procedures are included that produce results other than what you expect to see. You should be honest in your recording of data and resist the temptation to 'fudge' the data so that you can get a better result and therefore a better grade on your report.

Sometimes people in industry change their data to produce more desirable results. If this is discovered by independent investigations the corporation that allowed this to happen usually pays significant fines. Students who change data and are discovered are also subject to penalty which is usually far more severe than what they would receive due to the 'bad' data they think they might have gotten. In addition to being honest in the reporting of data you also want to make sure that your words are your own. Even though you may have performed an experiment word for word from this lab book or other source, you want to make sure that you do not plagiarize the material. You should paraphrase material if you want to have the same meaning but not the same words as someone else.

Your lab report should follow the format described here unless your instructor decides to make significant changes in the write-up. Lab reports tend to take a long time to write, but the analysis in the lab report is where you work towards understanding the experiment.

CREDITS

PHOTOGRAPHS

Chapter 5
5.4: David M. Philips/Visuals Unlimited.

Chapter 30
30.3: Courtesy of Clay Adams, Inc.
30.4: Courtesy of Coulter International Corporation.

Chapter 37
Page 428: From E. K. Markell and M. Vogue, Medical Parasitology, 5th ed., W. B. Saunders.

Chapter 40
40.2: Courtesy of Collins-Cybermedic, Inc.
40.3: Courtesy of Propper, Inc.

INDEX

Page numbers followed by *italic* letters *f* and *t* indicate figures and tables, respectively.